Essential Genetics

ESSENTIAL

SECOND EDITION

GENETICS

PETER J. RUSSELL PhD
Professor of Biology
Reed College, Portland, Oregon

BLACKWELL SCIENTIFIC PUBLICATIONS
OXFORD LONDON EDINBURGH
BOSTON PALO ALTO MELBOURNE

Editorial offices:
Osney Mead, Oxford, OX2 0EL
8 John Street, London, WC1N 2ES
23 Ainslie Place, Edinburgh, EH3 6AJ
52 Beacon Street, Boston
Massachusetts 02108, USA
667 Lytton Avenue, Palo Alto
California 94301, USA
107 Barry Street, Carlton
Victoria 3053, Australia

First published 1980
Reprinted 1982
Second edition 1987

Typeset by Enset (Photosetting),
Midsomer Norton, Bath, Avon
Printed in Great Britain by
Butler & Tanner Ltd,
Frome and London

DISTRIBUTORS

USA and Canada
Blackwell Scientific Publications Inc
PO Box 50009, Palo Alto
California 94301

Australia
Blackwell Scientific Publications
(Australia) Pty Ltd
107 Barry Street,
Carlton, Victoria 3053

British Library
Cataloguing in Publication Data

Russell, Peter J.
Essential genetics—2nd ed.
1. Genetics
I. Title
575.1 QH430

ISBN 0-632-01602-7

Library of Congress
Cataloging-in-Publication Data

Russell, Peter J.
Essential genetics.
Updated ed. of: Lecture notes on genetics. 1980.
Includes index.
1. Genetics. I. Russell, Peter J. Essential genetics. II. Title.
QH430.R867 1987 575.1 87–734

ISBN 0-632-01602-7

Contents

Preface

Essential Genetics, the second edition of *Lecture Notes on Genetics*, is a brief but balanced review of the important aspects of genetics. It is intended to be used in a one-semester college course in genetics either as the primary text or as a supplementary text. Prior completion of college courses in general biology and chemistry is assumed, although some material from general biology is reviewed in this text to make the discussion of topics complete.

Essential Genetics is organized as a series of concise chapters that present the principles of genetics in an up-to-date, readable way. Many of the classical and modern genetics experiments are described so that the reader can gain an appreciation of the methodology of genetics as well as the facts and concepts of the subject. It is hoped that this book will give the reader a solid understanding of genetics as well as a sense of the excitement that pervades this rapidly progressing field. For those readers who are stimulated to seek further information about genetics, historical and up-to-date references are provided at the end of each chapter.

The sequence of chapters in this book is the same as in *Lecture Notes on Genetics,* beginning with the molecular aspects of genetics, progressing to transmission genetics, and then discussing the regulation of gene expression, extranuclear inheritance, and population genetics. All chapters of the book have been revised, with particular attention directed toward updating the material on molecular genetics, while maintaining the book at a reasonable length. Significant revision of the molecular genetics sections of the book include: addition of a discussion of other forms of DNA, including Z-DNA (Chapter 1); updating the presentations of chromatin structure and repetitive DNA sequences (Chapter 2) and DNA replication and topoisomerases; addition of a discussion of transposable genetic elements in prokaryotes and eukaryotes (Chapter 2); expanding the discussion of the individual steps in mitosis and meiosis (Chapter 5); addition of a description of the molecular structure of centromeres and telomeres (Chapter 5); updating and additional detailing of the transcription process, particularly promoters, RNA synthesis, and processing (Chapter 7); updating of the material on protein synthesis and the signal hypothesis for the secretion of proteins (Chapter 8); presentation of new material on the molecular aspects of mitochondria and chloroplasts (Chapter 17); and updating the chapters on gene regulation in bacteria

(Chapter 19) and in eukaryotes (Chapter 20). In addition, Chapter 12, "Recombinant DNA," was revised extensively to reflect the advances that have been made in that area since the first edition. Chapter 12 now includes, for example, a discussion of rapid DNA sequencing and more complete discussions of gene cloning techniques. Chapter 18, "Biochemical Genetics," was revised to include a new section on the genetic control of protein structure. Chapter 20, "Regulation of Gene Expression in Eukaryotes," was revised to include a more complete discussion of the general aspects of gene regulation, the roles of nonhistones, and the action of steroid hormones. Chapter 21, "Population Genetics," has been greatly expanded for this edition. Another significant revision is the addition of a number of new illustrations and photographs throughout the text to reinforce the text discussions.

Two pedagogical features have been added to the text of this second edition to aid students and to enhance their understanding and appreciation of genetics principles. First, a glossary of terms has been added so that the reader can quickly learn the key genetics terms and their meanings. These terms are in bold face in the text. Second, a set of questions and problems has been provided for each chapter so that the reader can test his or her understanding of the concepts presented in the chapters. The answers to the questions and problems are provided at the end of the book.

Acknowledgements

I acknowledge my wife, Jenny, and my children, Steven and Kristie, for their support during this project. I am grateful to all of the people at Blackwell Scientific Publications, Inc. who have worked with me on the book, especially John Staples. I thank Little, Brown & Co. for giving permission to use the "Questions and Problems" from my text, **GENETICS,** © 1986. A number of those questions and problems were graciously provided by Rowland H. Davis (University of California, Irvine), H. Branch Howe (University of Georgia, Athens), and David D. Perkins (Stanford University).

I thank the following people for providing new photographs for this edition: M. Meyer (Universiteit van Amsterdam, The Netherlands); P. Oudet (Laboratoire de Génétique Moleculaire des Eucaryotes du C.N.R.S., Institut de Chimie Biologique, Strasbourg, France); M. Meselson (Harvard University); O. L. Miller, Jr. (University of Virginia, Charlottesville); G. Stoffler (Max-Planck Institut für Molekulare Genetik, Berlin-Dahlem, Federal Republic of Germany); and G. Morgan (University of Washington, Seattle).

I am grateful to the Literary Executor of the late Sir Ronald A. Fisher, FRS, to Dr. Frank Yates, FRS, and to Longman Group Ltd, London, for permission to reprint part of a table of chi-square probabilities from their book *Statistical Tables for Biological, Agricultural, and Medical Research* (6th edition, 1974).

Last, but by no means least, I thank the reviewers of the second edition manuscript.

Peter J. Russell
Reed College

Chapter 1 The Genetic Material

Outline
Requirements for the genetic material
Nucleic acid structure
The DNA double helix
Different DNA forms
Evidence that DNA is the genetic material

The central theme of this book is the genetic material: its nature, structure, organization, replication, expression, etc. The approach used will be to discuss the salient facts in the context of the current literature and the analytical methods used in the areas under discussion. The emphasis of the book will be on the interrelationships between genetics, molecular biology, and biochemistry.

Requirements for the genetic material

The genetic material is of central importance to cell function and therefore must fulfill a number of basic requirements:

1. It must contain the information for cell structure, function, and reproduction in a stable form. This information is encoded in the sequence of basic building blocks of the genetic material.

2. It must be possible to replicate the genetic material accurately such that the same genetic information is present in descendant cells and in successive generations.

3. The information coded in the genetic material must be able to be decoded to produce the molecules essential for the structure and function of cells.

4. The genetic material must be capable of (infrequent) variation. Specifically, mutation and recombination of the genetic material are the foundations for the evolutionary process.

The nucleic acids, **deoxyribonucleic acid** (DNA) and **ribonucleic acid** (RNA), meet all these requirements.

Nucleic acid structure

Both DNA and RNA are linear polymeric macromolecules. The monomeric unit is called a **nucleotide**; it is **deoxyribonucleotide** in the case of DNA and **ribonucleotide** in RNA. A nucleotide consists of three components: a nitrogenous base (which is a derivative of either purine or pyrimidine), a pentose sugar, and a phosphate group (Fig. 1.1).

The carbon positions in the pentose sugar ring are labeled 1′ to 5′ to distinguish them from the numbering of the positions in the ring structure of the bases. The phosphoryl groups may be attached to any hydroxyl group of

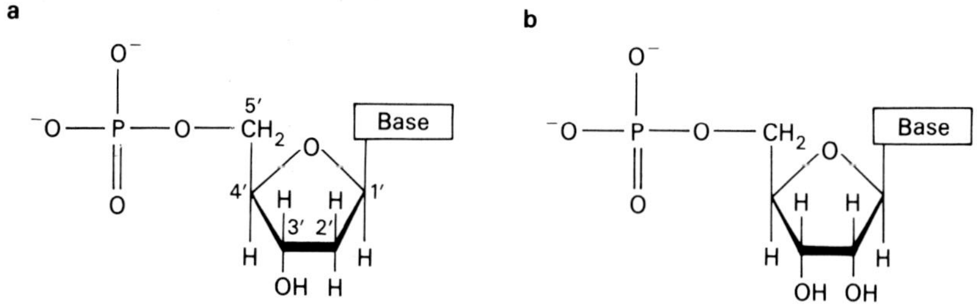

Fig. 1.1. Structure of the nucleotide components of DNA and RNA. (a) Deoxyribonucleoside 5′-monophosphate (monomeric unit of DNA). (b) Ribonucleoside 5′-monophosphate (monomeric unit of RNA).

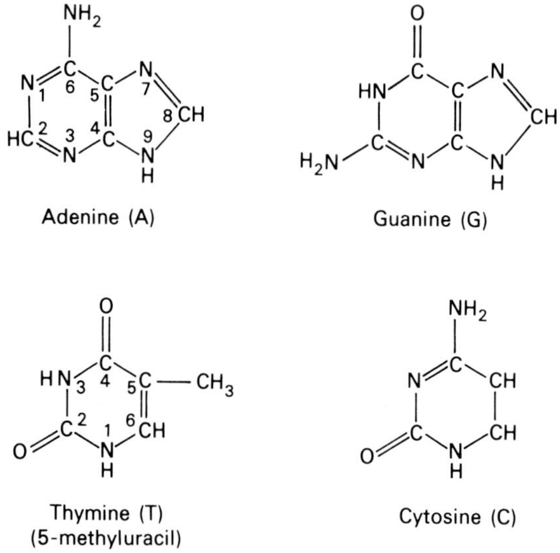

Fig. 1.2. Purine and pyrimidine nitrogenous bases found in DNA.

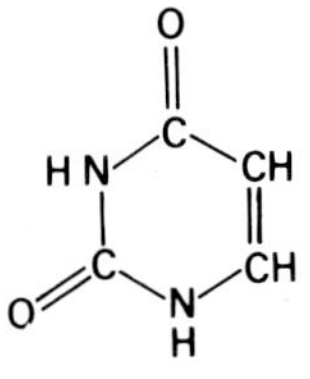

Fig. 1.3. The structure of uracil, the nitrogenous base found in RNA instead of thymine.

the sugar and the 5′- and 3′-nucleotides are of particular importance for the structure and function of DNA and RNA.

Four different deoxyribonucleotides are the major components of DNA. These are distinguished by the type of nitrogenous base they contain. The four bases characteristic of the deoxyribonucleotides are the purine derivatives, adenine (A) and guanine (G), and the pyrimidine derivatives, thymine (T) and cytosine (C) (Fig. 1.2).

Similarly, RNA is characterized by four different ribonucleotides that, like the monomeric units of DNA, contain the bases adenine, guanine, and cytosine. However, instead of thymine, RNA contains the pyrimidine derivative uracil (U), which has chemical and physical properties similar to those of thymine (Fig. 1.3).

In both DNA and RNA, the bases are attached to the pentose moiety by a covalent bond between the 1′ carbon of the sugar and the 9-position nitrogen of the purines or the 3-position nitrogen of the pyrimidines.

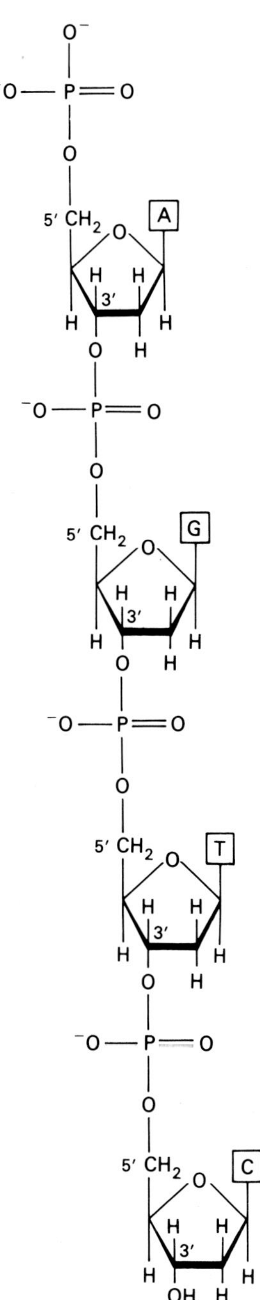

Fig. 1.4. An oligodeoxyribonucleotide chain showing the linkages between the monomeric units in a single DNA chain.

Another distinction between DNA and RNA is the nature of the pentose sugar each contains. Deoxyribonucleotides contain deoxyribose, whereas ribonucleotides contain ribose. As a result the two nucleic acids have different chemical properties, which are biologically important (e.g. enzymes can be specific for DNA or RNA) and may be exploited to separate the two molecules in the laboratory.

In DNA and RNA the mononucleotides are linked together by 3′, 5′-phosphodiester bonds. Thus the backbone of both molecules consists of alternating phosphate and pentose groups. The bases are not part of the backbone structure. An example of an oligodeoxyribonucleotide is shown in Fig. 1.4.

Polynucleotides have polarity. The pentose sugar at one end of the chain has a 5′-hydroxyl or phosphoryl group (5′ end), and the sugar at the other end has a 3′-hydroxyl group (3′ end). A shorthand way to represent a polynucleotide strand is depicted in Fig. 1.5.

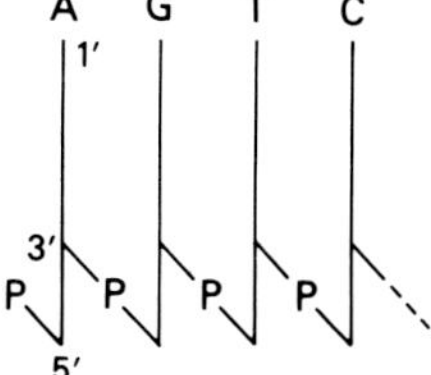

Fig. 1.5. A shorthand way to represent a polynucleotide chain.

The DNA double helix

In 1953 James D. Watson and Francis H. C. Crick proposed that DNA is in the form of a double-stranded, right-handed helix. (A right-handed helix is one that winds clockwise when viewed from the end.) The evidence for their hypothesis was as follows:

1. The DNA molecule consists of bases, sugars, and phosphoryl groups linked together as a polynucleotide chain as discussed earlier.
2. E. Chargaff analyzed the nucleotides released by chemical hydrolysis and found that the total amount of purines present is always equal to the total amount of pyrimidines present. More specifically, adenine always equals thymine (A = T), and guanine always equals cytosine (G = C). Thus the following equations hold for double-stranded DNAs:

$$A+G = C+T$$
$$A+G/C+T = 1$$
$$A+T/G+C \text{ does not } = 1 \text{ (in most cases)}$$

The last is called the base ratio of the DNA and is usually expressed as % GC. The base ratio varies widely among organisms but it remains constant for any one species.

R. Franklin, M. H. F. Wilkins and coworkers analyzed fibers of DNA by x-ray diffraction. The patterns they obtained indicated that DNA was a helical structure consisting of two or more chains wound around each other.

Watson and Crick fitted the chemical and physical data into a symmetric structure that was compatible with all the facts and also possessed the properties one would expect of genetic material. The Watson–Crick model of DNA involves two polynucleotide chains that are wound around each other to form a *double helix* (Fig. 1.6). The two chains are joined together by

Fig. 1.6. The Watson–Crick model of DNA. (**a**) Molecular model of DNA double helix. (After M. Feughelman et al. 1955. *Nature*, 175:834, courtesy of M. H. F. Wilkins). (**b**) Diagrammatic representation of the DNA double helix. (**c**) Diagram of one of the strands of the helix, viewed down the central axis. The sugar-phosphate backbone is outside, the bases (all pyrimidines here; in hatching) are inside. The tenfold symmetry is evident, thus enforcing a repeat in the helix after 10 nucleotide residues.

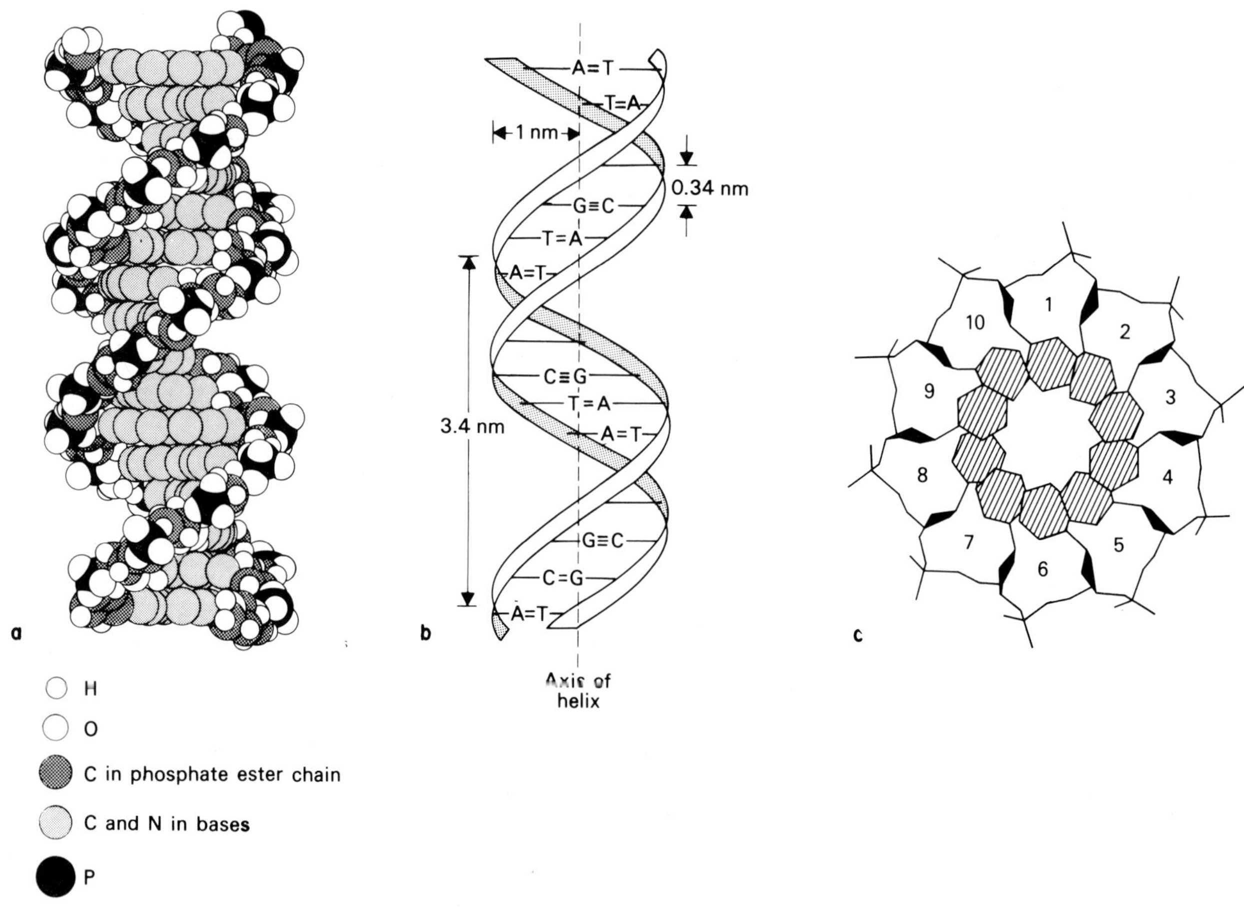

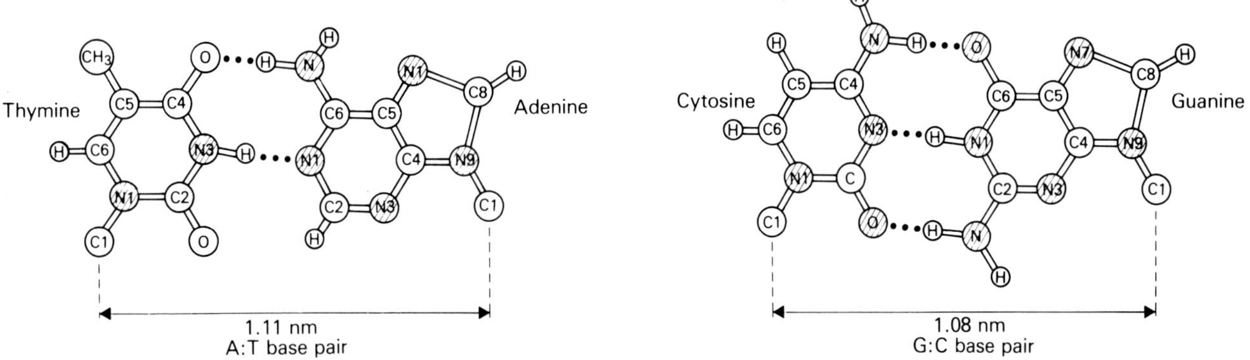

Fig. 1.7. Structures of the complementary base pairs in DNA—adenine:thymine and guanine:cytosine. In each case the bases are attached to the deoxyribose of the sugar-phosphate backbone by a covalent bond to the 1′ carbon of the sugar, labeled C1′ in the diagrams.

hydrogen bonding between the bases, which are flat structures stacked like coins and arranged at right angles to the long axis of the polynucleotide chain. The sugar-phosphate backbones are on the outside of the helix.

From the model it is possible to show that there are 10 base pairs per each complete turn of the polynucleotide chain. Since the distance between adjacent base pairs is 0.34 nm, it follows that the DNA helix has one turn each 3.4 nm of length. (1 nm = 1 nanometer = 10^{-9} m).

The most important feature of the model is the specific pairing of the bases. Only two complementary base pairs, A-T and G-C, can form stable bonds in the double-helical structure (Fig. 1.7). As a result of this, the nucleotide sequence in one strand dictates the nucleotide sequence of the other. In other words, the two strands are complementary. The A-T base pair has two hydrogen bonds, and the G-C pair has three hydrogen bonds. This specific complementary base pairing is of central importance for many functions of the nucleic acids (for example, DNA replication transcription, and translation).

Another property of the model is that the two chains of the double helix are oriented with opposite (antiparallel) polarity in terms of the 3′, 5′-phosphate-deoxyribose linkages (Fig. 1.8).

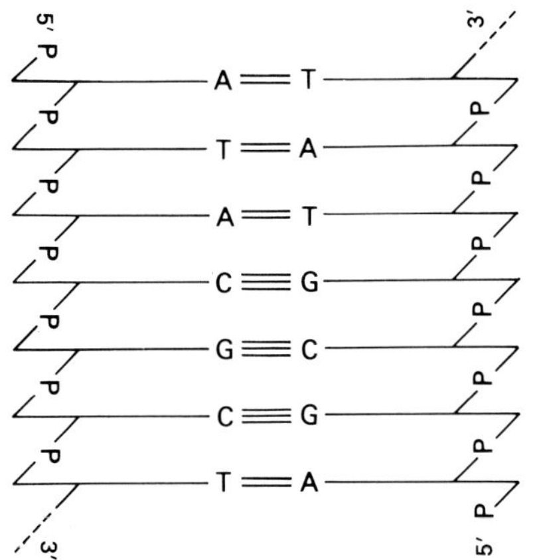

Fig. 1.8. A diagrammatic representation of double-stranded DNA showing the opposite polarity of the two strands.

Different DNA forms

The Watson and Crick model is a right-handed double helix of DNA. This is designated B DNA and is the most common of the configurations that DNA can assume. A stylized diagram of B DNA is shown in Fig. 1.9a. There are 10.0 base pairs per turn of the helix, and the base pairs are more-or-less perpendicular to the helix axis.

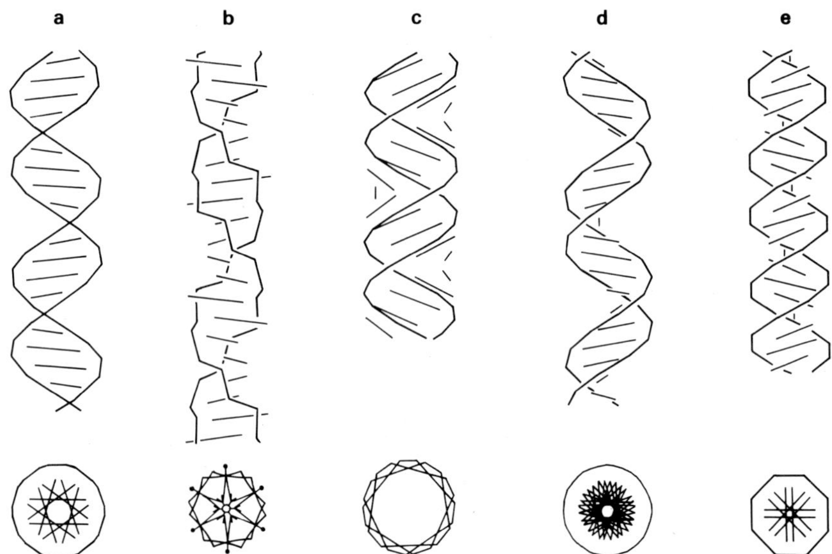

Fig. 1.9. Models for various forms of DNA. In each case a segment containing 20 base pairs is shown. The upper views are perpendicular to the helical axes and the lower views are "end on", looking along the helical axes. The continuous helical lines are the sugar-phosphate backbones, and the line segments indicate the positions of the base pairs: (**a**) B DNA; (**b**) Z DNA; (**c**) A DNA; (**d**) C DNA; (**e**) D DNA. (Reproduced with permission, from the *Annual Review of Biochemistry*, Vol. 51. 1982 by Annual Reviews Inc.)

Other forms of DNA have been described that differ from B DNA with respect to the direction of helical coiling and/or spacing and tilting of the base pairs. In Z DNA (named because of the zigzag appearance of the sugar-phosphate backbones), the backbones form a left-handed double helix in which the base pairs are arranged relatively peripherally to the central helix axis, and there are 12.0 base pairs per turn of the helix (Fig. 1.9b). This DNA form has been shown to exist in at least some eukaryotic chromosomes and may occur in some DNA locations in which gene function is being regulated. In vitro, the interconversion of B and Z forms of DNA has been demonstrated.

While B and Z DNA are the major forms of DNA encountered, other DNA forms have been described in natural and synthetic DNA. These are the right-handed double-helical A, C, and D forms (Fig. 1.9c, d, and e, respectively), which have 11.0, 7.9–9.6, and 8.0 base pairs per turn, respectively. In the A form the base pairs are arranged toward the outside of the helix when viewed along the helical axes (Fig. 1.9b, bottom), and they are inclined at about a 19° angle to the helix axis (the base pairs are perpendicular to the helix axis in B DNA). In the C form, the base pairs are arranged towards the middle of the helix (as is the case in B DNA) and, like A DNA, are at an inclined angle to the helix axis, although the angle of inclination is less extreme than in A DNA. In D DNA the helix as viewed along the helix axis is not circular but hexagonal in cross section. In this form of DNA, the base pairs are arranged toward the middle of the helix and are at an inclined angle to the helix axis. Note that the angles of inclination of the base pairs are approximately opposite in the C and D forms of DNA (compare Figs. 1.9d and e).

Evidence that DNA is the genetic material

Many lines of evidence strongly indicate that DNA is the genetic material in may organisms. Three examples are given here:

1. The nucleic acids show maximal absorbance of ultraviolet light at a wavelength of 260 nm (Fig. 1.10), and this correlates exactly with the wavelength at which maximal mutagenesis of cells can be achieved by ultraviolet irradiation. This observation provided further evidence that nucleic acids and not proteins (which show maximal absorbance of light at 280 nm) are the genetic material.

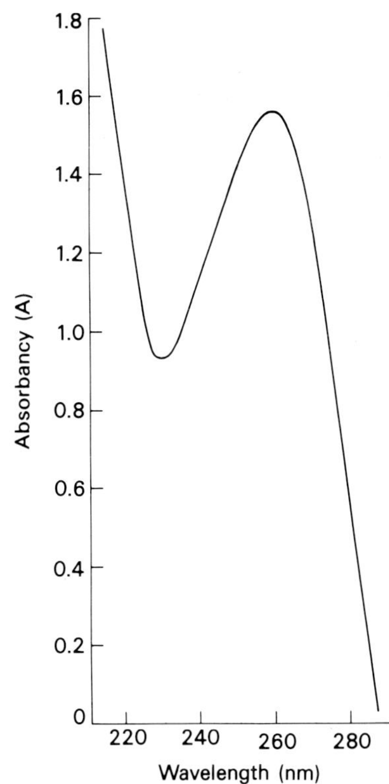

Fig. 1.10. Ultraviolet light absorbancy spectrum of DNA showing the maximal absorbancy.

2. In 1928 F. Griffith discovered that the S strain of the bacterium *Diplococcus pneumoniae* (pneumococcus), when injected into mice, causes death by septicemia (blood poisoning). Another strain, the R strain, had no effect on the same mice. The distinction between the two strains lies in the fact that the S strain bacteria have a polysaccharide capsule around them, resulting in a smooth colony appearance when they grow on solid medium in a culture dish (and hence the S designation). The R strain produces rough-appearing colonies, owing to the lack of the capsule. Griffith showed that the

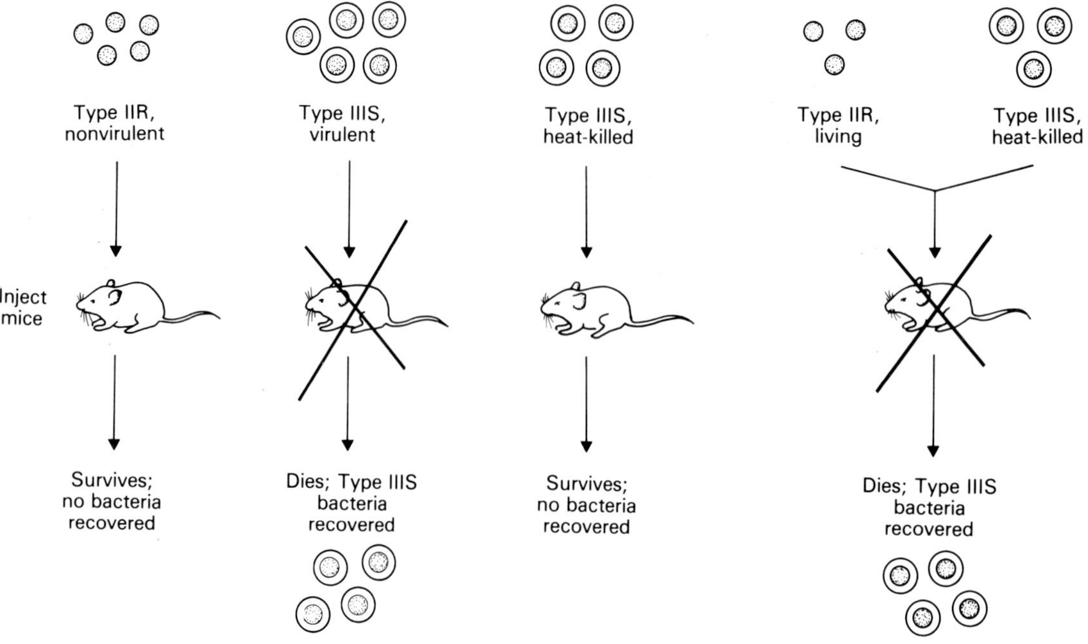

Fig. 1.11. Transformation experiment of Griffith. (After M. W. Strickberger, 1976. *Genetics*. Macmillan, New York.)

S bacteria could mutate spontaneously to give rise to the R type. Moreover, when mice were injected with a combination of live R bacteria and heat-killed S bacteria, the mice died from septicemia and live S bacteria could be isolated from their blood (Fig. 1.11). Thus something from the dead bacteria converted the R bacteria into S-type cells; this process is called transformation.

The transformation phenomenon received further scrutiny from O. T. Avery, C. M. Macleod, and M. McCarty in 1944. In some classic experiments they set out to determine the chemical nature of the substance (the so-called *transforming principle*) that induced the specific transformation of the pneumococcal types. They showed that a DNA fraction isolated from the S strain was capable of transforming unencapsulated R-type bacteria into fully encapsulated S-type cells. None of the other cell fractions, such as RNA, protein, lipid, carbohydrates, etc., was able to effect the transformation. Further, the transforming activity of the DNA fraction could be abolished by treatment with **deoxyribonuclease** (DNase), a DNA-degrading enzyme, but not by **ribonuclease** (RNase), an RNA-degrading enzyme. These results strongly indicated that DNA was the genetic material. However, Avery's work was criticized because the nucleic acids isolated were not completely pure and contained proteins.

The ultimate proof that DNA is genetic material came from other experiments such as that performed by A. D. Hershey and M. Chase, which is described next.

3. Hershey and Chase in 1952 studied the replication of bacteriophages (bacterial viruses) in their bacterial hosts. Bacteriophages consist of two components: DNA and protein. Hershey and Chase infected *Escherichia coli* cells growing in media containing either a radioactive isotope of phosphorus (^{32}P) or a radioactive isotope of sulfur (^{35}S) with bacteriophage T2 and collected the progeny phages that were produced (Fig. 1.12a). Since DNA

Fig. 1.12. The Hershey and Chase experiment. (**a**) Production of T2 phages either with (i) ^{32}P-labeled DNA or with (ii) ^{35}S-labeled protein. (**b**) Experiment that showed DNA and not protein was the genetic material of T2. In (i), *E. coli* is infected with ^{32}P-labeled T2, and in (ii), *E. coli* is infected with ^{35}S-labeled T2. (After P. J. Russell, 1986. *Genetics*. Little, Brown and Co., Boston.)

a

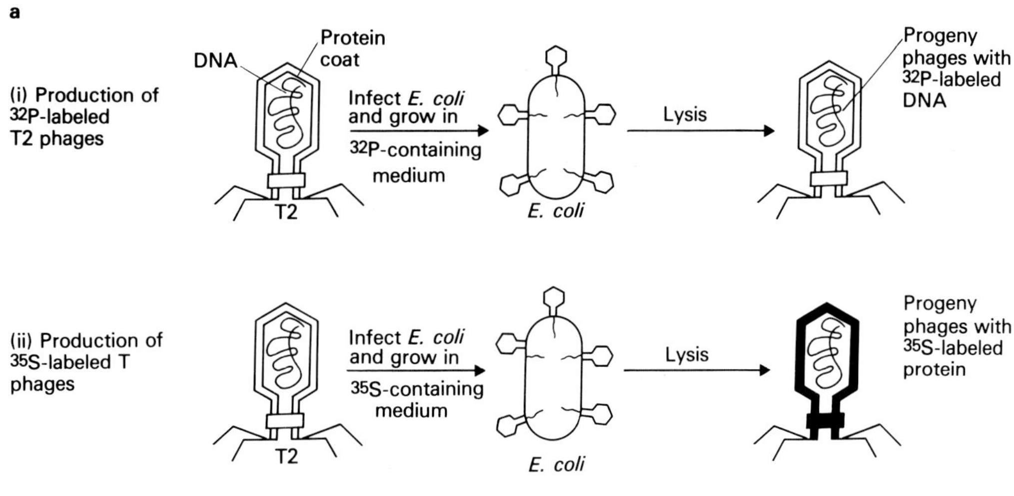

b

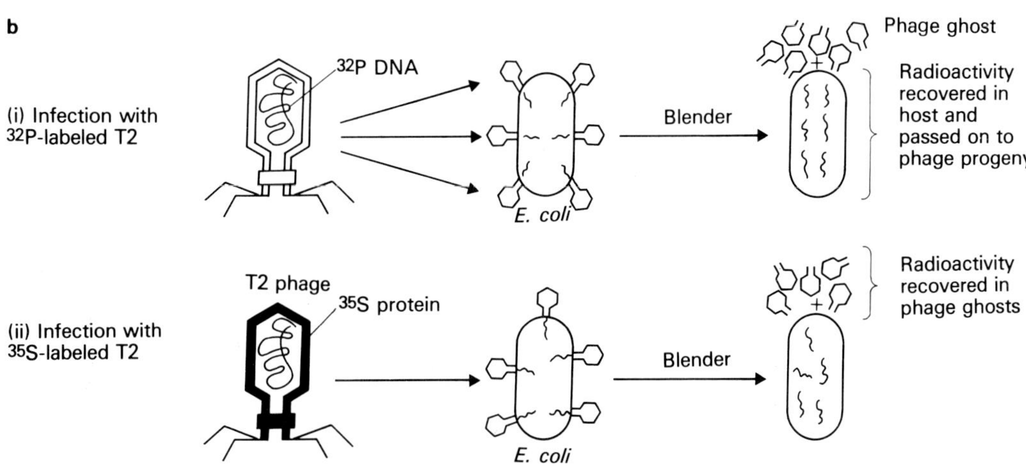

contains phosphorus but no sulfur and protein contains sulfur but no phosphorus, the DNA in the phages produced in ^{32}P-containing medium was radioactive and the protein in the phages produced in ^{35}S-containing medium was radioactive. Since each T2 phage consists of approximately half DNA and half protein, one of these two components must be the genetic material. To determine which component it was, Hershey and Chase next infected unlabeled *E. coli* with the two types of radioactively labeled T2 and obtained the results depicted in Fig. 1.12b. When the infecting phage was ^{32}P-labeled (Fig. 1.12b.i), most of the radioactivity was found within the bacteria soon after infection, and after lysis of the cells some of the label was found in the progeny phages. Very little label was found in the protein parts of the phages (called the phage ghosts) that were released from the cell surface by mixing the cells in a kitchen blender. When the infecting phage was ^{35}S-labeled (Fig. 1.12b.ii), essentially none of the radioactivity appeared either within the cell or in progeny phages, whereas most of the radioactivity was found in the phage ghosts released after the blender treatment.

We now know that the genetic material in most organisms is DNA. In some viruses, however, the genetic material is RNA, as will be discussed later.

Questions and problems

1.1 By differentially labeling the coat protein and the DNA of phage T2, Hershey and Chase demonstrated that (choose the right answer):
(a) only the protein enters the infected cell
(b) the entire virus enters the infected cell
(c) a metaphase chromosome is composed of two chromatids, each containing a single DNA molecule
(d) the phage genetic material is most probably DNA
(e) the phage coat protein directs synthesis of new progeny phage

1.2 In the 1920s, while working with *Diplococcus pneumoniae,* the agent that causes pneumonia, Griffith discovered an interesting phenomenon. In the experiments, mice were injected with different types of bacteria. For each of the following bacterial type(s) injected, indicate whether the mice lived or died:
(a) type IIR
(b) type IIIS
(c) heat-killed IIIS
(d) type IIR + heat-killed IIIS

1.3 Several years after Griffith described the transforming principle, Avery, MacLeod and McCarty investigated the same phenomenon.
(a) Describe their experiments.
(b) What did their experiments demonstrate beyond Griffith's?
(c) How were enzymes used as a control in their experiments?

1.4 The double helix model of DNA as suggested by Watson and Crick was based on a variety of lines of evidence gathered on DNA by other researchers. The facts fell into the following two general categories; give three examples of each:

(a) chemical composition
(b) physical structure.

1.5 For double-stranded DNA, which of the following base ratios always equals 1?
(a) (A+T)/(G+C)
(b) (A+G)/(C+T)
(c) C/G
(d) (G+T)/(A+C)
(e) A/G

1.6 If the ratio of (A+T) to (G+C) in a particular DNA is 1.00, does this result indicate that the DNA is most likely constituted of two complementary strands of DNA or a single strand of DNA, or is more information necessary?

1.7 Explain whether the (A+T)/(G+C) ratio in double-stranded DNA is expected to be the same as the (A+C)/(G+T) ratio.

1.8 The genetic material of bacteriophage ΦX174 is single-stranded DNA. What base equalities or inequalities might we expect for single-stranded DNA?

1.9 A double-stranded DNA molecule is 100,000 base-pairs (100 kilobase pairs) long.
(a) How many nucleotides does it contain?
(b) How many complete turns are there in the molecule?
(c) How long is the DNA molecule?

References

Avery, O.T., C.M. Macleod and M. McCarty, 1944. Studies on the chemical nature of the substance inducing transformation of pneumococcal types. Induction of transformation by a desoxyribonucleic acid fraction isolated from pneumococcus type III. *J. Exp. Med.* **79**:137–158.

Chargaff, E., 1950. Chemical specificity of nucleic acids and mechanism of their enzymatic degradation. *Experientia* **6**:201–209.

Chargaff, E., 1951. Structure and function of nucleic acids as cell constituents. *Fed. Proc.* **10**:654–659.

Davidson, J.N., 1972. *The Biochemistry of the Nucleic Acids,* 7th ed. Chapman and Hall, London.

Dickerson, R.E., H.R. Drew, B.N. Conner, R.M. Wing, A.V. Fratini and M.L. Kopka, 1982. The anatomy of A-, B-, and Z-DNA. *Science* **216**:475–485.

Fraenkel-Conrat, H. and B. Singer, 1957. Virus reconstitution: Combination of protein and nucleic acid from different strains. *Biochim. Biophys. Acta* **24**:540–548.

Franklin, R.E. and R. Gosling, 1953. Molecular configuration of sodium thymonucleate. *Nature* **171**:740–741.

Freifelder, D. 1978. *The DNA Molecule. Structure and Properties.* W.H. Freeman, San Francisco.

Hershey, A.D. and M. Chase, 1952. Independent functions of viral protein and nucleic acid in growth of bacteriophage. *J. Gen. Physiol.* **36**:39–56.

Pauling, L. and R.B. Corey, 1956. Specific hydrogen-bond formation between pyrimidines and purines in deoxyribonucleic acids. *Arch. Biochem. Biophys.* **65**:164–181.

Watson, J.D. 1968. *The Double Helix.* Atheneum, New York.

Watson, J.D. and F.H.C. Crick, 1953. A structure for desoxyribose nucleic acids. *Nature* **171**:737–738.

Wilkins, M.H.F., A.R. Stokes and H.R. Wilson, 1953. Molecular structure of deoxypentose nucleic acids. *Nature* **171**:738–740.

Zimmerman, S.B., 1982. The three-dimensional structure of DNA. *Annu. Rev. Biochem.* **51**:395–497.

Chapter 2 The Genetic Material and Chromosome Structure

Outline

Phage chromosomes
- Terminal redundancy
- Circular permutation

Bacterial chromosomes

Eukaryotic chromosomes
- Karyotype
- Chromatin composition
- Nucleosomes
- Euchromatin and heterochromatin
- Repetitive DNA sequences

Transposable genetic elements
- Prokaryotic transposable genetic elements
  - Insertion sequence elements
  - Transposons
  - Transposition of IS elements and transposons
- Eukaryotic transposable genetic elements
  - *Ty* elements in yeast
  - Human transposable genetic elements

Phage chromosomes

A number of phages can infect the intestinal bacterium *Escherichia coli* (*E. coli*). The T-series of phages all have double-stranded DNA as their genetic material (Fig. 2.1). When they are added to a culture of *E. coli*, they attach to the outer surface of the bacterium and then inject a single molecule of DNA into the host. Once inside, the DNA is replicated, progeny phages are assembled and eventually the bacterium lyses releasing progeny into the medium (Fig. 2.2). Phages like the T-even phages that always result in the lysis of the infected bacterial cell are called **virulent phages**.

The chromosomes of the T-phages are naked DNA in that no proteins are attached. In both the T2 and T4 phages, the chromosomes are longer than the complete genome. This results from the chromosomes being *terminally redundant* and *circularly permuted* as if each phage particle has a standard length of DNA randomly excised from a long chain of chromosomes joined end-to-end. On the other hand, the T3, T5, and T7 phages have terminally

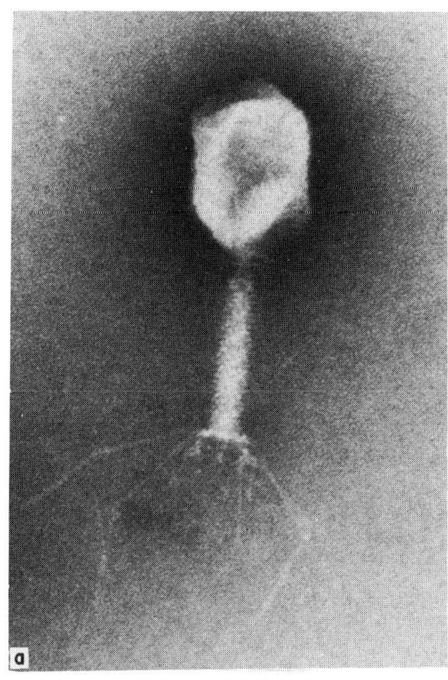

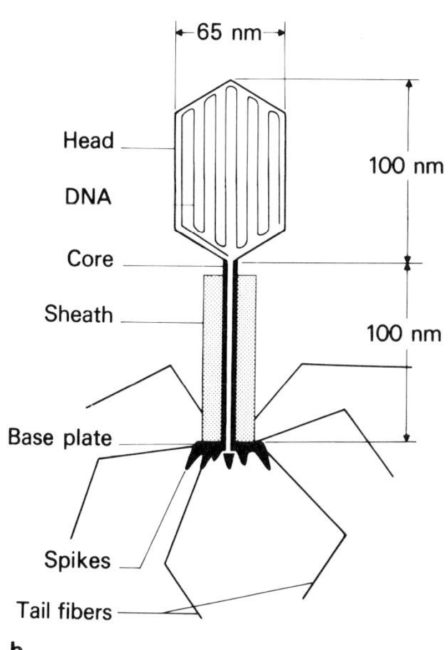

Fig. 2.1. Bacteriophage T4. (**a**) EM photograph of phage T4 (×264,000, negatively stained. Courtesy of M. Wurtz). (**b**) Diagrammatic representation of a T4 phage. (From W. H. Hayes, *Genetics of Bacteria and Their Viruses*. Copyright © 1968, Blackwell Scientific Publications Ltd., With permission from W. H. Hayes and Blackwell Scientific Publications, Oxford.)

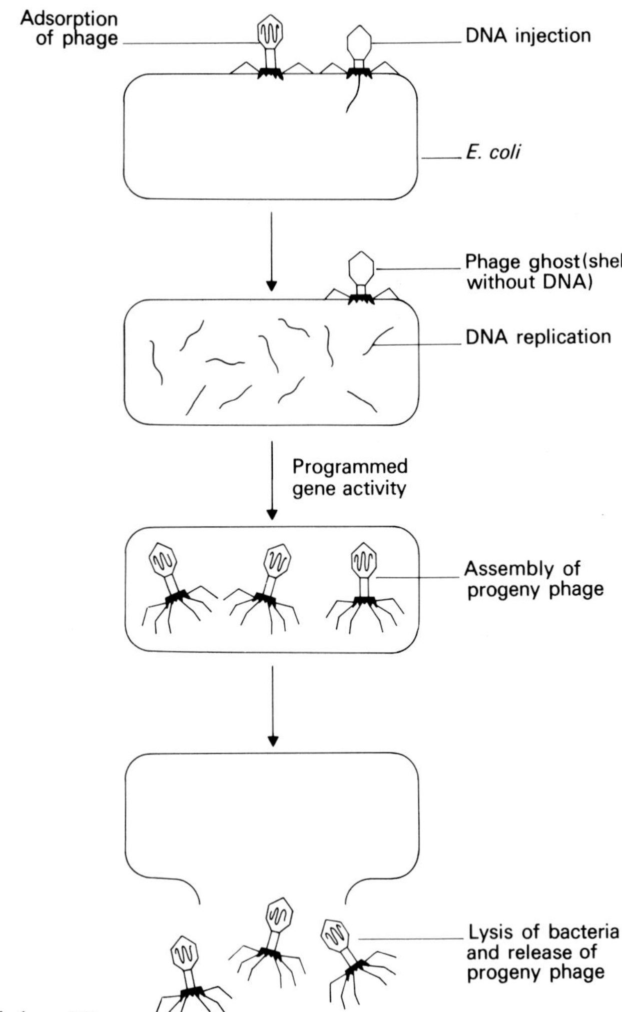

Fig. 2.2. Life cycle of phage T2.

redundant but not circularly permuted chromosomes. These different types of chromosome arrangements are shown in the following:

123456	Complete genome
12345612	Terminally redundant chromosome
123456 456123 345612	Circularly permuted chromosomes
12345612 34561234 61234561	Terminally redundant and circularly permuted chromosomes

The experimental evidence for these types of chromosome arrangements follows.

Terminal redundancy

If a chromosome is treated with the enzyme exonuclease III, nucleotides are removed from the 3′ end of each strand of the DNA. If the chromosome is terminally redundant, this treatment will expose complementary 5′-ended strands at the two ends of the linear molecule. These ends may then pair by hydrogen bonding to form circles, which may be visualized by electron microscopy (Fig. 2.3).

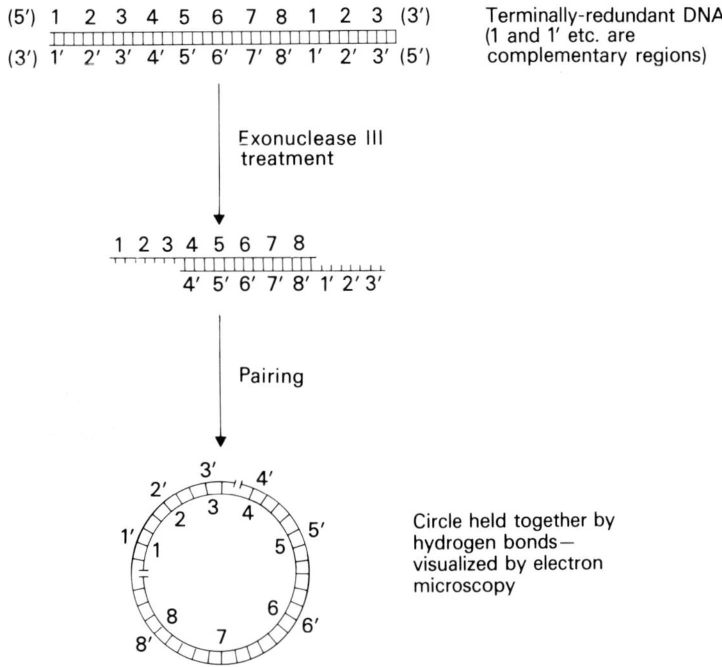

Fig. 2.3. Demonstration of terminally redundant DNA by exonuclease treatment. (After W. H. Hayes, 1968, *Genetics of Bacteria and Their Viruses*. Blackwell Scientific Publications, Oxford.)

Circular permutation

If a linear double-stranded DNA molecule is heated, eventually the two strands will separate (denature) by breakage of the hydrogen bonds between the base pairs. If allowed to cool, complementary strands of DNA can come together again (renature) to form double-stranded molecules. If a population of circularly-permuted DNA molecules is denatured and then allowed to renature, it is possible that the terminal region of one strand will be complementary to the central part of a second strand. If the complementary regions anneal, a molecule is produced that is double-stranded in the middle and has complementary, single-stranded ends. This condition can lead to the formation of circles of double-stranded DNA (Fig. 2.4).

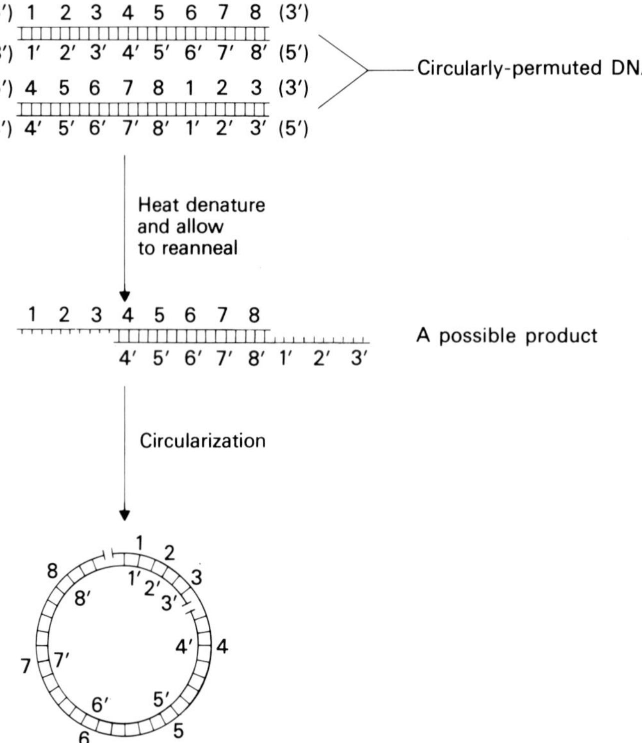

Fig. 2.4. Demonstration of circularly permuted DNA by heat denaturation and reannealing (After W. H. Hayes, 1968, *Genetics of Bacteria and Their Viruses.* Blackwell Scientific Publications, Oxford.)

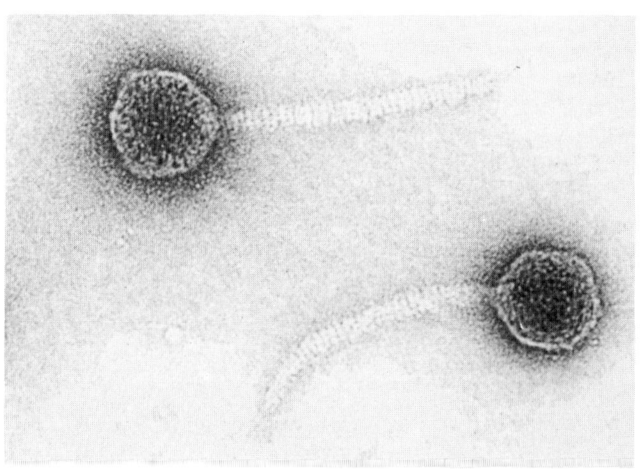

Fig. 2.5. Electron micrograph of phage lambda (λ). (×258,000; negatively stained with 2% uranyl acetate. Courtesy of R. B. Luftig.)

Another phage that infects *E. coli* is *lambda* (λ) (Fig. 2.5). Its chromosome is double-stranded DNA, and in the phage particle the chromosome is a linear molecule. If the chromosome is heated and then cooled, circular structures are seen under the electron microscope. The explanation for this is that the two ends of the chromosome have complementary single-stranded regions that can therefore form hydrogen bonds between them to produce circles. This property is exhibited during the λ life cycle; that is, when the λ DNA is injected into the host bacterium, the DNA rapidly becomes a covalently bonded circle as a result of the cohesive ends and with the aid of specific enzymes. This event is necessary to enable the phage chromosome either to replicate or to integrate into the bacterial chromosome.

Other viruses present us with examples of a variety of chromosome structures: for example, phage ΦX174 has a single-stranded, circular DNA chromosome; phage Qβ (Q-beta) and the plant virus tobacco mosaic virus (TMV) have single-stranded RNA chromosomes; and polyoma viruses have double-stranded RNA for their genetic material.

Bacterial chromosomes

The chromosomes of bacteria are circular, naked, double-stranded DNA molecules. The DNA is usually found attached to the cell membrane at some point or points. Although bacteria do not possess a nucleus, the DNA is localized in a distinct area within the cell called the nucleoid region. Within this region, the DNA is highly convoluted and folded, and very few cytoplasmic particles are found there. There is no membrane around the nucleoid region. An electron micrograph of an *E. coli* nucleoid is shown in Fig. 2.6.

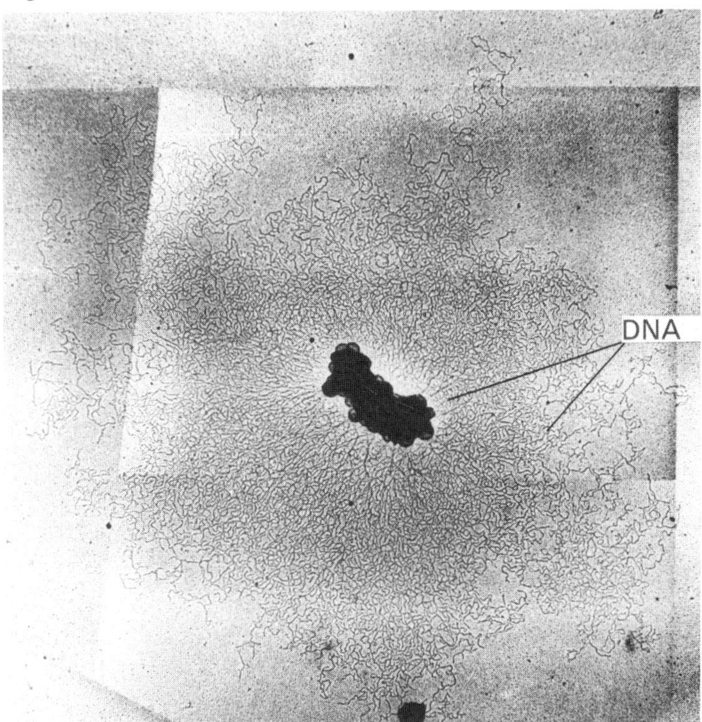

Fig. 2.6. Electron micrograph of an *E. coli* nucleoid. (Courtesy of M. Meyer, Universiteit van Amsterdam. From M. Meyer et al., 1976. *Eur. J. Biochem.* 63:469–475.)

The chromosomes of bacteria can be isolated in a highly folded conformation by gently lysing the cells at room temperature with nonionic detergents in 1.0 M NaCl. Electron microscopic analysis of the chromosomes reveals extensive packing of the DNA, which are folded into loops (10 to 80 per chromosome) and supercoils (Fig. 2.7). The supercoiling of the DNA is controlled by enzymes called **topoisomerases**. The folded chromosomes have been shown to contain all of the nascent RNA chains of the cell and the enzyme for RNA synthesis, RNA polymerase, but no ribosomes. Hence protein synthesis apparently does not occur in the proximity of the DNA.

In addition to the main chromosome, bacteria often contain other DNA material, found as small double-stranded DNA circles called **plasmids**. These plasmids replicate independently of the main chromosome. In natural populations of bacteria, the DNA present in plasmids may constitute up to 1–2% of the cellular DNA amount.

Fig. 2.7. Electron micrograph of the chromosome extracted from *E. coli*. Note the extensive supercoiling of the DNA. The bar is 2 μm. (Courtesy of A. Worcel. With permission from H. Delius and A. Worcel, 1974. Electron microscope visualization of the folded chromosome of *Escherichia coli*. *Journal of Molecular Biology* **82**:107–109. Copyright © 1974, Academic Press Ltd., London.)

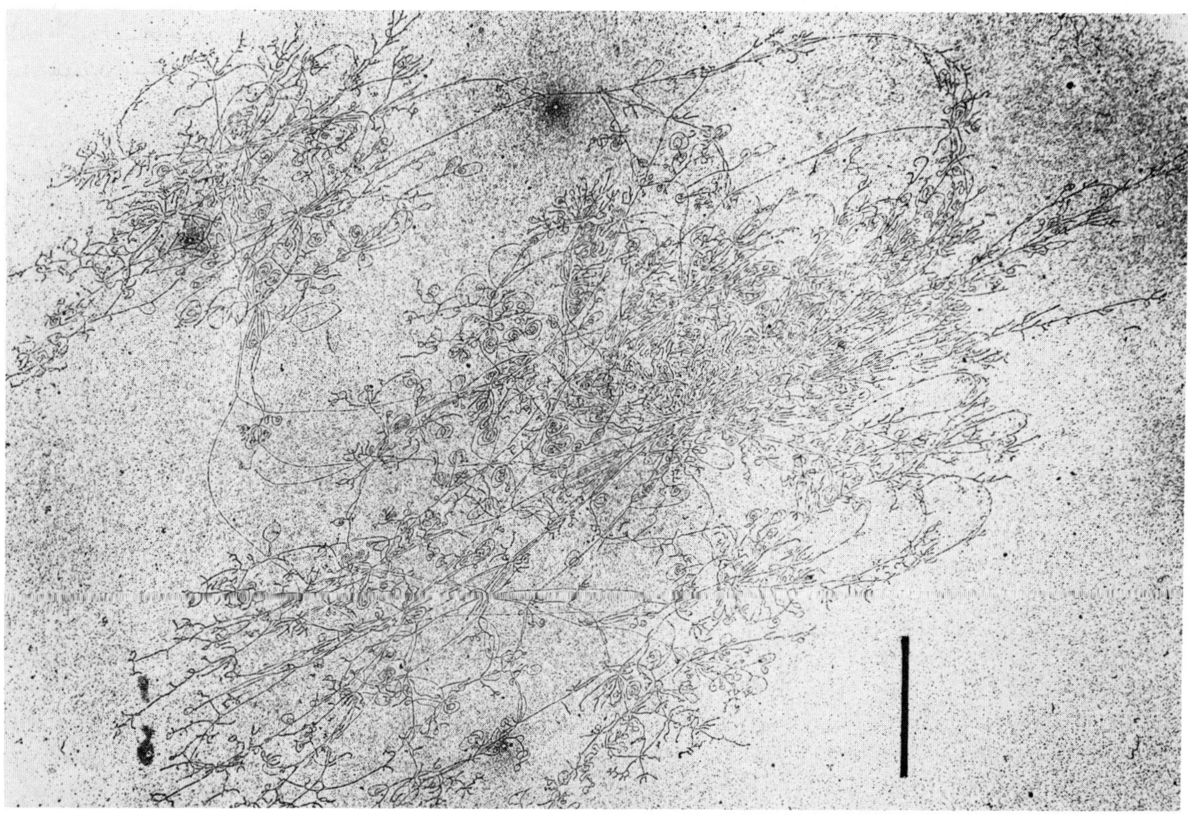

Eukaryotic chromosomes

In eukaryotes, most of the DNA is located in the chromosomes found in the nucleus, and the following discussion concentrates on these. However, some DNA is found in mitochondria and chloroplasts, and in these instances it is naked, double-stranded, and circular, that is, very similar to the organization of the genetic material in bacteria.

Karyotype

The genetic material of eukaryotes is distributed among multiple chromosomes, the number of which is characteristic of the organism. Most eukaryotes have two copies of each type of chromosome and those organisms are said to be **diploid** (2N) in terms of their chromosome complement. The gametes of all eukaryotes and the cells of some so-called lower eukaryotes (for example, the fungus *Neurospora crassa*) have a haploid (N) set of chromosomes. The complete haploid complement of genetic information is called the genome. In diploid cells, the members of a pair of chromosomes are called **homologous chromosomes**, while each member of a pair is called a homolog. Homologous chromosomes are identical with respect to the genetic loci they contain. Chromosomes from different pairs are called **nonhomologous chromosomes**.

In eukaryotic cells, the chromosomal complement is called its **karyotype** and is characterized both by the number and morphology of the chromosomes. The latter includes both the relative sizes of the chromosomes and the positions of the centromeres (where spindle fibers attach during the cell division processes). In animals the karyotype is generally different in males and females owing to the presence of different complements of X and Y chromosomes (**sex chromosomes**). Barring chromosomal aberrations, the karyotype for the **autosomes** (that is, the set of chromosomes other than the sex chromosomes) is invariant within species but differs from species to species. A karyotype for a normal human male is shown in Fig. 2.8: there are 46 chromosomes, which, because humans are diploid, involve 22 pairs of homologous autosomes and two sex chromosomes, in this case one X and one Y. In this organism the two sex chromosomes are quite distinct morphologically. Until recently it was possible to differentiate the chromosomes into only seven groups, designated A to G and ordered according to their size. Within each group the chromosomes were almost indistinguishable morphologically. Staining techniques (called banding techniques) have been developed that give rise to specific patterns of bands along the chromosomes that serve to differentiate each chromosome in the karyotype. In Fig. 2.9, for example, the G (acetic acid-saline-Giemsa) banding pattern of human chromosomes is shown.

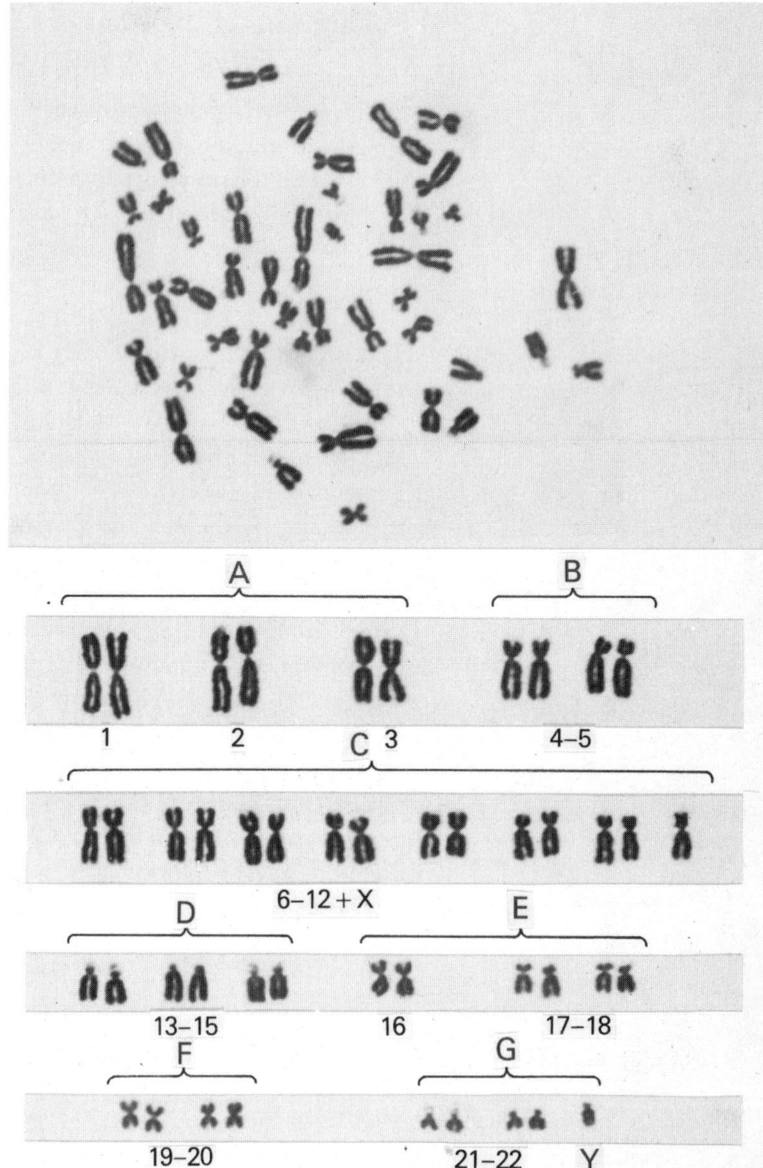

Fig. 2.8. Human karyotypes. Top: mitotic chromosomes of a normal human male (46,XY). Bottom: Unbanded karyotype: the same chromosomes arranged in homologous pairs. (Courtesy of A. T. Sumner, with permission from C. J. Bostock and A. T. Sumner, *The Eukaryotic Chromosome*. Copyright © 1978, North Holland Publishing Co., Amsterdam, The Netherlands.)

Chromatin composition

Most biochemical studies of chromosome structure and function have been done with chromosomes isolated from cells in interphase. (Interphase is the period in the cell division cycle of proliferating cells when the cell is not

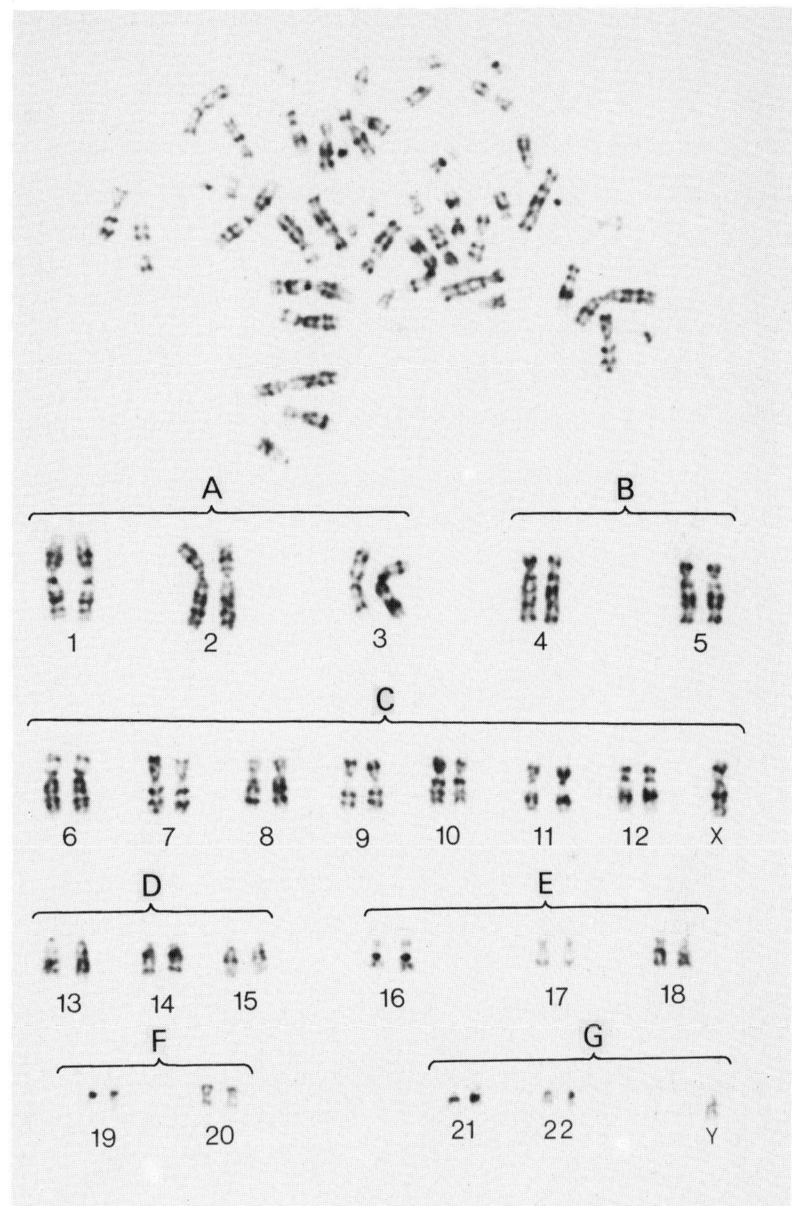

Fig. 2.9. Human karyotypes. Top: Human male mitotic chromosomes stained with acetic acid-saline-Giemsa (ASG) to show G-banding pattern. Bottom: The same G-banded chromosomes arranged in homologous pairs and numbered. (Courtesy of A. T. Sumner, with permission of A. T. Sumner et al. *Nature New Biology* **232**:31–32. Copyright © 1971. Macmillan Journals Ltd., England.)

undergoing either mitosis or meiosis.) When nuclei are isolated and then lysed, the chromosomes are released. Each chromosome contains a single, unbroken, double-stranded DNA running through its length. This DNA is associated with proteins in a complex called chromatin. When purified chromatin is analyzed, it is found that both basic proteins (**histones**) and acidic proteins (**nonhistones**) are associated with the DNA and this is characteristic of eukaryotic nuclear chromosomes. All eukaryotic chromosomes contain five distinct histone proteins attached to the DNA in a DNA–histone complex: H1, which is very rich in the basic amino acid lysine; H2A and H2B, which are lysine-rich; and H3 and H4, which are rich in the basic amino acid arginine (Table 2.1). These molecules have been conserved through evolution, as would be expected for the proteins since they have such a fundamental and universal role in determining chromosome structure. Weight for weight there is about an equal amount of histone and of DNA in chromatin.

Table 2.1. Characteristics of calf thymus histones. (After S. C. R. Elgin and H. Weintraub, 1975. *Annu. Rev. Biochem.* **44**:725.)

Type	Characteristics	No. of amino acids	Molecular weight
H1	Very lysine rich	~ 215	~ 21,500
H2A	Lysine rich	129	14,000
H2B	Lysine rich	125	13,775
H3	Arginine rich	135	15,320
H4	Arginine rich	102	11,280

The nonhistone acidic proteins are found more or less firmly attached to the DNA–histone complexes. In contrast to the histones, the molecules are numerous and heterogeneous. In most organisms that have been examined, there are more than 100 nonhistones, and these include the replication and transcription enzymes, DNA polymerase and RNA polymerase, and molecules involved in the control of DNA and RNA synthesis. Thus, as would be expected in view of this, the complement of nonhistones varies throughout the cell cycle and from one differentiated cell to another. This is in contrast to the histones, which remain constant in both these instances. Weight for weight there may be as much nonhistone protein present as DNA and histone combined.

At present, there is a large body of information on the organization of the histones along the DNA. The number and arrangement of nonhistones on the DNA have not been defined in as much detail.

Nucleosomes

Chromosomes are tight complexes between DNA and protein. It has long been known that the DNA in chromosomes is much longer than the chromosome length, and thus some means of compacting the genetic material must be available. Indeed, in chromatin the DNA is in a tightly packed state representing at least a 100-fold contraction. This is achieved by several levels of folding. The simplest level of packing involves the coiling of DNA around a core of histones to form a structure called the **nucleosome**, and the most complex level of packing involves higher-order coiling of the DNA–histone complexes to produce chromosome morphologies exemplified by chromosomes in the division cycle, such as those shown in the karyotypes of Figs 2.8 and 2.9.

When viewed under the electron microscope, the DNA–protein complex in chromosomes is seen as fibers, or nucleofilaments about 10 nm in diameter. If a nucleofilament is completely unraveled, the chromatin fiber looks like "beads on a string," where the beads are the nucleosome (Fig. 2.10a) and the thinner thread connecting the beads is naked DNA, called linker DNA.

The nucleosome was first described in the 1970s by A. Olins and D. Olins, and by C. L. F. Woodcock. They described the nucleosome as 10-nm diameter, flattened spheres in which the DNA was somehow associated with a core of histones. Our current information indicates that the nucleosome is found at about 200 base-pair intervals along the DNA molecule. If chromatin is treated for a short time with a bacterial endonuclease, micrococcal nuclease (which will digest any DNA not protected from its action by association with proteins), the linker DNA between nucleosomes is degraded and individual nucleosomes are released. (This procedure was pioneered by R. Kornberg.) Studies of isolated nucleosomes reveal them to be flat, cylindrical particles of dimensions $11 \times 11 \times 5.7$ nm (shaped like a hockey puck or a can of tuna) with the 2-nm diameter DNA wrapped around the histone core with 80 base pairs per turn (see Fig. 2.11). Since there are 146 base pairs of DNA in the average nucleosome particle, the DNA is wrapped around the core approximately one and three-fourths times. The association of the DNA with the core histone proteins is a tight one, since the DNA is well protected against enzyme attack.

By dissociating the nucleosomes with high salt concentration, it has been shown that each nucleosome consists of about 146 base pairs of DNA and a histone core (the nucleosome core particle: Fig. 2.10b) consisting of an octamer or two subunits each of the four histone types H2A, H2B, H3, and H4.

In the cell, chromatin exists in a more highly coiled state than the 10 nm nucleofilament. While the packaging of DNA into nucleosomes is well understood, understood less well is the higher orders of folding of the nucleosomes in the chromosomes of the cell. Above the level of the nucleosome and 10 nm nucleofilament, the next level of packing is the 30 nm chromatin fiber (Fig. 2.11). Examination of isolated 30-nm chromatin fiber reveals the compact nature of this level of chromosome structure, making it difficult to distinguish individual nucleosomes. As a consequence, the arrangement of nucleosomes

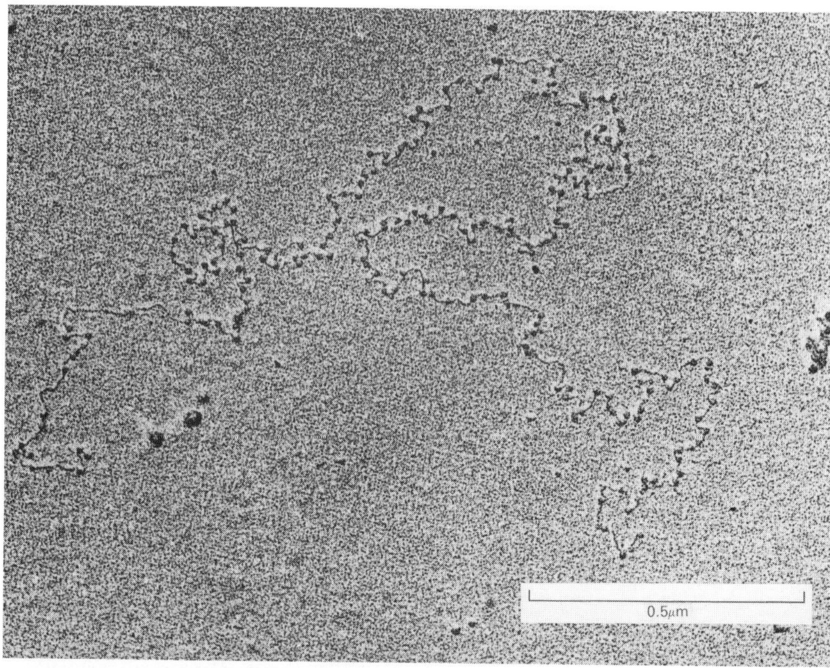

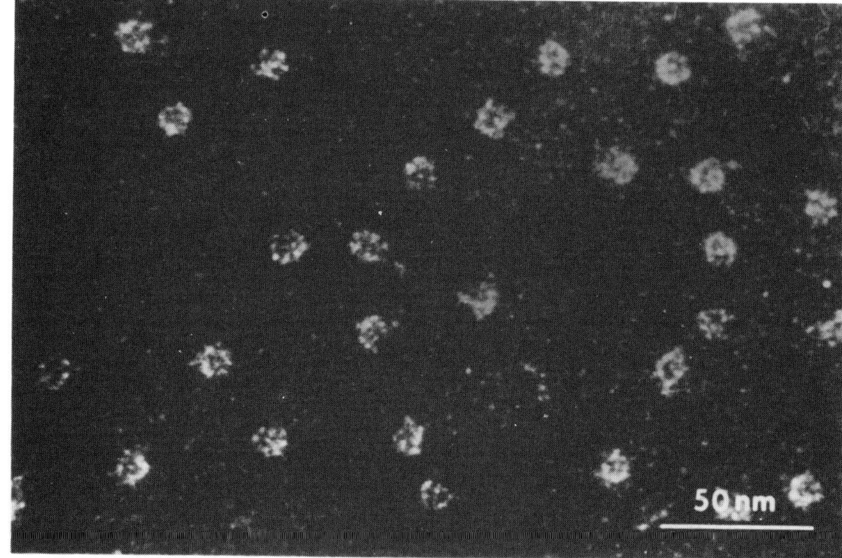

Fig. 2.10. Nucleosomes. (**a**) Electron micrograph of chick liver chromatin after removing histone H1. The chromatin resembles beads on a string, with each bead a nucleosome consisting of DNA wrapped around a core of two each of histones H2A, H2B, H3 and H4. (Courtesy of P. Oudet. From P. Oudet et al. 1975. Electron micrograph of chick liver chromatin showing nucleosome beads. *Cell* **4**:281–300. Copyright 1975, M.I.T. Press, Cambridge, Mass.) (**b**) An electron micrograph of nucleosome core particles from the chicken. (Courtesy D. Olins. From D. Olins and A. Olins, 1978, "Nucleosomes: the structural quantum in chromosomes". *American Scientist* **66**:704–711. Reprinted with permission of American Scientist, Journal of Sigma Xi, the Scientific Research Society.)

in the fiber is difficult to define. Based on biochemical and biophysical data, a recent model for the 30-nm fiber involves a zigzag ribbon of nucleosomes that is twisted to generate the 30-nm diameter chromatin fiber: the elements of this model are shown in Fig. 2.11. In the figure, the least compact form is the relaxed zigzag ribbon, which can make the transition to the compact zigzag

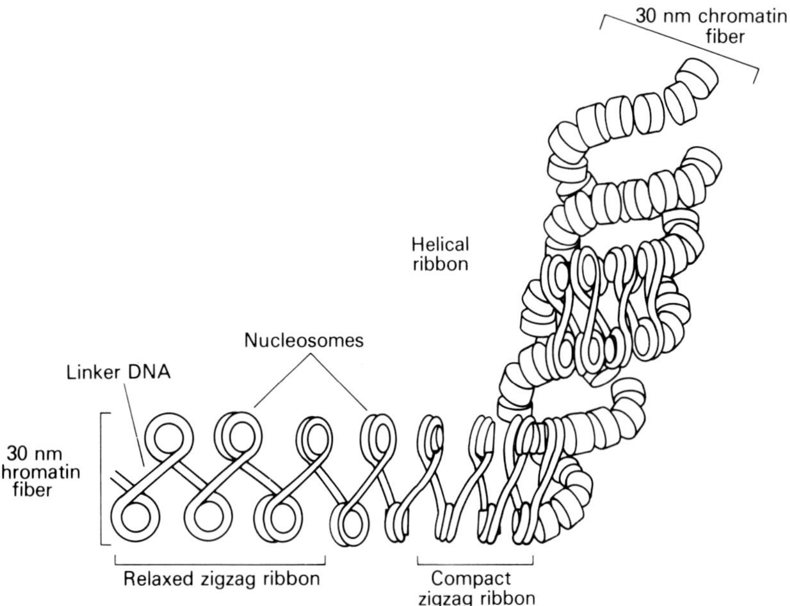

Fig. 2.11. Model for the levels of folding of a nucleosomal chain (the relaxed zigzag ribbon) via the (compact zigzag ribbon) into the helical ribbon proposed to be the basic form of the 30-nm chromatin fiber. (After C. L. F. Woodcock et al., 1984. *J. Cell Biol.* **99**:42–52.)

ribbon through no major changes in basic structure, especially the loss or addition of nucleosome-nucleosome contacts. Then the 30-nm chromatin fiber is constructed by folding the compact zigzag ribbon in such a way as to minimize changes in nucleosome-nucleosome contacts. This coiling of the compact zigzag ribbon creates a double-helical ribbon with a range of possible diameters and with the nucleosomes in face-to-face contact.

What of histone H1? There is very good evidence that H1 is located at the entry/exit point of the DNA on the nucleosome. Experiments have shown that chromatin without H1 can form 10-nm nucleofilaments but not 30-nm chromatin fibers; hence, H1 must function to pack the nucleosomes together into the 30-nm fibers. Nothing is known about the regulation of this phenomenon, however.

The next levels of packing beyond the 30-nm chromatin fiber are poorly understood. The diameter of an interphase chromosome may be 300 nm while the diameter of a metaphase chromosome may be 700 nm. To achieve these levels of compaction, the simplest concept is that the 30-nm fiber becomes looped and coiled (like a rope) in various ways to give the morphologies characteristic of these chromosomes.

Euchromatin and heterochromatin

When interphase chromosomes of metabolically active cells are chemically stained and examined under the microscope, it becomes apparent that there are two distinct types of organization of the chromatin material. One type is lightly staining and is called *euchromatin*, and the other type is darkly staining and is called *constitutive heterochromatin*. The former involves chromatin that is in a relatively uncoiled state, whereas the latter exhibits higher-order folding of the chromatin.

The distribution of constitutive heterochromatin varies from organism to organism, and examples are known where parts of chromosomes or whole chromosomes are heterochromatinized. In general, constitutive heterochromatin is found interspersed in short segments among euchromatin and is also located around the centromeres. Functionally, euchromatin contains DNA in an active or potentially active state (that is, capable of being transcribed), whereas heterochromatin contains DNA in a transcriptionally inactive configuration. Characteristically, heterochromatin replicates later than euchromatin in the cell cycle.

Repetitive DNA sequences

If the DNA of any organism is isolated, sheared to pieces several hundred nucleotide pairs long, denatured to single strands, and then the complementary strands are allowed to reassociate, the rate of reassociation for a particular nucleotide sequence will be related to the number of copies of that sequence in the genome.

From this type of DNA-DNA reassociation experiment, it was discovered that prokaryotic chromosomes consist almost entirely of unique DNA sequences. By contrast, eukaryotes were found to have some DNA sequences that are repeated from a few to many millions of times in the genome. The properties of these *repeated DNA sequences* and of **unique sequence DNA** will now be described.

Unique (single copy) sequence DNA is represented one to a few times in the genome. About 70% of the DNA in eukaryotes is unique sequence DNA. Most of the protein-coding genes of the cell are found in unique sequence DNA. Conversely, not all unique sequence DNA contains protein-coding genes.

Some repeated sequences are found clustered, that is, tandemly repeated in the genome. Among these sequences are **satellite DNA** sequences, which are simple sequences, some with repeat units no more than six base pairs long repeated as many as 10^6 to 10^7 times in the genome, and others hundreds of base pairs long and repeated millions of times. They are representative of **highly repetitive sequences** found in the genome. The amount of highly repetitive DNA varies from a few percent to as much as 50%. Commonly, highly repetitive DNA sequences are localized around centromeres and at the ends (telomeres) of the chromosomes. Characteristically these sequences replicate later than other DNA sequences.

Some clustered repeated DNA sequences contain genes, notably those for ribosomal RNA (rRNA), transfer RNA (tRNA), and histones. The number of copies in these cases ranges from tens to thousands, and these sequences are hence examples of **moderately repetitive sequences**.

Also among moderately repetitive sequences are *interspersed repeated sequences*. These are characteristic of eukaryotic DNAs, although there is no unifying description of their arrangement. Generally, there are 10^3 to 10^5 copies of these sequences dispersed among single copy sequences in the genome. Two common patterns are found. In the *short-period* pattern found in many eukaryotes, 100 to 300 base-pair repeated sequences are interspersed with 1000 to 2000 base-pair lengths of unique sequence DNA. Between 50 and 80% of the DNA may show this pattern. In the *long-period* pattern found in *Drosophila*, 5000 base-pair repeated sequences are dispersed among single copy sequences that may be up to 35,000 base pairs or more in length. Intermediate patterns have been found in other eukaryotes.

Transposable genetic elements

The classical picture of genes has been one in which the genes are at fixed loci on the chromosomes. It is clear, however, that certain genetic elements of chromosomes of both prokaryotes and eukaryotes have the ability to mobilize themselves and move from one location to another in the genome. These "jumping genes" are more formally called **transposable genetic elements (TGE)**, and they are defined as unique DNA segments that have the capacity to insert themselves at one or more of several sites in the genome. The existence of TGEs in both prokaryotes and eukaryotes suggests that they are general features of genomes. In eukaryotes, TGEs can move to new positions within the same chromosome or to a different chromosome. In both prokaryotes and eukaryotes, TGEs can produce genetic change; for example, by inserting into a gene or the region adjacent to a gene that regulates its expression, they can affect gene expression. These effects of TGEs have been established through genetic, cytological, molecular, and recombinant DNA (see Chapter 12) procedures.

Prokaryotic transposable genetic elements

The molecular nature of TGEs is best understood in bacteria and their viruses. There are four types of TGEs in prokaryotes: **insertion sequence (IS) elements, transposons (Tn), plasmids**, and certain **temperate bacteriophages**. The first two of these will be discussed in more detail here; information about plasmids and temperate bacteriophages is presented elsewhere in the book.

Insertion sequence elements. An IS element, the simplest TGE found in prokaryotes, is a mobile segment of DNA that contains sequences required for the process of insertion of the DNA segment into the chromosome and for the mobilization of the element to different locations. The IS elements have been detected at many locations in prokaryotic genomes.

The IS elements were first identified in *E. coli* as a result of studies of a particular set of mutations that did not have properties expected of typical gene mutations. These mutations, in genes involved with the fermentation of galactose, were shown to result from the insertion of an 800–base-pair (bp) DNA segment into a gene. The particular DNA segment involved is now named *insertion sequence I*, or *IS1*.

IS1 is one of a family of genetic elements capable of integrating into a chromosome at locations with which it has no homology. This event is an example of a *transposition* event. The integration event itself occurs independently of normal DNA recombination activity.

If an IS element "lands" within a gene, the function of that gene is usually altered, although the exact effect depends on the IS element involved. For example, IS1 always decreases the activity of the gene into which it integrates, while IS2 decreases gene expression if it integrates into the chromosome in one way and increases gene expression if it integrates in the other way.

An IS element can also excise itself from the chromosome. If the excision event is perfect, the function disrupted by the integration event should be fully restored. The excision event may be imprecise, however, so that some or all parts of the genes surrounding the IS element may be deleted. Alternatively, the IS element can transpose to another site in the chromosome while leaving a copy of itself at the original site.

In *E.coli* there are four major IS elements—IS1, IS2, IS3, and IS4—and each is present in 5–30 copies per genome. Collectively, they constitute approximately 0.3% of the cell's genome. Among prokaryotes as a whole, the ISs range from 768 bp to over 5000 bp in length and they are *normal* cell constituents. All IS elements that have been sequenced end in a perfect or nearly perfect inverted repeating segment of between 20 and 40 bp. As a result, if the stretch of DNA containing an IS is denatured and allowed to reanneal, complementary base pairing can occur within the same strand of DNA (Fig. 2.12). Between the complementary ends is a segment of unpaired DNA, which gives the reannealed molecule a lollipop appearance (called a *stem-and-loop* structure) under the electron microscope.

Transposons. The next most complex type of prokaryotic TGE is the transposon (Tn). A transposon is a mobile DNA segment that, like IS elements, contains genes for the insertion of the DNA segment into the chromosome and for the mobilization of the element to other locations on the chromo-

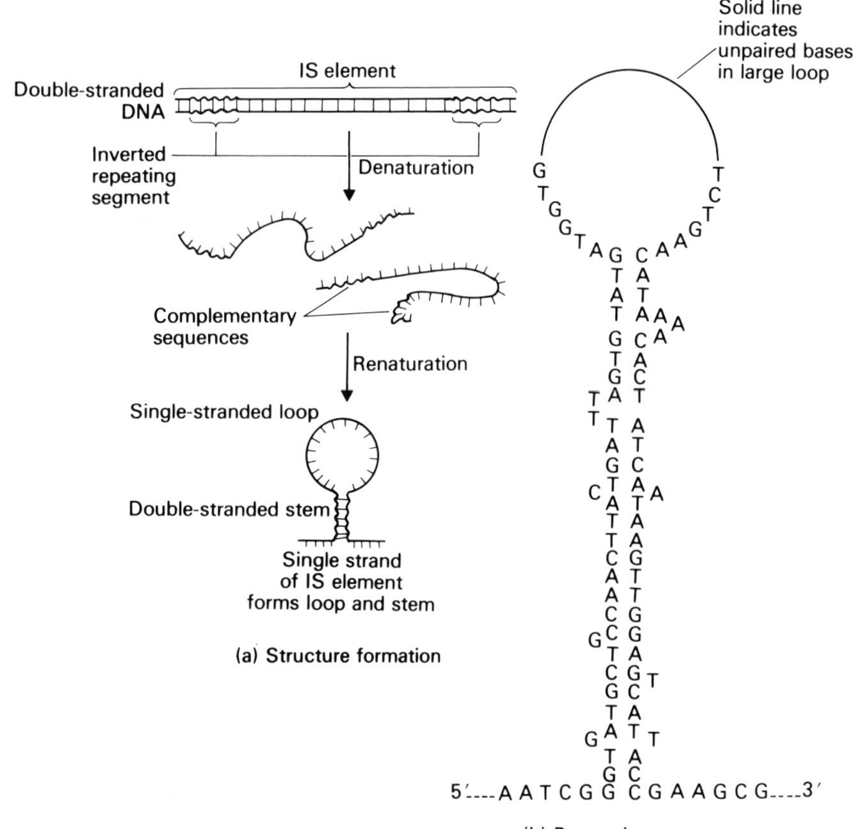

Fig. 2.12. Stem-and-loop structure of IS1 revealed by denaturing/reannealing experiment. (**a**) Formation of stem-and-loop after denaturing and reannealing; (**b**) actual base-pair sequence of the stem-and-loop structure of IS1 in *E. coli*.

some. Unlike IS elements, however, transposons also contain genes of identifiable function (e.g. for antibiotic resistance). Despite this difference, ISs and Tns are closely related; for example, Tns often end in long (800–1500 bp) direct or inverted repeats, which may be IS elements themselves or closely related to them. It seems that many Tns are mobile in the genome because they have IS elements (called *modules*) at their ends. Since IS elements themselves end in short, inverted repeats, a Tn flanked by two IS elements will also be flanked by inverted repeats. Thus, all IS elements and Tns are characterized by terminal inverted repeats.

Fig. 2.13 shows the genetic organization of Tn10. Tn10 is 9300 bp long and consists of 6500 bp of nonrepeated material flanked at each end with a 1400-bp IS-like module. The IS modules of Tn10 are designated IS10L (left) and IS10R (right) and are arranged in an inverted orientation. Since IS10L and IS10R are inverted repeats, then like IS elements, when Tn10 is denatured to single strands and allowed to reanneal, a stem-and-loop structure forms. Within the 6500-bp segment of nonrepeated material is a gene for tetracycline resistance. Thus, cells containing Tn10 are resistant to the antibiotic

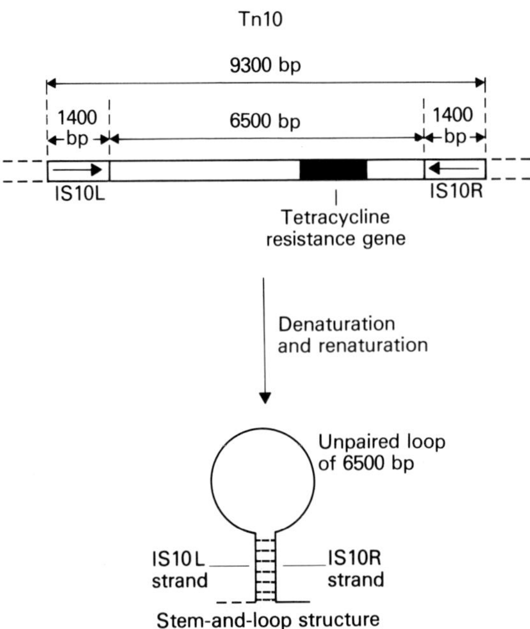

Fig. 2.13. Structure of the bacterial transposon Tn10.

tetracycline. In vivo, Tn10 promotes its own transposition. As a result of this event, a copy of the Tn element is placed in a new location while the original insert is retained. Deletion and insertion events also occur as a result of the gene activities of Tn10. Each of the three processes of transposition, deletion, and insertion involves specific transposon-coded sites and functions.

Transposition of IS elements and transposons. The most frequent event of a transposon's "life" is transposition, the integration of the transposon into new sites in the genome. The site for the integration of the transposon is called the *target site.*

The transposition of IS1 into the *E. coli* chromosome has been well characterized. The transposition event results in the repeat of a 9-bp segment of the target DNA on either side of the IS element, where only one copy was present prior to the integration of IS1 (Fig. 2.14). This phenomenon is a general one, since repeated sequences have been found for every IS element and Tn studied. Evidence suggests that the repeated sequence is generated during the integration process.

Eukaryotic transposable genetic elements

One of the earliest observations of a TGE came from B. McClintock's work with maize in the 1950s. She described a number of what she called *controlling elements* that modified or suppressed gene activity in maize. One of the

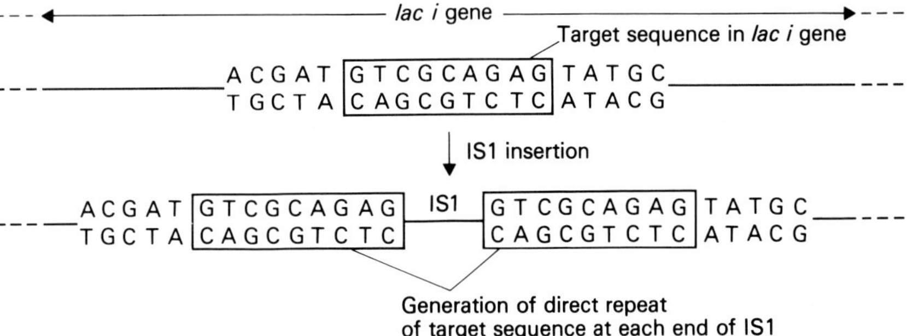

Fig. 2.14. Molecular consequences of IS1 integration into the *lac i* gene of *E. coli.* The integration generates a direct repeat of a 9-bp DNA sequence at each end of the inserted sequence, where only one copy of the sequence existed prior to integration.

characteristics she studied was the extent of pigmentation in the corn kernels. A number of different genes must function together for the synthesis of the red anthocyanin pigment in the kernel. Classical genetic experiments had shown that mutations in any one of these genes causes an unpigmented kernel. McClintock studied kernels that, rather than being red or unpigmented, had spots of red pigment in an otherwise white kernel. From her careful genetic and cytological studies she concluded that the spotted phenotype was not the result of conventional mutational events but was due to a controlling element. The explanation is that a mobile controlling element had transposed into one of the pigment-producing genes, causing that gene to become nonfunctional. As the kernel developed, the cells were unpigmented. The controlling element is unstable, however, and can excise from one chromosomal location and transpose to another location in the genome. When this element transposes out of the pigment-producing gene, that gene is again functional. All cells derived from that cell produce red pigment, and thus a red spot of the kernel results. Depending on the time during kernel development at which the transposition event occurs, the red spot may be large (indicating an early excision of the controlling element) or small (indicating a late time of excision).

The remarkable fact of McClintock's conclusion was that at the time of her work there was no precedent for the existence of transposable genetic elements. Only recently have TGEs been widely identified and studied, and only in 1983 was direct evidence obtained for the movable genetic element hypothesis proposed by McClintock. McClintock was awarded the Nobel Prize for medicine in 1983 for her pioneering work in this area.

Transposable genetic elements have also been studied in many other eukaryotes, mostly in yeast, *Drosophila*, and humans. The structure and function of these elements are very similar to those of prokaryotes. Eukaryotic TGEs have been shown to be capable of integrating into chromosomes at a number of sites and thus may be able to affect the function of virtually any gene, turning it on or off, depending on the element involved and how it integrates into the gene. The integration events, like those of prokaryotic TGEs, involve nonhomologous recombination (i.e. integration into DNA which does not have base pair sequence similarity with the element). Many of the eukaryotic TGEs carry genes and there is good evidence that these genes are transcribed from the integrated elements and that at least some of the resulting mRNA is translated. However, the role of the products of such mRNA is unknown.

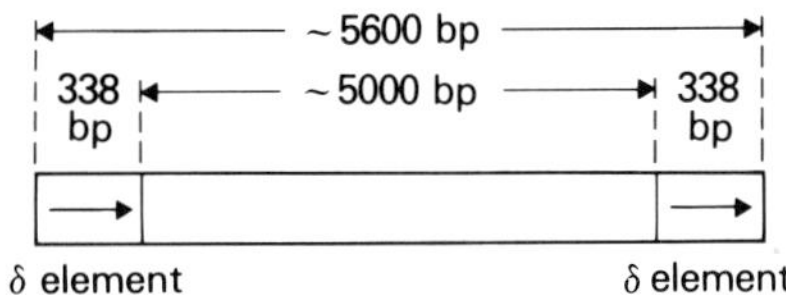

Fig. 2.15. Structure of the transposable genetic element *Ty1*, a transposon found in some strains of yeast.

Ty *elements in yeast.* The *Ty* elements are segments of DNA found in yeast that are capable of transposition around the yeast genome and that are similar in structure and function to bacterial transposons. Fig. 2.15 diagrams the *Ty1* element of yeast, which is about 5600 bp long and includes two 388-bp-long, directly repeated sequences called delta (δ), one at each end of the element. There are about 35 copies of the *Ty1* element in some yeast strains, although the number of copies of this element varies between strains. At least two other types of *Ty* elements occur in yeast—these differ not only in the unrepeated DNA region but also in the composition of the delta segments.

The *Ty* elements have a number of properties in common with bacterial transposons:

1. They are found inserted in many different target sequences that share no obvious homology to each other or to the ends of the TGEs.
2. Upon insertion, the elements generate a duplication of a few base pairs of the target sequence located immediately adjacent to the ends of the element.
3. The elements contain terminal repeated sequences flanking the main body of the element.

Experiments have shown that the transcription of *Ty* elements accounts for 10–15% of all the mRNAs in yeast. At least some of these messages are translated, although the function of their products is unknown. Recent experiments have shown that the transposition event involving *Ty1* occurs not by a direct DNA to DNA process but via an RNA intermediate. That is, in the transposition process, the integrated *Ty1* element produces a single-stranded RNA copy of itself, which, by a process called *reverse transcription* catalyzed by an enzyme *reverse transcriptase,* produces a new double-stranded DNA copy that can integrate elsewhere in the genome.

Human transposable genetic elements. The highly repetitive sequences of human DNA include a family of sequences containing a cleavage site for an enzyme called AluI. This family of sequences is called the Alu family, and since there are about 300,000 or more copies of the Alu sequences per haploid genome, it is the most abundant component of short-length, repetitive DNA in humans.

Each sequence in the Alu family is about 300 bp long, and there is about 80% homology between members of the family. Most of the Alu DNA members are flanked by tandem direct repeats, which suggests that they may be TGEs.

In conclusion, the picture we are left with is a genome that is dynamic rather than static in that DNA segments are moving around the genome, altering its basic organization and affecting the activities of genes.

Questions and problems

2.1 The nucleotide sequences of two DNA molecules from a population of T2 DNA molecules are as shown:

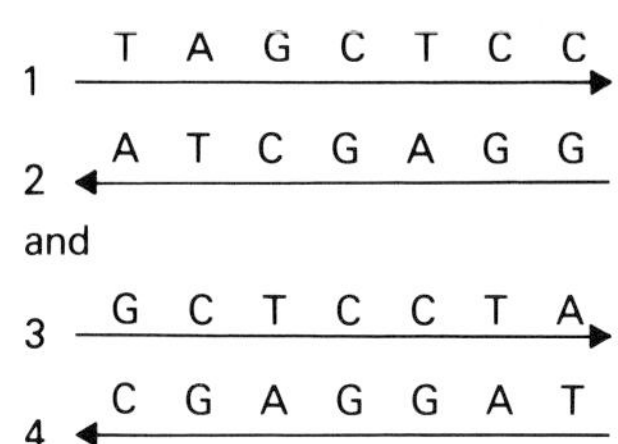

These molecules are heat-denatured, and then the separated strands were allowed to renature. Diagram the structures of the renatured molecules most likely to appear when (a) strand 2 renatures with strand 3 and (b) strand 3 renatures with strand 4. Mark the strands and indicate sequences and polarity.

2.2 Chromosomes contain (choose the best answer):
(a) protein
(b) DNA and protein
(c) DNA, RNA, histone, and nonhistone protein
(d) DNA, RNA, and histone
(e) DNA and histone

2.3 List five major features of eukaryotic chromosomes that distinguish them from prokaryotic chromosomes.

2.4 Discuss the structure and role of the nucleosomes.

References

Boeke, J.D., D.J. Garfinkel, C.A. Styles and G.R. Fink, 1985. *Ty* elements transpose through an RNA intermediate. *Cell* **40**:491–500.

Britten, R.J. and D.E. Kohne, 1968. Repeated sequences in DNA. *Science* **161**:529–540.

Bukhari, A.L., J.A. Shapiro and S.L. Adhya (eds.), 1977. *DNA Insertion Elements, Plasmids and Episomes*. Cold Spring Harbor Laboratory, New York.

Calos, S.N. and J.H. Miller, 1980. Transposable elements. *Cell* **20**:579–595.

Calos, S.N. and J.A. Shapiro, 1980. Transposable genetic elements. *Sci. Amer.* **242**:40–449.

Chambon, P., 1978. Summary: The molecular biology of the eukaryotic genome is coming of age. *Cold Spring Harbor Symp. Quant. Biol.* **42**:1209–1234.

Cold Spring Harbor Symposia on Quantitative Biology, vol. 38, 1973, *Chromosome Structure and Function*. Cold Spring Harbor Laboratory, New York.

Cold Spring Harbor Symposia on Quantitative Biology, vol. 42, 1977, *Chromatin*. Cold Spring Harbor Laboratory, New York.

Cold Spring Harbor Symposia on Quantitative Biology, vol. 45, 1980, *Movable Genetic Elements*. Cold Spring Harbor Laboratory, New York.

Comings, D.E., 1978. Mechanisms of chromosome banding and implications for chromosome structure. *Annu. Rev. Genetics* **12**:25–46.

Crick, F.H.C., 1976. Linking numbers and nucleosomes. *Proc. Natl. Acad. Sci. USA* **73**:2639–2642.

Dubochet, J. and M. Noll, 1978. Nucleosome arcs and helices. *Science* **202**:280–286.

DuPraw, E.J., 1970. *DNA and Chromsomes*. Holt, Rinehart and Winston, New York.

Elgin, S.C.R. and H. Weintraub, 1975. Chromosomal proteins and chromatin structure. *Annu. Rev. Biochem.* **44**:725–774.

Engels, W.R., 1983. The P family of transposable elements in *Drosophila*. *Annu. Rev. Genetics* **17**:315–344.

Foster, T.J., M.A. Davis, D.E. Roberts, K. Takashita and N. Kleckner, 1981. Genetic organization of transposon Tn10. *Cell* **23**:201–213.

Fredericq, E., 1982. Supramolecular aspects of chromatin structure. *Arch. Biol.* **93**:127–142.

Garfinkel, D.J., J.D. Boeke and G.R. Fink, 1985. Ty element transposition: reverse transcriptase and virus-like particles. *Cell* **42**:507–517.

Gottesfeld, J.M. and D.A. Melton, 1978. The length of nucleosome-associated DNA is the same in both transcribed and nontranscribed regions of chromatin. *Nature* **273**:317–319.

Jagadeeswaran, P., B.G. Forget and S.M. Weissman, 1981. Short interspersed repetitive DNA elements in eucaryotes: transposable DNA elements generated by reverse transcription of RNA pol III transcripts? *Cell* **26**:141–142.

Jelinek, W.R. and C.W. Schmid, 1982. Repetitive DNA sequences in eukaryotic DNA and their expression. *Annu. Rev. Biochem.* **51**:813–844.

Kennell, D.E., 1971. Principles and practices of nucleic acid hybridization. *Progr. Nucl. Acid Res.* **11**:259–302.

Kleckner, N., 1981. Transposable elements in prokaryotes. *Annu. Rev. Genetics* **15**:341–404.

Kornberg, R.D., 1977. Structure of chromatin. *Annu. Rev. Biochem.* **46**:931–954.

Lilly, D. and J. Purdon, 1979. Structure and function of chromatin. *Annu. Rev. Genetics* **13**:197–233.

Long, E.O. and I.B. Dawid, 1980. Repeated genes in eukaryotes. *Annu. Rev. Biochem.* **49**:727–764.

McClintock, B., 1961. Some parallels between gene control systems in maize and in bacteria. *Amer. Nat.* **95**:265–277.

McGhee, J.D. and G. Felsenfeld, 1980. Nucleosome structure. *Annu. Rev. Biochem.* **49**:1115–1156.

Noll, M. and R.D. Kornberg, 1977. Action of micrococcal nuclease on chromatin and the location of histone H1. *J. Mol. Biol.* **109**:393–404.

Singer, M.F., 1982. Highly repeated sequences in mammalian genomes. *Int. Rev. Cytol.* **76**:67–112.

Singer, M.F., 1982. SINEs and LINEs: Highly repeated short and long interspersed sequences in mammalian genomes. *Cell* **28**:433–434.

Streisinger, G., J. Emrich and M.M. Stahl, 1967. Chromosome structure in bacteriophage T4. III. Terminal redundancy and length determination. *Proc. Natl. Acad. Sci. USA* **57**:292–295.

Thomas, C.A., 1971. The genetic organization of chromosomes. *Annu. Rev. Genetics* **5**:237–256.

Thomas, C.A. and L.A. MacHattie, 1967. The anatomy of viral DNA molecules. *Annu. Rev. Biochem.* **36**:485–518.

Wang. J.C., 1982. The path of DNA in the nucleosome. *Cell* **29**:724–726.

Woodcock, C.L.F., L.-L.Y. Frado and J.B. Rattner, 1984. The higher-order structure of chromatin: evidence for a helical ribbon arrangement. *J. Cell Biol.* **99**:42–52.

Worcel, A. and C. Benajati, 1977. Higher order coiling of DNA in chromatin. *Cell* **12**:83–100.

Worcel, A. and E. Burgi, 1972. On the structure of the folded chromosome of *Escherichia coli. J. Mol. Biol.* **71**:127–147.

Wu, R. and E. Taylor, 1971. Nucleotide sequence analysis of DNA II. Complete nucleotide sequence of the cohesive ends of bacteriophage lambda DNA. *J. Mol. Biol.* **57**:491–511.

Wu, R., 1978. DNA sequence analysis. *Annu. Rev. Biochem.* **47**: 607–634.

Chapter 3 DNA Replication: Prokaryotes

Outline

Nucleotide synthesis
Purine biosynthesis
Pyrimidine biosynthesis
DNA synthesis in vitro
  Role of DNA in the reaction
DNA replication in vivo
  The Meselson and Stahl experiment
  Semidiscontinuous DNA replication
  Enzymes and proteins involved in DNA replication
  DNA chains are initiated by primers
  Model for DNA replication

Fig. 3.1. The structure of ribose 5′-phosphate.

Nucleotide synthesis

To be efficient, virtually all cells use two different kinds of pathway for the synthesis of nucleotides. One is a *de novo pathway*, in which a sugar (ribose 5′-phosphate), certain amino acids, carbon dioxide, and NH_3 are combined in a series of reactions to form the nucleotides directly. In this case neither the free purines or pyrimidines nor the nucleosides are intermediates in the pathway.

The second kind of pathway is called a *salvage pathway*, in which the purines, pyrimidines, and nucleosides released by the breakdown of nucleic acids can, by a variety of routes, be converted back to the nucleotides needed for nucleic acid synthesis. Both pathways are important to the cell, and how active each pathway is depends upon the resources available to the cell.

Purine biosynthesis

The purine biosynthetic pathway follows essentially the same sequence of reactions in a wide range of prokaryotic and eukaryotic organisms, including *E. coli*, yeast, plants, and man. The purine ring structure is assembled on the sugar, ribose 5′-phosphate (Fig. 3.1). A number of components are used to produce the purine ring, including some amino acids. Fig. 3.2 summarizes the origin of the purine ring atoms, and Fig. 3.3 presents the structures of the amino acids involved.

The primary product of the purine biosynthesis pathway is the ribonucleotide inosine 5′-monophosphate (IMP) and it is from this compound that the adenine and guanine ribonucleotides are derived (Fig. 3.4). As will be

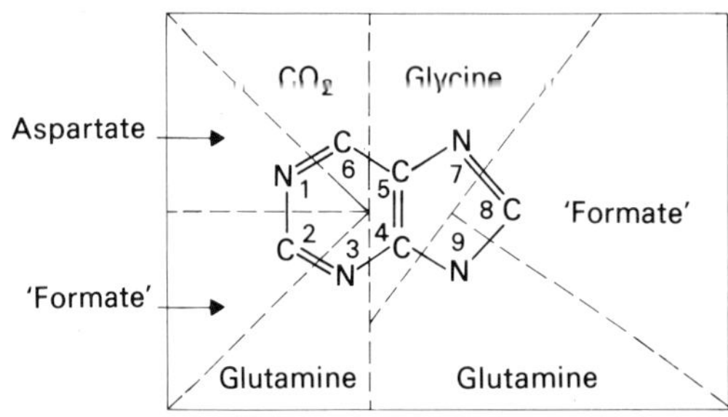

Fig. 3.2. Origin of the carbon and nitrogen atoms of the purine ring.

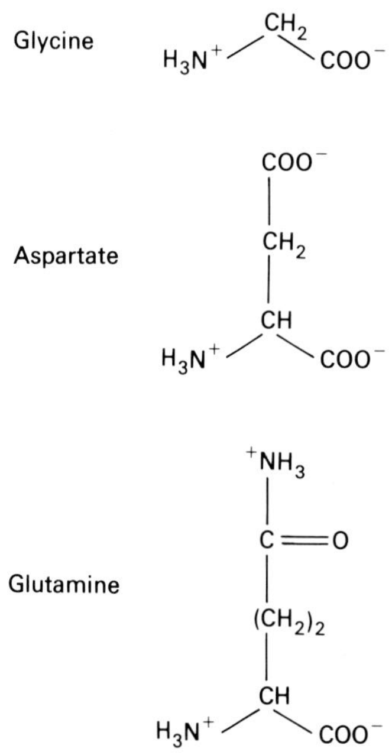

Glycine $H_3N^+-CH_2-COO^-$

Aspartate $H_3N^+-CH(CH_2-COO^-)-COO^-$

Glutamine $H_3N^+-CH((CH_2)_2-C(=O)-{}^+NH_3)-COO^-$

Fig. 3.3 Structures of amino acids used in the formation of purine (and pyrimidine) rings.

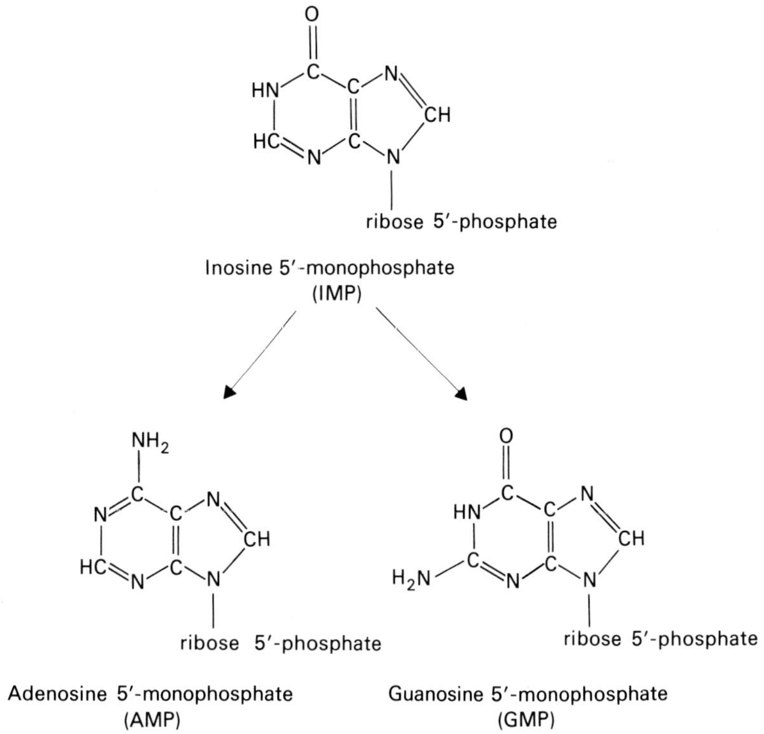

Inosine 5′-monophosphate (IMP)

Adenosine 5′-monophosphate (AMP)

Guanosine 5′-monophosphate (GMP)

ribose 5′-phosphate

Fig. 3.4. Purine biosynthesis: production of the adenine and guanine nucleotides from inosine 5′-monophosphate.

described later, these latter two compounds are phosphorylated further to convert them to the immediate RNA and DNA precursors.

Pyrimidine biosynthesis

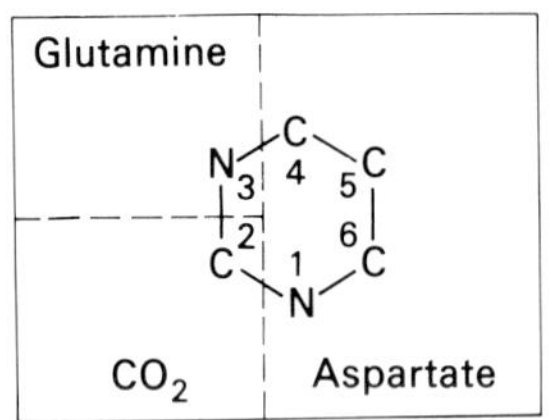

Fig. 3.5. Origin of the carbon and nitrogen atoms of the pyrimidine ring.

The biosynthesis of pyrimidine nucleotides differs in general from that of the purine nucleotides in that the pyrimidine ring is assembled first and then is attached to the ribose 5′-phosphate. As was the case with the purine ring, a number of compounds are used in the construction of the pyrimidine ring, and these are shown in Fig. 3.5. The pyrimidine biosynthesis pathway is depicted in Fig. 3.6. The key pyrimidine intermediate in this pathway is orotate, which becomes attached to an activated form of ribose 5′-phosphate, phosphoribosyl pyrophosphate (PRPP), to form a nucleotide, orotidine 5′-monophosphate (OMP). Removal of a carbon dioxide molecule generates uridine 5′-monophosphate (UMP). The cytosine nucleotide then is produced from UMP by the addition of an amino (NH_2) group (Fig. 3.6).

The nucleoside monophosphates generated by the de novo purine and pyrimidine biosynthesis pathways do not participate directly in nucleic acid biosynthesis. Instead, they are first converted to the triphosphate derivatives via diphosphate intermediates. The phosphate donor for all of these reactions is the ribonucleotide, adenosine 5′-triphosphate (ATP), and the reactions are catalyzed by enzymes called *kinases*. (These enzymes are sites for regulation of the production of DNA and RNA precursors.) The conversion of nucleoside monophosphates to diphosphates involves kinases that are specific for each base but nonspecific with regard to the ribose or deoxyribose sugar. Thus, they are involved in the synthesis of both RNA and DNA precursors. On the other hand, the synthesis of nucleoside triphosphates from the diphosphates is catalyzed by a kinase that is nonspecific for both the base and the sugars.

By the kinase-catalyzed reactions, the ribonucleoside 5′-triphosphates are produced from the ribonucleoside monophosphate end products of the purine and pyrimidine biosynthesis pathways (Fig. 3.7). These triphosphates are the immediate precursors of RNA synthesis, a process which will be described in a later chapter. The precursors for DNA synthesis are also derived from the ribonucleoside 5′-monophosphates (Fig. 3.7). First, the diphosphates are produced and then the ribose sugar is reduced to the deoxyribose sugar in a reaction catalyzed by the enzyme ribonucleoside phosphate reductase. With the exception of dUDP, the resulting deoxyribonucleoside 5′-diphosphates (dADP, dGDP, dCDP) are then phosphorylated to produce 5′-triphosphates. As was discussed earlier, the base uracil is specific for RNA and this pyrimidine is replaced in DNA by thymine, which is actually 5′-methyluracil. To produce the thymine deoxyribonucleotide,

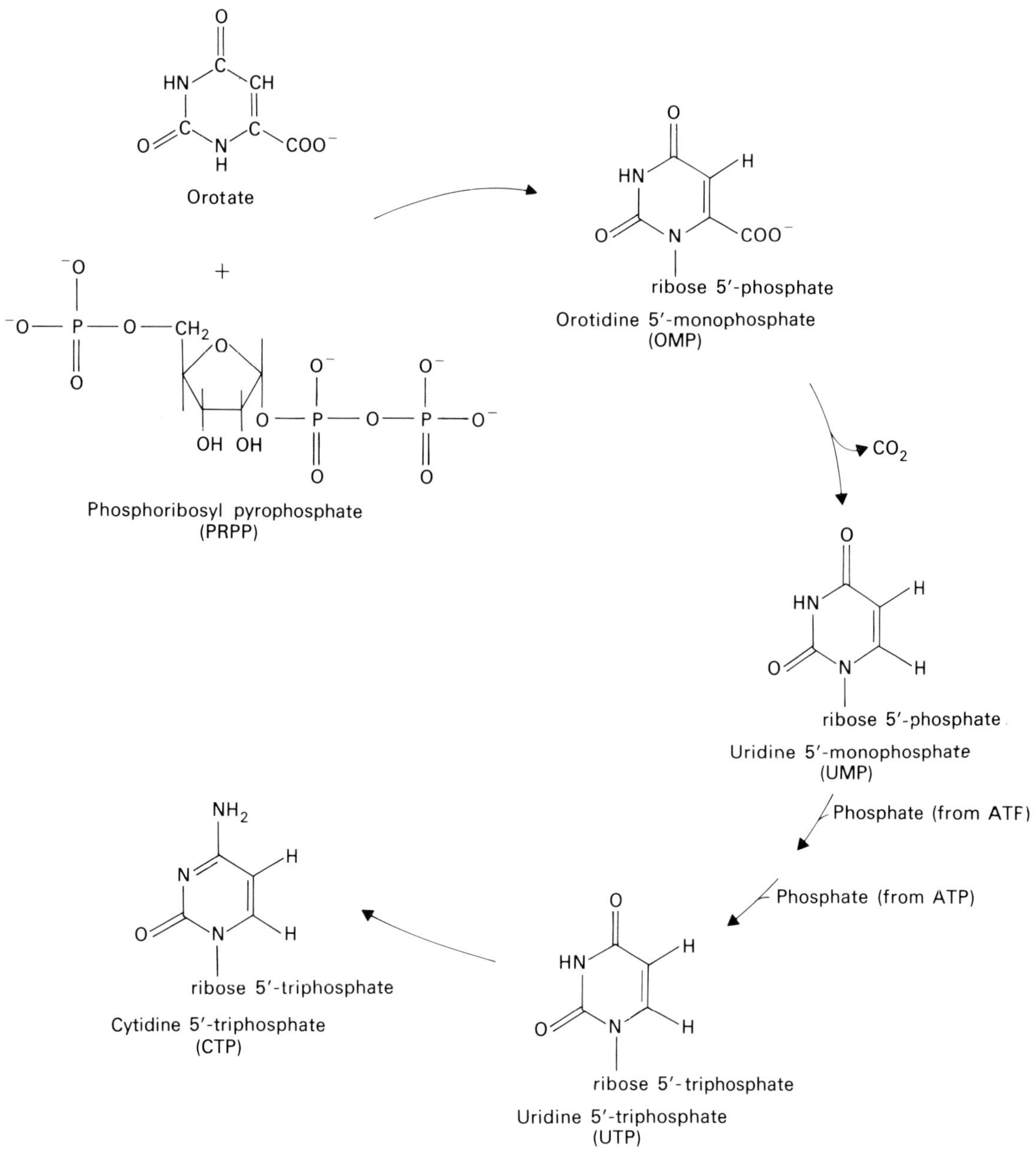

Orotate

+

Phosphoribosyl pyrophosphate (PRPP)

ribose 5′-phosphate

Orotidine 5′-monophosphate (OMP)

CO_2

ribose 5′-phosphate

Uridine 5′-monophosphate (UMP)

Phosphate (from ATF)

Phosphate (from ATP)

ribose 5′-triphosphate

Uridine 5′-triphosphate (UTP)

ribose 5′-triphosphate

Cytidine 5′-triphosphate (CTP)

Fig. 3.6. General outline of the pyrimidine biosynthesis pathway.

dUDP is dephosphorylated to dUMP and the 5-position carbon of the pyrimidine ring is then methylated in a reaction catalyzed by thymidylate synthetase. The resulting deoxythymidine 5′-monophosphate is phosphorylated to produce the triphosphate. In this way, the four DNA precursors,

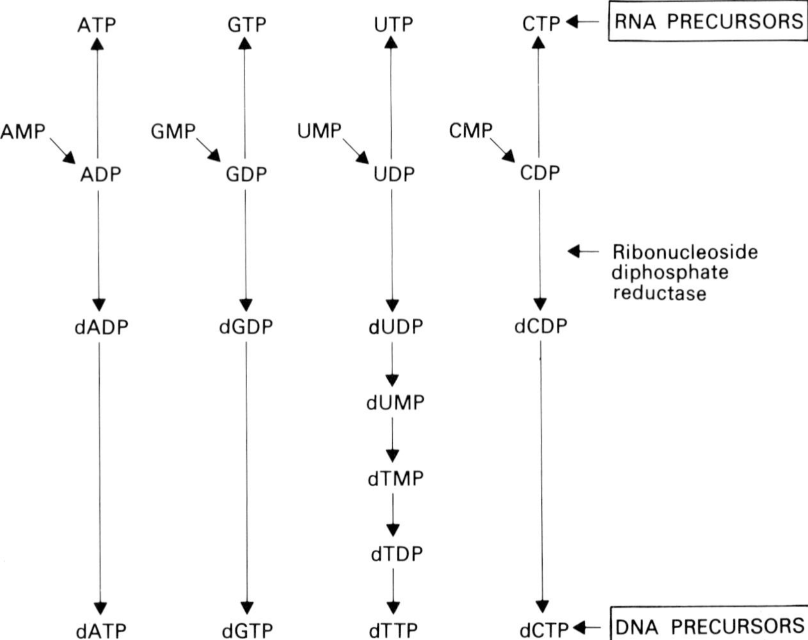

Fig. 3.7. Formation of ribonucleotide precursors for RNA and deoxyribonucleotide precursors for DNA.

dATP, dGTP, dTTP, and dCTP, are produced in reactions catalyzed by the kinases discussed previously.

Since the syntheses of RNA and DNA precursors are intimately related, one might expect that the regulatory elements of the system would be complex and indeed that is the case. For example, since cells must distribute their resources appropriately between RNA and DNA synthesis, one of the pivotal enzymes is ribonucleoside diphosphate reductase. This enzyme receives complex inhibitory and stimulatory signals by a variety of deoxyribonucleotides so as to provide a balanced supply of precursors for DNA synthesis. The synthesis of the enzyme can also be repressed by the deoxyribonucleotides if their pool sizes become large.

DNA synthesis in vitro

In Chapter 1, it was shown that the basic building blocks for DNA synthesis are the deoxyribonucleoside 5′-triphosphates, and the precursors are polymerized into DNA as shown in Fig. 3.8. In DNA synthesis, the polynucleotide chain grows from one end by the stepwise addition of deoxyribonucleoside triphosphates. In this, the direction of synthesis is determined by the fact that the 5′-triphosphate precursors must bond to the 3′-OH of the preceding deoxyribose. Before discussing recent information about DNA synthesis, we

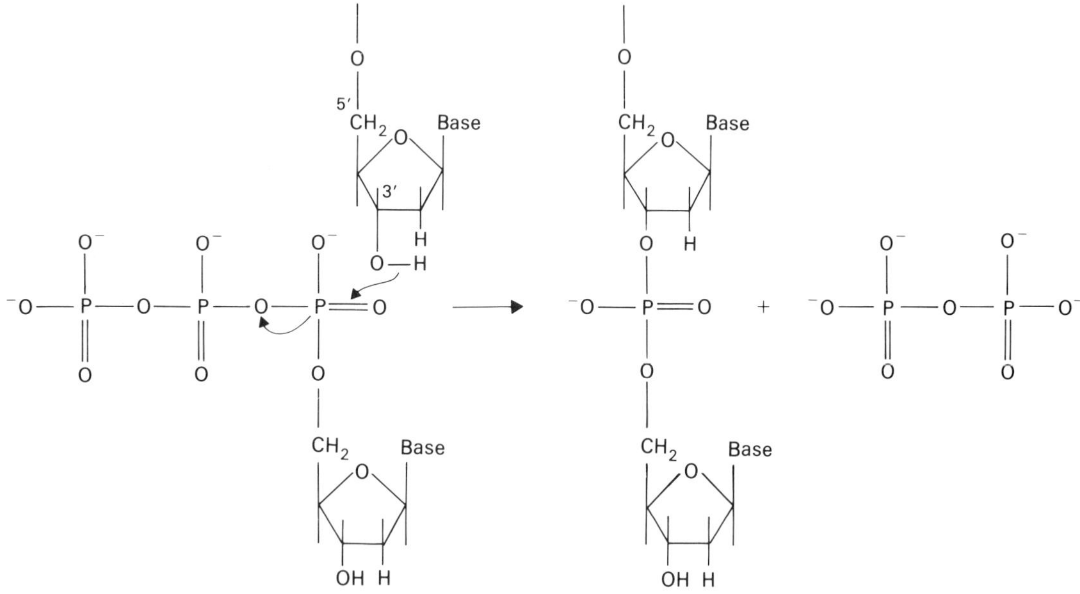

Fig. 3.8. Mechanism of DNA polymerization.

shall review the historical work of A. Kornberg and his colleagues that resulted in the in vitro synthesis of DNA under defined conditions.

In his experiments of the 1950s, Kornberg determined that the following ingredients were required for the in vitro synthesis of DNA:

1. A mixture of all four deoxyribonucleoside 5′-triphosphates.
2. Magnesium ions (Mg^{2+}).
3. A purified enzyme, **DNA polymerase**, obtained from cell-free extracts of *E. coli.*
4. High-molecular-weight DNA.

In his initial experiments with the above ingredients, Kornberg got an approximately 20-fold increase in the amount of DNA over that added. The reaction itself continued until one of the 5′-triphosphate precursors ran out.

That all four ingredients are required for DNA synthesis in vitro was shown by experiments in which one ingredient at a time was omitted from the reaction mixture. In each case no DNA synthesis occurred. Similar omission experiments showed that all four deoxyribonucleoside 5′-triphosphates must be present in order for DNA synthesis to proceed. The key to the reaction, though, is the DNA polymerase, which promotes synthesis by effecting the 3′-5′-phosphodiester linkages.

Role of DNA in the reaction

Two possibilities were proposed originally for the role of high-molecular-weight DNA in the in vitro synthesis of DNA:

1. The DNA acts as a **primer**, a growing point for the terminal random addition of nucleotides.

2. The DNA acts as a **template**; that is, the base sequence of the DNA chain specifies the complementary base sequence of the new DNA chain with which it is bound by hydrogen bonds in the double helix structure. The requirement for all four DNA precursors to be present for any DNA synthesis to occur suggests that DNA has a template role.

All the available evidence indicates that the second possibility is correct: DNA acts as a template. The initial evidence came from studies that showed that the (A+T)/(G+C) ratio of in vitro synthesized DNA was the same as the DNA used to start the reaction. However, an objection to this is that the results could be a coincidence since base ratios do not indicate the arrangement of bases along the DNA. A more exacting comparison of newly synthesized and template DNA can be made by *nearest neighbor analysis*, which shows the relative frequency with which each of the four bases is located adjacent to each of the four bases in the DNA chain (i.e. A next to A, G, C, and T; C next to A, G, C, and T; G next to A, G, C, and T; and T next to A, G, C, and T). Nearest neighbor analysis does not give information about the sequence of bases along the DNA, but it does give some information about the general arrangement of the bases in the DNA.

In nearest neighbor analysis DNA is isolated from the organism of interest and then is used as the template in Kornberg's in vitro synthesizing system (Fig. 3.9). In this case, only one of the four deoxyribonucleoside 5′-triphosphates (dATP in the figure) is labeled with ^{32}P, a radioactive isotope of phosphorus. The resulting DNA is treated with a mixture of micrococcal deoxyribonuclease and spleen diesterase. These enzymes break the backbone of DNA between the 5′-carbon and the phosphoryl group, producing deoxyribonucleoside 3′-monophosphates, which can be analyzed by paper electrophoresis to determine the amount of radioactivity each contains (Fig. 3.9).

In the example shown, the ^{32}P is now found in the form of dG-3′MP (deoxyguanosine 3′-monophosphate), which was the nearest neighbor to the 5′ side of the adenine nucleotide in the DNA. Extrapolating this discussion to the entire DNA, the method provides data concerning the relative amount of radioactive label transferred from ^{32}P-labeled dATP to the four possible neighbors. This gives information about the relative frequency with which A is next to A, G, C, and T in the DNA. These four nearest neighbors are usually written as 5′-ApA-3′, GpA, CpA, and TpA. The experiment is then repeated

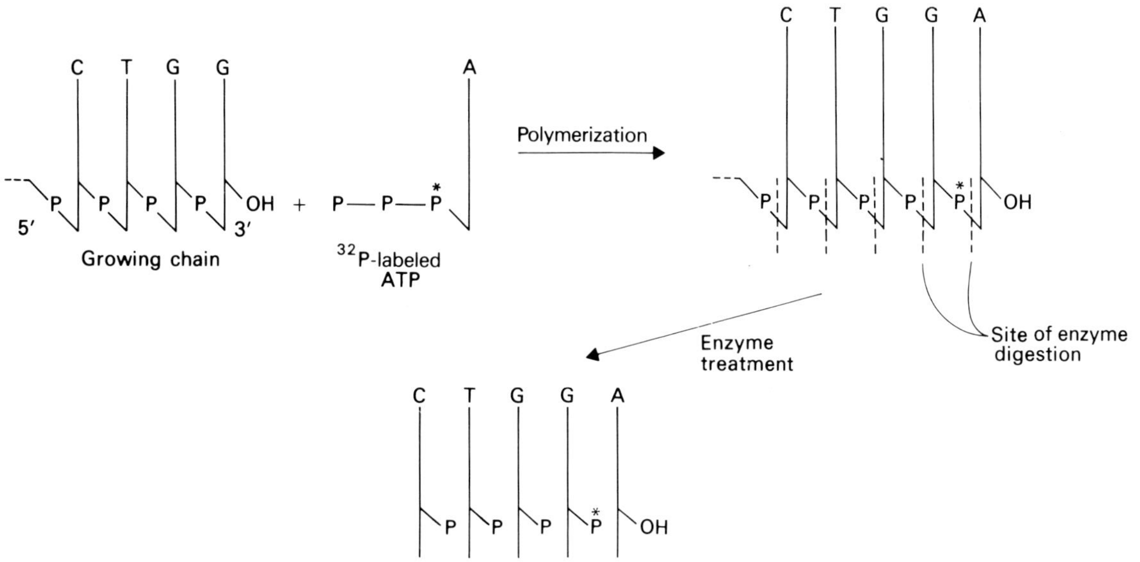

Fig. 3.9. The procedure of nearest neighbor analysis.

with each of the other three DNA precursors such that in the end a matrix of 16 nearest neighbor values is produced. The 16 frequencies found are generally unique to the DNA in question. An example of the type of data that is obtained from nearest neighbor analysis is shown in Table 3.1.

Table 3.1. Nearest neighbor frequencies of *Mycobacterium phlei* DNA. (With permission from J. Josse et al., 1961. *J. Biol. Chem.* **236**:804, ©1961, Journal of Biological Chemistry.)

Labeled triphosphate	Deoxyribonucleoside 3′-monophosphate isolated			
	Tp	Ap	Cp	Gp
dATP	TpA 0.012	ApA 0.024	CpA 0.063	GpA 0.065
dTTP	TpT 0.026	ApT 0.031	CpT 0.045	GpT 0.060
dGTP	TpG 0.063	ApG 0.045	CpG 0.139	GpG 0.090
dCTP	TpC 0.061	ApC 0.064	CpC 0.090	GpC 0.122
Sum	0.162	0.164	0.337	0.337

The data presented illustrate several points:

1. The sixteen possible nearest-neighbor frequences occur with a large number of frequencies.
2. The sums of the vertical columns show that the amount of A is equal to the amount of T and that G equals C, thus indicating the DNA was probably replicated correctly.
3. The two DNA strands are of opposite polarity. This is borne out by the frequency equivalence of the pairs of sequences: CpT and ApG, GpT and ApC, GpA and TpC, and CpA and TpG, as predicted by antiparallel DNA strands. Were the two strands of the same polarity, different matching sequences would have been predicted, for example, TpA and ApT, GpA and CpT, CpA and GpT, etc.

To show that newly synthesized DNA is made using a DNA template requires two rounds of nearest neighbor analysis. In the first round the originally isolated DNA is used in the reaction and a set of 16 nearest neighbor frequencies are obtained as described. In the second round the enzymatically synthesized DNA is used in the reaction and a second set of frequencies is produced. The results show good agreement between the two sets of frequencies, thus indicating that DNA plays a template role in DNA replication in vitro.

Since the early, comparatively crude in vitro DNA synthesis experiments, a large number of refinements have been made. For example, the reaction mixtures now duplicate the conditions that are present in growing cells such that it is possible to synthesize DNA in vitro that is identical in all respects with DNA produced in vivo. The reaction mixtures required for efficient DNA synthesis vary, depending on the DNA used as the template. In general, there is a basic set of enzymes and proteins required for the synthesis of all DNA sources. Beyond that, each DNA requires a number of specific proteins for new synthesis to occur, and the actual set of proteins needed depends on the DNA in question.

DNA replication in vivo

The Meselson and Stahl experiment

This experiment showed that when DNA replicates, the two strands separate and each serves as a template for the synthesis of a complementary strand. This is called the **semiconservative model** for DNA replication. The analytical method used by M. Meselson and F. W. Stahl was cesium chloride (CsCl) density gradient centrifugation. When a concentrated solution of CsCl is centrifuged in an ultracentrifuge, the opposing forces of sedimentation and diffusion produce a stable concentration gradient of the CsCl. The concentration gradient results in a continuous increase of density along the direction of the centrifugal force. If DNA, for example, is present in the gradient, the DNA will come to equilibrium in the region of the gradient where the solution density is equal to its buoyant density. It is possible, for example, to grow cells in a medium in which the sole nitrogen source is the "heavy" isotope ^{15}N. DNA produced in this medium will have a higher density than DNA produced in a medium containing the normal isotope ^{14}N. If heavy ^{15}N-DNA, and light (normal) ^{14}N-DNA, are analyzed in a CsCl gradient, at the completion of centrifugation they will have formed two distinct bands in the centrifuge tubes.

Meselson and Stahl grew a culture of *E. coli* in a medium in which the sole nitrogen source contained the heavy isotope ^{15}N. The DNA synthesized during this time therefore had a higher density than normal, since the ^{15}N becomes incorporated into the purine and pyrimidine rings. Then, at time zero the cells were collected and resuspended in fresh medium where only the normal nitrogen isotope (^{14}N) was present. Several rounds of replication were allowed to proceed in the normal medium. During this time samples of the cultures were taken; the DNA was extracted and analyzed in CsCl gradients to compare its density with the known densities of heavy and normal DNA: the results are shown in Fig. 3.10 in which part (a) shows an ultraviolet absorption photograph of the DNA bands in the centrifuge tube, and part (b) shows a tracing of such a photograph in a densitometer.

At the time these experiments were done, there were two main hypotheses for DNA replication. The first said that DNA replicated semiconservatively; that is, the two strands separate and act as templates for the synthesis of the new complementary strands. Thus each progeny double helix contains one parental strand and one new strand. The second hypothesis stated that DNA replicated conservatively; that is, the double helix remains intact while serving as a template for production of a new double helix. In this model both strands of the progeny DNA are newly synthesized. The predicted densities of DNA for thc two models in the Meselson and Stahl experiment are diagrammed in Fig. 3.11.

For the conservative replication hypothesis to be correct, some heavy DNA must be present at each generation and all new DNA must have normal density. Thus at generation number one, half of the DNA double helices should be heavy and half should be normal. On the other hand, in the first generation, the semiconservative replication model predicts that *all* DNA molecules will have one heavy and one normal strand. This would cause a hybrid DNA to form a band at a position intermediate between those for heavy and normal DNA. By contrast, no hybrid DNA will ever be formed by conservative DNA replication. Inspection of the banding data allowed Meselson and Stahl to conclude that the semiconservative model is correct.

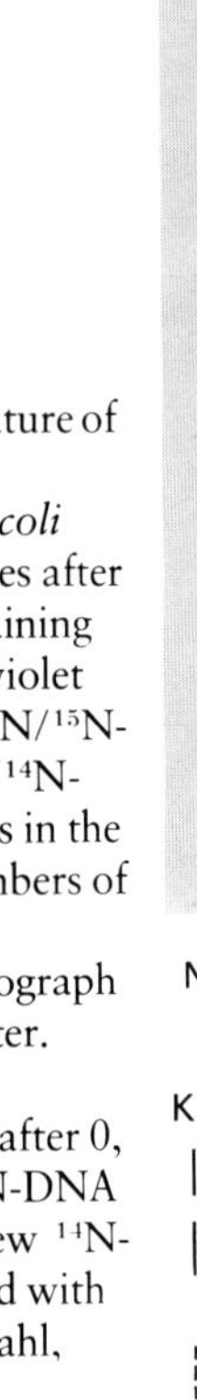

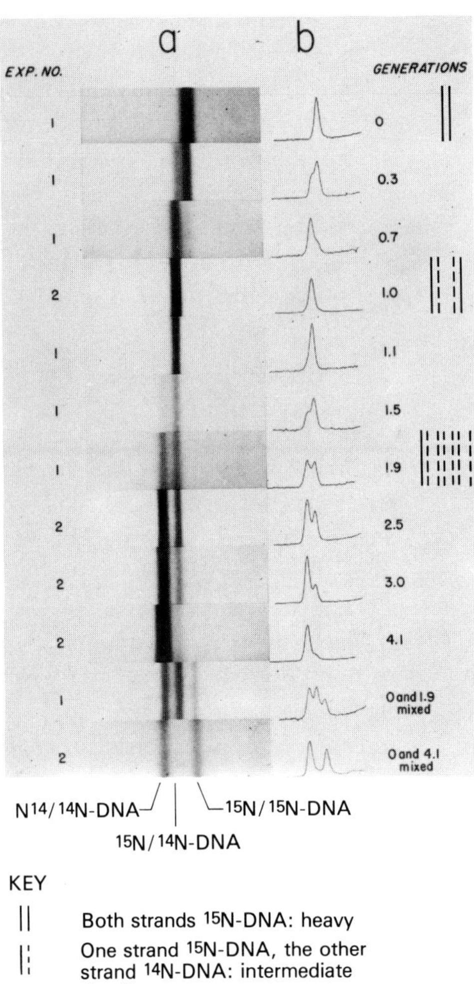

Fig. 3.10. The semiconservative nature of DNA replication: a tracing of the photograph in a densitometer. *E. coli* DNA was analyzed at various times after transferring from a medium containing $^{15}NH_4Cl$ to $^{14}NH_4Cl$. (a) An ultraviolet absorption photograph of the $^{15}N/^{15}N$-(heavy), $^{14}N/^{14}N$-(light), and $^{15}N/^{14}N$-(intermediate density) DNA bands in the centrifuge tubes after various numbers of generations in the ^{14}N-containing medium. (b) A tracing of the photograph shown in (a) made in a densitometer. (c) Schematic interpretation of double-stranded DNA molecules after 0, 1, and 2 generations: parental ^{15}N-DNA is depicted as solid lines, and new ^{14}N-DNA as dotted lines. (Reproduced with permission from Meselson and Stahl, 1958.)

Semidiscontinuous DNA replication

DNA replication proceeds in one overall direction such that both strands are replicated simultaneously. This unidirectional movement of the replication fork poses some problems since the two strands are of opposite polarity and the known DNA polymerases can only catalyze DNA synthesis in the 5′ to 3′ direction. As the DNA helix unwinds to provide the template for new synthesis, the DNA cannot be continuously polymerized on one of the strands. The first evidence for how cells get around this problem came from

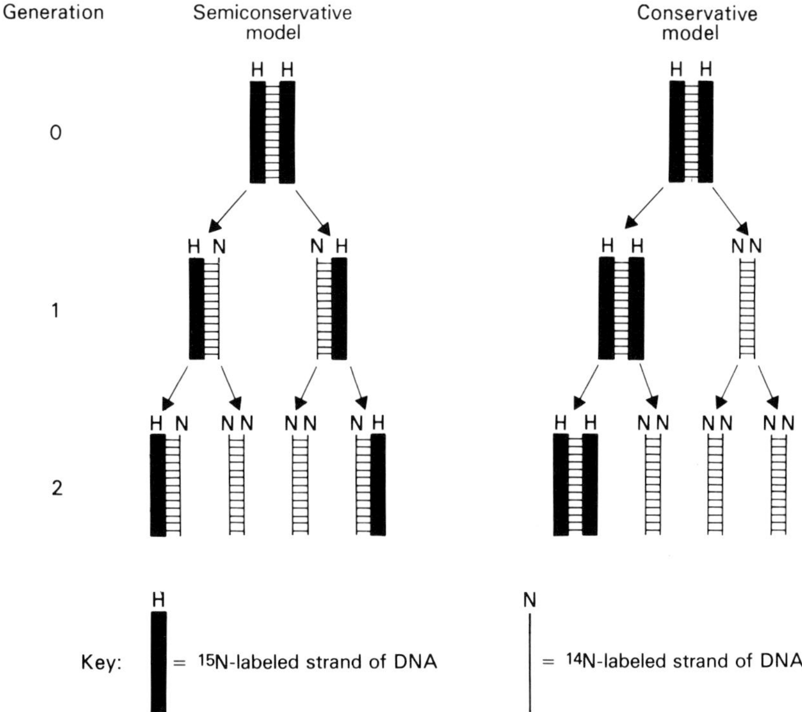

Fig. 3.11. Predicted densities for daughter DNAs produced by semiconservative and conservative methods of DNA replication.

the work of R. Okazaki and colleagues. They added radioactive DNA precursors to cultures of *E. coli* for very short time intervals (0.5% of a generation time) and determined the size of the newly labeled DNA by sedimentation analysis (Fig. 3.12). They found that most of the incorporated label was present in relatively low-molecular-weight DNA. Then, as the labeling time increased, a significant fraction of the label sedimented as high-molecular-weight DNA. The conclusion they drew was that DNA synthesis is discontinuous; that is, short segments (**Okazaki fragments**) are synthesized initially and later these become covalently bonded to the new high-molecular-weight DNA. The average length of the Okazaki fragments has been shown to be 1000 to 2000 nucleotides in prokaryotes.

We now know that the synthesis of the new DNA strand that is made using the 3′-to-5′ parental template strand is continuous while the synthesis of the other new strand that is made using the 5′-to-3′ parental template strand is discontinuous. This method of replication, in which one new strand is synthesized continuously and the other new strand is synthesized discontinuously, is called *semidiscontinuous replication.* A model for DNA replication that involves semidiscontinuous DNA synthesis will be described in a later section of this chapter.

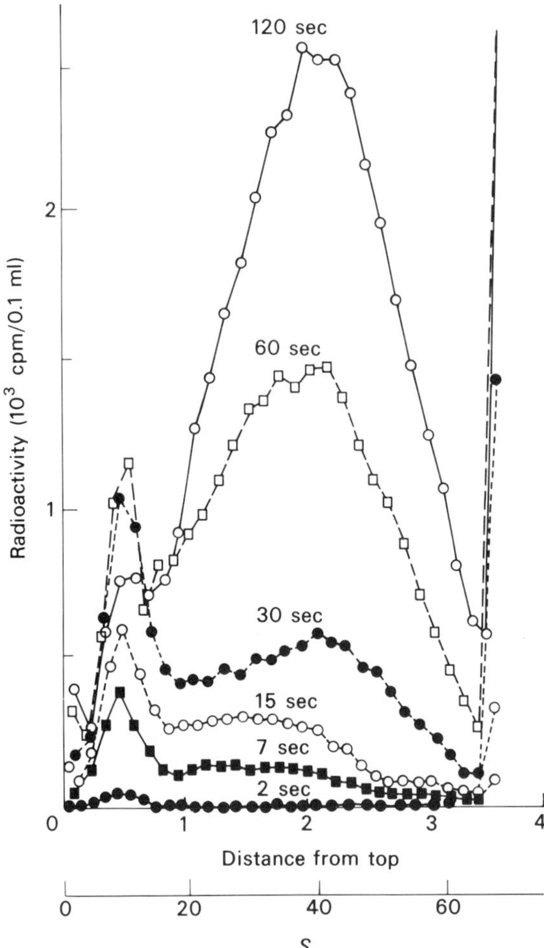

Fig. 3.12. Demonstration of discontinuous replication of DNA. Phage T4-infected *E. coli* cells were labeled with a radioactive DNA precursor. At various times samples were taken and the DNA was analyzed on sucrose gradients. At early times most of the label is in low-molecular-weight DNA (near top of gradient), whereas at later times label is found in high-molecular-weight DNA. (Reproduced with permission from R. Okazaki et al., 1968. *Proc. Natl. Acad. Sci. USA* **59**:598.)

Enzymes and proteins involved in DNA replication

The complicated process of DNA replication in *E. coli* requires many different proteins and enzymes, some of which also function in other cellular processes such as the repair of DNA damage and genetic recombination. The key enzyme in the synthesis of new DNA is DNA polymerase.

In *E. coli* there are three DNA polymerases. Each catalyzes the DNA template-directed condensation of deoxyribonucleoside 5′-triphosphates. There are significant differences in their activity in DNA synthesis and in the nuclease (DNA or RNA breakdown) activities associated with them (Table 3.2).

All three polymerases catalyze new DNA synthesis in the 5′-to-3′ direction. The rate of DNA synthesis catalyzed by the three enzymes varies

Table 3.2. Properties of *E. coli* DNA polymerases.

	Pol I	Pol II	Pol III
Molecular weight (daltons)	109,000	120,000	175,000
Number of polypeptides	1	1	3 (140,000+25,000+10,000)
Molecules per cell	400	50–100	10–20
5′-to-3′ exonuclease activity	yes	no	yes
3′-to-5′exonuclease activity	yes	yes	yes
Gene	*polA*	*polB*	*polC*

considerably, with Pol III being the most active and Pol II being the least active. Exonuclease activities are associated with each of these enzymes. (An exonuclease will degrade a strand of nucleic acid from a free end, whereas an endonuclease will make cuts within a strand.) Each enzyme has 3′-to-5′ exonuclease activity, suggesting that each can trim away newly synthesized polynucleotide chains. It seems that this exonuclease activity functions like a correcting typewriter so that incorrectly paired bases are detected and excised to allow accurate replication to continue. This is called *proofreading*, and it is responsible for the high degree of accuracy found in DNA synthesis.

DNA chains are initiated by primers

None of the three *E. coli* DNA polymerases (and none of the DNA polymerases known in prokaryotes and eukaryotes) can initiate new DNA chains. Instead, the initiation of DNA synthesis involves the synthesis of a short **primer** to which deoxyribonucleotides are added by the action of DNA polymerases. In *E. coli*, the primer is a short chain of RNA and is produced by the action of the enzyme **primase.** The primer's base-pair sequence is directed by the nucleotide sequence of the DNA template strand. The primer is later replaced by DNA (see later). Thus each Okazaki fragment is initiated by a short segment of RNA.

Model for DNA replication

A model for DNA replication that incorporates the facts already discussed is presented in Fig. 3.13. The evidence for this has come from biochemical, genetic, and physiological experiments.

The first step in DNA replication is the denaturation and unwinding of the double helix (Fig. 3.13a). This produces a Y-shaped structure called a **replication fork.** Fig. 3.13 diagrams one replication fork but many linear and

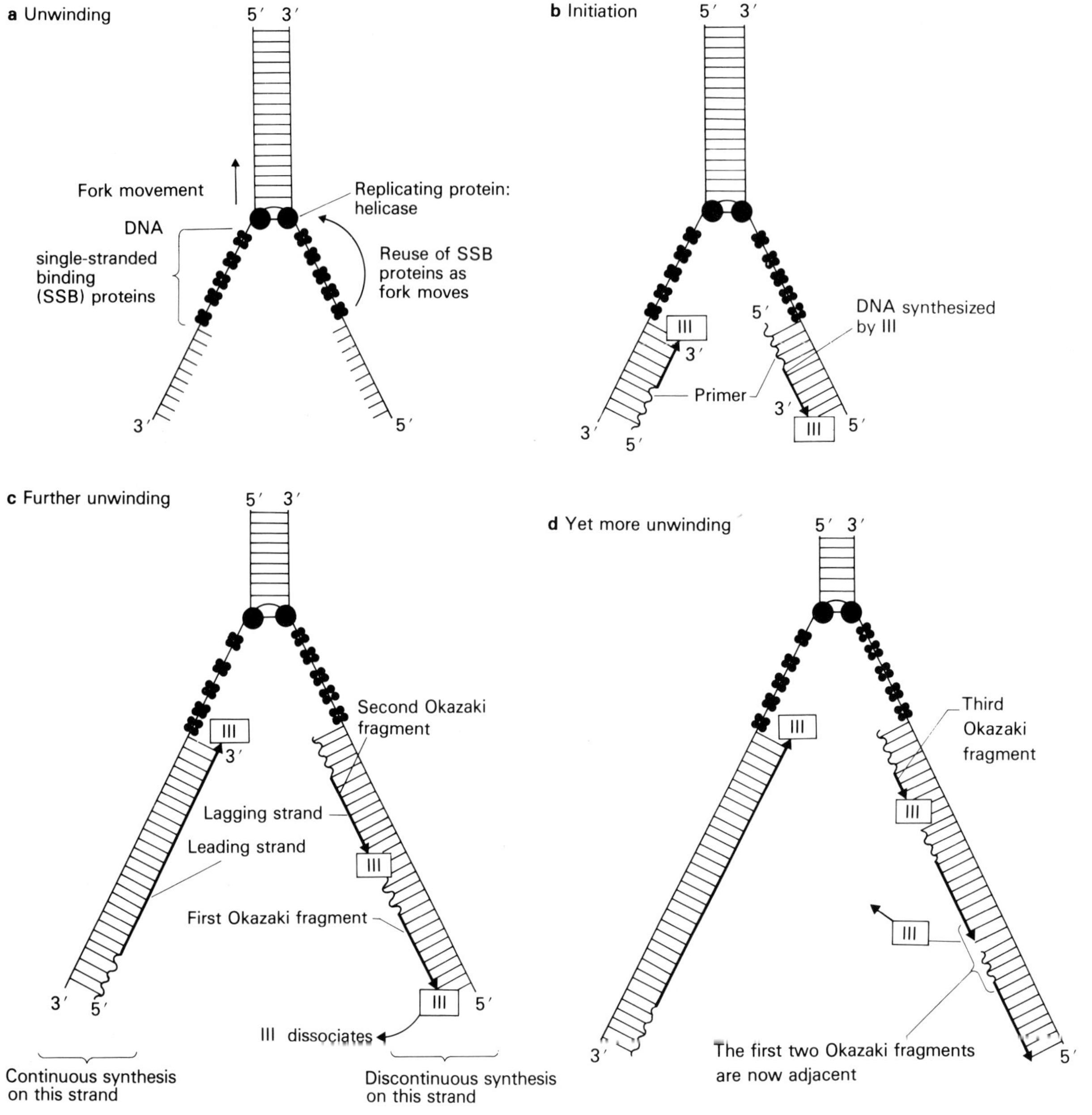

a Unwinding

5′ 3′

Fork movement

Replicating protein: helicase

DNA single-stranded binding (SSB) proteins

Reuse of SSB proteins as fork moves

3′ 5′

b Initiation

5′ 3′

III

3′

5′

DNA synthesized by III

Primer

3′

III

3′ 5′ 5′

c Further unwinding

5′ 3′

III

3′

Second Okazaki fragment

Lagging strand

Leading strand

III

First Okazaki fragment

III

3′ 5′ 5′

III dissociates

Continuous synthesis on this strand

Discontinuous synthesis on this strand

d Yet more unwinding

5′ 3′

III

Third Okazaki fragment

III

III

3′

The first two Okazaki fragments are now adjacent

5′

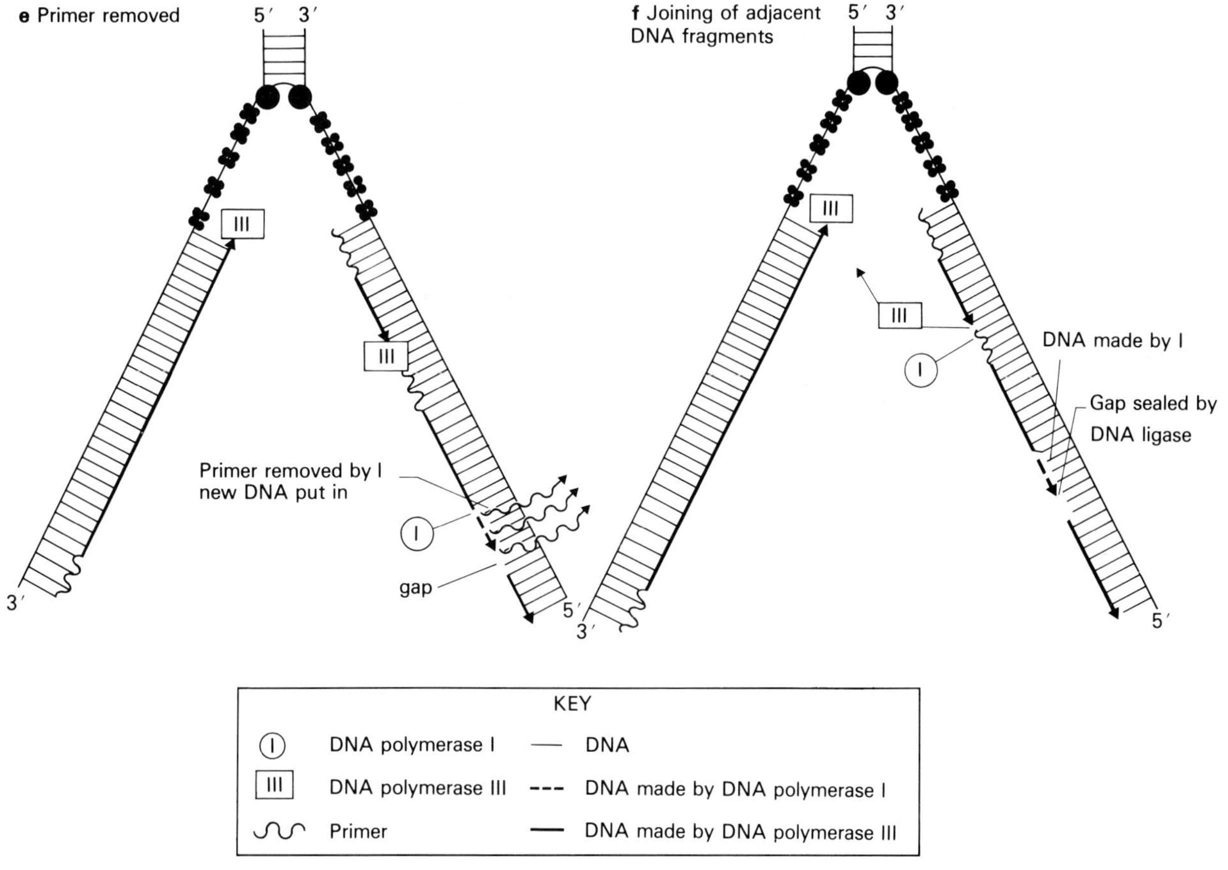

Fig. 3.13. Model for the events occuring around a replication fork of the *E. coli* chromosome. For details, see the text. (After P.J. Russell, 1968, *Genetics*. Little, Brown and Co., Boston.)

circular prokaryotic DNA molecules replicate bidirectionally (in both directions away from the origin of replication). In these cases there are two replication forks that are mirror images of one another: picture two Ys joined head-to-head by the two pairs of arms. The area of a double-stranded DNA molecule that has denatured for replication is called a *replication bubble*.

In *E. coli* the unwinding of the DNA is catalyzed by an enzyme called **helicase**, the product of a gene called *rep*. For each 10 base pairs of DNA unwound, one turn of the helix becomes untwisted. As a result, as the helix unwinds a torsional strain is imposed on the DNA. That is, since the DNA molecule is circular, the pulling apart of the helix at one location will cause increased tightening of the molecule elsewhere, much like what happens when the strands of a piece of rope that is fixed at each end are pulled apart.

This torsional strain is relieved by the action of enzymes called topoisomerases, which cause single-stranded breaks in DNA away from the replication forks and then allow one strand to rotate relative to the other strand.

The unwinding of the DNA produces single-stranded regions, which are stabilized by DNA *single-stranded binding* (SSB) *proteins*. Each SSB protein is a tetramer of a 74,000-dalton subunit, and this tetramer binds to eight bases of single-stranded DNA. Over 250 tetramers bind to each replication fork. Once the strands have started to unwind, the internal bases become available for the formation of bonds with bases in the new chain.

Initiation of DNA replication next takes place (Fig. 3.13b). In *E. coli*, primases bind to the single-stranded DNA of both arms of the replication fork and synthesize short primers. The primers are lengthened by the action of DNA polymerase III, which synthesizes the complementary DNA chains to the template strands. Note that, because of the opposite polarities of the two DNA template strands and the requirement for a 5′-to-3′ direction of new DNA synthesis, the primers are located at different relative positions on the two template strands (Fig. 3.13b).

The DNA helix continues to unwind (Fig. 3.13c). On the template on the left the new strand continues to be synthesized continuously (in the same direction as the direction of unwinding). However, because DNA polymerases can only synthesize DNA in the 5′-to-3′ direction, the new synthesis on the template on the right has gone as far as it can. Thus, a new initiation of DNA synthesis occurs on the right-hand template on the newly exposed single-stranded template close to the site of helicase activity. As before, a primer is made, which is lengthened by DNA polymerase III action. The whole process is repeated in Fig. 3.13d. The overall result is that the new DNA strand being made on the left-hand template is synthesized continuously while the new DNA being made on the right-hand template is synthesized discontinuously so that Okazaki fragments are produced.

Eventually the Okazaki fragments are joined together into a continuous DNA strand. This process requires the activities of the enzymes DNA polymerase I and **DNA ligase** (Figs. 3.13e, f). If we consider two adjacent Okazaki fragments, the 3′ end of the newer DNA fragment is adjacent to but not joined to the primer 5′ end of the previously synthesized fragment. The DNA polymerase III that has just completed the synthesis of the newer fragment now dissociates from the DNA and DNA polymerase I takes its place. This enzyme continues the 5′-to-3′ synthesis of the newer DNA fragment, at the same time removing the primer of the older fragment through the action of the 5′-to-3′ exonuclease function of DNA polymerase I (Fig. 3.13e). When DNA polymerase I is finished, there is a gap between the two new DNA fragments, and this gap is sealed in a reaction catalyzed by DNA ligase, thereby producing a longer DNA strand (Fig. 3.13f). The whole process is completed until all the DNA is replicated.

Questions and problems

3.1 Compare contrast the conservative and semiconservative models for DNA replication.

3.2 Describe the Meselson and Stahl experiment and explain how it showed that DNA replication is semiconservative.

3.3 In the Meselson and Stahl experiment, ^{15}N labeled cells were shifted to ^{14}N medium at what we can designate as generation 0.
(a) For the semiconservative model of replication, what proportion of ^{15}N-^{15}N, ^{15}N-^{14}N, and ^{14}N-^{14}N would you expect to find at generations 1, 2, 3, 4, 6, and 8?
(b) Answer the question in (a) but this time for the conservative model of DNA replication.

3.4 Describe the semidiscontinuous model for DNA replication? What is the evidence showing that DNA synthesis is discontinuous on at least one template strand?

3.5 Distinguish between a primer strand and a template strand.

3.6 An autoradiograph of the *E. coli* chromosome produced by J. Cairns showed its length to be 1100 μm.
(a) How many base pairs does the *E. coli* chromosome have?
(b) How many complete turns of the helix does this chromosome have?
(c) If this chromosome replicated unidirectionally and if it completed one round of replication in 60 minutes, how many revolutions per minute would the chromosome be turning during the replication process?
(d) The *E. coli* chromosome, like many others, replicates bidirectionally. Draw a simple diagram of a replicating *E. coli* chromosome that is half way through the round of replication. Be sure to distinguish new and old DNA strands.

Pure 14N DNA Pure 15N DNA

Generation

1

2

3

a Semiconservative model

Pure 14N DNA Pure 15N DNA

Generation

1

2

3

b Conservative model

3.7 Chromosome replication of *E. coli* commences from a constant point, called the origin of replication. The results of autoradiography experiments suggested to J. Cairns that chromosome replication was unidirectional. It is now known that DNA replication is bidirectional. Devise a biochemical experiment to prove that the *E. coli* chromosome replicates bidirectionally. (Hint: Assume that the amount of gene product is directly proportional to the number of genes.)

3.8 Compare and contrast the three *E. coli* DNA polymerases with respect to their enzymatic activities.

3.9 Distinguish between the activities of primase, single-stranded binding protein, helicase, DNA ligase, DNA polymerase I, and DNA polymerase III in DNA replication in *E. coli*.

3.10 Describe the molecular action of the enzyme, DNA ligase.

3.11 Suppose *E. coli* cells are grown on a ^{15}N medium for many generations. Then, they are quickly shifted to a ^{14}N medium and DNA is extracted from samples taken after 1, 2, and 3 generations. The extracted DNA is subjected to equilibrium density gradient centrifugation in CsCl. Using the reference positions of pure ^{15}N and pure ^{14}N DNA as a guide, indicate where the bands of DNA would equilibrate if replication were semiconservative or conservative.

3.12 Assume you have a DNA molecule with the base sequence T-A-T-C-A going from the 5′-to-3′ end of one of the polynucleotide chains. The building blocks of

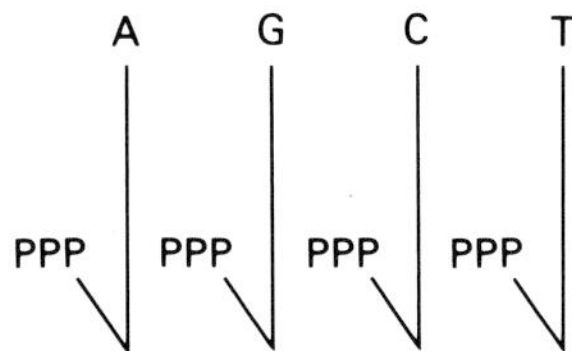

the DNA are as drawn in the accompanying figure. Use this shorthand system to diagram the completed molecule as proposed by Watson and Crick.

3.13 List the components necessary to make DNA in vitro by using the enzyme system isolated by Kornberg.

3.14 Give two lines of evidence that the Kornberg enzyme is not the enzyme involved in the replication of DNA for the duplication of chromosomes in growth of *E. coli.*

References

Cold Spring Harbor Symposia for Quantitative Biology, vol. 33, 1968. *Replication of DNA in Microorganisms.* Cold Spring Harbor Laboratory, New York.

Cozzarelli, N.R., 1980. DNA gyrase and the supercoiling of DNA. *Science* **207**:953–960.

Davidson, J.N. 1972. *The Biochemistry of the Nucleic Acids,* 7th ed. Chapman and Hall, London.

De Lucia, P. and J. Cairns, 1969. Isolation of an *E. coli* strain with a mutation affecting DNA polymerase. *Nature* **224**:1164–1166.

Dressler, D., 1975. The recent excitement in the DNA growing point problem. *Annu. Rev. Microbiol.* **29**:525–559.

Gefter, M.L., 1975. DNA replication. *Annu. Rev. Biochem.* **44**:45–78.

Gottesman, M.M., M.L. Hicks and M. Gellert, 1973. Genetics and function of DNA ligase in *E. coli. J. Mol. Biol.* **77**:531–547.

Goulian, M., P. Hanawalt and M. Fox, 1976. *DNA Synthesis and its Regulation.* Benjamin/Cummings, Menlo Park, California.

Gudas, L.J., R. James and A.B. Pardee, 1976. Evidence for the involvement of an outer membrane protein in DNA initiation. *J. Biol. Chem.* **251**:3470–3479.

Kornberg, A., 1960. Biologic synthesis of deoxyribonucleic acid. *Science* **131**:1503–1508.

Kornberg, A., 1980. *DNA Replication.* W.H. Freeman, San Fransisco.

Kornberg, A., I.R. Lehman, M.J. Bessman and E.S. Simms, 1956. Enzymic synthesis of deoxyribonucleic acid. *Biochim. Biophys. Acta* **21**:197–198.

Lehman, I.R., 1974. DNA ligase: structure, mechanism, and function. *Science* **186**:790–797.

Masters, M. and P. Broda, 1971. Evidence for the bidirectional replication of the *E. coli* chromosome. *Nature New Biol.* **232**:137–140.

Meselson, M. and F.W. Stahl, 1958. The replication of DNA in *Escherichia coli. Proc. Natl. Acad. Sci. USA* **44**:671–682.

Nossal, N.G., 1983. Prokaryotic DNA replication. *Annu. Rev. Biochem.* **58**:581–615.

Ogawa, T. and T. Okazaki, 1980. Discontinuous DNA replication. *Annu. Rev. Biochem.* **49**:424–457.

Okazaki, R.T., K. Okazaki, K. Sakobe, K. Sugimoto and A. Sugino, 1968. Mechanism of DNA chain growth. I. Possible discontinuity and unusual secondary structure of newly synthesized chains. *Proc. Natl. Acad. Sci. USA* **59**:598–605.

Sugino, A., S. Hirose and R. Okazaki, 1972. RNA-linked nascent DNA fragments in *Escherichia coli. Proc. Natl. Acad. Sci. USA* **69**:1863–1867.

Wickner, S.H., 1978. DNA replication proteins of *Escherichia coli. Annu. Rev. Biochem.* **47**:1163–1191.

Chapter 4 DNA Replication and the Cell Cycle in Eukaryotes

Outline

Outline of the eukaryotic cell cycle

In prokaryotes the process of DNA replication is well defined. Bacteria growing in a nutrient medium, for example, synthesize DNA throughout the cell cycle and then two daughter cells are produced by the formation of a transverse cell wall. By contrast the cell cycle in eukaryotes is much more complicated. In this chapter we will discuss the biochemical aspects of the cell cycle, and in Chapter 5 the cell division events of mitosis and meiosis will be discussed at the morphological level.

Both **asexual reproduction** and **sexual reproduction** are found in eukaryotes. Asexual reproduction is the development of a new individual from either a single cell or from a group of cells in the absence of any sexual process. This type of reproduction is found in both unicellular and multicellular organisms. An example is the propagation of plants by taking cuttings. Sexual reproduction, on the other hand, involves the fusion of two sex cells called gametes. This fusion produces a zygote cell from which the new individual develops. Usually, the two gametes are from different parents.

For both unicellular and multicellular eukaryotes, cellular reproduction is a cyclical process of growth, nuclear division, and (usually) cell division (cytokinesis): this is called the **cell cycle**. In proliferating somatic cells, the cell cycle consists of four phases (Fig. 4.1). The sequence of phases are G1 (gap 1), S (synthesis), G2 (gap 2), and M (**mitosis**), the first three collectively describing the interphase stage of the cell cycle. Following G2 the cell divides to produce two daughter cells by mitosis and each daughter cell then begins a

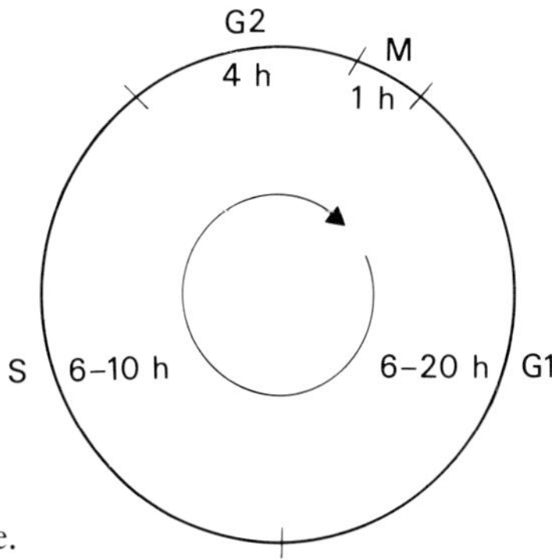

Fig. 4.1. Typical eukaryotic cell cycle.

new cell cycle. In mammalian cells in culture, the cell cycle takes approximately 24 hours.

Much of what we know about the molecular biology of the cell cycle has come from studies done with cell cultures. For eukaryotic cell cultures in general, the relative time spent in each of the four phases of the cell cycle varies, as does the cell cycle time itself (Fig. 4.2). At present knowledge of the S and M stages at the molecular level is incomplete, and much of what goes on in G1 and G2 is not known.

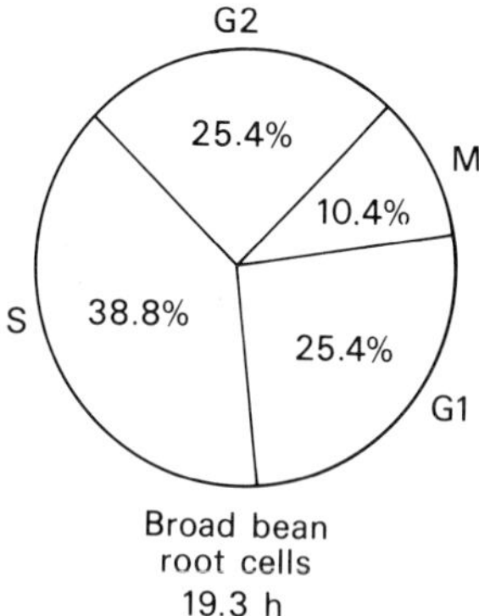

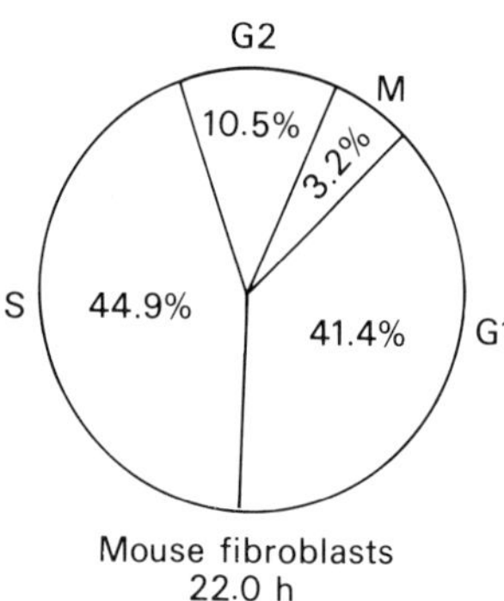

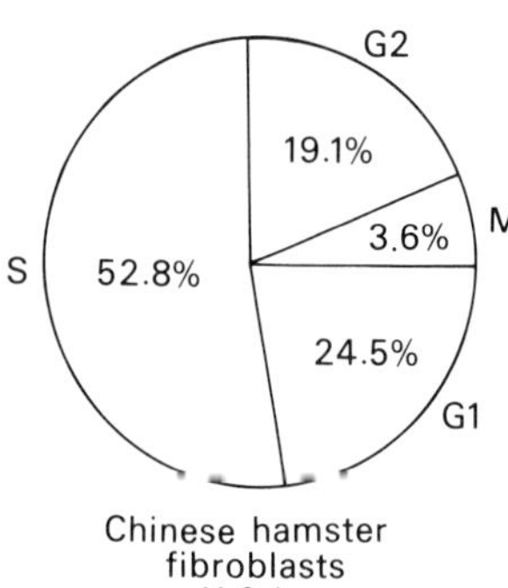

Fig. 4.2. Relative time in the four phases of the cell cycle for three cell types. (After B. Kihlman et al., 1966. *Hereditas* 55: 386.)

G1 phase

The G1 phase following mitosis is characterized by a change in the chromosomes from the condensed mitotic state of the more extended interphase state, and by a series of events leading to the initiation of DNA replication.

In a homogeneous population of cultured cells the cell cycle time is variable, and this is a major problem in attempts to work with synchronously dividing cells. The G1 phase of the cell cycle is far more variable in length from cell to cell of the same cell type than the other three phases, and it is this variability that is responsible for the variability in generation times within a cell population and for such variability in the different cell types of an organism. The cause of the variability of G1 is not known, although there is evidence to suggest that the length of this phase is related to protein content.

Apparently G1 is a very significant phase in the cell cycle since cells that stop dividing normally arrest in this phase. Apparently there is a point within G1 past which the cell is irreversibly committed to initiate DNA synthesis and to proceed through cell division once the biochemical event(s) associated with point has occurred. Whether a cell then continues to proliferate or whether it becomes specialized and nondividing presumably depends upon molecular regulatory signals acting at that point. One example of regulation at this level is that changes or differences in the rates of cell reproduction for cells of the same genetic constitution are brought about principally by alteration of the G1 phase. Thus, there may be a number of regulatory genes operating in G1, but we know very little about this area.

The same may be said of the specific molecular events occurring in G1. After mitosis, the chromosomes become less coiled and more transcriptionally active. Some of this transcriptional activity results in the production of a series of molecules required for the initiation of DNA synthesis. However, some organisms do not have a defined G1 phase (for example, the slime mold *Physarum polycephalum*, the fission yeast *Schizosaccharomyces pombe*, the cleavage stages of sea urchin embryos, *Xenopus* embryos, and mouse embryos). Thus the events leading to DNA synthesis may precede mitosis in the G2 phase in some systems rather than being in G1. These events include the synthesis of many of the enzymes and other proteins required for chromosome duplication.

S phase

Discontinuous DNA synthesis

During the S phase the chromatin material is duplicated. In animal cell cultures the S period lasts for 6 to 10 hours, whereas in vivo it may be as short as 10 minutes, as in the fission yeast *Schizosaccharomyces pombe*, or as long as 35 hours, as in mouse ear skin cells. J. H. Taylor in 1957 showed by autoradiographic studies of root tip cells of the broad bean pulse-labeled with tritium-labeled thymidine, that DNA replication was semiconservative (Fig. 4.3). He labeled cells during the DNA synthesis phase and showed by autoradiography that all chromatids were labeled. Then he allowed the cells to go through one more division cycle in the absence of label and found that at any one point along each chromosome, only one chromatid was labeled. There is also recent evidence from a number of workers that, as in prokaryotes, DNA replication in eukaryotes is semidiscontinuous. For example, when the animal virus SV40 (simian virus 40) replicates its DNA in vitro, newly synthesized fragments can be isolated that have the ability to self-anneal, that

Fig. 4.3. Semiconservative DNA replication in eukaryotes. (**a**) Autoradiogram of broad bean (*Vicia faba*) chromosomes showing semiconservative DNA replication. Cells were incubated with ^{3}H-thymidine to label all the DNA and then they were "chased" with cold thymidine for one generation (×1875). (**b**) Interpretative sketch of the autoradiogram showing that at any one point, only one chromatid is labeled (shaded area) thus indicating semiconservative DNA replication and coincidentally showing exchanges involving sister chromatids. (Both photographs courtesy of J. H. Taylor.)

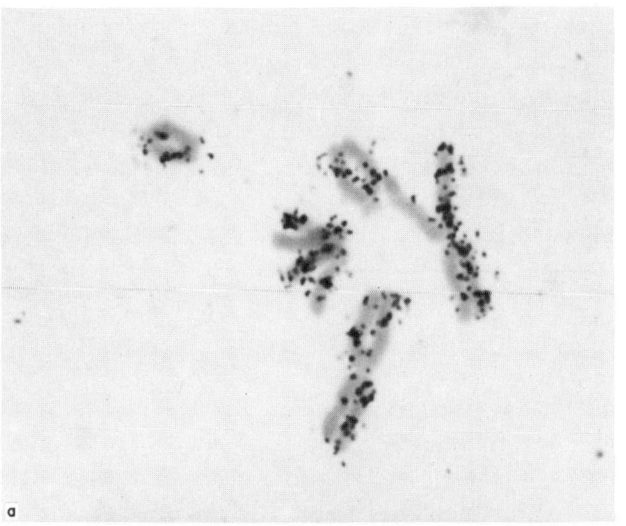

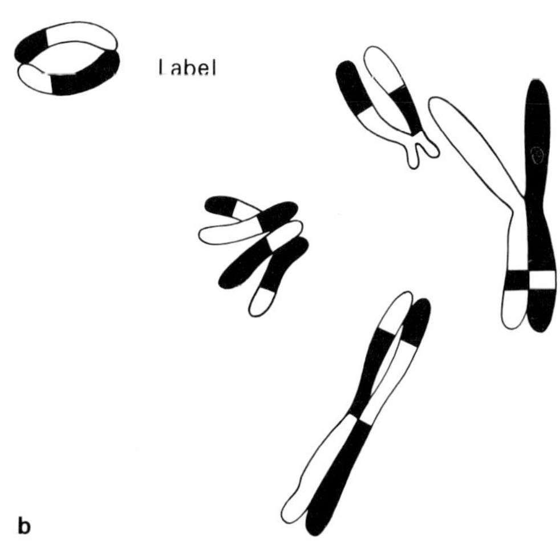

is, form short stretches of double helix. This is the result one would expect if Okazaki fragments were being generated on the two complementary template strands. Also, in mammalian cells there is evidence that nucleotides are polymerized into 100-nucleotide short-lived strands that are then joined into longer ones. In general, the fragments average 135 nucleotides in length in eukaryotes, with a range of 35 to 300 nucleotides.

DNA polymerases

As we discussed in Chapter 3, all prokaryotic DNA polymerases are unable to initiate DNA synthesis; rather they polymerize deoxyribonucleotides on a primer. At least in mammalian cells and in animal viruses, DNA synthesis is initiated by 9- to 10-nucleotide long RNA primers that are subsequently excised.

The enzymes required for DNA replication have been studied in a number of eukaryotes. In higher eukaryotes three DNA polymerases, alpha (α), beta (β), and gamma (γ), have been distinguished based on molecular weight, chromatographic properties, sensitivity to the inhibitor N-ethylmaleimide (NEM), sensitivity to salts, and the ability to copy various templates. These properties are summarized in Table 4.1. All of the α and β polymerase molecules are located in the nucleus, whereas the γ enzyme is apparently localized in the mitochondria. The α and β enzymes are usually found with other proteins in a replication complex. It is impossible to generalize too much about these enzymes, since there is much variability among the eukaryotic organisms that have been studied. For example, some organisms have multiple forms of the enzymes, whereas others have a single major enzyme. Plants, protozoa, and fungi lack the β polymerase, and eukaryotic microorganisms in general have an enzyme called α, although it is very different from the mammalian α polymerase. In many instances the enzymes have not been purified to homogeneity, and thus it is difficult to make comparisons or even to estimate the number of enzyme types or subtypes that a cell possesses.

Table 4.1. Properties of mammalian DNA polymerases. (After A. Weissbach, 1977. *Annu. Rev. Biochem.* **46,** 25.)

Polymerase type	Molecular weight	Inhibition by N-ethylmaleimide	Salt effect
α	120,000–300,000	yes	Inhibited by > 25 mM NaCl
β	30,000–50,000	no	Stimulated by 100–200 mM NaCl; inhibited by 50 mM phosphate
γ	150,000–300,000	yes	Stimulated by 100–250 mM KCl and 50 mM phosphate

Current models propose that in higher eukaryotes the α polymerase is required for nuclear DNA replication in general, and that the β polymerase may function as a repair enzyme. The γ polymerase is responsible for replication of the mitochondrial genome. None of these enzymes has proofreading activity. By contrast, lower eukaryotes (such as yeast) have nuclear DNA polymerases that more closely resemble those of prokaryotes, and many of those polymerases do have proofreading activity.

Replication units

In *E. coli* there is a single origin or replication in the chromosome, and DNA replication proceeds bidirectionally from that point. Electron microscopy studies of replicating chromosomes of eukaryotes show that each chromosome has a number of **replication units** (RUs) or **replicons** (Fig. 4.4), each of which has a specific origin and two termini for the replication process. This arrangement of chromosomes into replication units is necessary so that the enormous amount of DNA relative to a bacterial cell can be replicated within a reasonable time.

EM and autoradiographic studies show that DNA replication begins by the opening of a "bubble" in the DNA representing two replication forks (Fig. 4.4). These forks migrate bidirectionally as the DNA is synthesized until they reach specific termination points. Very little is known about the origin and termination signals, but it is presumed that specific nucleotide sequences provide recognition sites for the DNA polymerase-replication complex. (To date, specific sequences that signal the initiation of DNA replication have been identified in certain animal viruses [e.g. SV40] and in the yeast *Saccharomyces cerevisiae*.) The simplest model is that a specific initiator protein binds with a genetically determined origin sequence and facilitates

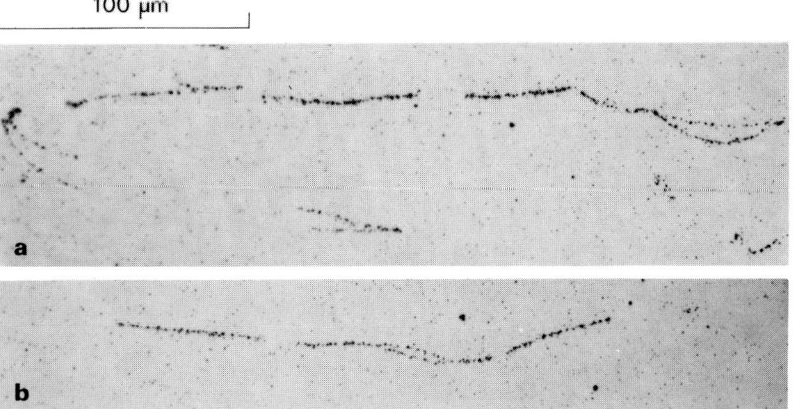

Fig. 4.4. Autoradiographs of replicating eukaryotic chromosomes showing distinct replication units. Both are from *Xenopus laevis* tissue culture cells labeled with ^{3}H-thymidine. (**a**) One replicating unit that had been initiated some hours before ^{3}H-thymidine was supplied. The divergent V-tracks give direct evidence that replication is bidirectional. (**b**) Evidence of distinct replication units is shown, and the middle unit shows sister strand separation. (Both photographs courtesy of H. G. Callan. Photograph [**a**] is with permission from the *Proceedings of the Royal Society of London B* **181:**19–41, 1972. The Royal Society, Copyright © 1972.)

the binding of the DNA replication complex and the initiation of DNA replication.

Careful examination of replicating chromosomes under the EM shows that the spacing between origins of replication varies between 7 to 100 μm (30,000 to 300,000 base pairs) in a wide range of eukaryotic organisms. Thus, for example, there are perhaps as many as 100 RUs per chromosome in HeLa cells (a type of tissue culture cell derived from a human cancer cell). The rate of movement of the replication complex in RUs is very similar in most eukaryotes that have been examined ranging from 1000–15,000 nucleotides of DNA per minute at 37°C. This rate of replication fork movement does vary within a single system, however, as a result of genetic controls or temporal controls or in response to environmental changes.

Considering DNA replication for all the chromosomes as a whole, there is a temporal ordering of replication through the S phase. Specifically, there seems to be a highly ordered and complex pattern of replication of RUs in the nucleus. A diagrammatic representation of this is shown in Fig. 4.5. Initiation

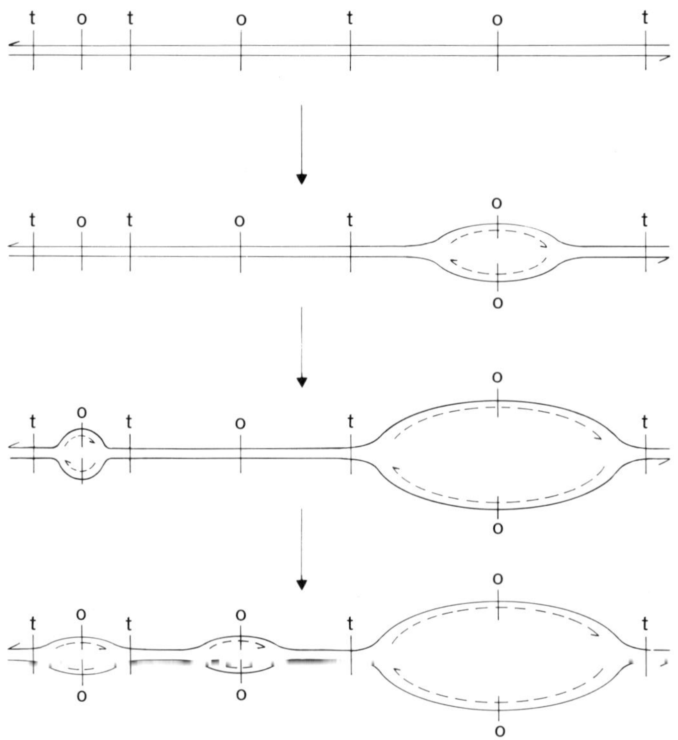

Fig. 4.5. Temporal sequencing of DNA initiation in replication units of eukaryotic chromosomes. (After J. A. Huberman and A. D. Riggs, 1968. *J. Mol. Biol.* **32**:327.)

of replication of RUs begins at the "o" regions. Replication proceeds bidirectionally producing the "bubbles" in the DNA. Termination of the process involves fusion of adjacent replication forks. Apparently the sequence of movement of replication complexes is species-specific and is repeated generation after generation. How the signals for these processes are coordinated is not known, but it is clear that protein synthesis is required for continued intitiation of replication of sets of RUs. There is also some evidence that the more GC-rich DNA is replicated first, and the AT-rich DNA regions are replicated later in the S phase.

DNA replication itself is dependent upon protein synthesis. This has been shown by the use of inhibitors of protein synthesis such as cycloheximide and puromycin. The addition of either of these to mammalian cells in culture causes an immediate, rapid decline in the initiation of DNA synthesis, but not in the elongation of DNA chains already begun. Since DNA replication is discontinuous, the overall replication of chromosomes cannot be completed in the absence of new protein synthesis. DNA synthesis is also dependent on RNA synthesis, but the inhibition of the latter process does not have an immediate effect on the former process. Thus the addition of the RNA synthesis inhibitor actinomycin D to cells early in the S phase blocks DNA replication late in S. Further, ribosomal RNA synthesis must continue to within an hour or so of the transition from G1 to S if DNA synthesis is to start. This is correlated with the genetic, physiological, and biochemical evidence that suggests that progression of cells into the S phase is determined by a regulatory protein synthesized during the G1 phase. This protein apparently acts as a positive effector to derepress DNA replication. Indeed a large number of DNA binding proteins have been identified in eukaryotes, and one or more of these may serve to facilitate initiation of RNA replication.

Assembly of replicated DNA into nucleosomes

Eukaryotic chromosomes consist of DNA complexed with histones in nucleosome structures. When the DNA is replicated, the nucleosome constant must be doubled through the synthesis of new histones and the assembly of new nucleosomes.

There are many copies of the histone genes in the eukaryotic genome. Recent evidence indicates that these genes are active at least in G1 and S and perhaps in G2 also, producing the large amounts of histone needed for the formation of new nucleosomes in the S phase.

Newly replicated DNA rapidly becomes complexed with histones to form nucleosomes. The distribution of old and newly synthesized histones in the nucleosomes of newly replicated DNA is not clearly understood. In the systems studied, evidence has been found that there is not any mixing of old and new histones in the nucleosomes. This indicates that the old histone octamers are conserved from generation to generation and that histone octamers consisting only of newly synthesized histones are used at the replication forks to form new nucleosomes. Once the nucleosomes are assembled at the replication fork, about 2–15 minutes are needed to convert the chromatin to the mature state, that is, the state indistinguishable from non-replicating chromatin.

G2 phase

In G2 the chromosomes condense in preparation for mitosis. The condensation process involves higher-order folding of the chromatin filament, but the mechanism for it is poorly understood. Inhibitor studies have shown that both RNA and protein synthesis are required for the completion of G2. The end of the G2 phase is delineated by the onset of mitosis. This transition is difficult to define by light microscopy.

One other significant event during G2 is the synthesis of the protein tubulin. Tubulin is polymerized to produce microtubules that make up the spindle apparatus used for the segregation of chromosomes into daughter nuclei mitosis and meiosis.

Molecular aspects of mitosis

Mitosis is the division of a cell into two genetically identical daughter cells. By the beginning of mitosis the chromosomes have duplicated and the chromosomes are then distributed to the progeny cells by the division process. The resulting cells are then in the G1 phase. The behavior of the chromosomes in mitosis will be described in the next chapter; here we shall discuss the general morphological and biochemical changes of mitosis.

Mitosis is characterized by large changes in cell structure and function, but the molecular bases of these events are not clearly understood. As mitosis commences, RNA synthesis decreases, then stops by the metaphase stage, and resumes again by late telophase. It is possible that this is related to the unavailability of transcription initiation sites owing to the highly condensed state of the chromosomes. The rate of protein synthesis also drops drastically at the beginning of mitosis and then resumes in late telophase concomitant with the rise in the rate of RNA synthesis. Along with the inhibition of RNA and protein synthesis, the nuclear membrane and nucleolus break down, and these structures are reformed in daughter nuclei in late telophase. The molecular signals for these events are not known.

Summary of cell cycle events

In summary, the cell cycle of a eukaryotic cell may be depicted as shown in Fig. 4.6. The sequence of events in the cycle are presumably dependent on the transcription and translation of cell cycle genes in a particular temporal order. Some progress has been made in identifying cell cycle genes in a number of organisms (for example, yeast and mammalian cells) by the selection for genetic mutants with conditional blocks in the cell cycle. Study of these mutants has provided valuable information about the ordering and regulation of cell cycle events at the biochemical level.

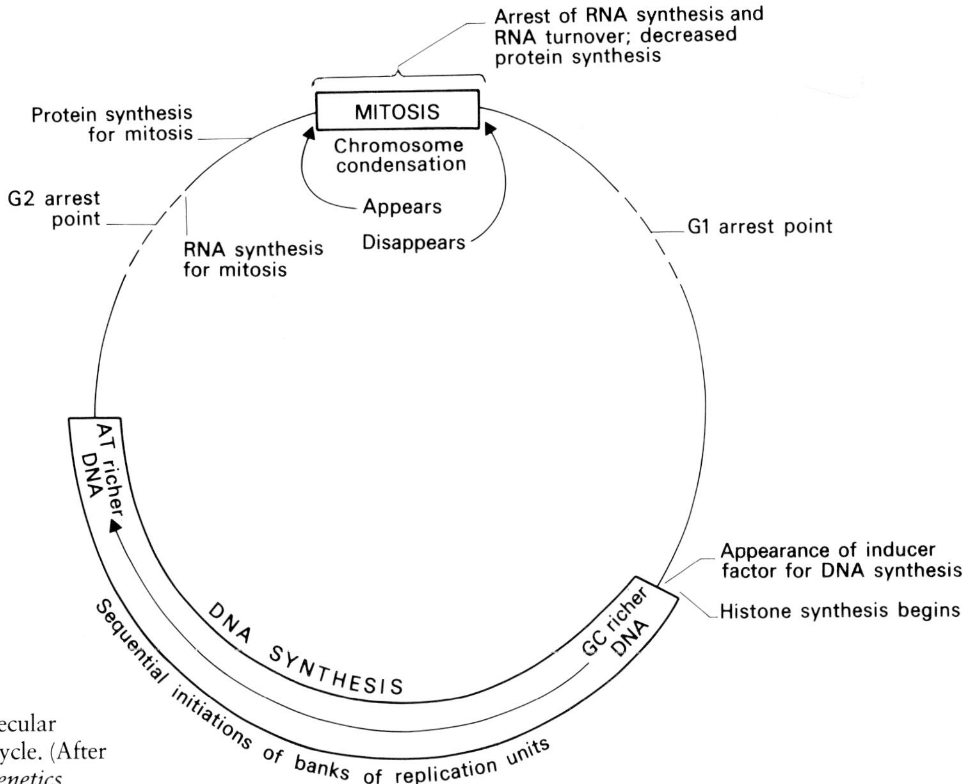

Fig. 4.6. Summary of the molecular events of the eukaryotic cell cycle. (After D.M. Prescott, 1976. *Adv. Genetics* **18**:99.)

Questions and problems

4.1 Compare and contrast eukaryote and prokaryote DNA polymerases.

4.2 When the DNA replicates, the nucleosome structure must duplicate. Discuss the synthesis of histones in the cell cycle and discuss the model for the assembly of new nucleosomes at the replication forks.

References

Bollum, F.J., 1975. Mammalian DNA polymerases. *Prog. Nucl. Acid Res. Mol. Biol.* **15**:109–144.

Challberg, M.D. and T.J. Kelly, 1982. Eukaryotic DNA replication: viral and plasmid model systems. *Annu. Rev. Biochem.* **51**:901–934.

Cold Spring Harbor Symposia for Quantitative Biology, 1977, vol. 42 *Chromatin.* Cold Spring Harbor Laboratory, New York.

DuPraw, E.J., 1970. *DNA and Chromosomes.* Holt, Rinehart, and Winston, New York.

Edenberg, H.J. and J.A. Huberman, 1975. Eukaryotic chromosome replication. *Annu. Rev. Genetics* **9**:245–284.

Gefter, M.L. 1975. DNA replication. *Annu. Rev. Biochem.* **44**: 45–78.

Groppi, V.E. and P. Coffino, 1980. G1 and S phase mammalian cells synthesize histones at equivalent rates. *Cell* **21**:195–204.

Hartwell, L.H., 1974. *Saccharomyces cerevisiae* cell cycle. *Bacteriol. Rev.* **38**:164–198.

Huberman, J.A. and H. Horwitz, 1973. Discontinuous DNA synthesis in mammalian cells. *Cold Spring Harbor Symp. Quant. Biol.* **38**:233–238.

Klein, A. and F. Bonhoeffer, 1972. DNA replication. *Annu. Rev. Biochem.* **41**:302–322.

Kreigstein, H.J. and D.S. Hogness, 1974. Mechanism of replication in *Drosophila* chromosomes: structure of replication forks and evidence for bidirectionality. *Proc. Natl. Acad. Sci. USA* **71**:135–139.

Loeb, L.A., 1974. Eucaryotic DNA polymerases, In *The Enzymes,* P.D. Boyer (ed.), vol. X, pp. 174–210. Academic Press, New York.

Pardee, A.B., R. Dubrow, J.L. Hamlin and R.K. Kletzien, 1978. Animal cell cycle. *Annu. Rev. Biochem.* **47**:715–750.

Prescott, D.M., 1970. The structure and replication of eukaryotic chromosomes. *Adv. Cell Biol.* **1**:57–117.

Prescott, D.M., 1976. The cell cycle and the control of cellular reproduction. *Adv. Genetics* **18**:99–177.

Sheinin, R., J. Humbert and R.E. Pearlman, 1978. Some aspects of eukaryotic DNA replication. *Annu. Rev. Biochem.* **47**:277–316.

Simchen, G., 1978. Cell cycle mutants. *Annu. Rev. Genetics* **12**:161–191.

Taylor, J.H., P.S. Woods and W.L. Hughes, 1957. The organization and duplication of chromosomes as revealed by autoradiographic studies using tritium-labeled thymidine. *Proc. Natl. Acad. Sci. USA* **43**:122–128.

Watson, J.D., 1971. The regulation of DNA synthesis in eukaryotes. *Adv. Cell Biol.* **2**:1–46.

Weissbach, A., 1977. Eukaryotic DNA polymerases. *Annu. Rev. Biochem.* **46**:25–47.

Chapter 5 Mitosis and Meiosis

Outline

As has already been discussed, DNA synthesis is continuous through the bacterial cell cycle. As the DNA content doubles, the cell enlarges and then begins to synthesize a dividing wall in the central region of the cell. This serves to segregate the two daughter chromosomes to different cell compartments and, when the wall is completed, the bacterium separates into two progeny cells. This process is repeated as long as the bacteria keep growing. In Chapter 4 we described the eukaryotic cell cycle in detail, and clearly it differs markedly from that of a bacterium. Dividing somatic (nongamete producing) cells go through four distinct but interrelated phases of a cell cycle described as G1, S, G2, and M. The chromosomes are replicated during the S phase of the cycle, and later, in a process called **mitosis** (the M phase), the duplicated chromosomes are segregated into two daughter cells that have the same genetic content as the parent cell. (An exception will be discussed in a later chapter.)

Before describing mitosis, we must review some information introduced in Chapter 2 that concerns the chromosome content of eukaryotes. The number of chromosomes per nucleus is generally constant for all individuals of a species and varies from one species to another. Thus, for example, man has 46, rat has 42, and pea has 14 chromosomes. For the somatic cells of these organisms and of other eukaryotes (except the lower eukaryotes), the chromosomes are present in pairs; that is, there are 23 pairs in man, and so on. Conventionally, the somatic cells of these organisms are described as being diploid (2N) in chromosome number. By contrast, the mature germ cells (gametes, produced by meiosis) of a sexually reproducing individual contain only half the somatic number of chromosomes: one member of each pair. The gametes are described as being haploid (N) in chromosome number. For later consideration the cells of the lower eukaryotes such as yeast and *Neurospora* also have a haploid chromosome number.

Mitosis

Mitosis is the mechanism by which the chromosome content of a somatic cell (haploid or diploid) is kept constant through successive cell divisions. When an interphase cell is stained with basic dyes, the **nucleus** becomes visible under the light microscope and is seen to be surrounded by a membrane. Within the nucleus one or two RNA-rich regions, the *nucleoli*, are apparent (Fig. 5.1).

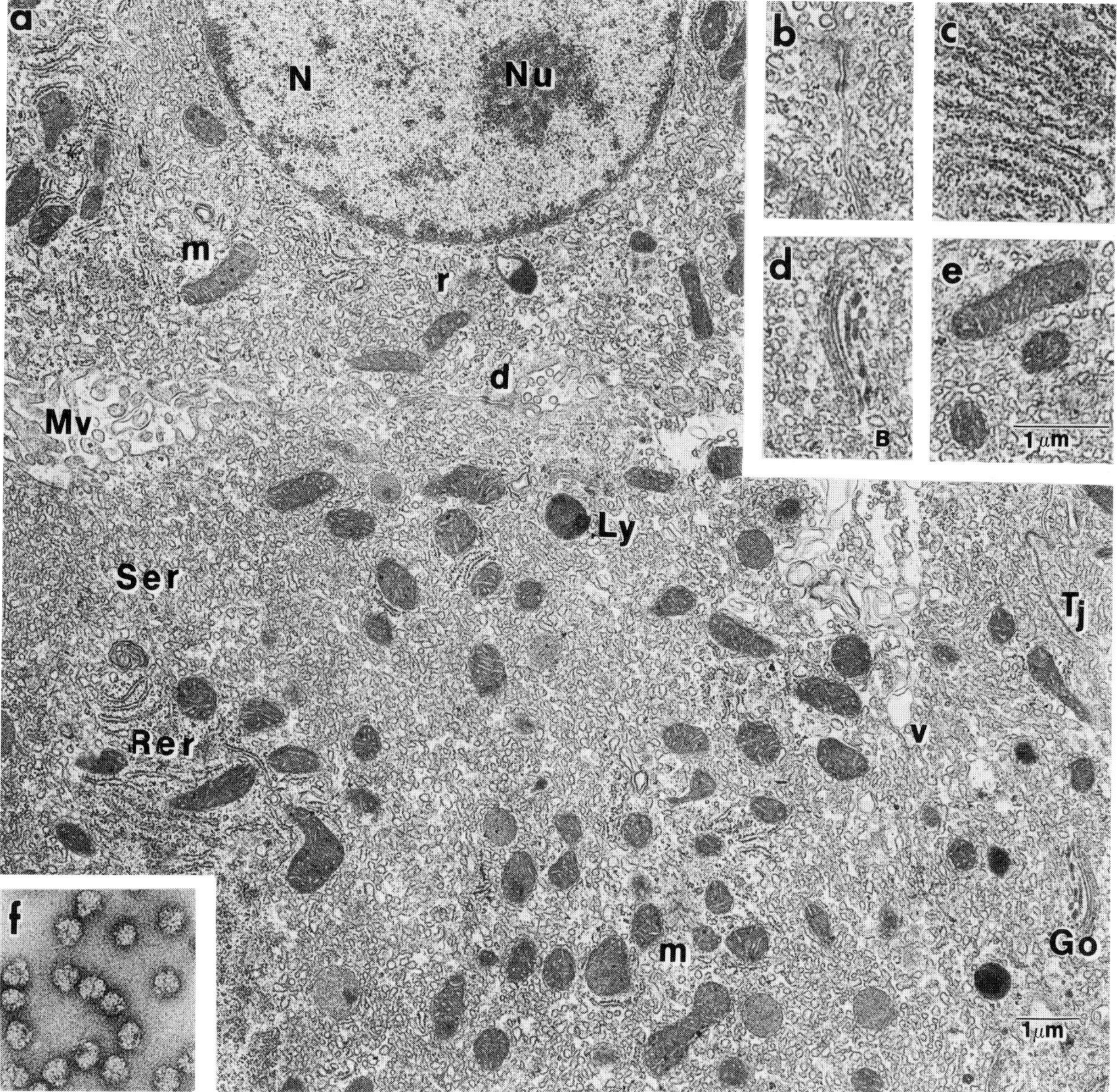

Fig. 5.1. An electron micrograph of a eukaryotic cell (from rat liver) with "close ups" of some of the major cell parts. (Courtesy of M. Boublik.) (**a**) General ultrastructure of rat liver cells (×9600). Key to parts: N, nucleus; Nu, nucleolus; Tj, tight junction (where the cell membranes of adjacent cells have fused); d, desmosome (a portion of the cell membrane specialized for adhesion to a neighboring cell); Mv, microvilli (invaginations of the cell surface); r, ribosomes (sites of protein synthesis); Rer, rough endoplasmic reticulum; Ser, smooth endoplasmic reticulum; m, mitochondria, v, vacuole; Ly, lysosome (small vesicles containing digestive enzymes); Go, Golgi apparatus (a stack of flattened vesicles). (**b**) Cell membrane with desmosome (×14,400). (**c**) Rough endoplasmic reticulum (×14,400). (**d**) Golgi apparatus (×14,400). (**e**) Mitochondria (×14,400). (**f**) Ribosomes (×160,000).

During the S phase of the cell cycle, each chromosome is replicated, but the **centromeres** do not appear to divide. Actually, the DNA of the centromere has duplicated but only one centromere structure is visible. The product of replication of a chromosome is two precise copies called **sister chromatids** that are held together by the unseparated centromere.

Mitosis itself is a continuous process, but for descriptive purposes it is broken down into four stages called **prophase, metaphase, anaphase**, and **telophase**. Photographs showing the typical appearances of the various phases of mitosis are in Fig. 5.2. For the purposes of simplifying the discussion, Fig. 5.3 shows the events of mitosis for a hypothetical, diploid cell with

Fig. 5.2. The stages of mitosis in *Trillium erectum*. Since this is a plant, there is no centriole and spindle fibers are present but not easy to see in photomicrographs. The mitosis shown is occurring in pollen, and the cell in this case is haploid. (**a**) Interphase. (**b**) Prophase. (**c**) Late prophase. (**d**) Metaphase. (**e**) Anaphase. (**f**) Telophase. (All photomicrographs ×2000.) (Courtesy of A. H. Sparrow and R. F. Smith, Brookhaven National Laboratory.)

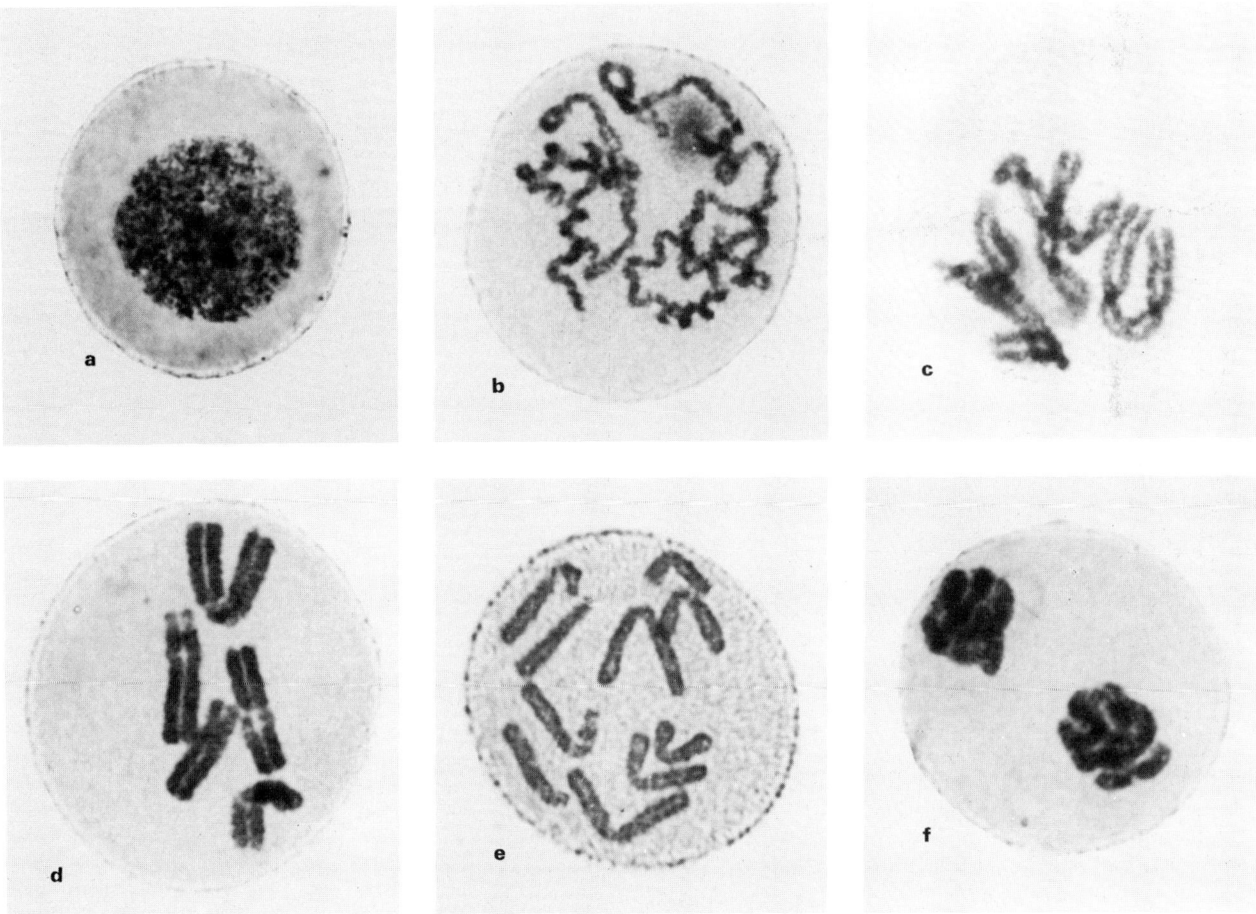

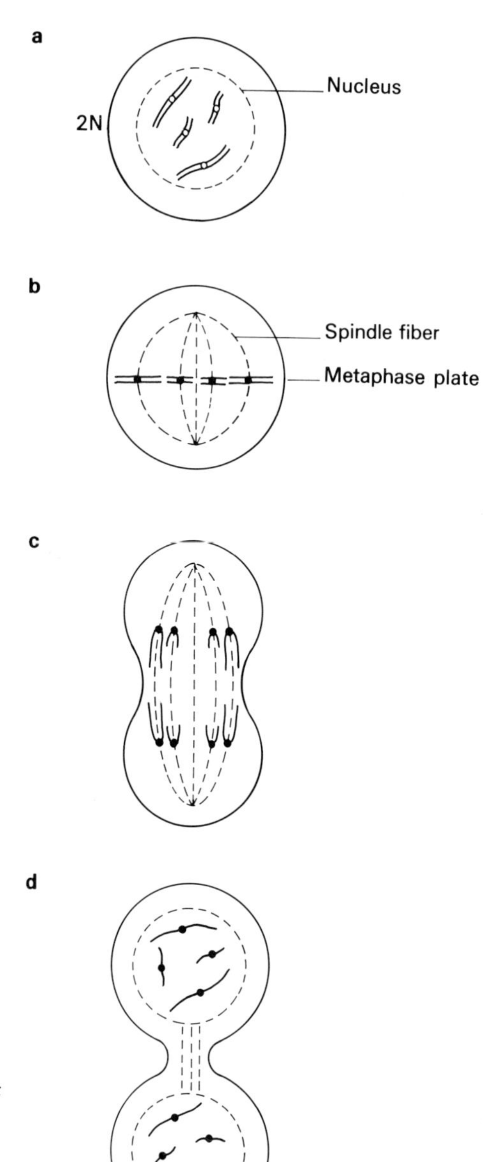

Fig. 5.3. Diagrammatic representation of mitosis in an hypothetical diploid animal cell with a haploid chromosome number of two. (**a**) Prophase. (**b**) Metaphase. (**c**) Anaphase. (**d**) Telophase.

two pairs of chromosomes and the main feature of each phase are listed here. The same events occur for haploid cells undergoing mitosis.

Prophase (Fig. 5.3a)

a. The chromosomes become visible as a result of coiling events.

b. Each chromosome is seen to consist of two sister chromatids.

c. Near the end of prophase, the nucleolus or nucleoli and the nuclear membrane are broken down.

Metaphase (Fig. 5.3b)

a. Spindle fibers appear and radiate from the opposite poles of the cells. In animals the fibers are attached to structures called *centrioles*, which are at the two poles; no such structures are visible in plant cells.

b. Some spindle fibers become attached to the centromere regions of the chromosomes.

c. The sister chromatids become aligned in one plane in the middle of the cell in a region called the *metaphase plate*.

Anaphase (Fig. 5.3c)

a. This stage is begun by division of the centromere of each chromosome.

b. Sister chromatids undergo **disjunction** (separation) during this phase, and the daughter chromosomes migrate toward the opposite poles of the cell. This migration occurs as a result of the properties of the spindle fibers. This serves to segregate the two identical copies of each chromosome, one to each pole.

Telophase (Fig. 5.3d)

a. The migration of the daughter chromosomes to the two poles is completed. Thus, the two identical replicates of each chromosome are separated and reformed into two groups in the cell.

b. A nuclear membrane forms around each set of chromosomes.

c. The nucleolus or nucleoli reform.

d. The spindle fibers disappear.

e. The chromosomes uncoil and become "invisible" under the light microscope. Two typical interphase nuclei are now apparent.

f. In most cases telophase is followed by **cytokinesis**, which involves the division of the cytoplasm. This compartmentalizes the two new nuclei into separate daughter cells. In animal cells, cytokinesis occurs by the production of a constriction in the center of the cell until two progeny cells are produced. Plant cells, however, divide by assembling a new cell membrane and cell walls between the two new nuclei so that two new cells are generated.

For studies of genetics, the key points of mitosis are:

1. Homologous chromosomes divide to give two chromatids each in the S phase of the cell cycle.

2. The homologous chromosomes (two sister chromatids each) align *independently* at the metaphase plate.

3. Mitosis provides for the maintenance of the genetic content of a cell from generation to generation.

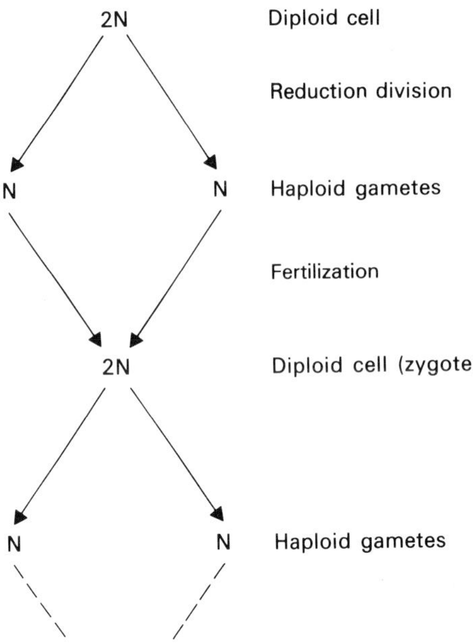

Fig. 5.4. The alternation of haploid (N) and diploid (2N) states from generation to generation.

Meiosis

The sexual cycle of a diploid organism involves the alternation of haploid and diploid states (Fig. 5.4). **Meiosis** is the process by which haploid gametes or spores are produced by two successive divisions of a diploid nucleus. During meiosis, homologous chromosomes pair, replicate once, and undergo assortment so that each of the four meiotic products receives one representative of each chromosome. The two nuclear divisions are called first (**meiosis I**) and second meiotic divisions (**meiosis II**). Photomicrographs of meiosis are shown in Fig. 5.5.

First meiotic division (meiosis I)

In meiosis I, the chromosome number is reduced from diploid to haploid.

Prophase I (Fig. 5.6a, b)

a. The diploid number of chromosomes become visible as they condense by coiling. Unlike the prophase stage in mitosis, each chromosome appears to be single.

b. Homologous chromosomes become paired.

c. Each chromosome later is seen to have divided to produce two sister chromatids (Fig. 5.6b).

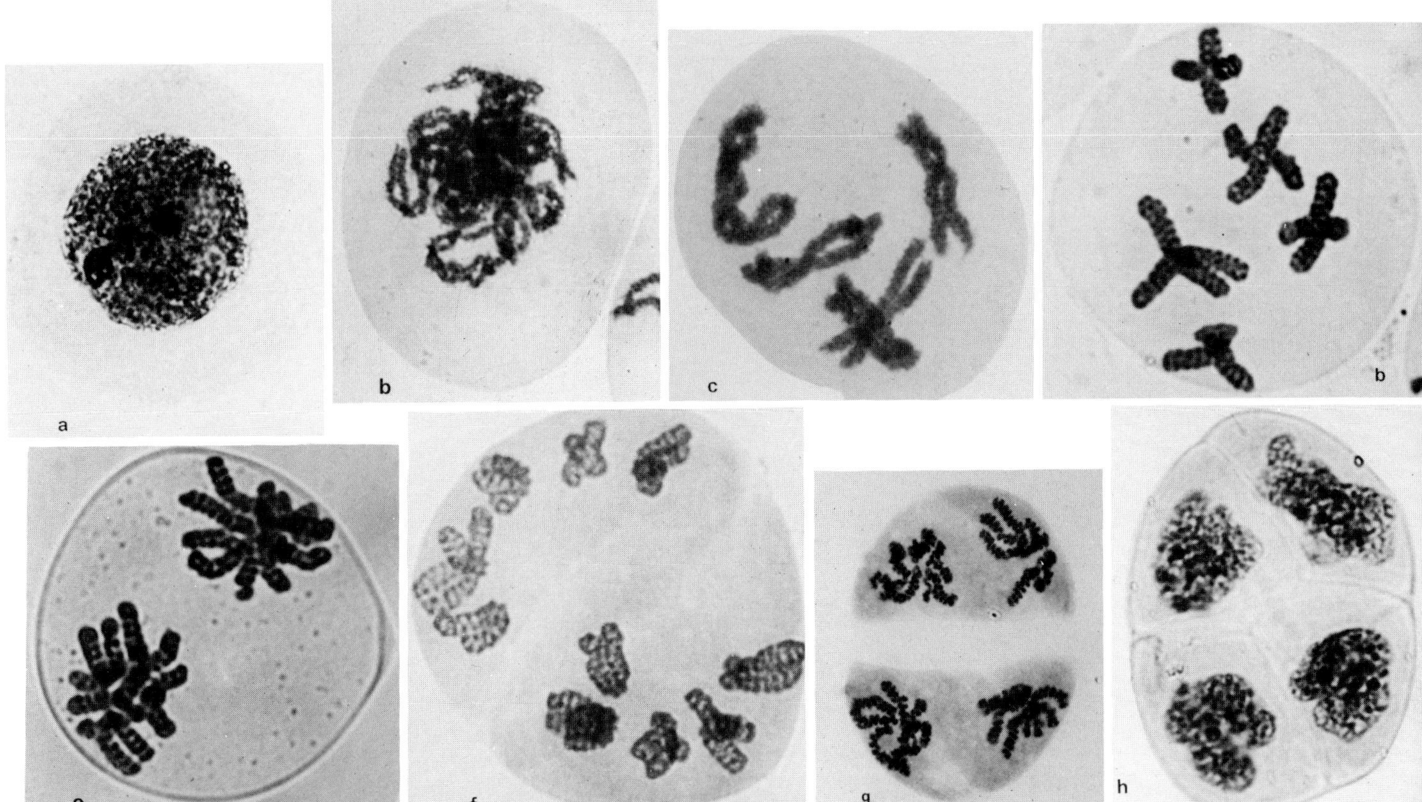

Fig. 5.5. The stages of meiosis in the plant, *Trillium erectum*. (**a**) Prophase I (early). (**b**) Prophase I (middle). (**c**) Prophase I (late). (**d**) Metaphase I. (**e**) Anaphase I. (**f**) Metaphase II. (**g**) Anaphase II. (**h**) Early interphase following the two meiotic divisions. Four cells are apparent. (All photomicrographs approximately × 1000.) (Courtesy of A. H. Sparrow and R .F. Smith, Brookhaven National Laboratory.)

d. It is at this stage of meiosis (the bivalent, or tetrad, stage) that genetic exchange occurs betwen paternally and maternally derived homologs. That is, the chromosomes become very closely aligned (or synapsed; the process is called **synapsis**), and physical exchanges of chromosome regions take place between homologous chromosomes. This phenomenon is called **crossing over**, and the place on the chromatids where crossing over has occurred is called a *crossover*. Later in prophase I, the result of crossing over becomes visible as a cross-shaped structure called a **chiasma** (plural = chiasmata).

Metaphase I (Fig. 5.6c)

a. At the onset of metaphase I the nucleolus and nuclear membrane disappear.

b. The undivided centromeres become aligned on the spindle fibers and the associated chromatids become oriented on the metaphase plate.

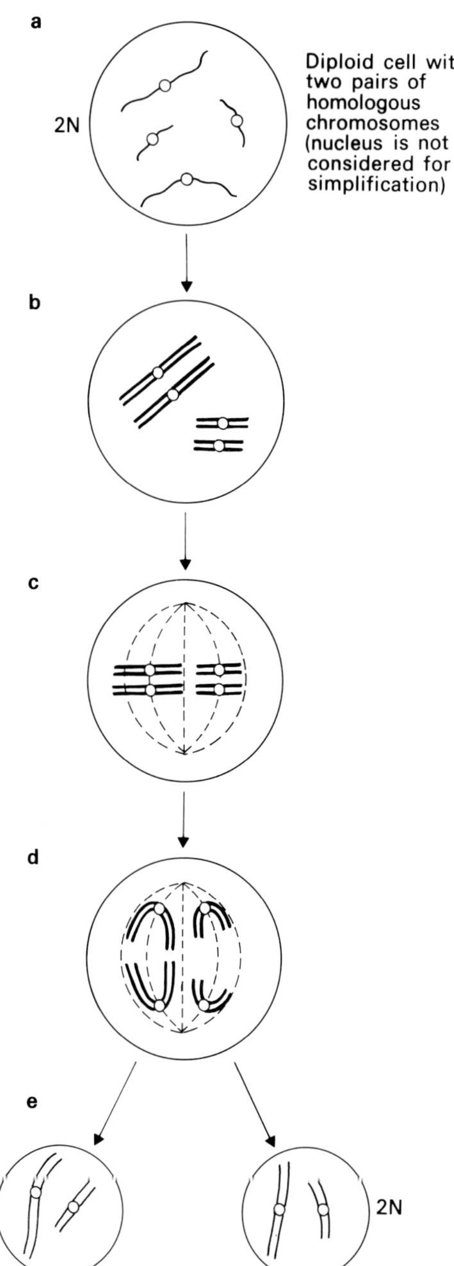

Fig. 5.6. Diagrammatic representation of the first meiotic division in an hypothetical animal cell with a haploid chromosome number of two. (**a**) Prophase I (early). (**b**) Prophase I (late). (**c**) Metaphase I. (**d**) Anaphase I. (**e**) Telophase I/interphase.

Anaphase I (Fig. 5.6d)
The chromosomes in each bivalent separate in this stage so that homologous pairs disjoin and migrate toward the opposite poles. As a result, the maternally and paternally derived homologs are segregated.

Telophase I/Interphase II (Fig. 5.6e)
a. Cell division (cytokinesis) usually occurs to produce two progeny cells. Each of these cells has only one complete haploid set of chromosomes (two chromatids each), and there is an equal probability that a particular chromosome will be paternal or maternal in origin.
b. In a brief interphase, the chromosomes elongate and a nuclear membrane reforms.

Second meiotic division (meiosis II)

In prophase II (Fig. 5.7a) the chromosomes condense and the centromeres divide. In metaphase II (Fig. 5.7b) a spindle apparatus is organized and the now-divided centromeres become attached to it so that the chromosomes become aligned at the equatorial plan.

Anaphase II (Fig. 5.7c)
The centromeres migrate to the opposite poles of the spindle, pulling the chromatids with them.

Telophase II (Fig. 5.7d)
a. Each of the two cells produced by the first division now divides. Thus four haploid cells are produced for each diploid cell that goes through meiosis.
b. The chromosomes become less condensed and a nuclear membrane forms.
Thus each of the four haploid cells produced contains half the number of chromosomes of a normal diploid cell, or one of each homologous pair. The chromosomes present in the haploid cell are randomly distributed with respect to paternal and maternal origin.

The behavior of chromosomes in the meiotic division is directly relevant to the segregation of genes and this relationship will be developed in detail in later chapters. Of particular relevance to genetics is the bivalent (tetrad) stage of the first meiotic division, which is the point where crossing over occurs, and the random segregation of each of the four chromatids of a homologous pair of chromosomes to the four haploid cells independently of the four chromatids of every other chromosome pair. Finally, by a cycle of meiosis and

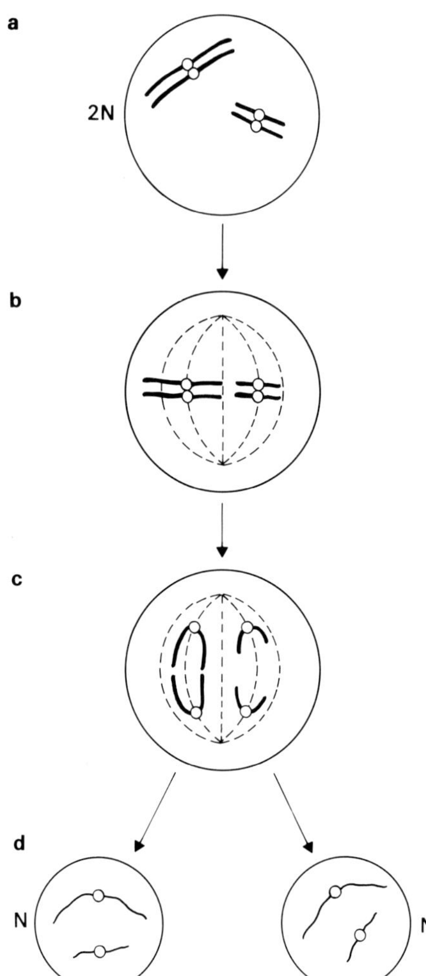

Fig. 5.7. Second meiotic division: a continuation from Fig. 5.6. (a) Prophase II. (b) Metaphase II. (c) Anaphase II. (d) Telophase II.

fusion of haploid cells (in fertilization), the chromosome number is maintained from generation to generation.

Molecular structure of centromeres and telomeres

We have learned in this chapter that the centromere plays an extremely important role in mitosis and meiosis. In this section, we will discuss some of the recent information about the molecular nature of centromeres, the points of attachment of chromosomes to the spindle, and *telomeres*, the ends of chromosomes.

Centromeres

The centromere is the site of attachment to the spindle in mitosis or meiosis. This attachment is necessary for the migration of the chromosomes to the spindle poles in mitosis and for the separation of homologous chromosomes in meiosis.

The DNA sequences of some centromeres from some lower eukaryotes have been determined. The sequences of the core regions I, II, and III of three yeast centromeres, CEN3, 4, and 11, are:

	I	II	III
CEN3	ATAAGTCACATGAT	- - - - - 88 bp, 93% AT - - - - -	TGATTTCCGAA
CEN4	AAAGGTCACATGCT	- - - - - 82 bp, 93% AT - - - - -	TGATTACCGAA
CEN11	ATAAGTCACATGAT	- - - - - 89 bp, 94% AT - - - - -	TGATTTCCGAA

As can be seen, even though different centromeres have the same function in the same cell, their sequences, while similar, are not identical. The functional meaning of this fact is not known. More specifically, region I is a conserved sequence among the yeast centromeres that have been examined, region II is a very AT-rich region with no conserved sequence, and region III is a short, highly conserved sequence.

Deletion experiments have shown that the three regions I, II, and III are important to the centromere's function. Deletions that removed all of regions I, II, and III completely abolished centromere activity. Deletions that went up to and included 5 bp of region I retained full activity, suggesting that region I is less important to centromere activity than regions II and III.

The nucleosomal organization of centromeres has been examined. Nuclease digestion experiments have shown that the core centromere regions I, II, and III are highly protected against digestion with DNase I or micrococcal nuclease. Nuclease digestion experiments have also shown that the DNA surrounding the centromeres is more uniformily packed into nucleosomes than is the rest of the genomic DNA.

Finally, it is clear that the complex structure of the centromere and the attachment of the spindle microtubules must be mediated by proteins, some of which presumably interact directly with centromeric DNA. Some progress is being made in identifying and characterizing centromere-binding proteins. One of the major problems to be solved is the mechanism of attachment of the centromere to the spindle.

Telomeres

A telomere is the region of DNA at the molecular end of a linear chromosome that is required for replication and stability of that chromosome.

Telomeric regions of chromosomes are characteristically but not necessarily heterochromatic in appearance. In most organisms that have been examined, the telomeres are positioned just under the nuclear envelope, and are often found associated with each other as well as with the nuclear envelope. There is some evidence for the existence of Z-DNA (see Chapter 1) in the telomeres of some organisms.

There is molecular evidence that all telomeres in a given species share a common sequence. Telomeric sequences may be divided into two types:

1. Regions near the ends of chromosomes often contain repeated but still complex DNA sequences extending for many thousands of base pairs from the molecular end of chromosomal DNA. These sequences, called *telomere-associated sequences*, may well mediate many of the telomere-specific interactions (e.g. among telomeres and between telomeres and the nuclear envelope).

2. Sequences at or very close to the extreme ends of the chromosomal DNA molecules consist of simple, tandemly repeated DNA sequences. These so-called simple telomeric sequences are the essential functional components of telomeric regions in that they are sufficient to supply a chromosomal end with stability. In the ciliate, *Tetrahymena*, for example, the repeated sequence consists of 5′-CCCCAA-3′ elements, and in the flagellate, *Trypanosoma*, the repeated sequence is CCTAA. Much remains to be learned about these sequences in general and the properties of chromosomal ends.

Questions and problems

5.1 Interphase is a period corresponding to the cell cycle phases of (choose the correct answer):
(a) mitosis
(b) S
(c) G1+S+G2
(d) G1+S+G2+M

5.2 The general life cycle has the sequence (choose the correct answer):
(a) 1N → meiosis → 2N → fertilization → 1N
(b) 2N → meiosis → 1N → fertilization → 2N
(c) 1N → mitosis → 2N → fertilization → 1N
(d) 2N → mitosis → 1N → fertilization → 2N

5.3 Which statement is true?
(a) Gametes are 2N, zygotes are 1N.
(b) Gametes and zygotes are 2N.
(c) Gametic and somatic chromosome numbers can be the same.
(d) The zygotic and the somatic chromosome numbers cannot be the same.
(e) Haploid organisms have haploid zygotes.

5.4 Chromatids joined together by a duplicated but unseparated centromere are called (choose the correct answer):
(a) sister chromatids

(b) homologs
(c) alleles
(d) bivalents

5.5 Mitosis and meiosis always differ in regard to the presence of (choose the correct answer):
(a) chromatids
(b) homologs
(c) bivalents
(d) centromeres
(e) spindles

5.6 The following does *not* happen in prophase I of meiosis (choose the correct answer):
(a) chromosome condensation
(b) pairing of homologs
(c) chiasma formation
(d) terminalization
(e) segregation

5.7 Give the name of the stages of mitosis or meiosis at which the following events occur:
(a) Chromosomes are located in a plane at the center of the spindle.
(b) The chromosomes move away from the spindle equator of the poles.

5.8 What are the important differences, from the genetic point of view, between mitosis and meiosis?

References

Blackburn, E.H., 1984. Telomeres: Do the ends justify the means? *Cell* **37**:7–8.

Blackburn, E.H. and J.W. Szostak, 1984. The molecular structure of centromeres and telomeres. *Annu. Rev. Biochem.* **53**:163–194.

Bloom, K.S., E. Amaya, J. Carbon, L. Clarke, A. Hill and E. Yeh, 1984. Chromatin conformation of yeast centromeres. *J. Cell Biol.* **99**:1559–1568.

Brachet, J. and A.E. Mirsky (eds.), 1961. *The Cell: Meiosis and Mitosis,* Vol. 3. Academic Press, New York.

Brinkley, B.R. and E. Stubblefield, 1970. Ultrastructure and interaction of the kinetochore and centriole in mitosis and meiosis. *Adv. Cell Cycle* **1**:119–186.

Carbon, J., 1984. Yeast centromeres: Structure and function. *Cell* **37**:351–353.

Fitzgerald-Hayes, M., L. Clarke and J. Carbon, 1982. Nucleotide sequence comparisons and functional analysis of yeast centromere DNAs. *Cell* **29**:235–244.

Henderson, S.A., 1970. The time and place of meiotic crossing over. *Annu. Rev. Genetics* **4**:295–324.

Jabs, E.W., S.F. Wolf and B.R. Migeon, 1984. Characterization of a cloned DNA sequence that is present at centromeres of all human autosomes and the X chromosome and shows polymorphic variation. *Proc. Natl. Acad. Sci. USA* **81**:4884–4888.

John, B. and K.R. Lewis, 1965. *The Meiotic System.* Springer-Verlag, New York.

Moses, M.J., 1968. Synaptinemal complex. *Annu. Rev. Genetics* **2**:363–412.

Stern, H. and Y. Hotta, 1969. Biochemistry of Meiosis. In *Handbook of Cytology,* C.A. Lima-de-Faria (ed.). North Holland, Amsterdam.

Westergaard, M. and D. von Wettstein, 1972. The synaptinemal complex. *Annu. Rev. Genetics* **6**:74–110.

Chapter 6 Mutation, Mutagenesis, and Selection

Outline

Mutagenesis

A gene is a specific sequence of nucleotides in the DNA and different genes have different sequences of nucleotides. **Mutations** are changes in the sequence of base pairs of the DNA such as **transitions** (where a purine-pyrimidine base pair is changed to the other purine-pyrimidine base pair; e.g. AT to GC or vice versa) and **transversions** (where a purine-pyrimidine base pair is changed to a pyrimidine-purine base pair; e.g. AT to TA or vice versa), or insertions or deletions of base pairs. Most single base-pair changes are reversible. The consequences of a mutation depend on its location within a gene, and thus not all mutations result in an altered (mutant) phenotype of the organism under study. The induction of mutations is called *mutagenesis*, and the agent involved is called a **mutagen**.

Spontaneous mutations

No mutagens are involved in the production of **spontaneous mutations**. Both

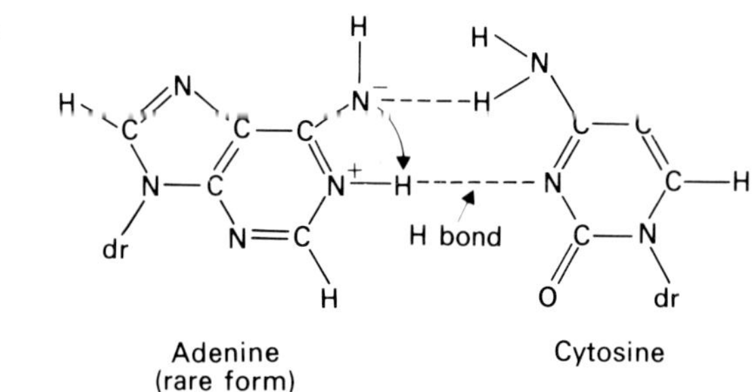

Fig. 6.1. Shift of adenine to rare form results in formation of unusual adenine-cytosine base pairing. (dr = deoxyribose.)

base-pair changes and chromosome aberrations can occur spontaneously. For example, the adenine molecule can exist in two forms. In its more stable configuration, adenine forms two hydrogen bonds with thymine in DNA but cannot hydrogen bond with cytosine (Fig. 6.1a). However, if adenine undergoes a tautomeric shift such that a hydrogen atom moves from the 6-amino group to the 1-N position, hydrogen bonding with cytosine can occur at two positions (Fig. 6.1b).

If the AC pairing occurs while DNA is replicating, then at the ensuing round of replication one of the two daughter DNA helices will have a GC pair instead of an AT at that position. This is an example of a **transition mutation** (Fig. 6.2). The consequences of this mutation to the organism depend on its location within the gene.

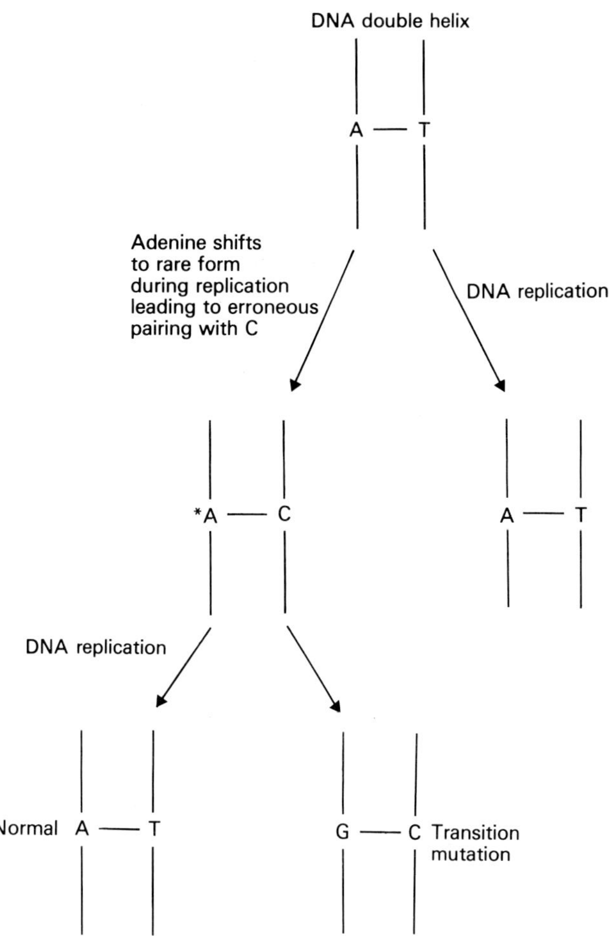

Fig. 6.2. Spontaneous mutation: erroneous pairing of adenine with cytosine during replication leads to transition mutation.

Induced mutations

Mutations can be induced by either physical or chemical means: such mutations are called **induced mutations**. Irradiation is an example of a physical mutagen, with x-rays, gamma-rays, and ultraviolet light being the most common examples used. One consequence of x- or gamma-ray irradiation is the breakage of chromosomes, which may result in chromosomal rearrangements, or the events may be lethal to the cell.

By contrast, chemical mutagens can act in a variety of ways, depending on the properties of the chemical and its reactions with the bases of the DNA.

5-Bromouracil

5-Bromouracil (5-BU) is a *base analog*; that is, its structure closely resembles one of the bases normally found in DNA. 5-BU can exist in two states. In its usual keto state it exhibits properties similar to those of thymine and thus will pair with adenine in DNA (Fig. 6.3a). Rarely it switches to the enol state, and in this form it will pair specifically with guanine (Fig. 6.3b).

Mutations can be induced by 5-BU (and in general by base-analog mutagens) in two ways. The first involves the incorporation of the usual form of

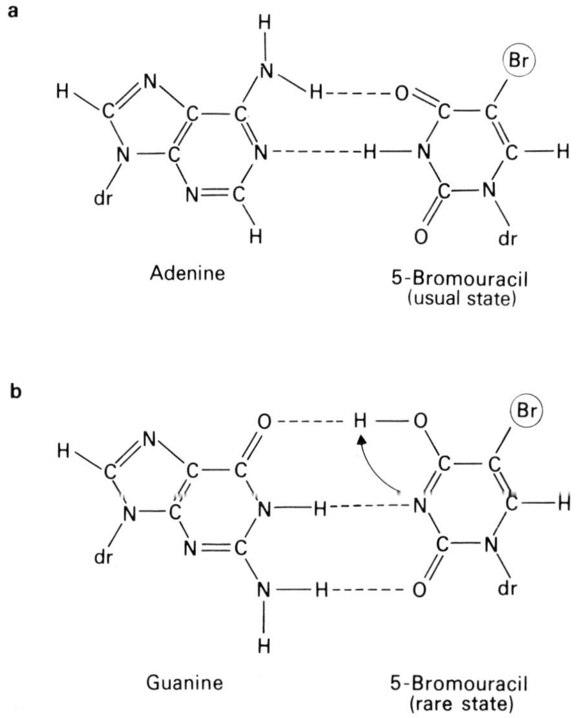

a

Adenine

5-Bromouracil (usual state)

b

Guanine

5-Bromouracil (rare state)

Fig. 6.3. Pairing properties of a 5-Bromouracil (5-BU): (**a**) In its usual keto sate, 5-BU pairs with adenine; (**b**) In its rare enol state, 5-BU pairs with guanine. (dr = deoxyribose.)

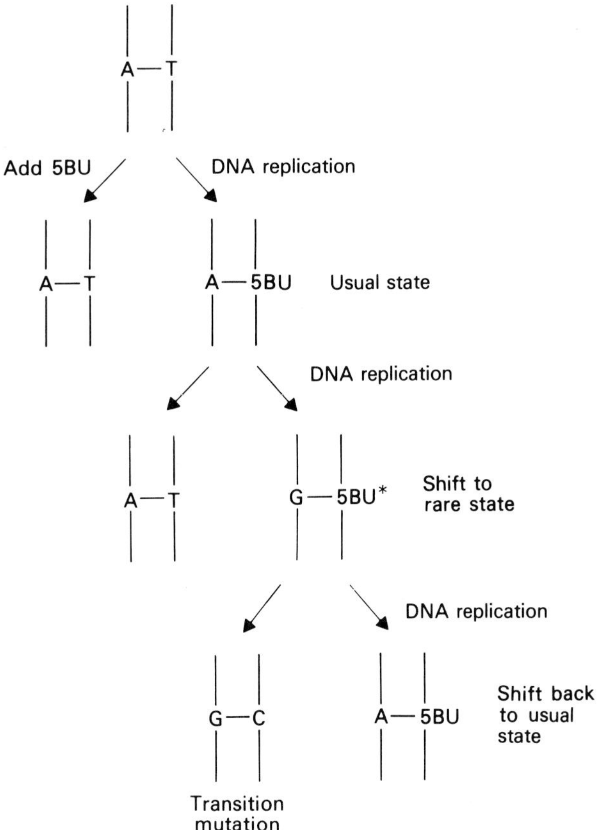

Fig. 6.4. Mutagenic action of 5-BU when it incorporates into DNA in its usual keto state and then shifts to its rare state during the next round of replication.

5-BU into DNA during replication. If 5-BU shifts to its rare enol state during the next round of replication, the result will be a transition mutation from AT to GC (Fig. 6.4).

The second way that 5-BU can induce mutations is if the base analog is incorporated into the DNA while it is in the rare enol state. This dictates insertion opposite a G on the complementary strand, and subsequent replication with a shift of the 5-BU to the usual keto state will result in a transition mutation from GC to AT (Fig. 6.5).

Thus 5-BU can induce either AT-to-GC or GC-to-AT transition mutations. (In the jargon of this area, 5-BU is said to induce two-way transition mutations.) Therefore it is possible to correct a 5-BU induced transition mutation by treating with 5-BU for a second time. This is called **reversion** of the mutation.

2-Aminopurine

2-Aminopurine (2-AP) is also a base analog and, like 5-BU, it can exist in two states (Fig. 6.6). In its usual state it behaves like adenine and will form two

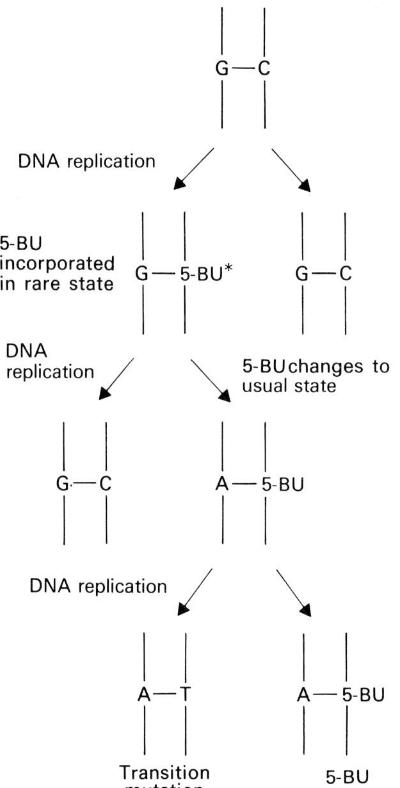

Fig. 6.5. Mutagenic action of 5-BU when it incorporates into DNA in the rare state and then shifts to the usual keto state during the next round of replication.

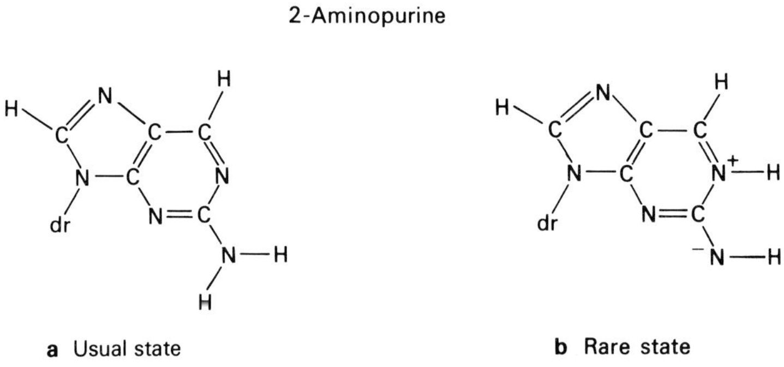

2-Aminopurine

a Usual state

b Rare state

Fig. 6.6. Structure of 2-Aminopurine: (**a**) In its usual state (pairs with thymine) and (**b**) in its rare imino state (pairs with cytosine). (dr = deoxyribose.)

hydrogen bonds with thymine. In its rare imino state 2-AP behaves like guanine and forms two hydrogen bonds with cytosine. Thus 2-AP can induce transition mutations both from AT-to-GC and from GC-to-AT. 2-AP induced mutations can therefore be reverted by 2-AP treatment.

Nitrous acid

Nitrous acid (NA: HNO_2) is a deaminating agent; it acts by removing amino groups (NH_2) from the bases. In some but not all instances this alters their base-pairing abilities and hence induces mutations. The three bases that have amino groups are adenine, guanine, and cytosine. When adenine is treated with NA, it is changed to hypoxanthine, which will pair with cytosine. This results in an AT-to-GC transition mutation (Fig. 6.7a).

Treatment of guanine with NA removes the amino group from the 2-carbon position and produces xanthine. However, since both guanine and xanthine pair with cytosine, no base-pair mutation results (Fig. 6.7b).

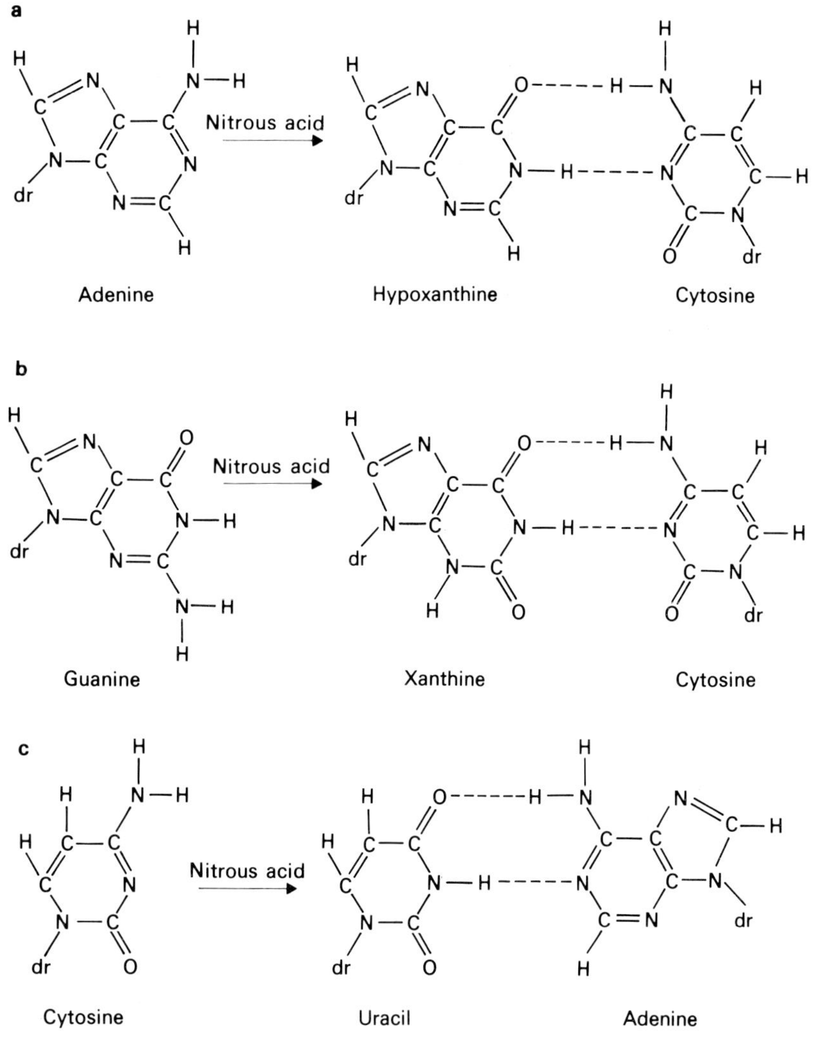

a

Adenine → Nitrous acid → Hypoxanthine Cytosine

b

Guanine → Nitrous acid → Xanthine Cytosine

c

Cytosine → Nitrous acid → Uracil Adenine

Fig. 6.7. Mutagenic action of nitrous acid. (**a**) Deamination of adenine by nitrous acid treatment produces hypoxanthine, which pairs with cytosine (transition mutation); (**b**) Deamination of guanine produces xanthine, which pairs with cytosine (no base-pair substitution); (**c**) Deamination of cytosine produces uracil, which pairs with adenine (transition mutation). (dr = deoxyribose.)

Deamination of cytosine by NA produces uracil, which, of course, pairs with adenine. This results in a GC-to-AT transition mutation—the opposite of NA's effect on adenine (Fig. 6.7c). Thus mutations induced by nitrous acid can be reverted by nitrous acid treatment: in other words, nitrous acid induces two-way transition mutations.

Hydroxylamine

Hydroxylamine (NH_2OH) reacts only with cytosine, hydroxylating it so that it can then pair only with adenine (Fig. 6.8). Thus hydroxylamine induces one-way transition mutations from GC to AT. Because of this, mutations induced by hydroxylamine cannot be reverted by treatment with the same mutagen, although they could be induced to revert by 5-BU, 2-AP, or NA treatment since these mutagens can bring about a GC-to-AT transition.

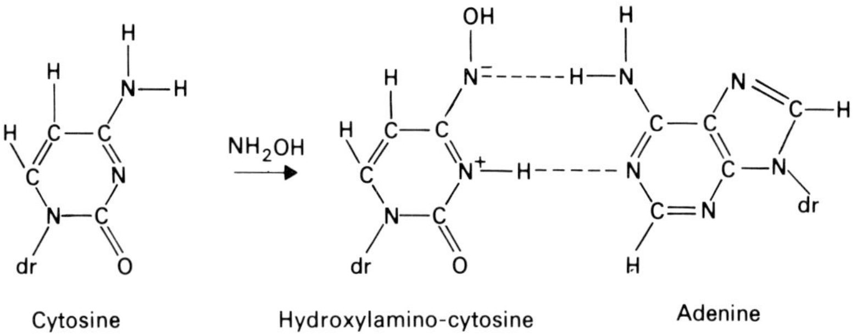

Fig. 6.8. Mutagenic action of hydroxylamine.

Acridines

Acridine treatment results in the addition or deletion of one base pair in the DNA. This has serious consequences, since the amino acid sequence of a protein coded for by a stretch of DNA altered in this fashion will be changed drastically. This will become more apparent in later discussions of messenger RNA translation.

When present at relatively low concentrations, acridine acts by becoming inserted between adjacent base pairs in the DNA. When this occurs, it "stretches" the distance between adjacent base pairs to 0.68 nm, which is precisely double the normal distance. The consequences of this depend on whether the acridine molecule is inserted into the template strand (the one being copied) or into the strand being synthesized. In the former case, a randomly chosen base is inserted opposite the acridine molecule when the replication fork passes by (Fig. 6.9a). At the next round of replication, the correct complementary base is paired with the inserted base, with the result

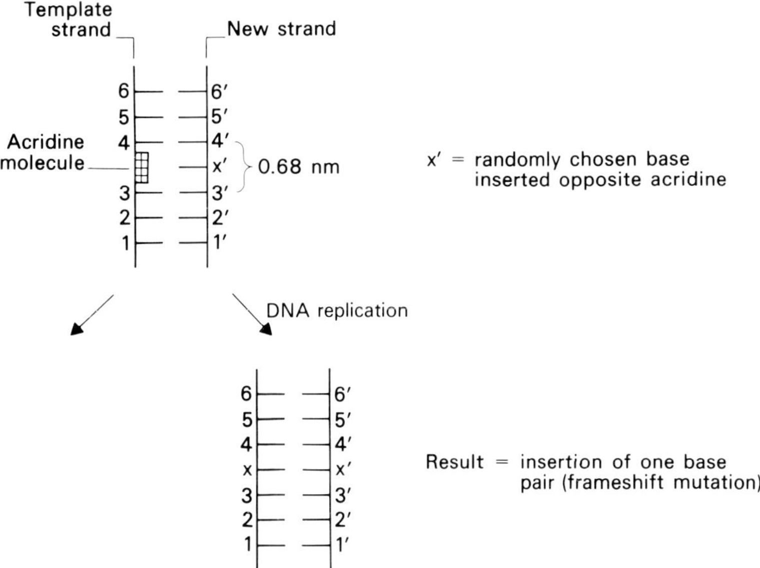

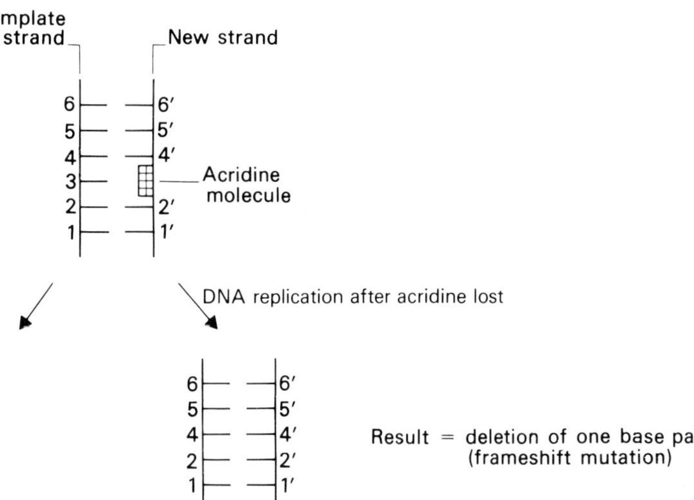

Fig. 6.9. Mutagenic action of acridines by intercalation into DNA. (**a**) Generation of addition mutation when acridine inserts in template strands. (**b**) Generation of deletion mutation when acridine inserts into newly synthesizing strand. (After W. Hayes, 1968. *The Genetics of Bacteria and their Viruses.* Blackwell Scientific Publications, Oxford.)

that one base pair is added to the DNA in that region. This is called an *insertion mutation*.

Alternatively, if the acridine becomes inserted into the newly synthesized strand, it blocks one of the bases on the template strand from having a complementary base (Fig. 6.9b). Then, if the acridine is lost before the next round of replication, the result will be a deletion of a base pair. Therefore it is possible to revert an acridine-induced mutation by a second treatment with acridine.

In summary, the mutagens described in this section have different modes of action and cause different mutational changes. These base-pair changes are summarized in Table 6.1.

Table 6.1. Summary of the modes of action of various chemical mutagens. (After W. Hayes, 1968. *The Genetics of Bacteria and Their Viruses*. Blackwell Scientific Publications, Oxford.)

Mutagen	Base-pair changes
5-Bromouracil	AT ↔ GC two-way transitions
2-Aminopurine	AT ↔ GC two-way transitions
Nitrous acid	AT ↔ GC two-way transitions
Hydroxylamine	GC → AT one-way transitions
Acridines	+1 or −1 insertion or deletion

Some of the chemical mutagens described in this section are commonly used in the laboratory. Many other chemicals appear to cause mutations, and public awareness of this is increasing as industrial effluents, cosmetic ingredients, food additives, etc. are examined carefully for any mutagenic activity in test organisms.

Repair of mutational damage

Throughout the lifetime of an organism, its cells are exposed to a number of agents that have the potential to damage the DNA and hence induce mutations. Examples of such agents are ultraviolet (UV) light (from sunlight) and chemicals in the environment. Accumulated damage to the DNA over a period of time is considered by some scientists to be a cause of transformation of cells to the neoplastic (cancerous) state. Thus to enhance a cell's chance for survival, a variety of inherent repair mechanisms have evolved that serve to reverse the effects of some spontaneous and induced mutations. A model system that has proved useful for the elaboration of two such mechanisms is *pyrimidine dimers* in which adjacent thymines or cytosines on the same DNA strand become bonded together when cells are irradiated with ultraviolet light (Fig. 6.10). The dimers distort the DNA so no pairing occurs with the purines of the opposite strand. Failure to remove the dimers may be a lethal event to the cell, or if the wrong nucleotides are inserted opposite the dimers during replication, a mutation may result. In this section two repair mechanisms, **photoreactivation** and **excision repair**, will be considered.

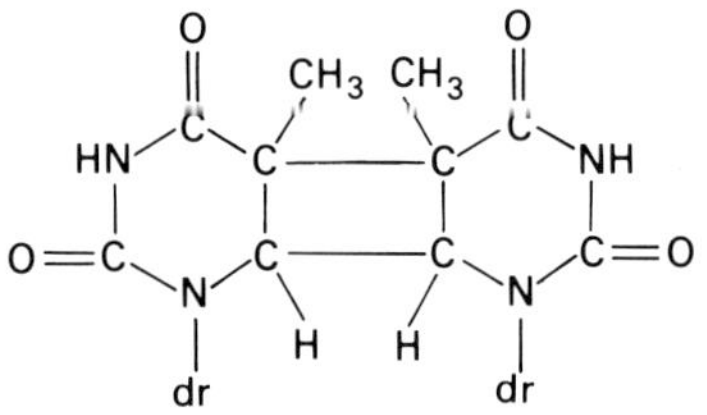

Fig. 6.10. Structure of thymine dimer induced by UV light.

Photoreactivation

In photoreactivation the UV-induced pyrimidine dimers are reversed directly to the original form. This reaction is dependent upon exposing the cells to visible light following irradiation, hence the name "photoreactivation" (Fig. 6.11). The reversal process is catalyzed by an enzyme called a *photolyase*, which monomerizes the dimers when activated by a photon of light in the wavelength range 320–370 nm. Examination of a large number of organisms leads us to the conclusion that the photolyase repair enzymes are probably ubiquitous.

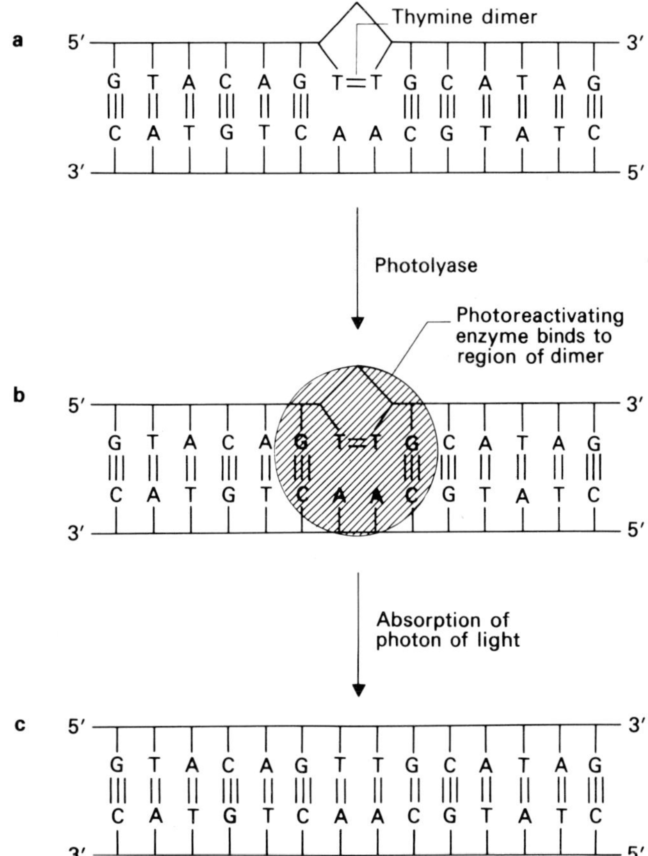

Fig. 6.11. Photoreactivation of thymine dimer induced by UV light. (**a**) Segment of DNA double helix distorted by a thymine dimer. (**b**) Attachment of photoreactivating enzyme (photolyase) to the thymine dimer region. (**c**) Absorption of photon of light in the blue end of the spectrum causes enzyme to split the thymine dimer, allowing the A-T base pairs to reform. The enzyme then dissociates from the DNA. (After M. W. Strickberger, 1976, *Genetics*. Macmillan, New York.)

Excision repair

The second repair mechanism was discovered separately by P. Boyce and P. Howard-Flanders, and by R. Setlow and W. Carrier. In this case the dimers are excised from the DNA by the action of nucleases, and the single-stranded gap is filled in by the action of polymerase and ligase enzymes (Fig. 6.12). Since this reaction does not require activation by light, it is also called "dark repair." In fact, the excision repair mechanism in *E. coli* is very complex and the details will not be given here. Generally, there are specific correcting nucleases involved with repair. The first step involves recognition of the dimer in the DNA, and this is followed by the formation of a single-stranded

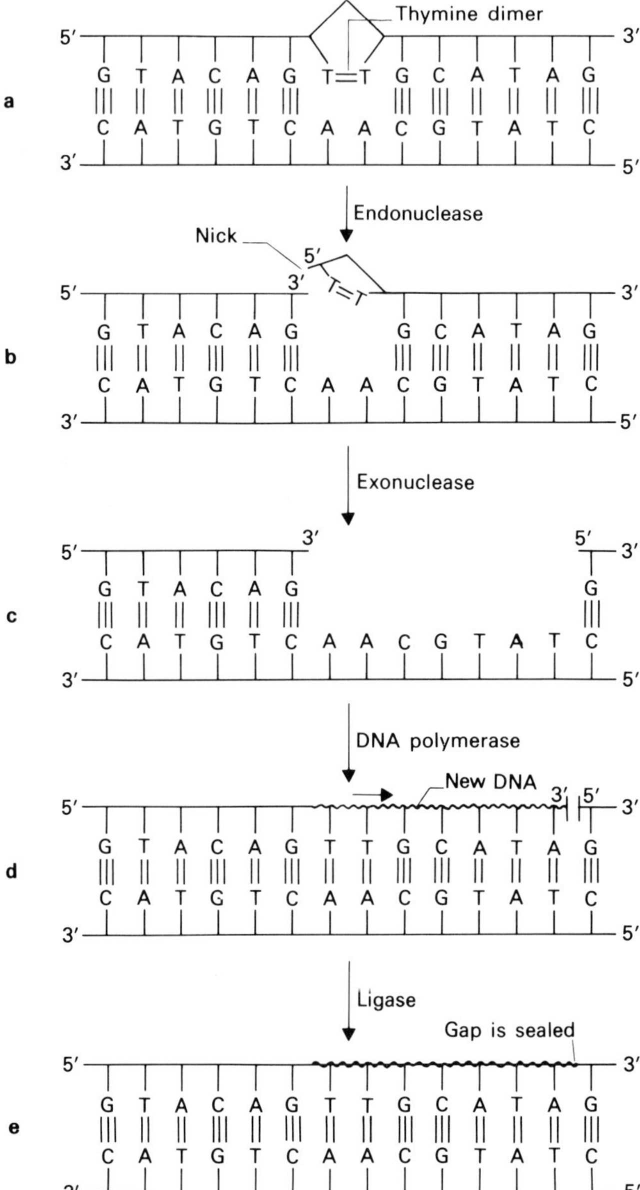

Fig. 6.12. Excision repair of thymine dimers in DNA. (**a**) Segment of DNA distorted by a UV-light induced thymine dimer. (**b**) Endonuclease activity "nicks" the DNA on the 5′ side of the dimer. (**c**) Exonuclease action removes the dimer and other nucleotides on the same strand in the 5′-to-3′ direction. (**d**) DNA polymerase I fills in the single-stranded gap, catalyzing the synthesis of new DNA in the 5′-to-3′ direction. (**e**) The gap between old and new DNA is sealed by the action of polynucleotide ligase. (After M. W. Strickberger, 1976. *Genetics*. Macmillan, New York.)

nick by the action of a correcting endonuclease. These enzymes are normally small proteins with a molecular weight of about 30,000 daltons. All of the ones that have been characterized make their incision close to the damage in the DNA strand. The result of the enzyme activity is a free 5′ end that is the substrate for the 5′-to-3′ exonuclease activity of DNA polymerase I. Concomitant with the removal of part of one strand of DNA, including the pyrimidine dimer, a new segment of DNA is polymerized into the gap in the 5′-to-3′ direction by the action of DNA polymerase I. This leaves a single-stranded nick in the DNA strand, which is sealed by polynucleotide ligase. The dark repair process also has been shown to occur in other organisms, and a number of the enzymes involved have been identified in several systems.

The genetic basis of excision repair has been studied in several organisms, for example, by selecting for UV sensitivity. Five genes, *uvrA-E*, have been shown to mutate in *E. coli* to give the UV-sensitivity phenotype (a phenotype is an observable physical or biochemical property of an organism), and the biochemical bases for the sensitivity have been examined in some cases. Thus *uvrA*, *uvrB*, and *uvrC* strains have very similar sensitivity to UV light, and although *uvrC* has normal levels of the dimer-specific endonuclease, *uvrA* and *uvrB* lack the enzyme activity and hence excision repair cannot proceed. The biochemical defect in the *uvrC* strain is not known. The *uvrD* mutant is less sensitive to UV irradiation than the three strains just discussed. In addition, for reasons not yet known, *uvrD* shows extensive degradation of the DNA after UV treatment. The *uvrE* strain is UV-sensitive and shows an increase rate of spontaneous mutation compared with the wild type.

A number of recombination-deficient (*rec*) mutations have also been found in *E. coli* and these are also UV-sensitive. These *rec* strains are not defective in excision repair, their primary role being in the recombination process.

Finally, as would be expected, mutations in the genes for polynucleotide ligase (*lig*) or DNA polymerase I (*polA*) result in decreased levels of excision repair. A genetic map showing the scattered locations of genes involved with dimer repair in *E. coli* is shown in Fig. 6.13. It is likely that similar enzymes play similar roles in other organisms also.

Mutant isolation

Spontaneous mutations occur very rarely among populations of any organism. Therefore, scientists use mutagens to increase the frequency of occurrence of mutations. Even so, as is apparent from the previous discussions, the mutagens generally used to induce mutations act at the base-pair level and they do not induce mutations in specific genes. Geneticists and biochemists usually want to study mutants that carry mutations affecting

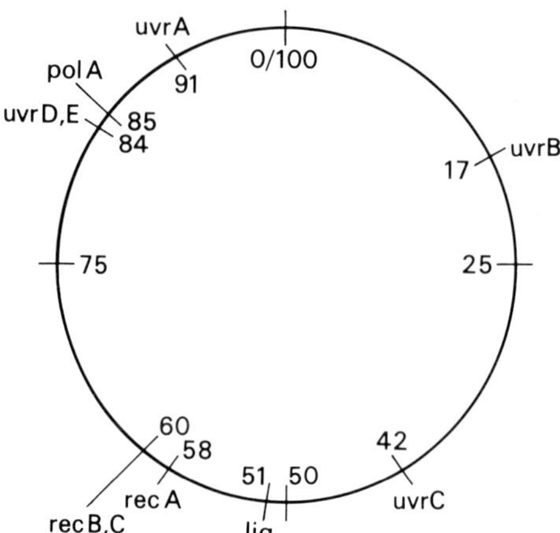

Fig. 6.13. Map location of the genes involved in the repair of DNA in *E. coli* (After U. Goodenough, 1978. *Genetics*. Holt, Rinehart, and Winston, New York.)

specific functions. Therefore a number of screening and enrichment procedures have been developed to help isolate specific mutants from among a heterogeneous mutagenized population of cells or organisms. These procedures will be described in the context of the types of mutants that are useful for genetic and/or biochemical types of analysis.

The first class of mutants to consider is the **visible mutants**. As their name suggests, these have phenotypes that are different from the normal (wild-type) organism at the gross macroscopic or microscopic level. In diploid organisms, mutations that give rise to a visible mutant phenotype are often recessive so the mutant phenotype is only seen when the mutation is homozygous. Thus, visible mutants in a diploid organism may only become apparent after further breeding of a mutagenized population of the organism. By contrast, visible mutants are immediately apparent in haploid organisms.

There are numerous examples of visible mutants. In diploid organisms, eye-color mutants, body-color mutants, wing-shape mutants in *Drosophila*, coat-color mutants in animals, and petal-color mutants in plants are some examples of this class of mutants. In haploid organisms, examples are mutants of *Pneumococcus* bacteria that produce rough instead of smooth colonies when grown on solid medium, mutants of yeast that produce colonies which are much smaller than the wild-type strain (*petite* mutants), and mutants of the fungus *Neurospora crassa* that grow with a colonial morphology instead of with the weblike growth habit characteristic of the wild-type strain (Fig. 6.14).

The second class of mutants affects the ability of the organism to grow. These types of mutants are most often isolated and studied in haploid organisms such as *E. coli*, yeast, and *Neurospora*. The reason for this is that these organisms can grow on a well-defined simple *minimal medium* that

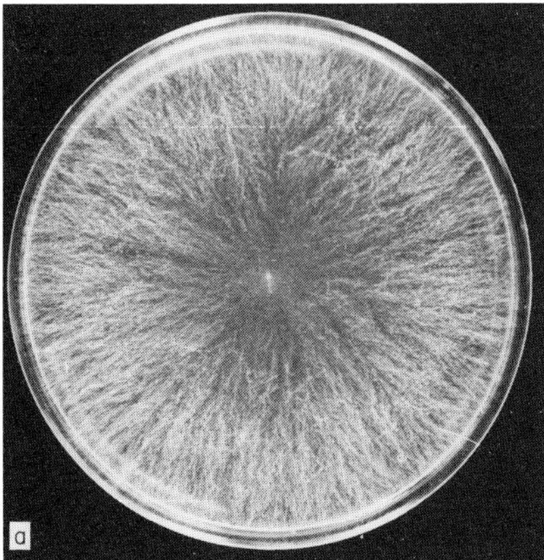

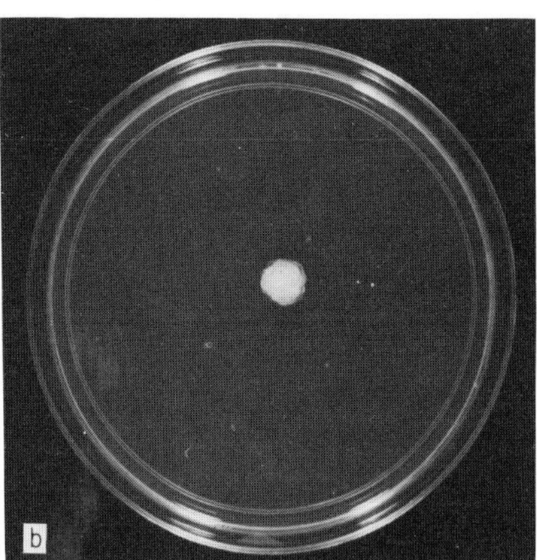

Fig. 6.14. (**a**) Morphology of wild-type *Neurospora crassa* growing on solid medium. (**b**) Morphology of a colonial mutant of *Neurospora*. (Photographs by P. J. Russell.)

provides a carbon source, salts, and trace elements (and sometimes vitamins) which the organisms use to make all the cell molecules such as amino acids, purines, pyrimidines, vitamins, etc. The standard laboratory strain of an organism that can grow on the minimal medium is called the wild-type or **prototrophic** strain. By mutagenesis of the wild-type strain, it is possible to isolate strains that cannot grow on the minimal medium alone but can grow on a supplemented minimal medium. These **auxotrophic mutants** (also called **nutritional** or **biochemical mutants**) are unable to make an ingredient essential for cell growth. Thus the mutation involved may affect the synthesis of an amino acid, purine, pyrimidine, vitamin, or other essential molecule. That ingredient must be present as a supplement to the minimal medium so that the mutant strain can grow.

For many organisms that will grow to discrete colonies on solid medium in a petri dish, auxotrophic mutants can be isolated using the **replica plating** technique developed by E. Lederberg and J. Lederberg (Fig. 6.15). In outline, samples of a cell culture (that has or has not been treated with a mutagen) are spread on the surface of a completely supplemented medium that is incubated to allow colonies (clones of the cell that landed on the medium) to grow. On this medium both prototrophic and auxotrophic cells will grow. The pattern of colonies can then be transferred onto velveteen cloth, which acts like thousands of tiny inoculating needles. By gently pressing a plate containing minimal medium onto the velveteen cloth, the pattern of colonies can be replicated onto that medium. Only prototrophic strains will grow on the minimal medium and thus, by comparing the patterns of colonies on the two media, auxotrophic strains can be detected and isolated from the original master plate. These auxotrophs can be tested as to the exact nature of their nutritional requirements by setting up a new master plate and replica plating onto a number of different media (e.g. minimal plus amino acids, minimal plus vitamins, and so on).

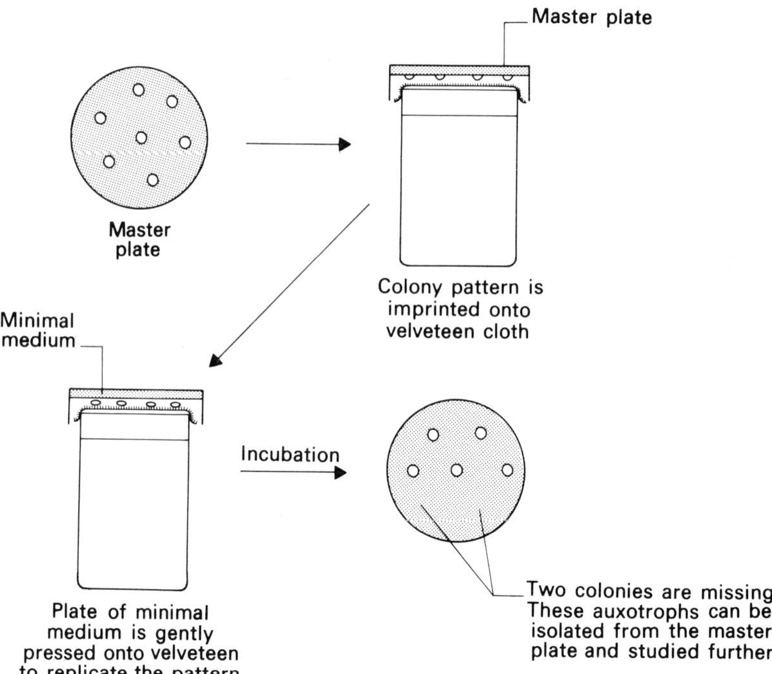

Fig. 6.15. The isolation of auxotrophic mutants of a colony-forming organism by replica plating.

Some degree of mutant enrichment can be achieved with replica plating by the choice of the media onto which the colony pattern is replicated. For example, one can specifically isolate mutants that require adenine for growth by replica plating onto a medium containing all possible supplements except adenine.

Auxotrophic mutants can be enriched for by using other procedures, such as the following:

1. *Antibiotic selection.* This procedure takes advantage of the fact that growing cells, but not nongrowing cells, of certain organisms can be killed by treating them with antibiotics. For example, if a mutagenized population of *E. coli* cells is placed in a medium that is restrictive (nonpermissive) for the type of auxotroph desired, only the strains not requiring the omitted supplement will be able to grow. If penicillin is added to the culture, the growing cells will be killed, leaving the auxotrophs surviving. The same principle applies for nystatin (an antibiotic named after or for New York state) enrichment in yeast.

2. *Filtration enrichment.* This procedure is used with filamentous fungi such as *Neurospora*. If a mutagenized population of cells is incubated in a selective medium, the mutant strains desired will not grow, whereas everything else will produce filamentous hyphae. The latter may be removed by

passing the culture through cheesecloth. The mutant, ungerminated cells pass through this material. When this is repeated over several days, there is significant enrichment for the auxotrophic mutant type desired.

Another class of mutants that is valuable in studies of macromolecular synthesis, cell function, and cell regulation is the **conditional mutant**. The most common examples are the *temperature-sensitive mutants*, which, unlike the wild type, do not grow well or at all at either high or low temperatures. These heat-sensitive (hs) and cold-sensitive (cs) mutants can be isolated easily following mutagenesis and the application of a suitable enrichment procedure. For example, the replica-plating method can be adapted by incubating the replicated colonies at either high or low temperature. Note that it is possible to isolate conditional auxotrophs in this way also. These can be avoided if the medium in all plates is fully supplemented.

Heat-sensitive and cold-sensitive mutants can be isolated following enrichment techniques such as antibiotic selection and replica plating described earlier. Another enrichment procedure that has been applied to the isolation of heat-sensitive and cold-sensitive mutants is *tritium suicide*. Tritium suicide is defined as the death of cells caused by the decay of the radioactive isotope of hydrogen, tritium (^3H), incorporated into their macromolecules, such as proteins and nucleic acids. This technique has been used with a number of organisms, including *E. coli*, yeast, *Neurospora*, and cultured mammalian cells, to enrich for particular mutant types in a heterogeneous mutagenized population of cells. Selection by tritium suicide can be made relatively specific by the choice of the tritiated precursor and the culture conditions used. For example, in some experiments in the author's laboratory, tritium suicide was used to enrich for hs mutants of *Neurospora*. Mutagenized asexual spores (conidia) were incubated in minimal medium at 35°C (the chosen nonpermissive temperature) in the presence of relatively high amounts of ^3H-amino acid mixture (protein precursor). After two hours the conidia were washed free of radioactivity and stored in a refrigerator. At various times thereafter, samples were taken to test for viability and for the presence of hs mutants. A culture that had not been incubated in the presence of the radioactive precurors was used as a control (Fig. 6.16).

The data indicated clearly that significant death of cells occurs when tritium is incorporated into their macromolecules. In addition, as time of storage increased, there was significant enrichment for hs mutants. The reason for this is that at 35°C the hs mutants are unable to grow, they are metabolically quiescent, and therefore much less of the tritiated precursor is incorporated into their macromolecules than is the case with nonmutants. The high level of radioactivity incorporated into the latter cells results in their death whereas the hs mutants have a better chance for survival.

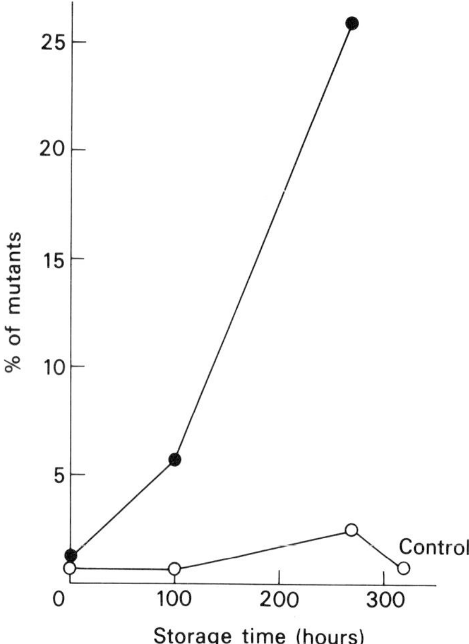

Fig. 6.16. Enrichment for heat-sensitive mutants of *Neurospora crassa* as a function of time after the incorporation of tritiated amino acids. (●) ^{3}H-labeled culture; (o) unlabeled control culture. (After P. J. Russell and M. P. Cohen, 1976. *Mut. Res.* **34**:359.)

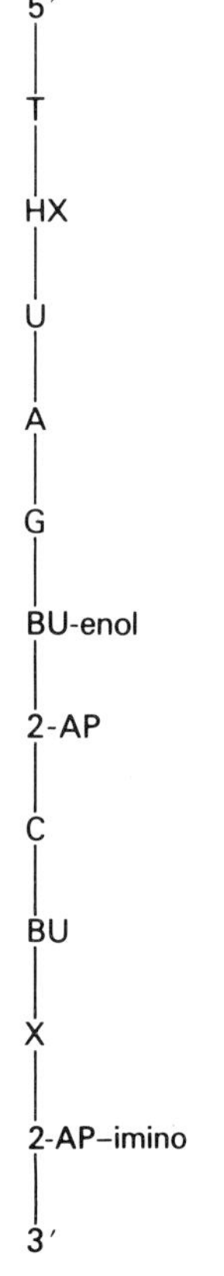

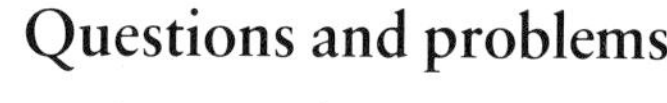

Questions and problems

6.1 Two mechanisms in *E. coli* were described for the repair of DNA damage (thymine dimer formation) after exposure to ultraviolet light: photoreactivation and excision (dark) repair. Compare and contrast these mechanisms, indicating how each achieves repair.

6.2 After a culture of *E. coli* cells were treated with 5-bromouracil (5-BU), it was noted that the frequency of mutants was much higher than normal. Mutant colonies were then isolated and grown and treated with nitrous acid; some of the mutant strains reverted to wild type.

(a) In terms of the Watson-Crick model, diagram a series of steps by which 5BU may have produced the mutants.

(b) Assuming the revertants were not caused by suppressor mutations, indicate the steps by which nitrous acid may have produced the back mutations.

6.3 A single, very hypothetical strand of DNA is composed of the base sequence indicated in the accompanying figure (left). In the figure, A indicates adenine, T indicates thymine, G indicates guanine, C indicates cytosine, U denotes uracil, BU is 5-bromouracil, 2-AP is 2-aminopurine, BU-enol is an enol tautomer of 5-BU, 2-AP-imino is an imino tautomer of 2-AP, HX is hypoxanthine, and X is xanthine; 5′ and 3′ are the numbers of the free, OH-containing carbons on the deoxyribose part of the terminal nucleotides.

(a) Opposite the bases of the hypothetical strand, and using the shorthand of the figure, indicate the sequence of bases on a complementary strand of DNA.

(b) Indicate the direction of replication of the new strand by drawing an arrow next to the "new" strand of DNA from part (a).

(c) When postmeiotic germ cells of a higher organism are exposed to a chemical mutagen before fertilization, the resulting offspring expressing an induced mutation are almost always mosaics for wild-type and mutant tissue. Give at least one reason why in the progenies of treated individuals these mosaics are found and not the so-called complete or whole body mutants.

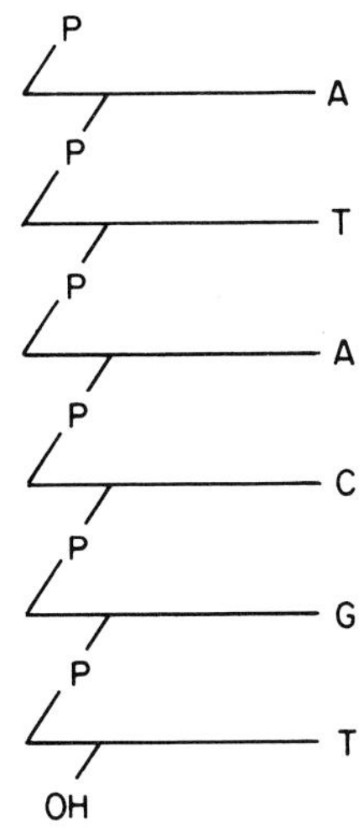

The following information applies to Questions 6.4–6.8. A solution of single-stranded DNA is used as the template in a series of reaction mixtures. The structure of the DNA is presented in the accompanying figure, where A = adenine, G = guanine, C = cytosine, and T = thymine. For Questions 6.4 through 6.8, use this shorthand system and draw the products expected from the reaction mixtures. Assume that a primer is available in each case.

6.4 The DNA template+purified Kornberg enzyme+dATP+dGTP+dCTP+dTTP+Mg^{2+}.

6.5 The DNA template+purified Kornberg enzyme+dATP+dGMP+dCTP+dTTP+Mg^{2+}.

6.6 The DNA template+purified Kornberg enzyme+dATP+dHTP+dGMP+dTTP+Mg^{2+}.

6.7 The DNA template is pretreated with HNO_2 (nitrous acid)+purified Kornberg enzyme+dATP+dGTP+dCTP+dTTP+Mg^{2+}.

6.8 The DNA template+purified Kornberg enzyme+dATP+dGMP+dHTP+dCTP+dTTP+Mg^{2+} (where H = hypoxanthine).

6.9 A strong experimental approach to determine the mode of action of mutagens is to examine the revertibility of the products of one mutagen by other mutagens. The following table represents collected data on revertibility of various mutagens on *rII* mutations in phage T2; + indicates that the majority of mutants reverted; − indicates that virtually no reversion occurred; BU = 5-bromouracil; AP = 2-aminopurine; NA = nitrous acid; HA = hydroxylamine. Fill in the empty spaces.

Mutation induced by	Proportion of mutations reverted by				Base-pair substitution inferred
	BU	AP	NA	HA	
BU	−			−	
AP		−		+	
NA	+	+		+	
HA			+	−	GC → AT

6.10 Three *ara* mutants of *E. coli* were induced by mutagen X. The ability of other mutagens to cause the reverse change (*ara* to *ara*$^+$) was tested, with the results shown in the accompanying table.

	Frequency of *ara*$^+$ cells among total cells after treatment with mutagen				
Mutant	None	BU	AP	HA	Frameshift
ara-1	1.5×10^{-3}	5×10^{-5}	1.3×10^{-4}	1.3×10^{-8}	1.6×10^{-3}
ara-2	2×10^{-7}	2×10^{-4}	6×10^{-5}	3×10^{-5}	1.6×10^{-7}
ara-3	6×10^{-7}	10^{-5}	9×10^{-6}	5×10^{-6}	6.5×10^{-7}

Assume all *ara*$^+$ cells are true revertants. What base changes were probably involved in forming the three original mutations? What kind(s) of mutations are caused by mutagen X?

References

Auerbach, C. and B.J. Kilbey, 1971. Mutation in eukaryotes. *Annu. Rev. Genetics* **5**:163–218.

Beadle, G.W. and E.L. Tatum, 1945. *Neurospora*. II. Methods of producing and detecting mutations concerned with nutritional requirements. *Am. J. Bot.* **32**:678–686.

Boyce, R.P. and P. Howard-Flanders, 1964. Release of ultraviolet light-induced thymine dimers for DNA in *E. coli K12*. *Proc. Natl. Acad. Sci. USA* **51**:293–300.

Drake, J.W., 1969. Mutagenic mechanisms. *Annu. Rev. Genetics* **3**:247–268.

Freese, E., 1959. The specific mutagenic effect of base analogues on phage T4. *J. Mol. Biol.* **1**:87–105.

Hanawalt, P.C., 1972. Repair of genetic material in living cells. *Endeavour* **31**:83–87.

Howard-Flanders, P., 1968. DNA repair. *Annu. Rev. Biochem.* **37**:175–200.

Kelley, R.B., M.R. Atkinson, J.A. Huberman and A. Kornberg, 1969. Excision of thymine dimers and other mismatched sequences by DNA polymerase of *E. coli*. *Nature* **224**:495–501.

Lederberg, J. and E.M. Lederberg, 1952. Replica plating and indirect selection of bacterial mutants. *J. Bacteriol.* **63**:399–406.

Lester, H.E. and S.R. Gross, 1959. Efficient method for selection of auxotrophic mutants of *Neurospora*. *Science* **129**:572.

Littlewood, B.S. and J.R. Davies, 1973. Enrichment for temperature-sensitive and auxotrophic mutants in *Saccharomyces cerevisiae* by tritium suicide. *Mut. Res.* **17**:315–1322.

Moat, A.G., N. Peters and A.M. Srb, 1959. Selection and isolation of auxotrophic yeast mutants with the aid of antibiotics. *J. Bacteriol.* **77**:673–681.

Russell, P.J. and M.P. Cohen, 1976. Enrichment for auxotrophic and heat-sensitive mutants of *Neurospora crassa* by tritium suicide. *Mut. Res.* **34**:359–366.

Setlow, R.B. and W.L. Carrier, 1964. The disappearance of thymine dimers from DNA; an error-correcting mechanism. *Proc. Natl. Acad. Sci. USA* **51**:226–231.

Tatum, E.L., R.W. Barratt, N. Fries and D. Bonner, 1950. Biochemical mutant strains of *Neurospora* produced by physical and chemical treatment. *Am. J. Bot.* **37**:38–46.

Woodward, V.W., J.R. De Zeeuw and A.M. Srb, 1954. The separation and isolation of particular biochemical mutants of *Neurospora* by differential germination of conidia, followed by filtration and selective plating. *Proc. Natl. Acad. Sci. USA* **40**:192–200.

Chapter 7 Transcription

Outline

The genetic material of a cell has a major function to direct protein synthesis. The genome contains all of the information for the structure and function of an organism, but not all of the genes are active at any one time.

The DNA itself is not a direct template for protein synthesis. Rather, the genetic information of DNA is first transferred to molecules of RNA in a process called **transcription.** The RNA template is then used to produce the sequence of amino acids of proteins in a process called *translation*. The relationship of DNA to protein is summarized by the central dogma (Fig. 7.1). In this chapter, the transcription of RNA from a DNA template will be discussed, and in the following chapter, the translation of messenger RNA to produce a polypeptide chain will be described.

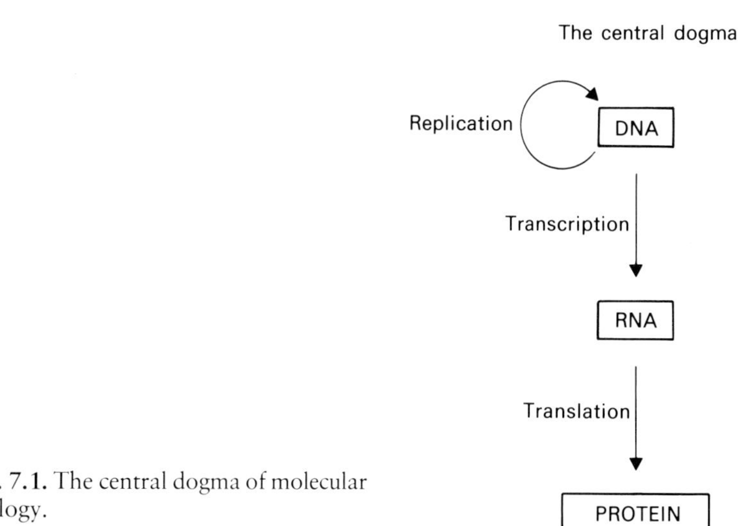

Fig. 7.1. The central dogma of molecular biology.

Classes of RNA

Three classes of RNA are produced by transcription of DNA in both prokaryotes and eukaryotes. The following is a brief description of the three classes:

1. *Messenger RNA (mRNA).* **Messenger RNA** serves as a template for protein synthesis. The size of a mRNA molecule is a function of the length of

the polypeptide chain for which it codes. Thus mRNAs of a cell represent a heterogeneous population of molecules. Messenger RNAs are also heterogeneous with respect to nucleotide sequence.

2. *Transfer RNA (tRNA)*. Each **transfer RNA** molecule is able to bind covalently with a specific amino acid and to form hydrogen bonds with a three-nucleotide sequence (**codon** or *triplet*) on an mRNA. The latter event occurs on the ribosome that is the site of protein synthesis. All tRNA molecules have a molecular weight of about 25,000–30,000 daltons and a sedimentation coefficient of 4S. (The S value is related to the rate of sedimentation of a molecule or particle in a centrifugal field.)

3. *Ribosomal RNA (rRNA)*. **Ribosomes** are made up of **ribosomal** RNA (rRNA) molecules and proteins. The rRNA molecules have discrete sizes (which are described with generalized S values) for prokaryotes and eukaryotes. Thus prokaryotic ribosomes are considered to contain 23S, 16S, and 5S rRNA molecules, and the cytoplasmic (nonorganellar) ribosomes of eukaryotes (here usually referring to the "higher" eukaryotes) contain 28S, 5.8S, 18S, and 5S rRNA molecules. We will use these S values in general discussions of eukaryotic ribosomes, but we will refer to specific values when dealing with "lower" eukaryotes in later sections.

Transcription in prokaryotes

The important features of transcription in both prokaryotes and eukaryotes are:

1. The product of transcription is single-stranded RNA (Fig. 7.2).
2. For transcription to occur, the DNA must unwind locally. For each gene, only one of the two strands serves as the template strand for RNA synthesis; this strand is called the **sense strand.**

Fig. 7.2. Schematic of the transcription of RNA from a DNA double helix.

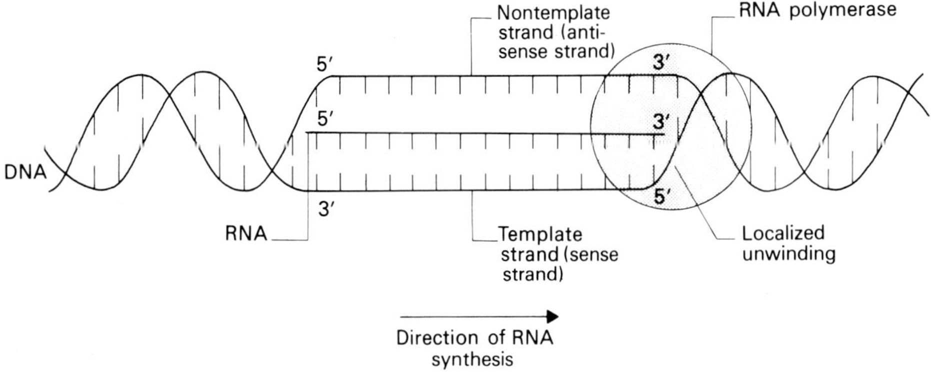

3. The RNA precursors (the molecules polymerized into the RNA chain) are the ribonucleoside triphosphates ATP, GTP, CTP, and UTP, collectively called NTPs. The polymerization reaction is catalyzed by the enzyme **RNA polymerase.**
4. The RNA polymerization reaction (Fig. 7.3) is similar to the DNA polymerization reaction: the next ribonucleotide to be added to the chain is selected by the RNA polymerase for its ability to form a complementary base pair with the exposed DNA base on the template strand.
5. RNA is synthesized in the 5′-to 3′ direction. For example, if the template strand is 3′-TCGGAT-5′, the RNA chain will be 5′-AGCCUA-3′.

Prokaryotic RNA polymerases

Each prokaryotic organism has its own specific RNA polymerase. Bacteriophages either use the host bacterium's RNA polymerase or code for their own, depending upon the bacteriophage.

RNA polymerases require a double-stranded DNA template, the four ribonucleoside triphosphates, and magnesium ions to catalyze RNA synthesis. Unlike prokaryotic DNA polymerases, RNA polymerases are able to initiate new RNA chains, and they lack proofreading functions.

Among prokaryotic RNA polymerases, the enzyme from *E. coli* has been studied most extensively. A stylized diagram of *E. coli* RNA polymerase is shown in Fig. 7.4. This enzyme has a sedimentation coefficient of 11–13S and a molecular weight of approximately 500,000 daltons. Relatively gentle treatment of the soluble enzyme dissociates a 95,000-dalton polypeptide called the sigma factor (σ), leaving the *core polymerase*. The latter can be further dissociated into four polypeptide chains. These are the beta-prime (β') subunit (165,000 daltons), beta (β) subunit (155,000 daltons), and two copies of the alpha (α) subunit (41,000 each).

The core enzyme alone can make an RNA copy of DNA, and the roles of the subunits in this process have been studied extensively. Isolated β' subunits, for example, have been shown to bind to DNA in vitro, whereas the α and β subunits do not. Thus apparently β' is the DNA-binding subunit. Our knowledge of the function of the β subunit has come from studies of antibiotic-resistant mutants. Rifampicin inhibits the initiation of transcription and streptolydigin inhibits the elongation of the RNA chain. These antibiotics bind to the β subunit of wild-type cells but do not bind to the β subunit isolated from mutants resistant to these drugs. Thus the β subunit may be involved in the catalysis of phosphodiester bond formation. The role of the α subunit is not known.

The sigma factor plays a very important role in RNA synthesis. This 95,000-dalton subunit is necessary for the core enzyme to initiate RNA

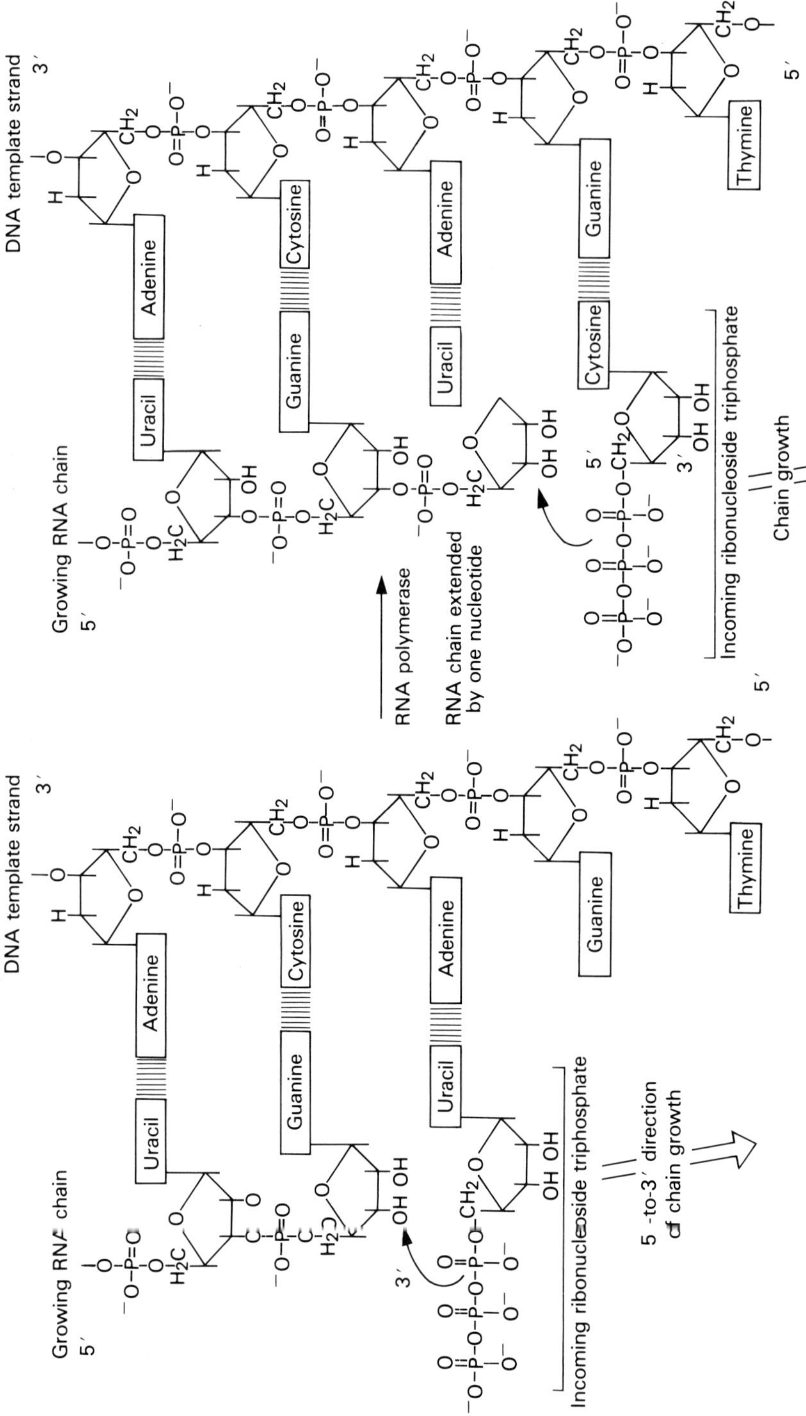

Fig. 7.3. Chemical reaction involved in the RNA polymerase-catalyzed synthesis of RNA on a DNA template.

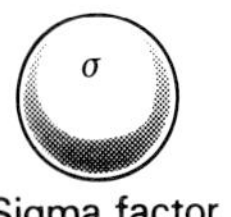

Fig. 7.4. Stylized diagram of the *E. coli* RNA polymerase showing the four subunits of the core enzyme and the dissociable sigma factor.

synthesis at specific sites along the DNA. These sites are called **promoter sequences**, or more simply, **promoters**. If σ is not bound to the core enzyme, the latter initiates RNA synthesis at random sites along the DNA and, moreover, both strands are copied instead of one.

The promoter recognition and transcription events catalyzed by RNA polymerase occur very rapidly, often within 0.2 second. The key process is the recognition of the promoter sequence in the DNA by the enzyme. Prokaryotic promoters that are recognized by the *E. coli* RNA polymerase have in common (with minor variations) a seven-nucleotide sequence called a **Pribnow box** found about five or six bases upstream of the actual start of transcription. The Pribnow box sequence on the antitemplate strand has the general formula 5′-TAT purine AT purine-3′.

A second sequence called a *recognition sequence* is in a region about 35 base pairs upstream from the start of the RNA transcript. The RNA polymerase probably first binds there and then slides to the Pribnow box to initiate transcription.

After RNA synthesis has begun, the σ factor dissociates from the core enzyme, which then continues the transcription process. The σ factor may then associate with another core enzyme to initiate another RNA transcription at the same or a different promoter. Thus the σ factor acts catalytically in terms of the initiation of transcription at specific sites.

There is evidence that the core RNA polymerase undergoes conformational changes. Specifically, its sensitivity to salt, proteases (protein degrading enzymes), and inhibitors changes when it becomes bound to DNA and when RNA synthesis is being catalyzed. And, as was noted before, the enzyme is no longer sensitive to the RNA synthesis inhibitor, rifampicin, once chain initiation has begun.

When a gene has been transcribed, the RNA polymerase is released, and the newly synthesized RNA is displaced by the reformation of double-helical DNA. Transcription stops when a **transcription termination sequence** after a gene is recognized by a complex of the RNA polymerase core enzyme and a protein factor coded for by the *nusA* gene. When transcription is correctly terminated, the *nusA* factor dissociates from the core enzyme and can be used for other termination reaction.

Initiation, elongation, and termination steps of transcription

Three types of base-pair sequence are needed for a gene to be transcribed: the DNA sequences that code for the base sequence of the RNA transcript (the gene), a sequence that signals the start point for transcription (the promoter sequence), and a sequence that signals the stopping point for transcription (the terminator sequence), 5′-promoter-gene-terminator-3′. The promoter is considered to be "upstream" from the gene and the terminator is considered to be "downstream".

Fig. 7.5. Process of initiation, elongation, and termination of transcription in *E. coli*.

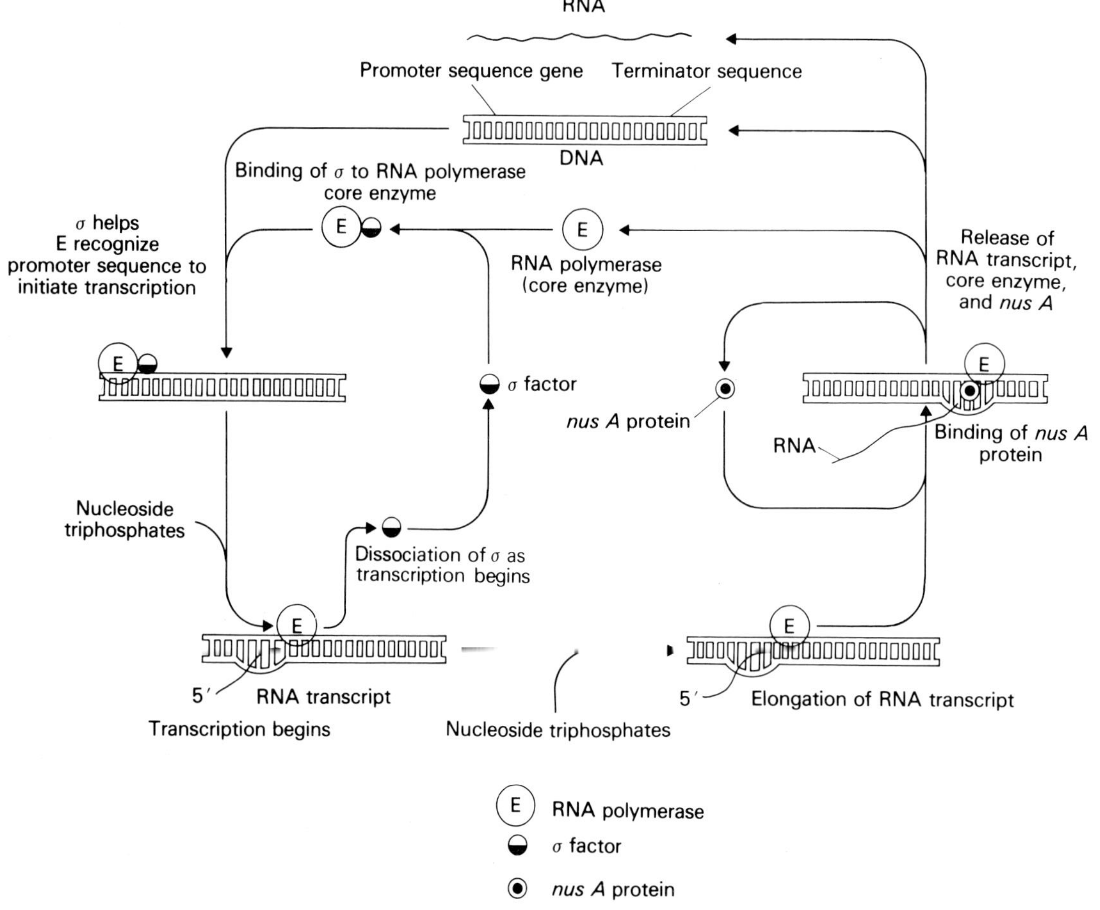

Fig. 7.5 shows the general processes of initiation, elongation, and termination of transcription. The important features are:

1. RNA polymerase core enzyme plus σ factor recognizes and binds to the promoter sequence.
2. RNA polymerase slides toward the gene and catalyzes the localized denaturation of the two DNA strands, exposing the template strand for the initiation of RNA synthesis.
3. RNA synthesis commences and the RNA chain is made in the 5′-to-3′ direction. Once a few polymerization reactions have been completed, the σ factor is released.
4. The transcription termination sequence signals the RNA polymerase to cease transcribing the DNA. The terminator sequence consists of two parts: (a) a region with a high proportion of GC base pairs (a GC-rich region) and (b) a series of T bases on the nontemplate strand in an AT-rich region. Both of these regions are transcribed by RNA polymerase, and base pairing between two parts of the transcript in this region causes an altered relationship between the RNA polymerase and the DNA. As a result, the RNA polymerase cannot elongate the DNA chain any further, and it and the RNA chain both dissociate from the DNA. The termination reaction involves a complex between the RNA polymerase and the *nusA* protein factor, and the release of the RNA transcript from the DNA requires the activity of a ρ (rho) protein factor.

Transcription in eukaryotes

Transcription in eukaryotes occurs in a very similar way to transcription in prokaryotes. Two differences are that the transcription controlling sequences are different, and more than one RNA polymerase enzyme is used in eukaryotic organisms.

Eukaryotic RNA polymerases

Three types of nuclear RNA polymerases (I, II, and III) have been found in a number of eukaryotic cells. These have been shown to have distinct specialized roles in RNA transcription, and they vary in their sensitivity to the RNA synthesis inhibitor α-amanitin. None is sensitive to the prokaryotic RNA synthesis inhibitor rifampicin.

RNA polymerase I. This enzyme is not inhibited at all by α-amanitin. It is found within the nucleolus and is responsible for the transcription of the genes for the 28S, 5.8S, and 18S rRNA molecules.
RNA polymerase II. This polymerase is inhibited at low concentrations of

α-amanitin and is required for the synthesis of most other RNA species, notably mRNA.

RNA polymerase III. This enzyme is inhibited at high concentrations of α-amanitin and functions in the transcription of tRNA (4S RNA) and 5S ribosomal RNA.

All of these RNA polymerases are complex, each consisting of a number of polypeptide subunits.

The mitochondria and chloroplasts contain their own RNA polymerases, which resemble bacterial RNA polymerases in their sensitivity to rifampicin and their insensitivity to α-amanitin.

Transcription controlling sequences

Analysis of base-pair sequences upstream from eukaryotic protein-coding genes has shown the presence of the consensus sequence TATAAAA, reading 5′-to-3′ on the nontemplate strand. This sequence, called the TATA box or the **Goldberg–Hogness box**, is found approximately 30 base pairs upstream from the first base pair transcribed into RNA, and is considered to be the likely promoter sequence. Other sequences further upstream from the TATA box also appear to play a role in the regulation of transcription, in particular determining the rate of transcription initiation of the gene. There is also evidence that several different types of promoter sequences are used in eukaryotes.

To date, relatively little has been learned about the nature of transcription termination sequences in eukaryotes, in part because of the rapid processing that occurs at the 3′ ends of transcripts for protein-coding genes. In yeast, part of the signal for transcription termination is sequence TTTTTATA, and there is evidence that the complete signal is no more than 21 base pairs. It appears that the complete signal requires the presence of other sequences downstream for full effectiveness.

Messenger RNA

General properties

Genes coding for proteins are transcribed into mRNA molecules, which become associated with ribosomes where protein synthesis occurs. In prokaryotic organisms where no nuclear membrane exists it is possible for mRNA to associate with ribosomes before the mRNA chain has been completed so that polypeptide synthesis can commence. This coupled transcription and translation are shown in Fig. 7.6. In eukaryotes, owing to the cellular compartmentation, the mRNAs must migrate out of the nucleus to the cytoplasm where the ribosomes active in protein synthesis are found. The

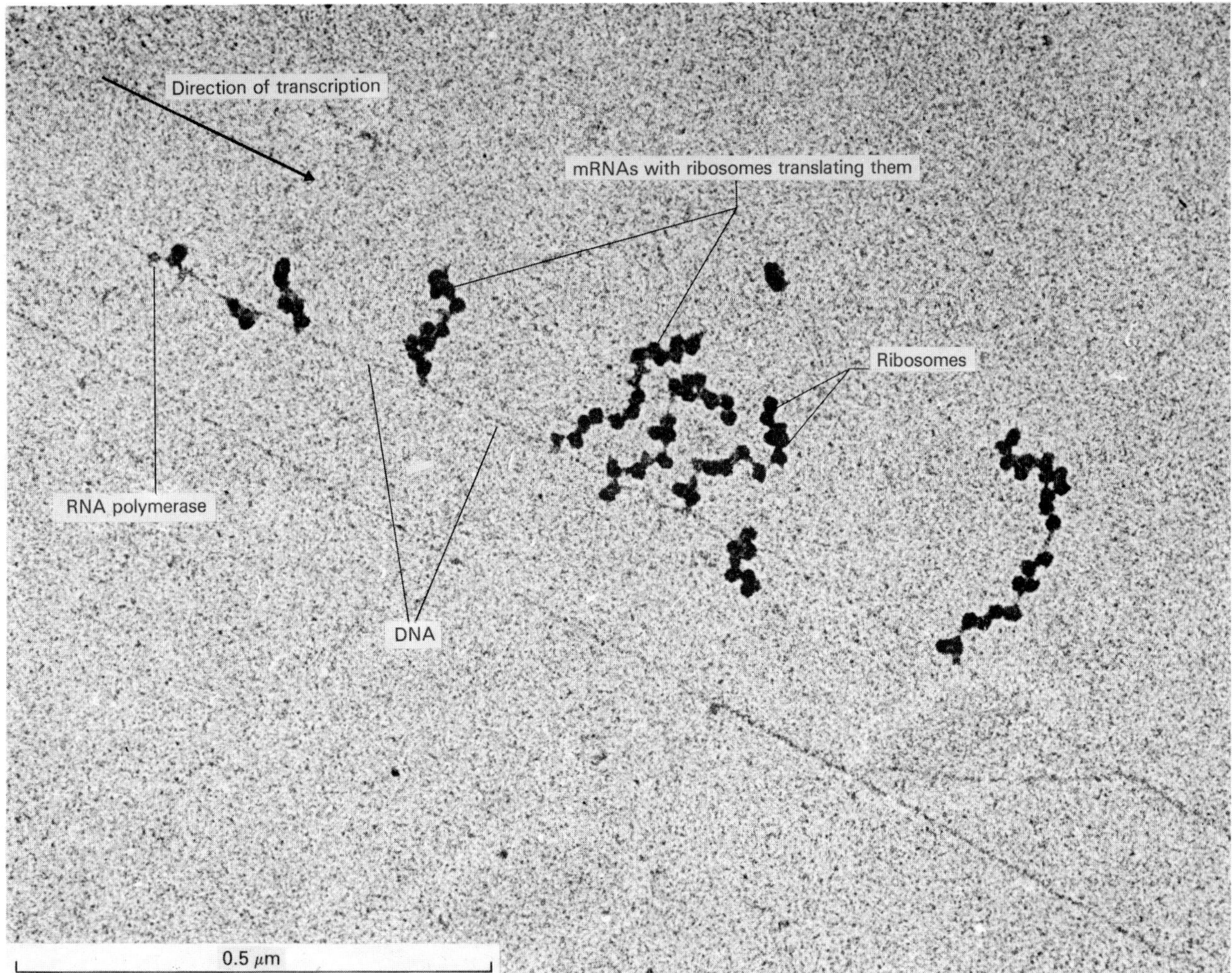

Fig. 7.6. Electron micrograph of coupled transcription and translation in *E. coli* (Based on data from O. Miller et al. 1975. Reproduced with permission, © 1975 by the American Association for the Advancement of Science.)

other main differences between prokaryotic and eukaryotic mRNAs are:

1. Many prokaryotic mRNAs are transcripts of several adjacent genes (these mRNAs are called **polygenic** or **polycistronic mRNAs**), whereas all eukaryotic mRNAs are transcripts of single genes (these mRNAs are called *monogenic* or *monocistronic mRNAs*).

2. Prokaryotic mRNAs are not processed and have short lifetimes, whereas eukaryotic mRNAs are processed and exhibit a range of lifetimes.

3. Most of the prokaryotic genome codes for mRNA, whereas only a small fraction of the eukaryotic genome may code for mRNA. At any one time, less than 10–20% of the total cellular RNA in either prokaryotic or eukaryotic cells is present as mRNA.

When the mRNA population of a prokaryotic or eukaryotic cell is examined, it is found to be heterogeneous in size, with a fairly wide range of S values. This results from the fact that the number of nucleotide pairs in a gene is related to the size of the protein for which they code. However, since a mRNA molecule contains variable-length sequences other than those that specify the amino acid sequence of a protein, this size relationship is not absolutely direct.

The general structure of a mature mRNA molecule is shown in Fig. 7.7.

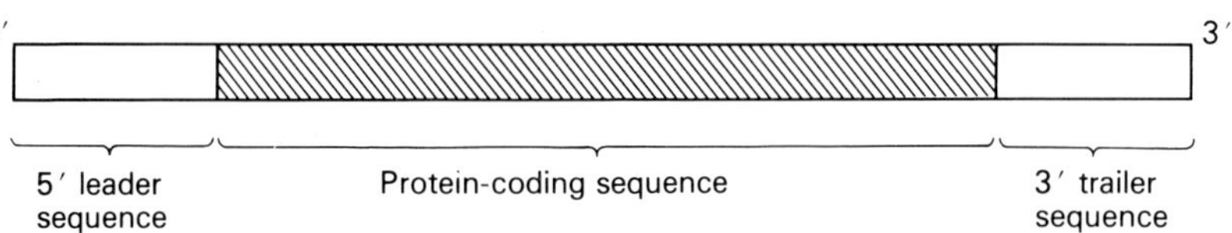

Fig. 7.7. General structure of mature prokaryotic and eukaryotic mRNA molecules.

All mRNAs are transcribed from **structural genes**, that is, genes that code for proteins. The three main parts of an mRNA molecule are:

1. **Leader sequence** at the 5′ end. The length of the leader sequence varies from mRNA to mRNA, and within this sequence is the information that the ribosome uses to initiate protein synthesis at the correct position. The leader sequence is not translated.
2. **A coding sequence** following the leader sequence. The coding sequence is translated into the amino acid sequence of a polypeptide.
3. **A trailer sequence** at the 3′ end. The trailer sequence is not translated and varies in length from mRNA to mRNA. Note that the polycistronic mRNAs in prokaryotes have leader and trailer sequences and spacer sequences between the coding sequences.

Prokaryotic mRNA

In prokaryotes there is generally *colinearity* between a protein-coding gene and the mRNA for which it codes. That is, there is a contiguous sequence of nucleotide pairs in a gene and this sequence is transcribed into a contiguous sequence of RNA nucleotides, which are then translated to give the corresponding contiguous sequence of amino acids in the protein. We will see later that this is often not the case in eukaryotic mRNAs.

Prokaryotic mRNAs have relatively short lifetimes; they are quite labile. By this we mean that a newly synthesized mRNA molecule may only function for two or so minutes in organisms such as *E. coli*. After this time, which is approximately 10% of a cell cycle, the RNA will be destroyed. Thus to continue to synthesize a particular protein, the gene for that protein must be transcribed continuously. This affords the cell a very rapid means of conserving resources because, if a protein is no longer needed for a cell to function, the gene can be blocked from being transcribed and any extant mRNA will soon be destroyed. This allows the organism to respond rapidly to changing needs when its external environment changes by turning off the transcription of some genes and turning on the transcription of a different set of genes.

While most prokaryotic mRNAs are short-lived, some mRNA species in prokaryotes have relatively long lives. These appear to be resistant to the nucleases that are responsible for the rapid destruction of the short-lived mRNAs. In *E. coli* for example, at least five ribonuclease (RNase) activities have been demonstrated, and some of these, if not all, may be involved in mRNA breakdown.

Eukaryotic mRNA

Longevity of mRNAs. Eukaryotic mRNAs are generally long-lived. Compared with prokaryotic mRNAs, many mRNAs in eukaryotic organisms are functional for only an hour or so, which is on the order of 5–10% of the cell cycle in actively dividing cells. These mRNAs could be considered short-lived in these terms, in much the same way as were the short-lived prokaryotic mRNAs.

In general, the lifetimes of eukaryotic mRNAs, when expressed as a fraction of the normal cell cycle time, are highly varied. This is a practical consequence of the functional characteristics of a eukaryotic cell as compared with a prokaryotic cell. In particular, eukaryotic organisms are differentiated into various cell types that have different functions. For example, some cell types are specialized to produce one type of protein almost entirely. Characteristically, the mRNA for that protein will be stable.

Chromatin structure and transcription. As discussed in Chapter 2, the DNA of eukaryotic nuclear chromosomes is organized into nucleosome structures. A number of studies have shown that transcriptionally active chromatin is more sensitive to deoxyribonuclease attack than is nontranscribing chromatin. This result indicates that the transcribed segments of the chromatin must have a different structural conformation than the inactive chromatin. However, this is not due to any major alteration in nucleosome structure.

5′- and 3′-end modifications. Unlike prokaryotic mRNAs, eukaryotic mRNAs are modified at both the 5′ and 3′ ends. The modifications occur post-transcriptionally and involve the action of specific enzymes.

The 5′ ends of most eukaryotic mRNAs do not have a free 5′-nucleoside triphosphate, but instead they are *capped*. In other words, after transcription of the mRNA, a modification occurs in which a guanine nucleotide is added to the terminal 5′-nucleotide by a 5′-5′ linkage. In addition, methyl groups are added to the now terminal guanine and the 2′-hydroxyl group of the adjacent nucleotide (Fig. 7.8). The capping phenomenon is ubiquitous among eukaryotes, although there are slight variations of the "cap" structure itself. At the functional level, the cap structure appears to be essential for the formation of the mRNA–ribosome complex and therefore for the initiation of protein synthesis.

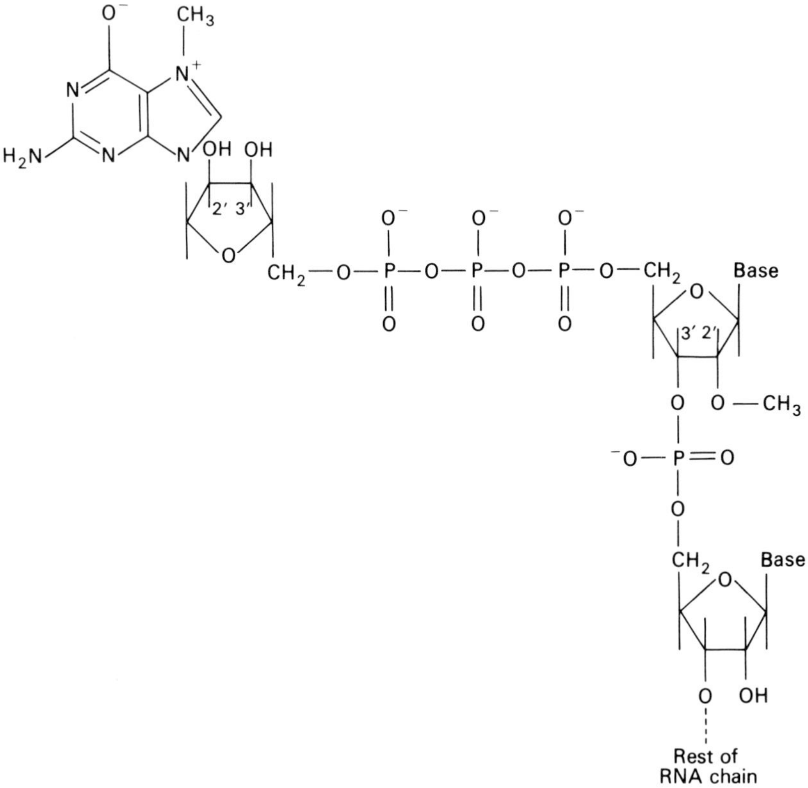

Fig. 7.8. The cap structure at the 5′ end of eukaryotic mRNAs.

The 3′ ends of most eukaryotic mRNAs (histone mRNA is a general exception) is modified by the addition of 50 to 200 adenine nucleotides to give what is called a **poly(A) tail**. These poly(A) segments are added after transcription has been completed with the aid of the enzyme poly(A)-polymerase. There appear to be multiple forms of the enzyme in cells, and they all catalyze the reaction shown in Fig. 7.9.

Fig. 7.9. The synthesis of poly(A) sequences at the 3′ end of many eukaryotic mRNAs catalyzed by polyadenylate polymerase.

$$\text{RNA} + \text{nATP} \xrightarrow[\text{Mg}^{2+}]{\text{Polyadenylate polymerase}} \text{RNA}-(\text{A})\text{n} + \text{nPPi}$$

The exact role of the poly(A) segment is not clear, although it is suggested that the poly(A) segment stabilizes the mRNA against nuclease attack. Some evidence to support this comes from studies in which the translatability of a mRNA from which the poly(A) had been removed by enzymatic means was tested in an in vitro system. Compared with the control, the mRNA in this instance was much more labile.

The polyadenylation of mRNAs is an aid to the isolation of such RNAs from a mixture of cellular RNAs. Chemical beads can be made that have poly(U) or poly(T) sequences bound to them. These beads are poured into a column, and a mixture of RNAs is poured through. In this column chromatographic procedure, the mRNAs with poly(A) tails (the poly[A]+ RNAs) bind to the column by the formation of complementary base pairs between poly(A) and poly(U) or poly(T). The poly(A)− RNAs do not bind and are washed through. The bound poly(A)+ RNAs can subsequently be dissociated and collected.

Noncoding sequences. Not all the nucleotide sequence between the 5′-cap and the 3′-poly(A) segment in eukaryotic mRNAs is used to code for the amino acid sequence in proteins. At the 5′ end, a short untranslated segment, called a leader sequence, is involved with ribosome recognition for the initiation of protein synthesis. This is also the case for prokaryotic mRNAs. Comparisons of the amino acid sequences of proteins and the nucleotide sequences of some eukaryotic mRNAs has shown that there is a segment at the 3′ end of the mRNAs that is not translated. This segment is called the trailer sequence. The number of nucleotides in the noncoding sequence (this excludes the poly[A]) may be as many as one-third to one-half of those found in the coding region itself. Thus mature mRNAs may be significantly longer than is required for their coding capacity. The function of the nontranslated segment is unknown at present, but it may contain binding sites that recognize the proteins that must be involved in the synthesis, processing, transport, association with ribosomes, and degradation of mRNAs.

It is known that the 3′ ends of polyadenylated mRNAs in at least higher eukaryotes are generated by cleavage from much larger transcripts. The specificity of cleavage and polyadenylation has been shown to be at least partly determined by the sequence AAUAAA (or related sequences) located 11–30 bases 5′ to the actual cleavage site. Although the necessity for the AAUAAA sequence has been clearly shown, this sequence alone cannot be sufficient to signal the processing because the sequence has also been shown to occur within coding sequences of some messages. Presumably other sequences or mRNA conformations are necessary in addition to the specific AAUAAA sequence to signal correct processing.

"Split" genes and mRNA production from precursors. In the nucleus of eukaryotic cells, a rapidly labeled RNA species can be isolated that is distinct from ribosomal and transfer RNA or their precursors. This RNA is called **heterogeneous nuclear RNA (hnRNA)**, and in general it has a relatively short lifetime in the nucleus. The hnRNA is heterodisperse in length, is much longer than cytoplasmic mRNA, and has both the 5′-cap and 3′-poly(A) modifications that are characteristic of the cytoplasmic mRNAs. The last two properties have lended credence to the hypothesis that hnRNA molecules are precursors to the functional mRNAs, the former being "processed" to the latter in the nucleus. However, in most eukaryotes, a large fraction (approximately 90%) of the hnRNA that is made is rapidly degraded, and thus only a small portion of the synthesized material is transported to the cytoplasm in the form of mRNA. The reason for this "wastage" is not known.

What is the relationship between the DNA, hnRNA, and mRNA? In 1977 A. J. Jeffreys and R. A. Flavell discovered that there is a 600 base-pair (bp) segment inserted within the coding sequence for the 146-amino acid β-globin chain in rabbits. (β-globin with α-globin and a prosthetic group constitutes hemoglobin.) The insert is called an **intervening sequence** (ivs), or **intron**, and the coding sequences are called *coding sequences* or *exons*. The introns are transcribed but they are not translated.

A large number of eukaryotic protein-coding genes contain introns. Higher eukaryotes have more of these interrupted ("split") genes, and the introns are more extensive than is the case in lower eukaryotes. For example, the chicken ovalbumin gene has seven introns (Fig. 7.10), collagen genes have at least 51 introns, the human preproinsulin gene contains two introns, and the yeast actin gene has one intron. Not all eukaryotic protein-coding genes have introns, however. Histone genes of invertebrate animals are uninterrupted, as is the interferon gene. In general, there is no general pattern as to what types of genes are interrupted and what types are not.

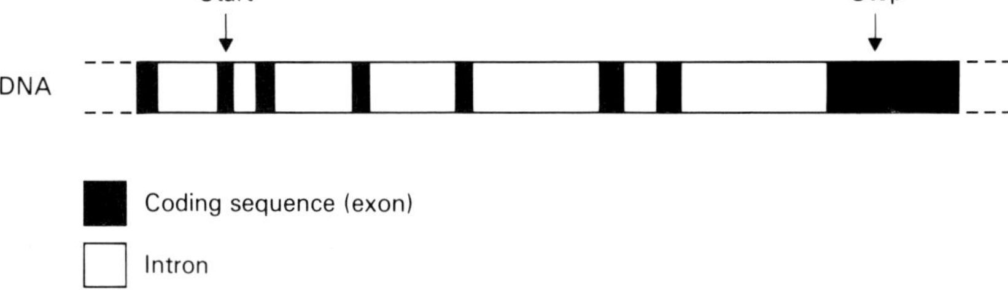

Fig. 7.10. Organization of the chicken ovalbumin gene showing the seven introns separating the coding sequences. Also shown are the positions of the start and stop signals for translation when mature mRNA is produced. (From the work of B. W. O'Malley.)

Given the fact that some eukaryotic genes contain introns, a simple model can be proposed for the production of functional mRNA (Fig. 7.11). In the example, the hypothetical gene contains two introns. Genes without introns are expressed in the same way, although, obviously, there are no introns to be removed from the RNA transcript. The steps are:

1. RNA polymerase binds to the promoter and transcribes the DNA continuously until a termination sequence is read; that is, leader, trailer, and coding sequences as well as introns are transcribed.

Fig. 7.11. Model for the production of a mature mRNA from a hypothetical gene with introns.

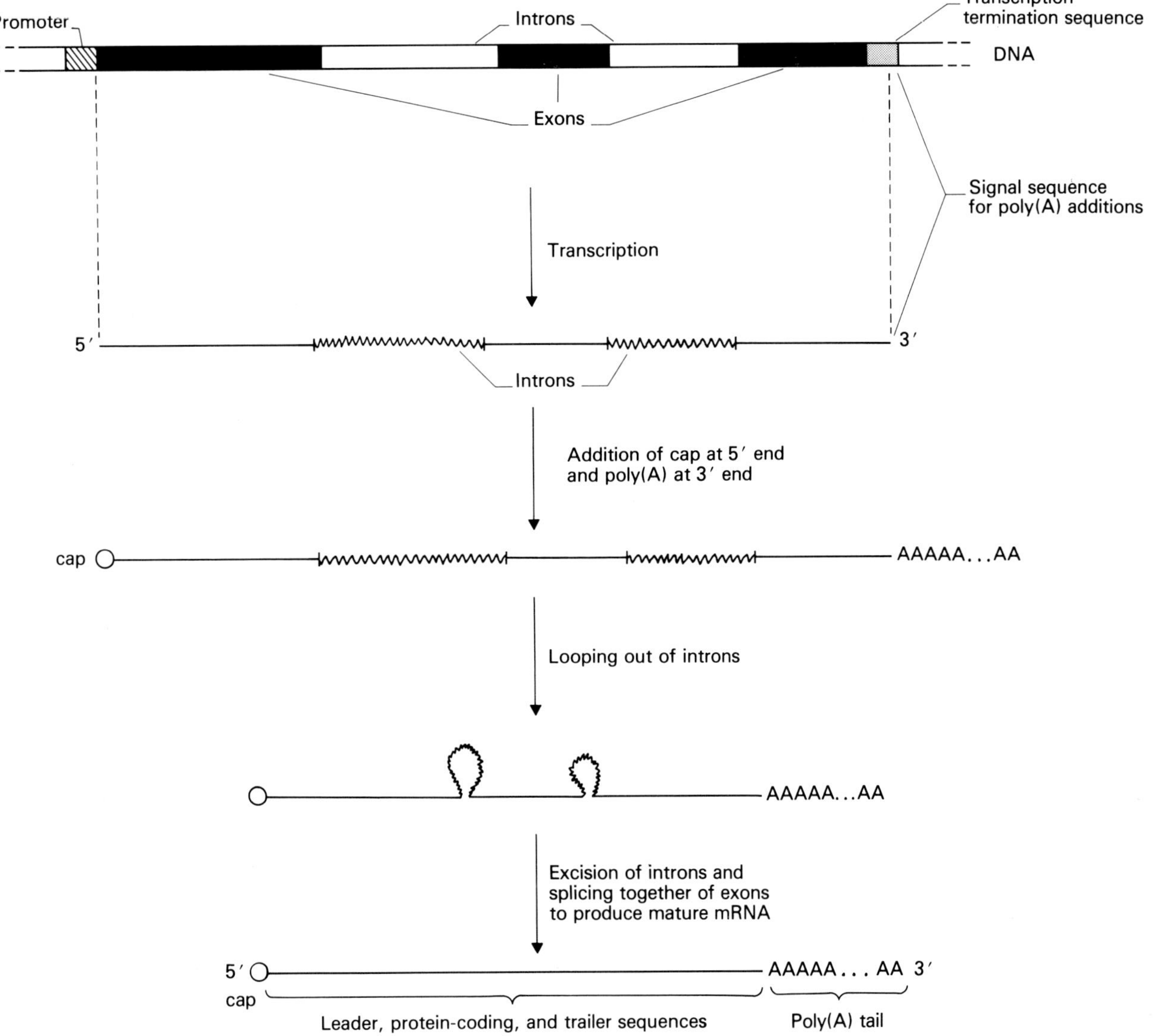

2. The transcript is capped at the 5′ end, the transcript is cleaved near the 3′ end, and a poly(A) tail is added to the new 3′ end (for most RNAs) to produce a *precursor-mRNA (pre-*mRNA), which is longer than the mature mRNA because of the introns. These pre-mRNA molecules constitute or at least are included in the hnRNA population discussed earlier.
3. The pre-mRNA molecule is *processed* to remove introns. The introns are looped out, and the loop is removed by nuclease cleavage. The now-adjacent coding sequences are then ligated together. These events are called *splicing*. Little is known about the enzymes involved in splicing. It is known that all introns (at the RNA level) start with GU and end with AG, and these sequences are presumably crucial if correct splicing is to occur.

Recent work has revealed the details of the splicing of introns out of mRNA precursors (Fig. 7.12). Consider a pre-mRNA molecule with two coding sequences, 1 and 2, separated by an intron (Fig. 7.12a). The first step in splicing is an endonucleolytic cleavage at the 5′ splice junction that results in the separation of coding sequence 1 from an RNA molecule that contains the intron and coding sequence 2. The resulting 5′ end of the intron becomes joined to an A that is part of a sequence about 20 to 40 nucleotides before the 3′ splice junction (Fig. 7.12b). A branch point results in the RNA because the A is in 2′-5′ linkage to the first G and simultaneously in 3′-5′ linkage to the downstream nucleotide.

Next, mRNA and a precisely excised intron form appear in parallel as a result of an endonucleolytic cleavage at the 3′ splice junction and ligation of the two coding sequences (Fig. 7.12c). The excised intron RNA molecule retains the branch point and is termed a *lariat structure* because of its shape. The lariat is subsequently converted to a linear molecule by a debranching enzyme.

The processing of pre-mRNA molecules just described appears to require a very large structure, which presumably contains several protein and RNA molecules. Such splicing complexes, called *spliceosomes* have been identified, but not completely characterized, in yeast and mammalian cells. It is speculated that the intron in lariat form holds the spliceosome together and that debranching of the lariat is a signal for recycling spliceosome components.

Since there are processing events involved with tRNA and rRNA production, we must be clear about the definition of an intron. An intron is a sequence in a gene that separates coding sequences of the gene. Both coding sequences and introns are transcribed, but each intron is subsequently removed such that the adjacent coding sequences are spliced together to form a contiguous sequence.

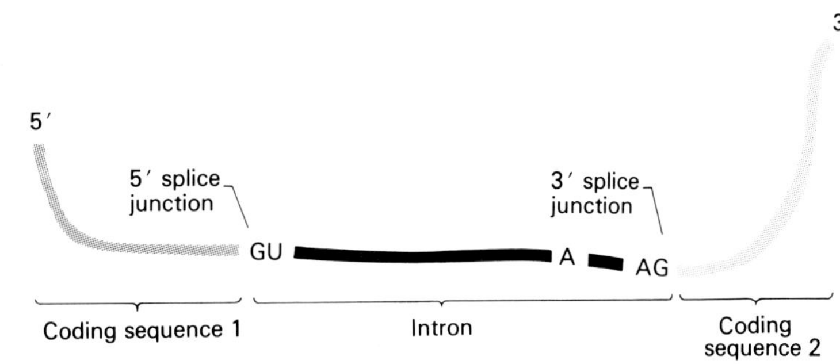

Fig. 7.12. Model for splicing of introns out of mRNA precursor molecules. (**a**) Theoretical pre-mRNA molecule containing two coding sequences, 1 and 2, separated by an intron. (**b**) An endonucleolytic cleavage event separates coding sequence 1 from the rest of the RNA molecule and is followed by a covalent pairing of the G at the 5′ end of the intron with an A residue located about 20 to 40 bases before the 3′ splice junction. (**c**) An endonucleolytic cleavage event at the 3′ junction releases the intron as a characteristic lariat structure, and the two coding sequences become ligated together to produce the mature mRNA molecule.

Transfer RNA

Function

Transfer RNA molecules have a central role in protein synthesis and interact with a wide range of other molecules. This class of molecules is predominantly located in the soluble portion of the cytosol (hence its earlier name, soluble RNA) and represents about 10–15% of the total cellular RNA in both prokaryotes and eukaryotes. Each molecule is able to combine specifically with one of the amino acids in a reaction catalyzed by one of the set of enzymes called **aminoacyl-tRNA synthetases**. The resulting **aminoacyl-tRNAs** migrate to specific sites on the ribosome and interact with specific three-nucleotide sequences (codons) on the mRNA so that the correct amino acid can be inserted into the growing polypeptide chain. These events will be described in detail in the next chapter.

Transfer RNA genes

There are 61 codons in mRNAs that code for amino acids in proteins, and each codon must be read by an anticodon on an appropriate tRNA molecule. Hence, a large number of tRNA types is found in the cell, meaning that there is a large number of tRNA genes in the genome.

In the *E. coli* chromosome, the typical case is that there is one gene for a given tRNA type. Some genes are represented two or more times, and this is called *gene redundancy*. Both nonredundant and redundant tRNA genes may be clustered or scattered in the chromosome.

Eukaryotes have many more tRNA genes than do prokaryotes. For example, in *Xenopus* there are more than 200 copies of each gene per haploid genome. All possible arrangements of tRNA genes can be found; that is, genes may be arranged singly or in clusters, and those genes found in clusters may be redundant or nonredundant tRNA genes.

Structure

Transfer RNAs have an S value of about 4, and a molecular weight of about 25,000–30,000 daltons. The nucleotide chain length is remarkably uniform for all the tRNAs of prokaryotes and eukaryotes, ranging from 76 to 85.

All the tRNA molecules sequenced to date (more than 40) appear to conform to the cloverleaf model with a stem, three large arms consisting of a stem and a loop, and occasionally an extra arm. One general form for this is shown in Fig. 7.13. Some of the modified bases are shown in Fig. 7.14.

Certain features of the primary sequence of tRNAs appear to be constant. For example, the 3′-terminal sequence –CCA.OH (which is added to all tRNAs post-transcriptionally) and the sequence TΨC (where Ψ is pseudouridine) in loop IV appear universal. The anticodon is a sequence of three nucleotides that must bond with the codon of mRNA during polypeptide elongation, and hence this region varies according to the tRNA in question. However, the nucleotide to the 5′ side of the anticodon is always U and that to the 3′ side is always a modified purine. The nucleotides present in the stem regions are variable, but the number of base pairs in a particular stem is fairly constant.

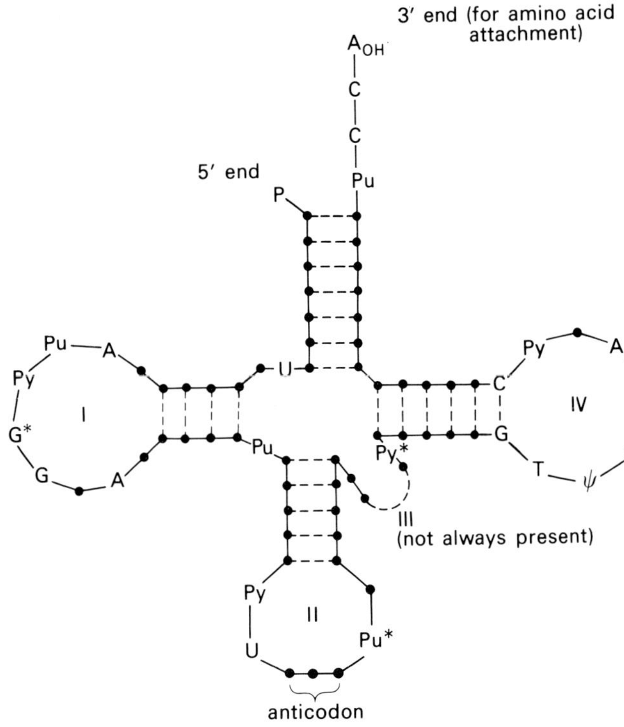

Pu = purine
Py = pyrimidine
P = phosphate
A,U,G,C = normal RNA bases
* = modified base
T = ribothymidine
ψ = pseudouridine
---- = hydrogen bonding

Fig. 7.13. A schematic of a generalized tRNA molecule shown in the cloverleaf configuration. (After A. L. Lehninger, 1975. *Biochemistry*. Worth Publishers, New York.)

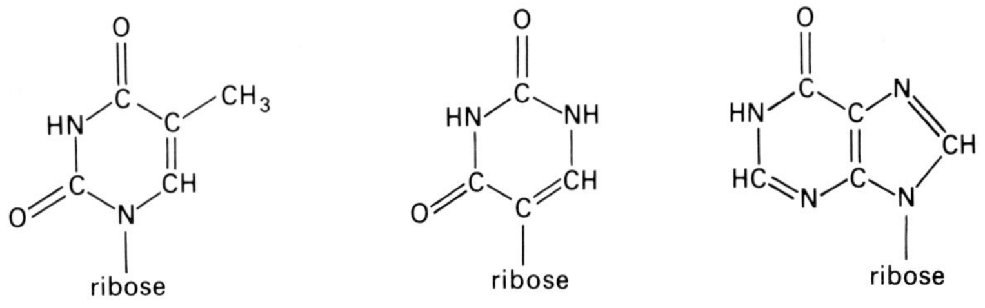

Ribothymidine (T)   Pseudouridine (ψ)   Inosine (I)

Fig. 7.14. Structures of some of the modified bases found in tRNA.

In spite of having more than 40 tRNA nucleotide sequences to compare, no general conclusions can be drawn regarding the functions of the various parts of the molecules. One exception to this is the sequence T-Ψ-C-G, which probably represents the common ribosome binding site of tRNA. Thus, it is likely that the functions of the tRNA are dependent upon the tertiary structure of the molecules.

Tertiary structure of tRNAs

As was indicated earlier, all of the sequenced tRNA molecules can be arranged in a similar cloverleaf structure. However, this model is derived entirely from analysis of the primary nucleotide sequence and the maximization of hydrogen bonding. The application of x-ray crystallography to stable crystals of tRNA (yeast phenylalanine tRNA was the first) has permitted the elucidation of its tertiary structure at the 0.3 nm resolution level, including the location of the major groups (Fig. 7.15). From the data, it was possible to conclude that all of the double-helical stems predicted by the cloverleaf model do exist. In addition, other hydrogen bonds bend the cloverleaf into a stable tertiary structure that has a rough L-shaped

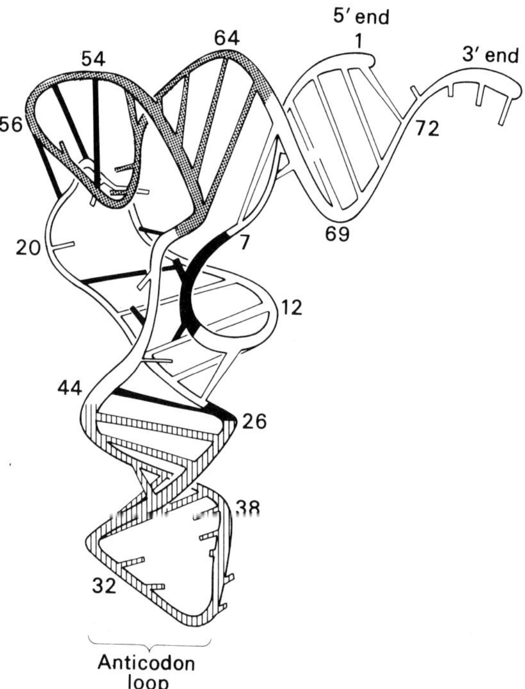

Fig. 7.15. Schematic model of yeast phenylalanine-tRNA. The ribose phosphate backbone is drawn as a continuous cylinder with bars to indicate hydrogen-bonded base pairs. The positions of single bases are indicated by short rods. The TΨC arm is heavily stippled, and the anticodon arm is marked by vertical lines. Tertiary structure interactions are illustrated by black rods. The numbers indicate the nucleotide position starting at the 5′ end. (With permission from S. H. Kim et al., *Science* **185**, 435–440. © 1974 by the American Association for the Advancement of Science.)

appearance. In this structure, the amino acid acceptor CCA group at the 3′ end of the chain is located at the opposite end from the anticodon loop.

tRNA biosynthesis

It is apparent that, unlike mRNAs, the tRNA molecules are extensively modified. These modifications occur post-transcriptionally. Indeed, the mature tRNAs not only have modified bases but also are considerably shorter (up to 40 nucleotides or so) than the primary gene transcripts. Thus, in both prokaryotes and eukaryotes there is evidence for **precursor-tRNAs (pre-tRNAs)**, which must be clipped and trimmed to produce the mature molecules. In prokaryotes, the precursors either contain only one tRNA molecule with extra leader and trailer sequences at the 5′ and 3′ ends, or they contain several species of tRNA molecules within a long precursor. In the latter case, the tRNA molecules are separated by spacer sequences. These multi-tRNA precursors reflect the tandem arrangement of tRNA genes from which they were transcribed, and they may be described as 5′-leader-(tRNA-spacer)n-tRNA-trailer-3′.

At least two enzymes are necessary for the processing of these pre-tRNAs. One of these, RNase P, catalyzes the removal of the 5′ leader sequence, and the other, RNase Q, catalyzes the removal of the 3′ trailer sequence. Evidence for this has come from studies of mutant strains of *E. coli* with temperature-sensitive defects in the enzymes. At the nonpermissive temperature, partially processed pre-tRNAs accumulate and can be isolated for sequence comparison with mature tRNAs. Another enzyme (or enzymes) is involved in removing the spacer sequences between tRNAs when they are present in clusters.

In eukaryotes, there is also good evidence for the existence of pre-tRNAs, although studies have generally been done with mixtures of molecules rather than with purified individual species as was done in prokaryotes. Pre-tRNAs from several eukaryotes sediment at about 4.8S compared with 3.8S for mature tRNAs. These molecules are transcribed by RNA polymerase III, and the data obtained from studies with denaturing agents indicate that there are an additional 15–35 nucleotides on the pre-tRNAs. Processing of the pre-tRNA molecules presumably occurs in the nucleus, although, as yet, the enzymes involved and their sites of action have not been defined in any detail.

Interestingly, about 10% of the approximately 400 yeast tRNA genes contain an intron. In every case, the intron in the pre-tRNA is just to the 3′ side of the anticodon and is usually less than 15 bp long. After leader and trailer sequences have been removed from the tRNA, the intron is spliced out of the transcript in a reaction catalyzed by *RNA ligase*. This reaction is different from the splicing reaction involved in pre-mRNA processing, and no lariat structures are produced.

Ribosomes and ribosomal RNA

Ribosome structure

Ribosomes are the sites of protein biosynthesis within the cell. Most of the reactions that take place on the ribosome are similar in both prokaryotic and eukaryotic cells. The major difference between bacterial and eukaryotic ribosomes is their size.

Ribosomes are usually prepared by breaking open the cells in buffered solutions containing magnesium ions. After centrifuging the cell lysate to sediment cell debris and subcellular organelles, the supernatant is centrifuged at 250,000 *g* for 60 minutes. This yields a pellet of ribosomal material from which pure ribosomes can be isolated in a few more steps.

An important property of ribosomes is that under certain ionic conditions, in vitro they dissociate into subunits. This is commonly brought about by substantially reducing the magnesium ion concentration. This potentiates studies of the complete ribosome and the ribosomal subunits.

Ribosomes are usually characterized by their sedimentation coefficients obtained by analytical ultracentrifugation experiments. In eukaryotic cells, ribosomes sediment at about 80S, whereas in bacteria the ribosomes sediment at about 70S. Both of these types of ribosomes have very similar structures. They are complexes of rRNA and proteins with molecular weights ranging from 2.7 million daltons (Mdal) for bacterial ribosomes to about 4 Mdal for mammalian ribosomes. Among eukaryotic ribosomes, the molecular weight of an undissociated (monomeric) ribosome becomes greater the more complex the organism is.

Each ribosome consists of two unequally sized subunits. Both subunits must associate to function in protein synthesis. Fig. 7.16a presents electron micrographs of 30S subunits, 50S subunits, and 70S ribosomes from *E. coli*; and Fig. 7.16b shows three-dimensional models of 30S subunits, 50S subunits, and 70S ribosomes in different orientations. The relative S values for the ribosomes and their subunits and the molecular weights of the rRNA molecules they contain are summarized in Fig. 7.17.

More than half of the mass of each ribosomal subunit consists of rRNA, the rest being protein. In both prokaryotes and eukaryotes, the small subunit contains only one species of rRNA complexed with a number of proteins. The large ribosomal subunit of prokaryotes contains two species of rRNA, 23S and 5S, complexed with proteins. In eukaryotes, the large subunit contains three rRNA species, 28S, 5.8S, and 5S, and a large number of proteins. The

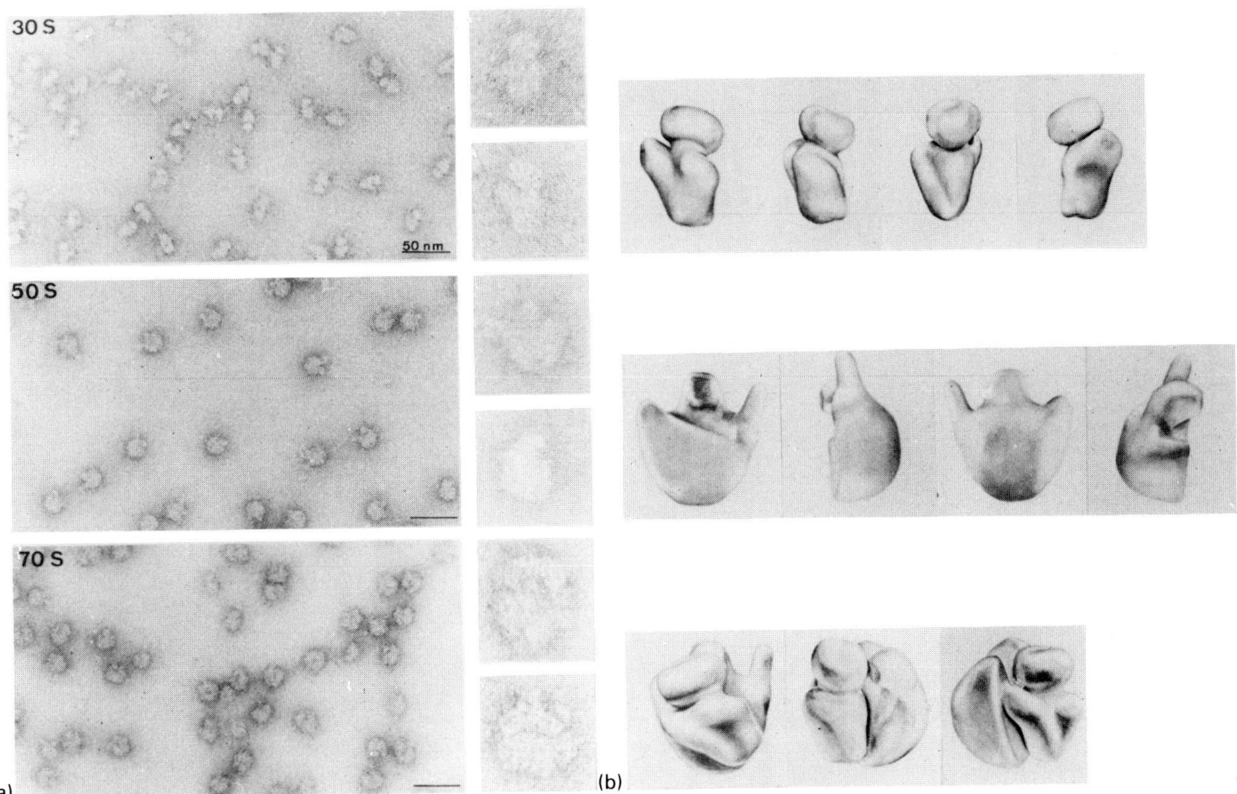

Fig. 7.16. *E. coli* ribosomes. (**a**) Electron micrographs of 30S subunits, 50S subunits, and 70S ribosomes. (**b**) Three-dimensional models of 30S subunits, 50S subunits, and 70S ribosomes seen in different orientations. (Courtesy of G. Stoffler.)

molecular weight of the 28S rRNA varies from 1.3 Mdal in eukaryotes such as higher plants, algae, protozoa, and fungi, to 1.75 Mdal in higher eukaryotes. When the rRNAs are extracted from the large subunit, the 5.8S is found to be hydrogen bonded to the 28S rRNA. It can be released by gentle heat treatment.

The ribosomal proteins of *E. coli* have been studied extensively. Since a number of protein factors necessary for protein biosynthesis associate transiently with ribosomes, there is a question of how one defines a ribosomal protein. One operational definition is that a ribosomal protein is a protein that remains attached to the ribosome after washing with high-salt solutions (which removes the transiently associated proteins), provided it is present in approximately molar yield. Most of the ribosomal proteins isolated from the ribosomes are basic proteins. In the small subunit of *E. coli* there are 20 proteins, and in the large subunit there are 34 proteins. All 54 proteins are immunologically distinct, with the exception of two that differ only by an acetyl group. The proteins exhibit enough differences in mass and charge that

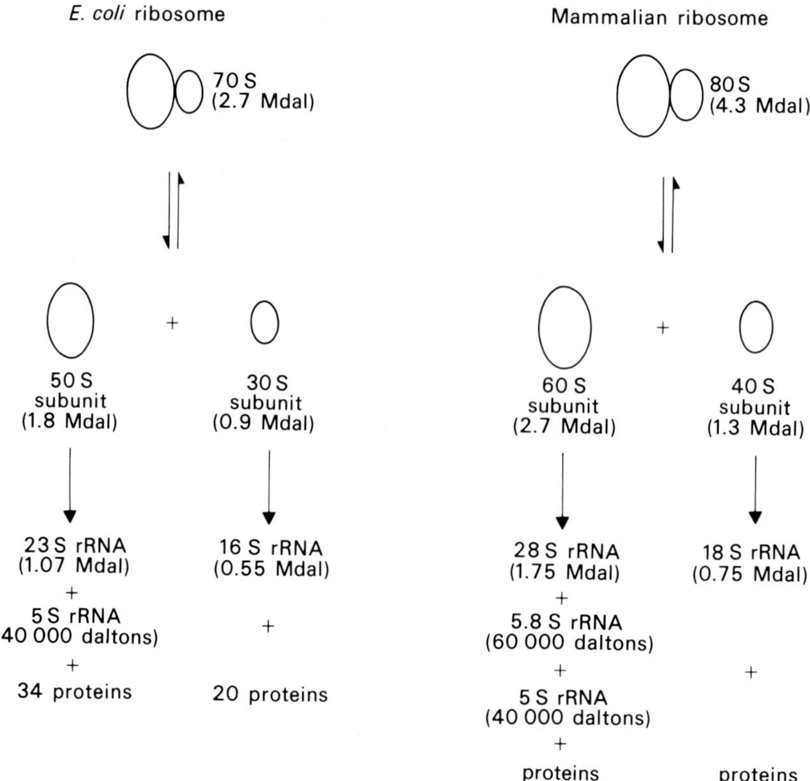

Fig. 7.17. A comparison of the subunit molecular weights and composition of ribosomes from *E. coli* and a mammal. (Mdal = 1 million daltons.)

they can be displayed by two-dimensional acrylamide gel electrophoresis. The exact role of each ribosomal protein in ribosome structure and function is not known.

In eukaryotic ribosomes, there are a correspondingly larger number of ribosomal proteins, and it is fair to say that the same generalizations that have been made for prokaryotic ribosomal proteins apply also to these.

Biosynthesis of prokaryotic ribosomes

Much work has been done to investigate how the protein and RNA components of the ribosome are assembled. One approach, taken by M. Nomura and colleagues, was to take purified ribosomal subunits of *E. coli* and to dissociate them chemically into their component rRNAs and proteins, and then to permit them to reassociate under appropriate ionic conditions. The 30S subunit was the first to be studied in this way, and it was found that the 20 proteins and the 16S rRNA could interact to form a complete, functional subunit at 37°C, the normal physiological temperature for *E. coli*. Since no other factors were present in the solution, the process was called *self-assembly*. This experiment, reported by Nomura's group in 1969, is summarized in Fig. 7.18.

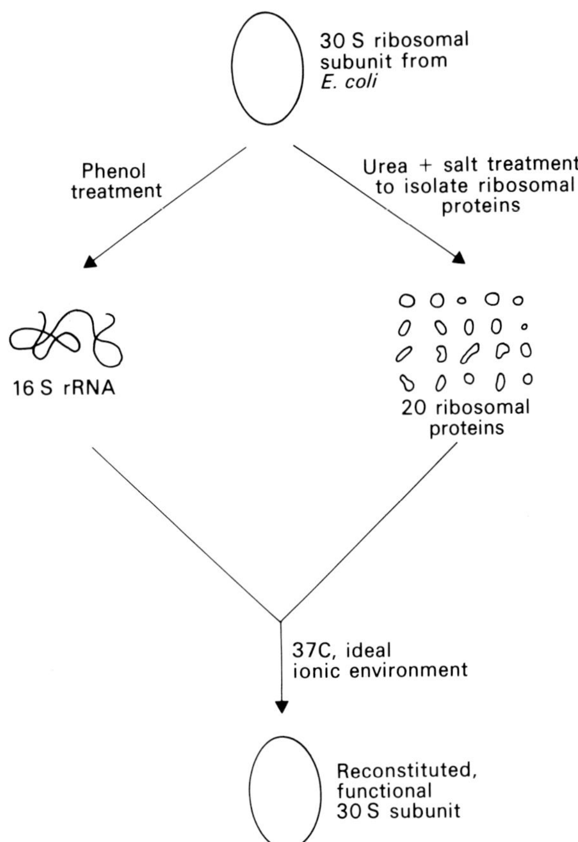

Fig. 7.18. Schematic of experiment that demonstrated the reconstitution of functional 30S ribosomal subunits of *E. coli* from the component RNA and protein molecules. (After M. Nomura, 1969. *Sci. Am.* **221**:28.)

Similar results were obtained more recently with the 50S subunit of *E. coli*. Thus, both subunits have the ability to self-assemble. Further reconstitution experiments by Nomura's group, in which one protein at a time is omitted from the reaction mixture, has provided information about the sequence of steps involved in the assembly reaction and about the necessity for all proteins to be present for the subunit to be functional.

The elegant in vitro reconstitution experiments do not completely reflect the assembly of *E. coli* ribosomal subunits in vivo, however. In *E. coli* the rRNA genes are arranged in seven repeats, each containing the 16S, 23S, and 5S rRNA sequences (Fig. 7.19). The primary transcripts of each repeat unit is a **precursor-rRNA (pre-rRNA)**, a precursor 30S (p30S) RNA that is cleaved

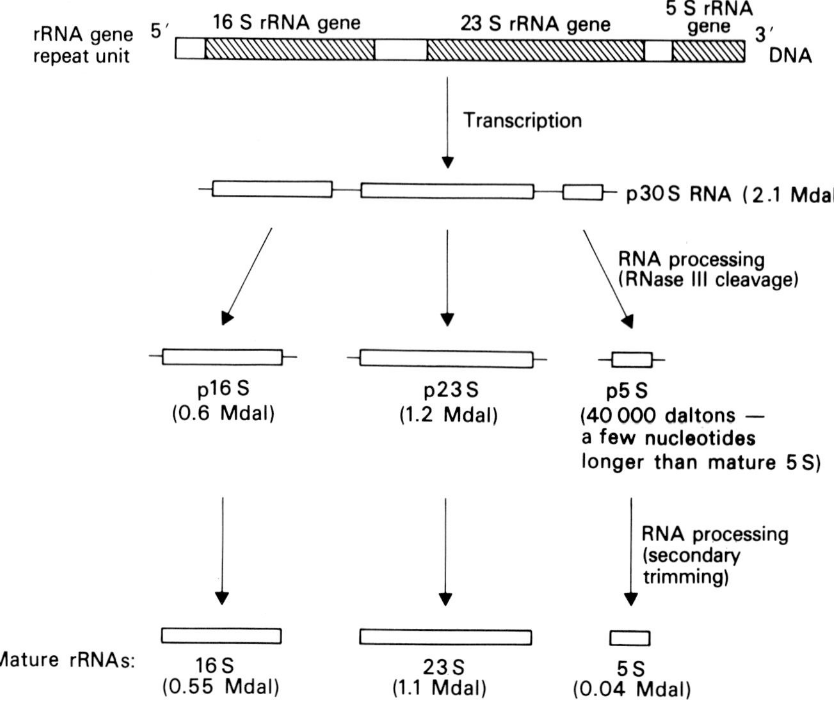

Fig. 7.19. Diagram of an *E. coli* rRNA gene repeat unit, and proposed scheme for the cleavage of a precursor-rRNA (p30S) to the mature rRNAs in *E. coli*.

to produce the p16S, p23S, and p5S RNAs, which are the immediate precursors to the mature 16S, 23S, and 5S rRNAs, respectively. Normally, the p30S molecule is not observed in wild-type cells, since the cleavages occur while p30S is still being transcribed. However, this cleavage is partially blocked in mutant strains that are deficient in RNase III, and the p30S molecule accumulates. Analysis of this RNA revealed the presence of the 16S, 23S, and 5S sequences within it. The polarity of the mature rRNAs within the p30S RNA is 5′-16S-23S-5S-3′, and a model for the processing steps based on in vitro cleavage studies with RNase III is as shown in Fig. 7.19.

Enzymes other than RNase III are required for the endonucleolytic cleavages that convert intermediate precursors to the mature RNAs. These final cleavages occur when the precursor rRNAs are associated with ribosomal proteins in preribosomal particles. In view of this, one must be cautious in extrapolating Nomura's in vitro assembly information to the in vivo state, since the conformations of the precursor rRNAs may facilitate different protein–protein or protein–RNA interactions than would the maturer RNAs. In other words, the end product is the same in the two, but the sequence of steps may be different.

For proper ribosome assembly to occur, the 16S and 23S RNAs must be methylated. This methylation occurs post-transcriptionally and most of the methyl groups are added to the bases, with only a few added to the 2′-OH of the ribose moiety. All of the methylations of the 23S rRNA occur on the p30S RNA component, whereas most if not all of the 16S rRNA methyl groups are added when it is in a mature form. Thus methylation of 23S rRNA is an early event in RNA processing and methylation of 16S rRNA is a late event.

Biosynthesis of eukaryotic ribosomes

In outline, eukaryotic ribosomes are made as follows: The rRNA genes are transcribed into a high-molecular-weight ribosomal precursor RNA (pre-rRNA), which is then modified and cleaved at specific sites to yield a number of discrete intermediate RNAs and finally the mature 18S, 5.8S and 28S rRNAs. Fig. 7.20 is an electron micrograph of transcription of pre-rRNA from an rRNA gene to a *Xenopus laevis* oocyte cell. In the figure, transcription is from left to right, and the "Christmas tree" appearance reflects the increasing length of the pre-rRNA transcript. The figure also shows that a number of RNA polymerase I molecules are transcribing each rRNA gene at any given time during its transcriptional activity. Assembly with ribosomal proteins and 5S rRNA (which is transcribed from separate rRNA genes) occurs in the nucleolus during pre-rRNA processing. The resultant ribosomal subunits are then released from the nucleus into the cytoplasm.

Ribosomal RNA maturation has been studied in a number of eukaryotes, including mammalian cells, amphibians, insects, higher plants, and fungi. The results indicate that all eukaryotes have similar pathways for the biosynthesis of ribosomes. In the following section, ribosome production in higher eukaryotes and fungi (yeast) will be described for comparison purposes and to point out the generalities involved.

Ribosomal DNA

In most eukaryotes studied to date, the genes for 18, 5.8, and 28S rRNA are multiple and clustered at the sites of the nucleolar organizer regions. Saturation hybridization experiments between purified rRNA and nuclear DNA have shown the extent of this gene multiplicity, and this varies from about 100 to 1000, depending on the organism being studied. In general there are more copies the more complex the eukaryote is in an evolutionary sense. In most organisms, the 5S rRNA is coded by extranucleolar genes whose multiplicity is usually higher than that for the other rRNA genes. The 5S rRNA genes may be clustered in the genome as in humans or scattered throughout the genome as in *Xenopus laevis*, the South African clawed toad. At least in yeast and slime moulds, the 5S rRNA genes are interspersed with the other rRNA genes. In a few eukaryotes, an intron is found in the 28S

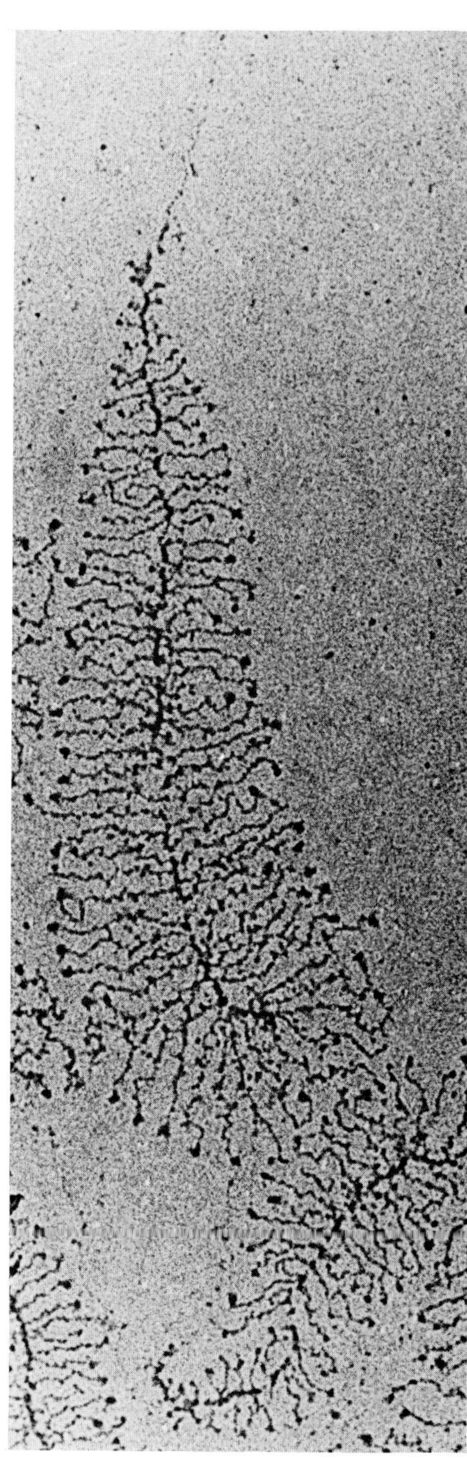

Fig. 7.20. Electron micrograph of transcription of pre-rRNA from an rRNA gene to a *Xenopus laevis* oocyte cell. Transcription of the rRNA gene is from left to right. (Courtesy of G. Morgan.)

rRNA coding sequence, and it is removed by a splicing reaction that is distinct from those reactions reported for pre-mRNA and pre-tRNA processing.

The DNA comprising the 18S+5.8S+28S gene clusters consists of three basic elements:

1. Sequences corresponding to the mature rRNAs.
2. Sequences transcribed as part of an initial precursor molecule but not found in the mature rRNAs (*transcribed spacer (TS) sequences*).
3. Additional *nontranscribed spacer (NTS) sequences* interspersed among the transcribed sequences.

Molecular studies of isolated rDNA have produced a generalized picture for the repeating unit of rDNA in eukaryotes (Fig. 7.21). Here A-G is the repeating unit. Within that unit B-C codes for 18S, D-E for 5.8S, and E-F for 28S rRNA. Regions A-B and C-D are TS sequences that are homogeneous from repeat to repeat, and F-G is a NTS sequence that is heterogeneous in length between repeats in higher eukaryotes such as *Xenopus*, is slightly heterogeneous in length in *Drosophila*, and is homogeneous in length in lower eukaryotes such as yeast. Differences in the lengths of the TS sequences and slight differences in the length of the 28S rRNA sequence accounts for the different sized pre-rRNAs seen in various organisms.

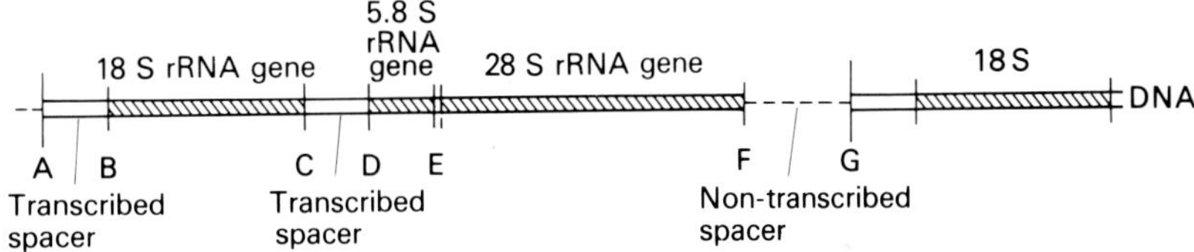

Fig. 7.21. Generalized diagram for the repeating unit in the ribosomal DNA of eukaryotes. A-G is the repeating unit. B-C, D-E, and E-F code for 18S, 5.8S, and 28S rRNAs, respectively. A-B and C-D are transcribed spacer sequences, and F-G is a nontranscribed spacer sequence.

Pre-rRNA transcription and processing

Ribosome formation commences with the transcription of pre-rRNA. In HeLa (mammalian tissue culture) cells it is a 45S molecule with a molecular weight of about 4.5 Mdal. In yeast (a lower eukaryote) it has a sedimentation coefficient of 35S with a molecular weight of about 2.5 Mdal. Within the pre-rRNA molecule, the arrangement of the rRNA sequences has been determined to be 5′-18S-5.8S-28S-3′, which is somewhat analogous to the arrangement of the rRNA sequences in the prokaryotic pre-rRNA molecule.

To produce mature ribosomes, the pre-rRNA molecule undergoes modifications (methylation and pseudouridylation), associates with ribosomal proteins and 5S rRNA, and is processed enzymatically to remove TS sequences. These events occur in the nucleolus, and many of the nonribosomal proteins involved are presumably stable, specific nucleolar proteins.

In HeLa cells over 100 methyl groups are added to 45S RNA. This

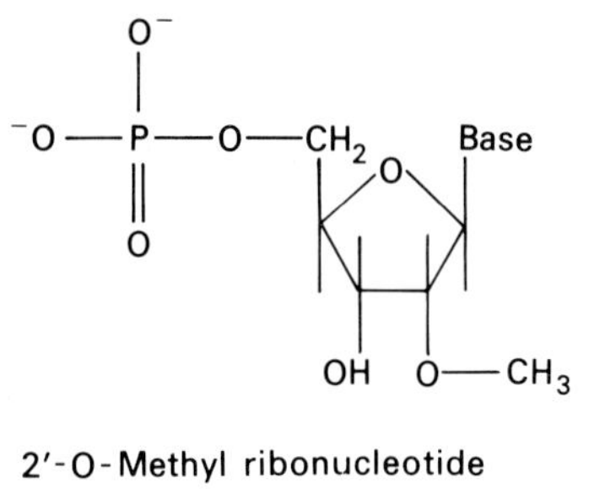

2′-O-Methyl ribonucleotide

Fig. 7.22. Site of methylation of the ribose moiety in rRNA.

methylation occurs close to the point of polymerization of ribonucleotides by the RNA polymerase, that is, during transcription. The significance of methylation is poorly understood, but it is clear that methylation is essential for ribosome maturation. In HeLa cells all the methylation sites are within the parts of the 45S RNA corresponding to the mature rRNA sequences, and all except for about six are on the 2′-OH of ribose moieties (Fig. 7.22).

Mature 18S, 28S, and 5.8S rRNAs of eukaryotic cells also contain many pseudouridine residues (c.f. tRNA modification). This modification also occurs in the nucleolus on pre-rRNA, but the significance of these residues is poorly understood.

The nucleolar 45S pre-rRNA molecule also associates with ribosomal proteins, which have been synthesized in the cytoplasm and then transported to the nucleolus, and with 5S rRNA which in most organisms is transcribed elsewhere in the nucleus. Thus nascent ribosomal particles can be extracted from nucleoli of actively growing cells.

Ribosome maturation, then, involves specific cleavages of the pre-rRNA, resulting in the elimination of TS regions and leaving only mature rRNA. These cleavages take place in the nascent ribosomal particles and are accompanied by the formation of specific protein–protein and protein–nucleic

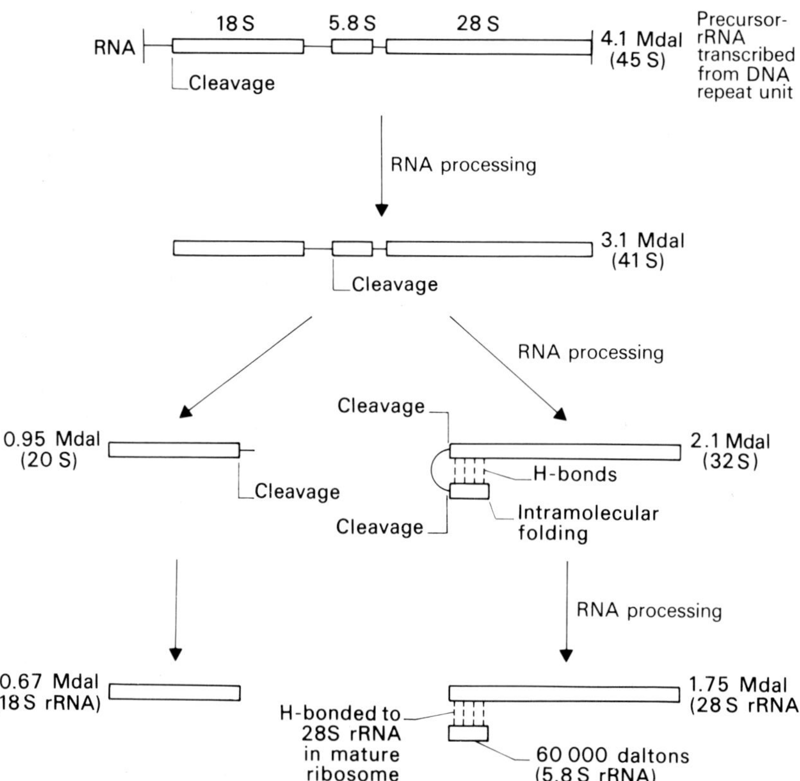

Fig. 7.23. Proposed scheme for the processing of 45S precursor rRNA in HeLa cells to produce the mature rRNA species. (After B. E. H. Maden, 1971. *Prog. Biophys, Mol. Biol.* **22:** 127.)

acid interactions as the mature ribosomal subunits are formed. Based on continuous-labeling and pulse-chase experiments with a radioactive RNA precursor and other supportive experiments, the maturation pathway for rRNA production in HeLa cells is proposed to be as shown in Fig. 7.23. In this case the TS sequences represent about 50% of the primary transcript, and these are nonconserved, being eliminated by the specific cleavages indicated.

The maturation scheme is essentially similar in all other eukaryotes examined to date. The details of rRNA processing vary owing to differences in the sizes of the initial transcript and the mature rRNA components, differences in the extent of the RNA modifications, and differences in the number of stable (and hence detectable) intermediates. In general the proportion of the primary transcript that is spacer decreases as one descends the evolutionary scale. In yeast, for example, the rRNA maturation scheme is as depicted in Fig. 7.24. Here about 20% of the primary transcript is nonconserved.

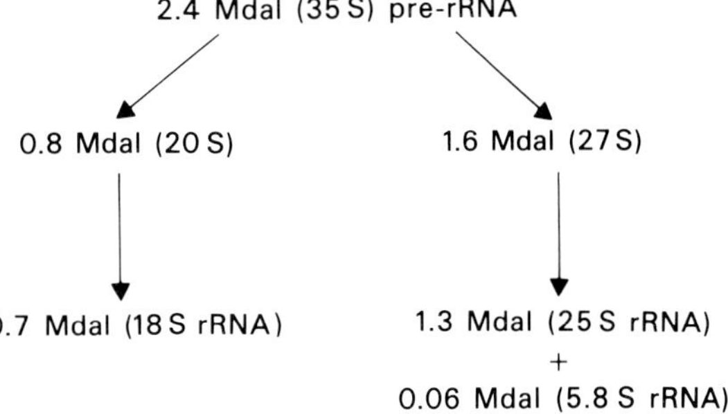

Fig. 7.24. Proposed scheme for the production of mature rRNAs from 35S pre-rRNA in the yeast, *Saccharomyces cerevisiae*. (After S. A. Udem and J. R. Warner, 1972. *J. Mol. Biol.* **65**:227.)

It should be pointed out that not all organisms exhibit an intermediate between the pre-rRNA and the 18S rRNA. Examples of these organisms are plants, *Xenopus* and *Neurospora*.

The entire process of ribosome production, in this case in HeLa cells, is summarized in Fig. 7.25. The numbers refer to the sedimentation coefficient of the RNAs.

Very little is known in eukaryotes about the enzymes that must be present for RNA processing to occur. Some potential processing enzymes have been identified, including a specific endoribonuclease and a 3′-OH-specific exonuclease.

Clearly, the synthesis of ribosomes is a highly complex process. Eukaryotic cells must possess elaborate mechanisms for regulating ribosome production according to needs. This control could be at the level of transcription of pre-rRNA and/or at post-transcriptional levels. Comparatively little specific information is available on this point.

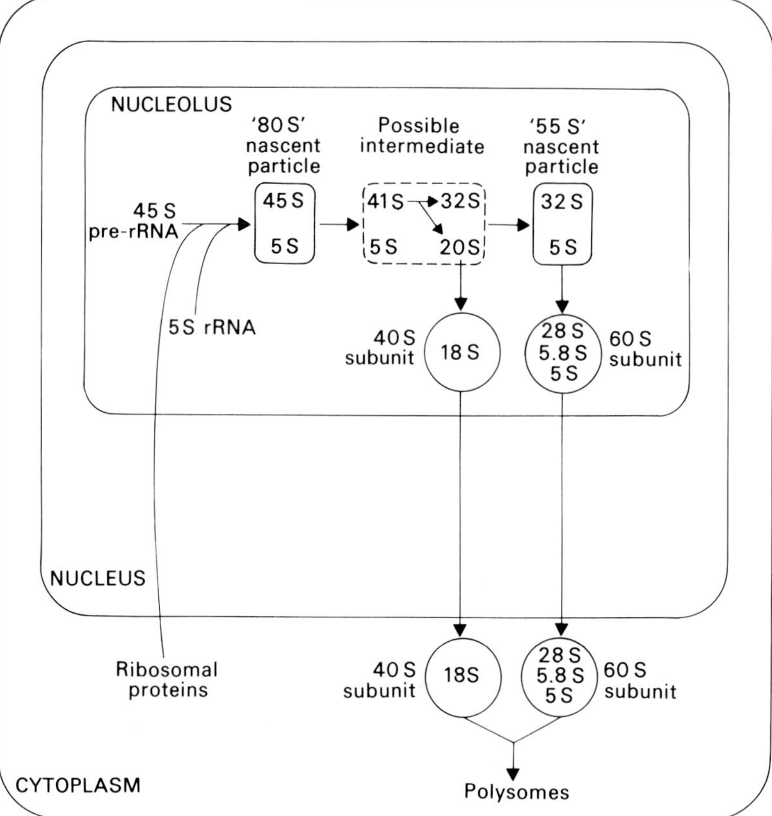

Fig. 7.25. Schematic representation of ribosome formation in HeLa cells showing the formation of ribonucleoprotein particles in the nucleolus and their processing to mature ribosomal subunits. In the diagram the rectangular particles are precursors and the circular particles are mature subunits. The numbers outside the particles indicates their S values, and the numbers inside indicate the S values for the rRNA they contain. (After B. E. H. Maden, 1971. *Prog. Biophys. Mol. Biol.* **22**:127.)

Questions and problems

7.1 Describe the differences between DNA and RNA.

7.2 Compare and contrast DNA polymerases and RNA polymerases.

7.3 Discuss the structure and function of the *E. coli* RNA polymerase. In your answer be sure to distinguish between core polymerase, RNA polymerase–sigma factor complex, and sigma factor.

7.4 Discuss the similarities and differences between the *E. coli* RNA polymerase, and eukaryotic RNA polymerases.

7.5 The RNA polymerases bind to promoter sequences in the DNA.
(a) Compare and contrast promoter sequences from prokaryotes and eukaryotes.
(b) Apart from the sequence information available for promoters for a number of genes, what evidence is there that promoters do not all have the same base sequence?

7.6 What is the role of each of the *E. coli* RNA polymerase subunits in the transcription process?

7.7 Describe an experiment for determining the nucleotide sequence of a eukaryotic promoter.

7.8 Discuss the molecular events involved in the termination of RNA transcription in prokaryotes and eukaryotes.

7.9 Compare and contrast the structures of prokaryotic and eukaryotic mRNAs.

7.10 Compare the structures of the three classes of RNA found in the cell.

7.11 Many eukaryotic mRNAs, but not prokaryotic mRNAs, contain intervening sequences. What is the evidence for the presence of intervening sequences in genes? Describe how these sequences are removed during the production of mature mRNA.

7.12 Discuss the posttranscriptional modifications that take place on the primary transcripts of tRNA, rRNA, and protein-coding genes.

7.13 Distinguish between leader sequence, trailer sequence, coding sequence, intervening sequence (ivs), transcribed spacer, and nontranscribed spacer sequence. Give examples of actual molecules in your answer.

7.14 Describe the organization of the ribosomal DNA repeat unit of a higher eukaryotic cell.

References

General

Adhya, S. and M. Gottesman, 1978. Control of transcription termination. *Annu. Rev. Biochem.* **47**:217–249.

Brenner, S., F. Jacob and M. Meselson, 1961. An unstable intermediate carrying information from genes to ribosomes for protein synthesis. *Nature* **190**:576–581.

Chamberlin, M.J., 1974. The selectivity of transcription. *Annu. Rev. Biochem.* **43**:721–775.

Chambon, P., 1977. Summary: the molecular biology of the eukaryotic genome is coming of age. *Cold Spring Harbor Symp. Quant. Biol.* **42**:1209–1234.

Darnell, J.E., 1977. Gene regulation mammalian cells: Some problems and the prospects for their solution. In *Cell Differentiation and Neoplasia,* 30th Annual Symposium on Fundamental Cancer Research, G. Saunders (ed.). M.D. Anderson Hospital and Tumor Institute, Houston, Texas.

Perry, R.P., 1976. Processing of RNA. *Annu. Rev. Biochem.* **45**:605–629.

Sirlin, J.L., 1972. *The Biology of RNA.* Academic Press, New York.

Stewart, P.R. and D.S. Letham (eds.), 1977. *The Ribonucleic Acids,* 2nd ed. Springer-Verlag, New York.

Weissbach, H. and S. Pestka (eds.), 1977. *Molecular Mechanisms of Protein Biosynthesis.* Academic Press, New York.

RNA polymerase

Burgess, R.R., 1971. RNA polymerase. *Annu. Rev. Biochem.* **40**:711–740.

Chambon. P., 1975. Eukaryotic nuclear RNA polymerase. *Annu. Rev. Biochem.* **44**:613–638.

Losick, R. and M. Chamberlin (eds.), 1976. *RNA Polymerase*, Cold Spring Harbor Laboratory, New York.

Pribnow, D., 1975. Nucleotide sequence of an RNA polymerase binding site at an early T7 promoter. *Proc. Natl. Acad. Sci. USA* **72**:784–788.

Travers, A.A. and R.R. Burgess, 1969. Cyclic reuse of the RNA polymerase sigma factor. *Nature* **222**:537–540.

Messenger RNA

Banerjee, A.K., 1980. 5′-terminal cap structure in eukaryotic messenger ribonucleic acids. *Microbiol. Rev.* **44**:175–205.

Berget, S.M., A.J. Berk, T. Harrison and P.A. Sharp, 1977. Spliced segments at the 5′ termini of adenovirus-2 late mRNA: a role for heterogeneous nuclear RNA in mammalian cells. *Cold Spring Harbor Symp. Quant. Biol.* **42**:523–529.

Brawerman, G., 1974. Eukaryotic messenger RNA. *Annu. Rev. Biochem.* **43**:621–642.

Brawerman, G., 1976. Characteristics and significance of the polyadenylate sequence in mammalian messenger RNA. *Progr. Nucl. Acid. Res. Mol. Biol.* **17**:117–148.

Breathnach, R. and P. Chambon, 1981. Organization and expression of eucaryotic split genes coding for proteins. *Annu. Rev. Biochem.* **50**:349–383.

Brody, E. and J. Abelson, 1985. The "spliceosome": yeast pre-messenger RNA associates with a 40S complex in a splicing-dependent reaction. *Science* **228**:963–967.

Corden, J., B. Wasylyk, A. Buchwalder, P. Sassone-Corsi, C. Kedinger and P. Chambon, 1980. Promoter sequences of eukaryotic protein-coding genes. *Science* **209**:1406–1411.

Crick, F.H.C., 1979. Split genes and RNA splicing. *Science* **204**:264–271.

Darnell, J.E., 1978. Implications of RNA. RNA splicing in evolution of eukaryotic cells. *Science* **202**:1257–1260.

Edmonds, M. and M.A. Winters, 1976. Polyadenylate polymerases. *Prog. Nucl. Acid Res. Mol. Biol.* **17**:149–179.

Furuichi, Y., M. Morgan, A.J. Shatkin, W. Jelenik, M. Salditt-Georgieff and J.E. Darnell, 1975. Methylated, blocked 5′ termini in HeLa cell mRNA. *Proc. Natl. Acad. Sci. USA* **72**:1904–1908.

Grabowski, P.J., R.A. Padgett and P.A. Sharp, 1984. Messenger RNA splicing in vitro: an excised intervening sequence and a potential intermediate. *Cell* **37**:415–427.

Henikoff, S. and E.H. Cohen, 1984. Sequences responsible for transcription termination on a gene segment in *Saccharomyces cerevisiae*. *Mol. Cell. Biol.* **4**:1515–1520.

Jeffreys, A.J. and R.A. Flavell, 1977. The rabbit β-globin gene contains a large insert in the coding sequence. *Cell* **12**:1097–1108.

MacCumber, M. and R.L. Ornstein, 1984. Molecular model for messenger RNA splicing. *Science* **224**:400–405.

Nevins, J.R., 1983. The pathway of eukaryotic mRNA formation. *Annu. Rev. Biochem.* **52**:441–466.

Padgett, R.A., M.M. Konarska, P.J. Grabowski, S.F. Hardy and P.A. Sharp, 1984. Lariat RNAs as intermediates and products in the splicing of messenger RNA precursors. *Science* **225**:898–903.

Perry, R.P., D.E. Kelley, K. Frederici and F. Rottman, 1975a. The methylated constituents of L cell messenger RNA: Evidence for an unusual cluster of the 5′ terminus. *Cell* **4**:387–394.

Perry. R.P., D.E. Kelley, K. Frederici and F. Rottman, 1975b. Methylated constituents of heterogeneous nuclear RNA: Presence of blocked 5′ terminal structures. *Cell* **6**:13–19.

Ruskin, B. and M.R. Green, 1985. A processing activity that debranches RNA lariats. *Science* **229**:135–140.

Sadofsky, M. and J.C. Alwine, 1984. Sequences on the 3′ side of hexanucleotide AAUAAA affect efficiency of cleavage at the polyadenylation site. *Mol. Cell. Biol.* **4**:1460–1468.

Salditt-Georgieff, M. and J.E. Darnell, 1982. Further evidence that the majority of primary nuclear RNA transcripts in mammalian cells do not contribute to mRNA. *Mol. Cell. Biol.* **2**:701–707.

Sharp, P.A., 1985. On the origin of RNA splicing and introns. *Cell* **42**:397–400.

Tilghman, S.M., D.C. Tiemeir, J.G. Seidman, B.M. Peterlin, M. Sullivan, J.V. Maizel and P. Leder, 1978. Intervening sequence of DNA identified in the structural portion of a mouse beta-globin gene. *Proc. Natl. Acad. Sci. USA* **78**:725–729.

Tonegawa, S., A.M. Maxam, R. Tizard, O. Bernhard and W. Gilbert, 1978. Sequence of a mouse germ-line gene for a variable region of an immunoglobulin. *Proc. Natl. Acad. Sci. USA* **75**:1485–1489.

Winicov, I. and R.P. Perry, 1976. Synthesis, methylation, and capping of nuclear RNA by a subcellular system. *Biochemistry* **15**:5039–5046.

Transfer RNA

Holley, R.W., J. Apgar, G.A. Everett, J.T. Madison, M. Marquisee, S.H. Merrill, J.R. Penswick and A. Zamir. 1965. Structure of a ribonucleic acid. *Science* **147**:1462–1465.

Nishimura, S., 1974. Transfer-RNA: structure and biosynthesis. In *Biochemistry of Nucleic Acids*. K. Burton (ed.), vol. 6, *MTP International Review of Science*, pp. 289–322. Butterworths, London.

Smith, J.D., 1972. Genetics of transfer RNA. *Annu. Rev. Genetics* **6**:235–256.

Smith, J.D., 1967. Transcription and processing of transfer RNA precursors. *Progr. Nucl. Acid Res. Mol. Biol.* **16**:25–73.

Sussman, J.L. and S.H. Kim, 1976. Three-dimensional structure of a transfer RNA in two crystal forms. *Science* **192**:853–858.

Ribosomes

Brimacombe, R., G. Stoffler and H.G. Wittmann, 1978. Ribosome structure. *Annu. Rev. Biochem.* **47**:217–249.

Chambliss, G., G.R. Craven, J. Davies, K. Davis, L. Kahan and M. Nomura, 1980. *Ribosomes: Structure, Function, and Genetics*. University Park Press, Baltimore, MD.

Craig, N.C., 1974. Ribosomal RNA synthesis in eukaryotes and its regulation. In *Biochemistry of Nucleic Acids*. K. Burton (ed.), vol. 6, *MTP International Review of Science*, pp. 255–288. Butterworths, London.

Davies, J. and M. Nomura. 1972. The genetics of bacterial ribosomes. *Annu. Rev. Genetics* **6**:203–234.

Kurland, C.G., 1977. Structure and function of bacterial ribosomes. *Annu. Rev. Biochem.* **46**:173–200.

Nierhaus, K.M., 1980. The assembly of the prokaryotic ribosome. *Biosystems* **12**:273–282.

Noller, H.F., 1984. Structure of ribosomal-RNA. *Annu. Rev. Biochem.* **53**:119–162.

Nomura, M., 1969. Ribosomes. *Sci. Amer.* **221**:28–35.

Nomura, M., A. Tissieres and P. Lengyel (eds.), 1974. *Ribosomes.* Cold Spring Harbor Laboratory, New York.

Russell, P.J., J.R. Hammett and E.U. Selker, 1976. *Neurospora crassa* cytoplasmic ribosomes: ribosomal ribonucleic acid synthesis in the wild type. *J. Bacteriol.* **127**:785–793.

Russell, P.J. and W.M. Wilkerson, 1980. The structure and biosynthesis of fungal cytoplasmic ribosomes. *Exp. Mycol.* **4**:281–337.

Udem, S.A. and J.R. Warner, 1972. Ribosomal RNA synthesis in *Saccharomyces cerevisiae. J. Mol. Biol.* **65**:227–242.

Chapter 8 Protein Biosynthesis (Translation)

Outline

Protein components
Peptide bond
Protein structure
Protein synthesis
Polypeptide chains are made in N-terminal to C-terminal direction
Protein synthesis in prokaryotes
  Initiation
  Elongation
  Termination
  Polysomes
  Relationship of transcription and translation
Protein synthesis in eukaryotes
  Initiation
  Elongation
  Termination
  Protein synthesis and cellular compartmentation

In the previous chapter, we discussed the transcription of the genes in DNA into RNA molecules. All three of the RNA classes are involved in the protein synthesis process. The mRNAs are transcripts of the structural genes that code for the production of proteins, which are composed of amino acids. The mRNA attaches to the ribosome upon which it is translated; the genetic content of the mRNA contained in the sequence of ribonucleotides is converted into a linear sequence of amino acids. The resulting protein has enzymatic, structural, or regulatory function within the cell. The conversion process from the four nucleotide language of nucleic acids to the 20 amino acid language of proteins is mediated by the *genetic code*. Thus the sequence of three nucleotides (a **codon** or triplet) in the mRNA codes for the insertion of one amino acid into a growing polypeptide chain. This ribosome-localized event involves the tRNA molecules to which the amino acids are covalently attached.

Protein components

The basic building blocks of proteins (polypeptides) are **amino acids**. There are 20 naturally occurring amino acids, and their structures are shown in Fig. 8.1.

With the exception of proline, all of the amino acids have a common structure consisting of a central carbon atom (the α-carbon) to which is bonded the α-amino group ($—NH_3^+$), an α-carboxyl ($—COO^-$), and a hydrogen atom (proton). The other part of the amino acid is called the *R group*, which varies from one amino acid to another. It is the R group that gives the amino acid its chemical properties, and the sequence of amino acids in a protein gives the protein its overall properties. Thus the general structure of an amino acid is as depicted in Fig. 8.2.

Peptide bond

Proteins consist of long chains connected together to form **polypeptide chains**. The amino acids are attached by **peptide bonds**, which are formed by the interaction of an α-carboxyl group of one amino acid with an α-amino group of another, resulting in the elimination of one molecule of water (Fig. 8.3).

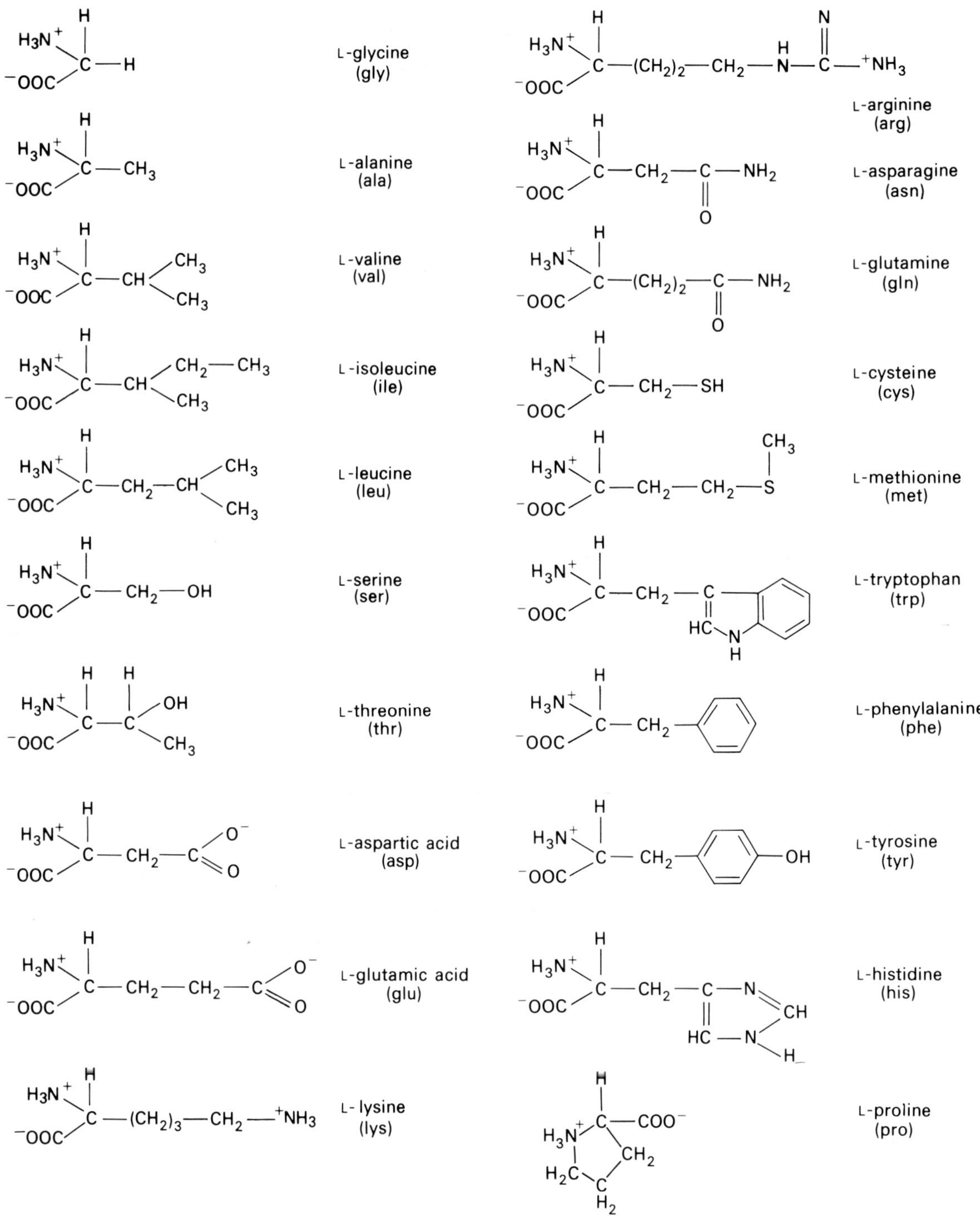

Fig. 8.1. Structures of the 20 naturally occurring amino acids.

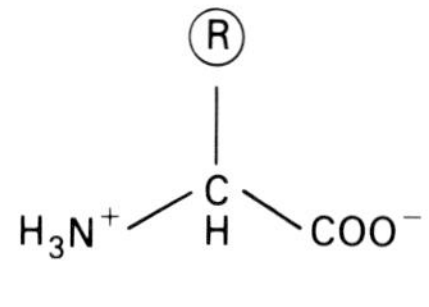

Fig. 8.2. General formula for an amino acid.

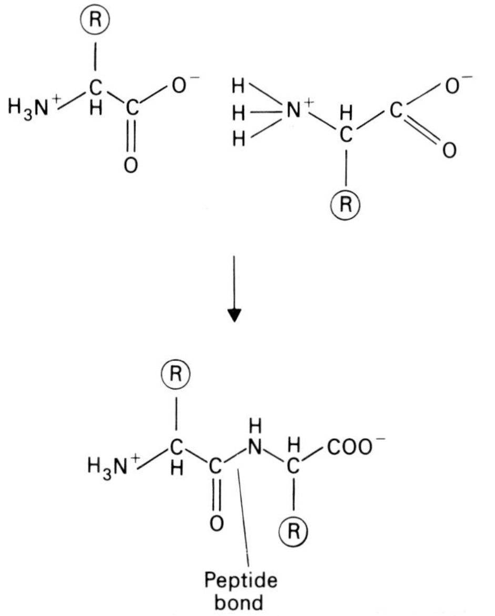

Fig. 8.3. The formation of a peptide bond.

Each polypeptide, then, has an *amino end*, with a free amino group, and a *carboxyl end*, with a free carboxyl group. These are often called the N- and C-terminals of the molecule, respectively. All of the other amino groups and carboxyl groups of the amino acids have been subsumed into the peptide bonds (Fig. 8.4).

As was the case with the nucleic acids, the backbone of the polypeptide chain carries no information; rather it is the order of the R groups and their chemical properties that gives the protein its characteristics.

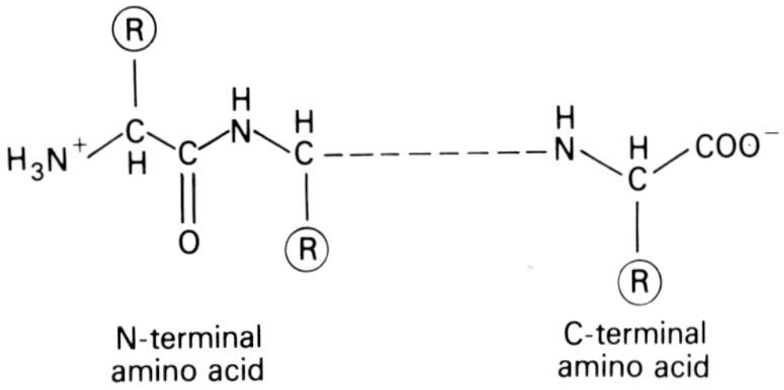

Fig. 8.4. General structure of a polypeptide chain.

Protein structure

There are four levels of protein structure:

Primary structure. This is the amino acid sequence of the polypeptide chain. This sequence determines the secondary and tertiary structure of the polypeptide chain.

Secondary structure. This is the folding of the chain into simple patterns as a result of the formation of electrostatic bonds (e.g. between carboxyl and amino groups) and hydrogen bonds between amino acids relatively close in the chain. The so-called α-helix found in many parts of the polypeptide chains is an example of secondary structure (Fig. 8.5).

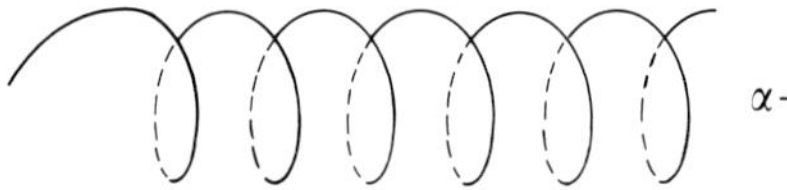

Fig. 8.5. Diagrammatic representation of an α-helix segment in a polypeptide chain.

Tertiary structure. This is the way the helices and other parts of the polypeptide are folded to make a compact globular molecule. The tertiary structure of a polypeptide (protein) often places hydrophobic (literally "water-hating") groups on the inside and hydrophilic ("water-loving") groups on the outside.

Quaternary structure. This is the way polypeptide chains are packed into the whole protein molecule if a number of chains is involved. For example, the hemoglobin molecule consists of four polypeptide chains, two α and two β, associated in the quaternary structure required for the molecule to function. (From this we can come to the realization that a protein is considered in a functional sense and may, in fact, contain more than one polypeptide chain.)

Protein synthesis

A polypeptide chain is synthesized from the N-terminal to the C-terminal end. The mRNA coding for the polypeptide is moved through the ribosome (the site of protein synthesis) starting with the 5′ end. The nucleotide sequence is read in groups of three (called codons or triplets) such that one amino acid is inserted into the polypeptide for each codon on the mRNA. A given codon is complementary (in a base-pairing sense) to an anticodon of a tRNA molecule, and each tRNA molecule carries a specific amino acid such that the correct amino acid is put into the polypeptide chain when a particular codon is exposed at the ribosome. As with other nucleic acid–nucleic acid complementary base pairings, the codon–anticodon pairing of the mRNA and tRNA involves strand oriented with opposite polarity.

As discussed in the next chapter, there are 64 possible codons, and 61 of these code are for amino acids. Thus there is usually more than one codon for a given amino acid, and there must be at least 61 tRNA molecules with the appropriate anticodons. All tRNA molecules have a similar tertiary structure that is suited for their function in protein synthesis. Nonetheless the correct amino acid becomes attached to a given tRNA so that "reading" of the codons is of high fidelity. There are 20 enzymes, called aminoacyl-tRNA synthetases, involved in the attachment of amino acids to their respective tRNAs, a reaction called *charging*. Each enzyme is specific for the attachment of one amino acid to a tRNA molecule. Thus all of the tRNA molecules for a particular amino acid, even though different anticodons may be involved, must have a common nucleotide sequence that can be recognized by the requisite, specific, aminoacyl-tRNA synthetase.

The link between an amino acid and a tRNA molecule is a covalent bond between the α-carboxyl group of the amino acid and the terminal ribose of the 3′ adenine nucleotide of the tRNA (Fig. 8.6). This linkage involves a high-energy bond, and hence it is common to talk about "activated" or "charged" tRNA. The energy in this bond is used in the formation of a

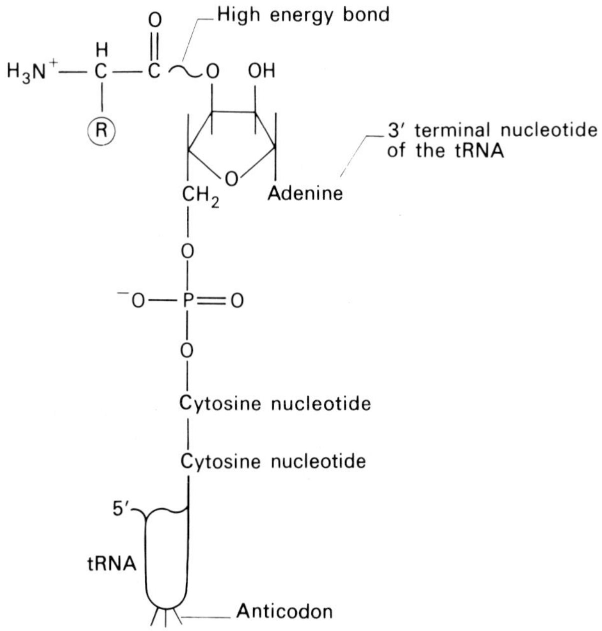

High energy bond
H_3N^+—C—C～O OH
H
O
R
3′ terminal nucleotide of the tRNA
CH_2
O
Adenine
O
^-O—P=O
O
Cytosine nucleotide
Cytosine nucleotide
5′
tRNA
Anticodon

Fig. 8.6. An aminoacyl-tRNA.

peptide bond during polypeptide chain growth. The source of energy for the bond is ATP, and the sequence of steps in forming the aminoacyl-tRNA is summarized in Fig. 8.7.

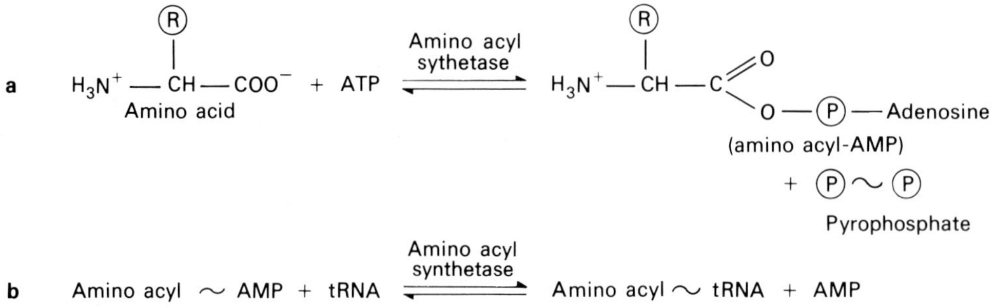

a H_3N^+ — CH(Ⓡ) — COO^- + ATP ⇌ (Amino acyl sythetase) H_3N^+ — CH(Ⓡ) — C(=O) — O — Ⓟ — Adenosine (amino acyl-AMP) + Ⓟ∼Ⓟ Pyrophosphate

Amino acid

b Amino acyl ∼ AMP + tRNA ⇌ (Amino acyl synthetase) Amino acyl ∼ tRNA + AMP

Fig. 8.7. Reactions catalyzed by aminoacyl-tRNA synthetase in the synthesis of aminoacyl-tRNA.

Polypeptide chains are made in N-terminal to C-terminal direction

A polypeptide chain is synthesized starting at the amino-terminal end. This was shown by H. Dintzis in 1960 in his work with reticulocytes of the rabbits. Reticulocytes are young red blood cells that make hemoglobin as their sole protein product. In radioactive-labeling experiments, Dintzis showed that an in vitro culture of reticulocytes made the β polypeptide of hemoglobin in about one minute. Then in a separate experiment he added radioactive leucine to the cell suspension for about 30 seconds, after which he stopped protein synthesis by quickly cooling the cells. He isolated the complete hemoglobin molecules ($\alpha_2\beta_2$) and purified them, thus removing incomplete chains from consideration. Then he separated the two types of polypeptides and broke the chains into fragments (peptides) with the aid of the enzyme, trypsin. The fragments were separated electrophoretically and the distribution of radioactivity among the fragments was examined. The reasoning here was that, since the label was present for only about half the time it took to make a complete chain, and since only whole chains were examined, then the radioactivity will only be found in peptides corresponding to the end of the chain synthesized last. Thus, if synthesis is from N to C terminus, the results depicted in Fig. 8.8 would be expected.

Here the prediction is that the label will be found toward the C-terminal end, and *no* label found at the N-terminal end. Data were obtained of this kind, thus supporting the N-to-C-terminal polarity of polypeptide synthesis.

More recent work, using in vitro protein-synthesizing systems, has confirmed Dintzis' conclusions.

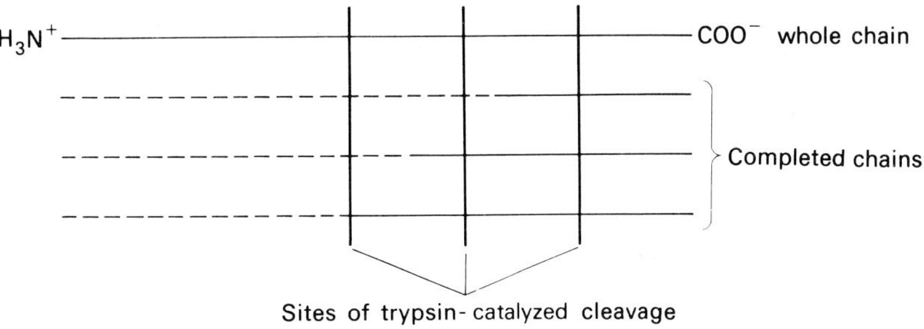

---- = unlabeled
——— = labeled

Fig. 8.8. Demonstration that polypeptides are made in the N-terminal to C-terminal direction.

Protein synthesis in prokaryotes

There are three major steps in protein synthesis: initiation, elongation, and termination. These will be discussed in turn.

Initiation

Initiator tRNA. The first amino acid in the synthesis of all bacterial polypeptides is N-formyl-methionine (fmet), which is a modified methionine amino acid in which the α-amino group is "blocked" and therefore cannot participate in peptide bond formation. The formyl group is added to the methionine after the amino acid has become attached to a specific tRNA, called tRNA.fmet. This reaction is catalyzed by the enzyme *transformylase* (Fig. 8.9). In many cases the fmet that starts a polypeptide chain is subsequently removed by enzymatic action.

Biochemical analysis has shown that at least two species of tRNA that can be charged with methionine are present in all prokaryotic organisms: one is the tRNA involved with initiation, and the other species is responsible for the insertion of methionine elsewhere in the polypeptide chain. The latter tRNA is designated tRNA.met. Both of these tRNAs in bacteria are aminoacylated by the same enzyme, but only tRNA.fmet is a substrate for the transformylase-catalyzed reaction. Both tRNAs read AUG (the only methionine codon), but in addition tRNA.fmet can recognize GUG and UUG codons. RNA-sequencing studies have shown that both molecules have an anticodon that is complementary to AUG. The two tRNA molecules do differ in some other properties. For example, the binding of fmet-tRNA.fmet to ribosomes is catalyzed by an initiator factor (discussed later), whereas the binding of met-tRNA.met is catalyzed by elongation factors. The two tRNAs apparently bind to the ribosome at different sites. Thus it is clear that fmet.-tRNA.fmet must have a structure that is specific for its role in initiation.

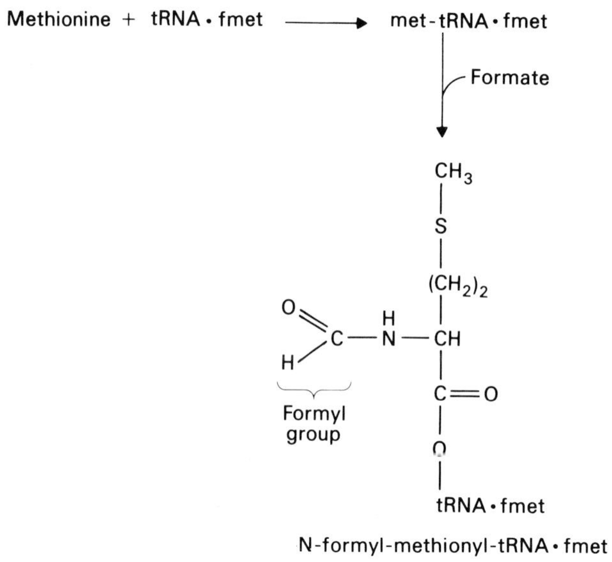

Methionine + tRNA • fmet ⟶ met-tRNA • fmet

Formate

CH_3 – S – $(CH_2)_2$ – CH(NH–C(=O)H) – C(=O) – O – tRNA • fmet

Formyl group

N-formyl-methionyl-tRNA • fmet

Fig. 8.9. Synthesis of N-formyl-methionyl-tRNA.

Ribosome binding sites. In bacteria, the first step in initiation is the formation of a complex between the 30S ribosomal subunit, fmet-tRNA, and an mRNA molecule. The 50S subunit is added later to form the active 70S ribosome (monosome). The mRNA may contain information for one to several distinct polypeptide chains. For each of the segments coding for a polypeptide, there is a specific nucleotide sequence for orienting the mRNA correctly and in the right reading frame on the ribosome. These sequences are called the **ribosome binding sites.**

Fig. 8.10 shows the sequence of some prokaryotic ribosome binding sites.

Fig. 8.10. Some prokaryotic ribosome binding sites. The initiation codon, AUG, is boxed. The larger boxed regions indicate the regions of contiguous complementarity (including allowable G-U base pairs) to the 3′ end of 16S rRNA.

Message origin	Ribosome binding site sequence
E. coli lac Z	UUC ACA CAG GAA ACA GCU AUG ACC AUG AUU
E. coli trp B	AUA UUA AGG AAA GGA ACA AUG ACA ACA UUA
E. coli RNA polymerase β	AGC GAG CUG AGG AAC CCU AUG GUU UAC UCC
Phage λ *cro*	AUG UAC UAA GGA GGU UGU AUG GAA CAA CGC

Most of the ribosome binding sites have a purine-rich sequence about 8 to 12 bases upstream from the AUG start codon. This sequence and other bases in this region are complementary to a pyrimidine-rich region, including at least CCUCC at the 3′ end of 16S rRNA. The mRNA region that binds in this way is called the *Shine-Dalgarno sequence* after the discoverers of this relationship. Thus it appears that the formation of complementary base pairs between mRNA and 16S rRNA in the 30S ribosomal subunit allows the ribosomes to locate and bind to the initiator regions in the mRNA.

Initiation factors and initiation. In addition to mRNA, fmet-tRNA, and ribosomal subunits, three protein **initiation factors** (IF-1, IF-2, and IF-3) and GTP are required for the initiation process to occur. First the properties of the initiation factors are discussed and then the scheme proposed for the initiation process in protein synthesis is presented.

1. *IF-3.* The IF-3 factor weighs 23,000 daltons and functions in binding mRNA to the 30S subunit. It also acts as a dissociation factor for separating the 30S and 50S subunits after polypeptide synthesis is complete. Like all the IFs, IF-3 is found bound to free 30S subunits and can be released by washing the subunits in 0.5 M ammonium chloride.

Experiments with radioactive IF-3 have shown that it is capable of binding to both 30S subunits and to mRNA molecules. In an in vitro protein-synthesizing system, IF-3 enhances the binding of fmet-tRNA to mRNA.30S subunit complexes. It is attractive to suppose that IF-3 recognizes mRNAs by the AUG or GUG initiation codons, but there is no solid evidence on this point.

In summary, the initiation reaction in which IF-3 is involved is:

$$\text{IF-3} + \text{mRNA} + \text{30S subunit} \rightarrow \text{(IF-3.mRNA.30S) complex}$$

2. *IF-2.* The 80,000 dalton IF-2 protein is involved with the binding of the initiator tRNA to the IF-3.mRNA.30S complex. The high-energy molecule GTP is used in this reaction. In vitro experiments have shown that IF-2 and GTP will bind to form a complex that is stabilized when it in turn forms a complex with fmet-tRNA. This latter complex then binds with the IF-3.mRNA.30S complex and the IF-1 protein factor (9000 daltons) to form the 30S initiation complex (Fig. 8.11).

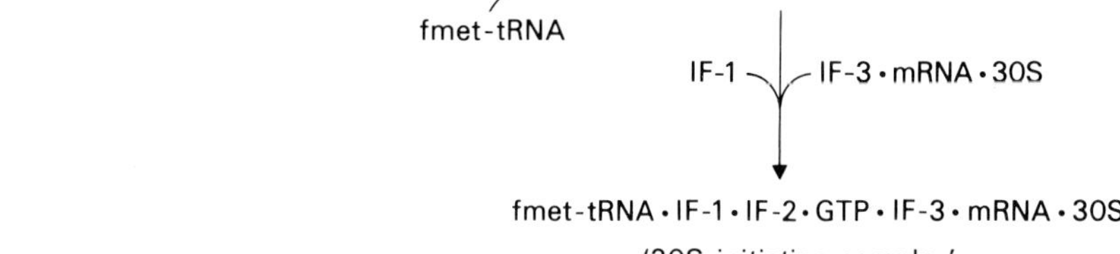

Fig. 8.11. Initiation of protein synthesis: steps in the formation of the 30S initiation complex.

3. *Dissociation of initiation factors from the initiation complex.* The initiation factors function to bring fmet-tRNA, mRNA and 30S subunits into a stable association. The next step is the addition of a 50S subunit to form a 70S initiation complex. This leads to the hydrolysis of GTP to GDP+ Ⓟ and the release of three initiation factors (Fig. 8.12). The factors can then be used for further initiation reactions on the same or different mRNA.

A summary of the initiation steps is given in Fig. 8.13.

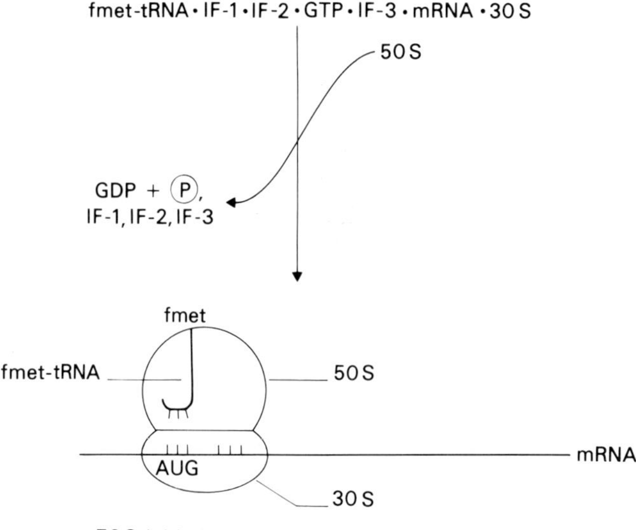

Fig. 8.12. Initiation of protein synthesis: addition of 50S ribosomal subunit to 30S initiation complex leads to formation of 70S ribosome in frame on the mRNA.

Elongation

The 70S ribosome has two sites for binding aminoacyl-tRNA. In protein synthesis, charged tRNA binds first to a site called the A (aminoacyl) site. Then the amino acid it carries becomes joined to the growing polypeptide chain carried by the tRNA at the site called the P (peptidyl) site by the formation of a peptide bond.

It is not known whether the fmet-tRNA enters the A site and then moves to the P site or whether it enters the P site directly. Before further protein synthesis can occur, however, the fmet-tRNA must become located in the P site hydrogen-bonded to the start codon on the mRNA. Once this has occurred a cyclic sequence of events commences in which one amino acid at a time is added to the growing polypeptide chain. This is called *elongation* and is summarized in Fig. 8.14. These steps will now be discussed in more detail.

Binding of aminoacyl-tRNA. The charged tRNA with the complementary anticodon to the codon in the reading frame of the A site becomes bound to that site of the ribosome in a reaction requiring **elongation factor** T (EF T) and GTP. This factor can be isolated from the soluble proteins of *E. coli*, and by column chromatography it can be separated into two polypeptides: Ts, which is stable and weighs about 30,000 daltons, and Tu, which is unstable and weights 42,000 daltons.

EF-T has been shown to bind with GTP and this is postulated to bring about the dissociation of the factor into the two polypeptides, resulting in the

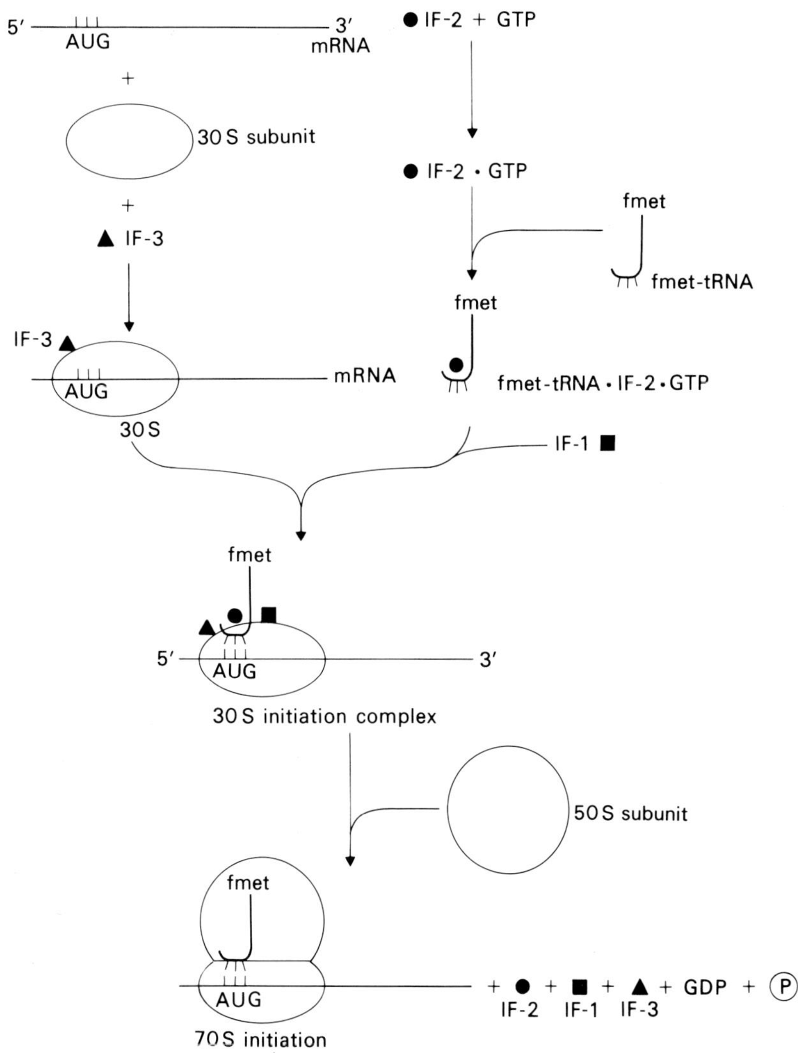

Fig. 8.13. Summary of the steps in the initiation of protein synthesis in prokaryotes. (After J. D. Watson, 1977, *The Molecular Biology of the Gene*. Benjamin/Cummings, Menlo Park.)

formation of an EF-Tu.GTP complex and releasing free EF-Ts. The next step in the elongation process is the binding of aminoacyl-tRNA to the complex to produce an aminoacyl-tRNA.Tu.GTP complex. There is evidence that this complex is an intermediate in aminoacyl-tRNA binding to ribosomes. Once the charged tRNA is bound in the A site, GTP is hydrolyzed as a result of the enzymatic action of one or more 50S ribosomal proteins. This hydrolysis causes the release of EF-Tu in a complex with GTP. The latter is released and the elongation factor can reassociate with EF-Ts. The process can then be repeated with another aminoacyl-tRNA. These events are summarized in Fig. 8.15.

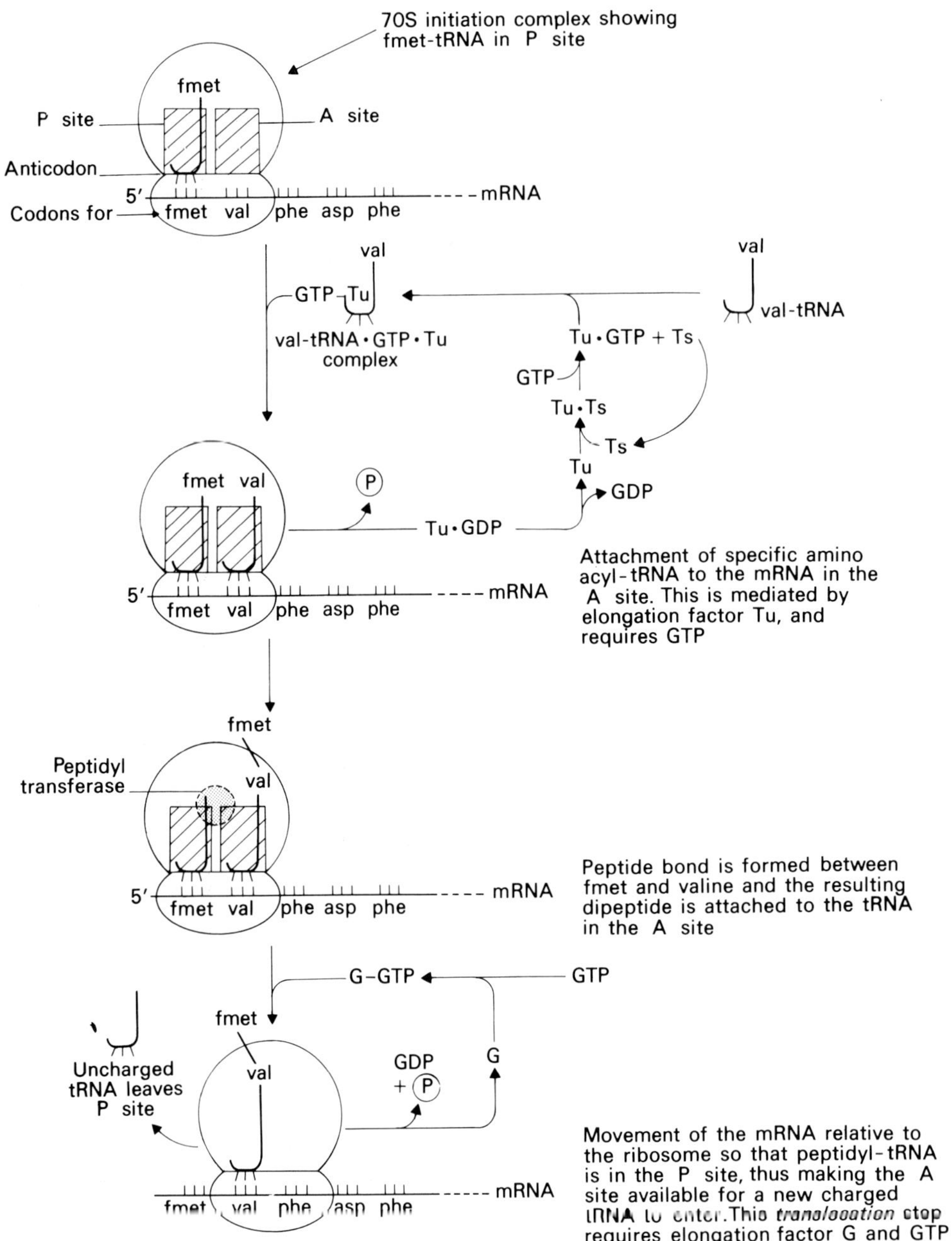

Fig. 8.14. Summary of the elongation (peptide bond formation) and translocation steps in protein synthesis. (After J. D. Watson, 1977. *The Molecular Biology of the Gene*. Benjamin/Cummings, Menlo Park.)

a Tu·Ts (Elongation factor T) + GTP ⇌ Tu·GTP + Ts

b Tu·GTP + aa-tRNA (amino acyl-tRNA) ⟶ aa-tRNA·Tu·GTP complex

c aa-tRNA·Tu·GTP + active 70S ribosome ⟶ aa-tRNA·70S (charged tRNA enters A site) + Tu·GDP + P_i (released from ribosome)

d Tu·GDP + Ts ⟶ Tu·Ts

Fig. 8.15. The reactions in protein synthesis involving the prokaryotic elongation factors EF-Tu and EF-Ts.

Experiments with an analog of GTP that cannot be hydrolyzed have shown that GTP hydrolysis is required for release of EF-Tu from the ribosome but it is not needed for aminoacyl-tRNA binding to the ribosome. Other experiments have shown that binding of fmet-tRNA to the ribosome does not require EF-Tu.

Peptide bond formation. At the beginning of this stage, a tRNA carrying the growing polypeptide chain is located in the P site, and an aminoacyl-tRNA is located in the A site. These tRNAs are maintained in positions conducive for peptide bond formation of the hydrogen bonds between the respective codons and anticodons and by the tertiary structure of the ribosome. The peptide bond is formed with the aid of the enzyme, *peptidyl transferase*, which is a ribosomal protein of the 50S subunit. The end result of the reaction is that the polypeptide chain is one amino acid longer, and the growing polypeptide chain has been transferred from the tRNA in the P site to the tRNA in the A site. The tRNA in the P site, which now has no amino acid bound to it, is called an *uncharged* tRNA. The reactions are presented schematically in Fig. 8.16.

Translocation. Once the peptide bond has been formed and the polypeptide chain is on the tRNA in the A site, the next step is advancement of the ribosome precisely one codon (three nucleotides) down the mRNA, a process called *translocation*. During this translocation event, the peptidyl-tRNA remains attached to the mRNA by codon-anticodon–pairing properties and thus becomes located in the P site. The A site is then vacant, and the aminoacyl-tRNA specified by the new codon there becomes bound by the process already described. The uncharged tRNA left in the P site after peptide bond formation is also released from the ribosome during translocation.

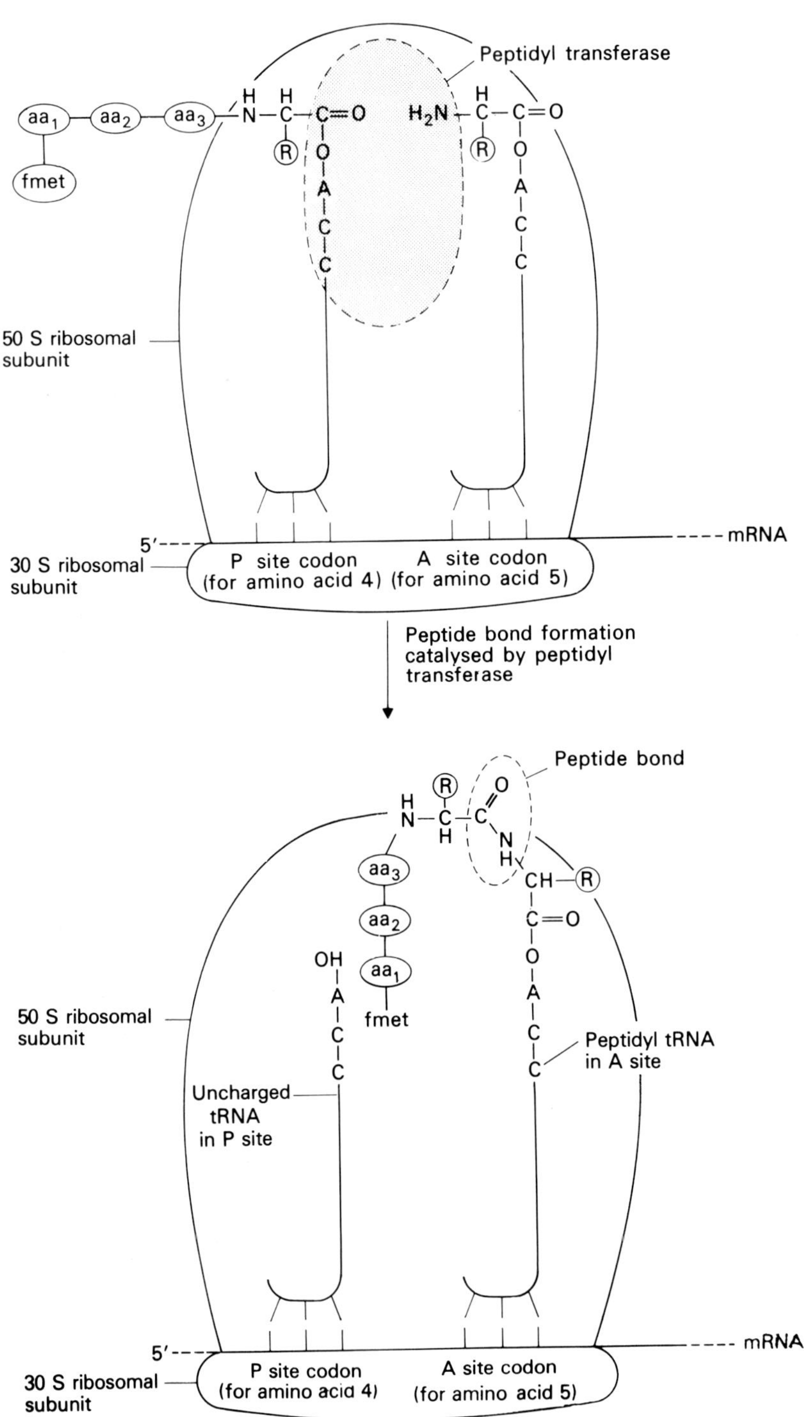

Fig. 8.16. Diagrammatic representation of peptide bond formation on ribosomes catalyzed by peptidyl transferase.

Elongation factor G (EF-G), a 72,000–84,000 dalton protein, and GTP hydrolysis are needed for translocation to occur, but it is not yet known how the translocation mechanism works. One GTP molecule is hydrolyzed for each translocation event. It appears that EF-G leaves the ribosome after translocation, since EF-Tu and EF-G cannot interact with the ribosome at the same time.

Termination

The end of the polypeptide chain is indicated on a mRNA molecule by a specific **chain-terminating** (stop) **codon**. Three such codons are known: UAA, UAG, and UGA. No naturally occurring tRNA has an anticodon for any of these stop codons, and therefore no amino acid can be put into the polypeptide. Three specific termination factors have been shown to be involved in reading the stop codon. They differ in their codon specificity and GTP requirement (Table 8.1).

Table 8.1. Properties of prokaryotic termination factors.

Termination factor	Molecular weight (daltons)	Stop codons recognized	GTP requirement
RF1	44,000	UAA and UAG	No
RF2	47,000	UAA and UGA	No
RF3	46,000	None	Yes

RF1 and RF2 have overlapping specificities for the stop codons. They have been shown to interact with the termination codons by interaction at the A site. The RF3 factor apparently plays a stimulatory role in RF1 and RF2 activity. There is some evidence for a GTP requirement in the RF3 factor's activity. In any event, chain termination, as mediated by these factors, involves the cleavage of the carboxyl group of the C-terminal end of the polypeptide chain from the tRNA in the P site. This results in the release of the polypeptide and the now uncharged tRNA. The ribosome will then move along the mRNA until a new initiation sequence is encountered (as it may be in polycistronic mRNAs), or it will dissociate from the mRNA. If none is found, when the ribosome is released from the mRNA, IF-3 functions to keep the two subunits apart. Thus, when a new 70S initiation complex is formed, the two subunits are drawn randomly from the free pools of 30S and 50S subunits.

While the polypeptide chain is being synthesized, the primary sequence of amino acids directs the three-dimensional shape. In other words, the elongating chain begins to assume its final shape as it is being made. Indeed, some enzyme activity can be detected on ribosomes that have not yet completed the synthesis of an enzymatic polypeptide.

Polysomes

Efficient translation of an mRNA molecule cannot be achieved by a single ribosome moving along it. The amount of space a ribosome takes up on a mRNA is relatively small, and thus several ribosomes can work on the mRNA at once. The association of a number of ribosomes on a single mRNA chain is called a *polyribosome* or *polysome*, and this allows several polypeptide chains to be made from each mRNA. The length of the polypeptide chain on a given ribosome will be directly proportional to how far the ribosome has moved along the mRNA from the 5′ end of the molecule. The existence of polysomes explains why a cell needs so little mRNA, while at the same time it contains so much more protein (Fig. 8.17).

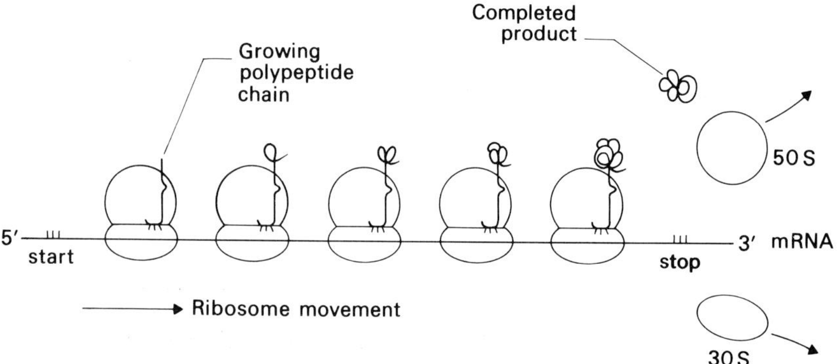

Fig. 8.17. Diagrammatic representation of a polysome engaged in protein synthesis.

Relationship of transcription and translation

In bacteria the mRNA typically becomes associated with ribosomes while synthesis of the mRNA molecule is continuing (see Fig. 7.6). This is possible owing to the lack of a nuclear membrane so that as the 5′ end of the growing mRNA molecule is displaced from the DNA as the double helix reforms, the ribosome binding site becomes available. Ribosomes then load onto the mRNA in rapid sequence, the first being close behind the RNA polymerase (Fig. 8.18).

To give some idea of the rates of these processes, the mRNA for the tryptophan biosynthetic operon (see Chapter 19) is transcribed at a rate of about 1000 nucleotides per minute and the translation process proceeds at about the same rate. Thus approximately 350 amino acids can be polymerized into polypeptide chains each minute.

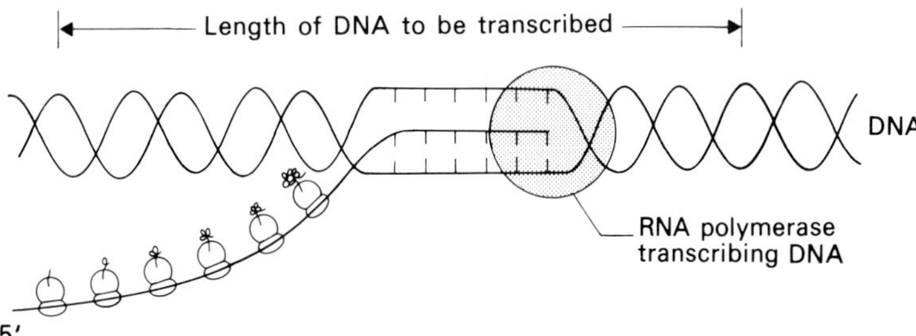

Fig. 8.18. Schematic of the possible translation of an mRNA while it is still being transcribed in prokaryotes.

Another important aspect of translation is the lability of the mRNA. As mentioned in an earlier chapter, prokaryotic mRNAs are considered to be short-lived in that degradation of the molecule by 5′-exonuclease action competes with the ribosome-mediated initiation of protein synthesis. Continued mRNA synthesis, then, is necessary for continued polypeptide synthesis.

Protein synthesis in eukaryotes

The steps and mechanisms of protein synthesis are similar in both eukaryotes and prokaryotes. As previously discussed, the ribosomes are different, and this is also the case with the soluble protein factors.

Initiation

Initiation of protein synthesis involves the binding of mRNA to the ribosomes. In eukaryotes it has now been established that the 5′-cap structure is necessary to get efficient binding but the 3′ poly(A) sequence apparently is not needed. There is good evidence, however, that the poly(A) sequence stabilizes the mRNA during the translation process. The precise mechanism whereby the eukaryotic message binds to the ribosome is not known, although it is most likely that RNA–RNA and RNA–protein interactions are involved. To date the nucleotide sequence has been determined to the 5′ side of the AUG start codon for a number of eukaryotic mRNAs and there are very few common features. Strikingly, while the last 50 nucleotides of *E. coli* 16S rRNA and eukaryotic 18S rRNA are highly homologous and similar secondary structure models can be built, the eukaryotic rRNA does *not* have the CCUCC sequence characteristic of prokaryotic rRNA. Also, there is no Shine–Dalgarno sequence in the eukaryotic mRNA. Thus it appears that, unlike the case in prokaryotes, there is not a fixed nucleotide sequence that plays a role in ribosome binding to the message.

As in prokaryotes the initiation codon is AUG, and a special initiator methionyl-tRNA recognizes that signal in the message. Unlike its prokaryotic counterpart the methionine carried by the tRNA does not become formylated, since the appropriate enzyme system does not exist in eukaryotic cells. The initiator tRNA can be distinguished from the met-tRNA that reads AUG codons elsewhere in the message, however, by the fact that it can be formylated in vitro in the presence of an *E. coli* extract. Hence in eukaryotes it is also appropriate to define the two methionine-accepting tRNAs as tRNA.fmet and tRNA.met.

At least in mammals, there are many more initiation factors (labeled eIFs for eukaryotic initiation factors) than in prokaryotes. In many cases the proteins have not been purified to homogeneity, and thus their absolute roles in protein synthesis are uncertain. Between them all, they carry out the initiation events performed by the three prokaryotic IFs. It remains to be seen to what extent the situation in mammals is generalizable throughout the eukaryotes.

Elongation

As in prokaryotes, there are two elongation factors: eEF-1 (equivalent to prokaryotic EF-T) and eEF-2 (equivalent to prokaryotic EF-G). eEF-1 has been studied in a number of systems, and in general it exists in multiple forms. Purified eEF-1 from rabbit reticulocytes, for example, has a molecular weight of 186,000 and consists of three subunits weighing 62,000 daltons each. In some systems the subunits aggregate to produce molecules of greater than 1 million daltons. Regarding the function of eEF-1, much less is known than in the prokaryotes. The eEF-1 from rabbit reticulocytes, for example, has been shown to bind to aminoacyl-tRNA and to GTP, and thus to facilitate binding of the aminoacyl-tRNA to the A site in ribosomes. During this step, GTP is hydrolyzed to GDP as a result of GTPase activity of the elongation factor, and an eEF-1.GDP complex is released from the ribosome.

The eukaryotic eEF-2 is similar to prokaryotic EF-G, although the two are not interchangeable in in vitro systems. The factor from rabbit reticulocytes has a molecular weight of 96,500–110,000 daltons and, after binding with GTP, binds to the ribosome. This event results in hydrolysis of GTP to GDP, translocation of the ribosome one codon down the message, and release of an eEF-2.GDP complex. All these steps are similar to the events that take place in prokaryotes, one exception being that the eukaryotic factor forms a stable complex with GTP whereas the prokaryotic factor does not.

Termination

The same chain termination codons are functional in eukaryotes as in prokaryotes. One release factor has been identified in rabbit reticulocytes, and this has a molecular weight of 115,000 daltons and may be a dimer. This factor recognizes all three chain termination codons and it requires GTP to carry out its function. No stimulatory factor analogous to the prokaryotic RF-3 has been found in eukaryotes.

Protein synthesis and cellular compartmentation in eukaryotes

In prokaryotes there is no nuclear membrane to separate the transcription process from the translation process. The presence of a nuclear membrane and the various modification processes peculiar to mRNAs in eukaryotes present many levels at which the regulation of gene expression can be affected. These include transcription itself, the processing of the primary transcript to produce mature mRNA, RNA-RNA splicing, and the movement of mRNA from the nucleus to the cytoplasm.

In eukaryotic cells, the cytoplasm contains a network of interconnecting channels bounded by membranes called the *endoplasmic reticulum* (ER). The membranes involved are continuous and may in fact connect to the nuclear membrane and the cell membrane. Close examination of the ER reveals that it is differentiated into two types, *smooth* (SER) and *rough* (RER), which are distinguished by the fact that the latter has ribosomes bound to it (hence the rough appearance) whereas the former does not. Thus, ribosomes in the cytoplasm are either membrane bound or free. The membrane-bound ribosomes synthesize proteins that are either secreted from the cell or that are packaged in lysosomes (where the proteins degrade other proteins). Thus, for example, pancreatic cells that secrete enzymes into the intestine are extremely rich in RER. The free ribosomes synthesize all other proteins found in the cell, that is, those in the cytoplasm, nucleus, mitochondria and chloroplasts (if present). We shall now consider the cellular compartmentation of proteins in more detail.

Fig. 8.19 presents a diagrammatic view of the secretion system of a eukaryotic cell. The proteins to be secreted are made on the ribosomes of the RER and then transferred across the membrane into the channel system. The mRNAs for the secreted proteins must somehow become associated specifically with the ribosomes of the RER. In 1975, G. Blobel and B. Dobberstein proposed a **signal hypothesis** to explain this. They suggested that there is a unique sequence of codons located to the 3′ side of the initiator AUG codon which is present only in mRNAs for proteins that must be transferred across membranes. Translation of these codons results in a specific amino acid sequence at the N-terminal end of the protein. Then they postulated that the special end of the protein facilitates the attachment of the ribosome to the membrane so that the protein can be transferred across it. Once the protein has been completed, they proposed that the ribosome dissociates from the ER.

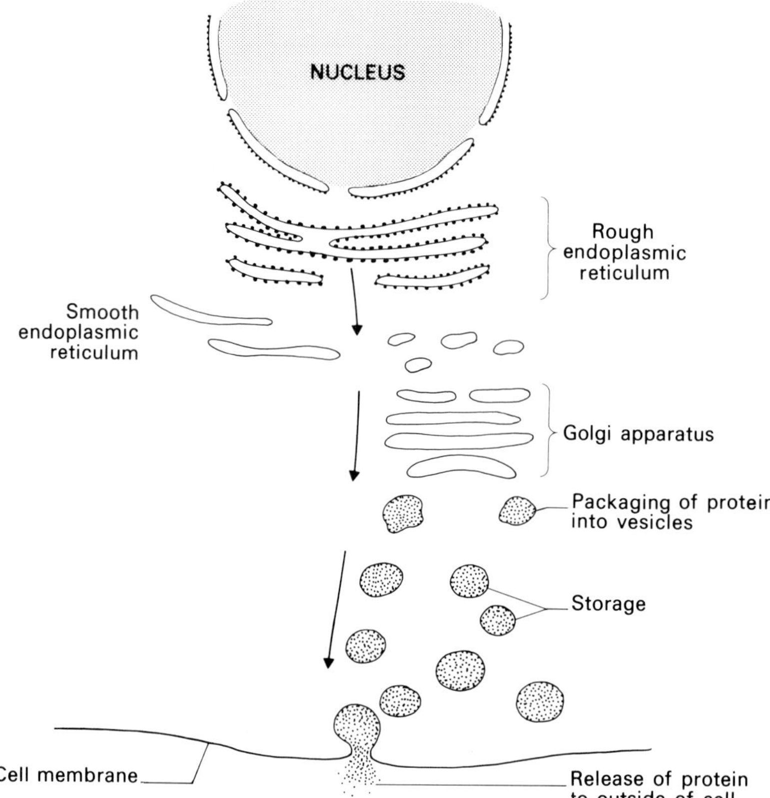

Fig. 8.19. Diagrammatic representation of a cell showing the steps involved in the secretion of particular proteins. The steps are discussed in the text. The overall path of the protein to be secreted is shown by the arrows. In brief, the protein is synthesized on the RER and is transported into the channel network. Next it is packaged into vesicles in the Golgi apparatus. The vesicles are stored in the cell and the contained proteins can be released from the cell by fusion of the vesicle with the cell membrane.

Recent work has shed more light on the processes involved. The growing (nascent) proteins that are to be secreted or inserted into membranes all have an N-terminal extension of about 15–30 amino acids called the *signal sequence*. Only proteins that have signal sequences can be cotranslationally transferred across or inserted into the membrane of the RER. When a nascent secretory or membrane protein exposes its signal sequence, a cytoplasmic receptor protein called the *signal recognition protein* (SRP) recognizes it, binds to it, and blocks its further translation. This translation arrest persists until the nascent protein-SRP-ribosome complex reaches and binds to the ER. A membrane protein called the *SRP receptor* (also called the *docking protein*) is the site for binding of the SRP-blocked ribosome with the ER. The direct interaction between the SRP on an arrested ribosome and the SRP receptor then causes a release of the translation arrest. The growing polypeptide is then translocated through the membrane into the lumen of the ER. This translocation step may require membrane proteins, although it is not clear if the polypeptide goes through a porelike structure or if a localized membrane alteration is involved. Once inside the ER, the signal sequences are removed by the action of *signal peptidase*. Once the protein is completely synthesized and translocated into the ER, the protein is then modified as necessary to facilitate its secretion. This usually involves the addition of specific carbohydrate residues.

Once in the channels of the RER the signal sequence is removed and the proteins to be secreted by the cell move toward the perimeter of the cell, becoming concentrated in the Golgi apparatus, which are stacks of flat membranous sacs. In the Golgi apparatus, the aggregated proteins are wrapped by a single membrane and the resulting vesicles migrate toward the cell surface where, by fusing with the cell membrane, the contents (the protein to be secreted) are released to the outside of the cell. Lysosomal proteins are synthesized and translocated into the ER in the same way as for secreted proteins. In this case, however, the proteins become packaged into vesicles—the lysosomes—which remain in the cytoplasm.

Mitochondrial proteins. Most proteins destined for mitochondria are synthesized as precursor proteins with specific signal sequences that are recognized by the organelles and that allow the proteins to be taken up by mitochondria. However, unlike secreted proteins, mitochondrial proteins are synthesized on free ribosomes and no ribosomes associate with the mitochondria themselves. Once inside the mitochondria, the signal sequences are cleaved from the proteins.

Interestingly, 30 to 40 inherited metabolic disorders are characterized by specific deficiencies of imported mitochondrial enzymes. It may be hypothesized that some or all of these disorders result from mutations affecting the signal sequences.

Nuclear proteins. The uptake of proteins by nuclei is extremely selective. Like the other cases we have discussed, nuclear proteins have simple signal sequences that are specific for translocation into the nucleus. There is good evidence that the recognition of the signal sequence is by some component of the nuclear envelope. The nuclear envelope itself is a double membrane in which there exists nuclear pore structures about 60 to 100 nm in diameter. There is some evidence that nuclear proteins enter the nucleus via the nuclear pore complex.

Unlike the other compartmentalized proteins that we have just discussed, however, nuclear proteins are distinguished by the fact that their signal sequences are not removed once they enter the nucleus. The reason for this is as follows: Each time the cell divides, the nuclear envelope is degraded and it reforms prior to cytokinesis. Thus, during cell division the nuclear proteins are free in the cytoplasm, and they must retain their ability to reenter the nucleus selectively once the nuclear envelope reforms.

Questions and problems

8.1 Antibiotics have been very useful in elucidating the steps of protein synthesis. If you have an artificial messenger of the sequence AUGUUUUUUUUUUUU . . . , this will produce the following polypeptide in a cell-free protein–synthesizing system: fMet-Phe-Phe-Phe- . . . In your search for new antibiotics, you find one called putyermycin, which blocks protein synthesis. When you try it with your artificial mRNA in a cell-free system, the product is fMet-Phe. What step in protein synthesis does putyermycin affect? Why?

8.2 Describe the reactions involved in the aminoacylation (charging) of a tRNA molecule.

8.3 Compare and contrast the following in prokaryotes and eukaryotes:
(a) protein-synthesis initiation
(b) protein-synthesis elongation
(c) protein-synthesis termination

8.4 Discuss the two species of methionine tRNA and describe how they differ in structure and function. In your answer, include a discussion of how each of these tRNAs binds to the ribosome.

8.5 Antibiotics have been useful in determining whether cellular events depend on transcription or translation. For example, actinomycin D is used to block transcription, and cycloheximide (in eukaryotes) is used to block translation. In some cases, though, surprising results are obtained after antibiotics are administered. The addition of actinomycin D, for example, may result in an increase and not a decrease in the activity of a particular enzyme. Discuss how this might come about.

References

General

Blobel, G. and B. Dobberstein, 1975. Transfer of proteins across membranes. I. Presence of proteolytically processed and unprocessed nascent immunoglobulin light chains on membrane-bound ribosomes of murine myeloma. *J. Cell. Biol.* **67**:835–851.

Chambliss, G., G.R. Craven, J. Davies, K. Davis, L. Kahan and M. Nomura (eds.), 1980. *Ribosomes: Structure, Function, and Genetics.* University Park Press, Baltimore, MD.

Gassen, H.G., 1982. The bacterial ribosome: a programmed enzyme. *Angew. Chem. Int. Engl.* **21**:23–36.

Haselkorn, R. and L.B. Rothman-Denes, 1973. Protein synthesis. *Annu. Rev. Biochem.* **43**:397–438.

Meyer, D.I., 1982. The signal hypothesis—a working model. *Trends Biochem. Science* **7**:320–326.

Pestka, S., 1976. Insights into protein biosynthesis and ribosome function through inhibitors. *Progr. Nucl. Acid Res. Mol. Biol.* **17**:217–245.

Revel, M. and Y. Groner, 1978. Post-transcriptional and translational controls of gene expression in eukaryotes. *Annu. Rev. Biochem.* **47**:1079–1126.

Shatkin, A.J., A.K. Banerjee, G.W. Both, Y. Furuichi and S. Muthukrishnan, 1976.

Dependence of translation on 5′-terminal methylation of mRNA. *Fed. Proc.* **35**:2214–2217.

Smith, A.E., 1976. *Protein Biosynthesis*,. Chapman and Hall, London.

Walter, P., R. Gilmore and G. Blobel, 1984. Protein translocation across the endoplasmic reticulum. *Cell* **38**:5–8.

Weissbach, H. and S. Ochoa, 1976. Soluble factors required for eukaryotic protein synthesis. *Annu. Rev. Biochem.* **45**:191–216.

Weissbach, H. and S. Pestka (eds.), 1977. *Molecular Mechanisms of Protein Biosynthesis*. Academic Press, New York.

Initiation

Kozak, M., 1978. How do eucaryotic ribosomes select initiation regions in messenger RNA? *Cell* **15**:1109–1123.

Kreibich, G., M. Czako-Graham, R.C. Grebenau and D.D. Sabatini, 1980. Functional and structural characteristics of endoplasmic reticulum proteins associated with ribosome binding sites. *Annu. N.Y. Acad. Sci.* **343**:17–33.

Maitra, U., E.A. Stringer and A. Chaudhuri, 1982. Initiation factors in protein biosynthesis. *Annu. Rev. Biochem.* **51**:869–900.

Steitz, J.A. and K. Jakes, 1975. How ribosomes select initiator regions in mRNA: Base pair formation between the 3′ terminus of 16S rRNA and the mRNA during initiation of protein synthesis in *E. coli*. *Proc. Natl. Acad. Sci. USA* **72**:4734–4738.

Elongation

Brot, N., 1977. Translocation. In *Molecular Mechanisms of Protein Biosynthesis*. H. Weissbach and S. Pestka (eds.), pp. 375–411. Academic Press, New York.

Caskey, C.T., 1977. Peptide chain formation. In *Molecular Mechanisms of Protein Biosynthesis*. H. Weissbach and S. Pestka (eds.), pp. 443–465. Academic Press, New York.

Termination

Brot, N., W.P. Tate, C.T. Caskey and H. Weissbach, 1974. The requirement for ribosomal proteins L7 and L12 in peptide-chain termination. *Proc. Natl. Acad. Sci. USA* **71**:89–92.

Shine, J. and L. Dalgarno, 1974. The 3′-terminal sequence of *Escherichia coli* 16S ribosomal RNA; complementarity to nonsense triplet and ribosome-binding sites. *Proc. Natl. Acad. Sci. USA* **71**:1342–1346.

Tompkins, R.K., E.M. Scolnick and C.T. Caskey, 1970. Peptide chain termination, VII. The ribosomal and release factor requirements for peptide release. *Proc. Natl. Acad. Sci. USA* **65**:702–708.

Chapter 9 The Genetic Code

Outline

- Evidence for three-letter code
- Elucidation of the genetic code
- Characteristics of the genetic code
- "Wobble"
- Mutations and the genetic code

Evidence for three-letter code

The information for the amino acid sequence in polypeptides is coded in the nucleotide sequence of mRNA. The first information about this *genetic code* came from the work of F. Crick and coworkers in 1961. The complete code was not determined until several years later.

It was reasoned *a priori* that the code must be at least a triplet code. Specifically, a one-letter code in which one nucleotide codes for one amino acid could not deal with the 20 amino acids known to occur in proteins. A two-letter code would have $4 \times 4 = 16$ "words", which is still insufficient coding capacity. A three-letter code, however, generates $4 \times 4 \times 4 = 64$ code words, which is more than enough to code for the 20 amino acids. There is irrefutible evidence that the code is a triplet code. The evidence came from studies with phage T4, which, as we have discussed before, infects and lyses *E. coli.* The experiments involved two types of strains of T4, wild type and *rII*, which can be distinguished by their *plaque morphology* and their *host-range phenotype.* When phages are added to a lawn of bacteria grown on solid medium in a petri dish, the successive rounds of infection and lysis produce cleared areas in the lawn called **plaques**. Wild-type T4 (r^+) produces small turbid plaques, whereas *rII* mutants produce large clear ones (Fig. 9.1).

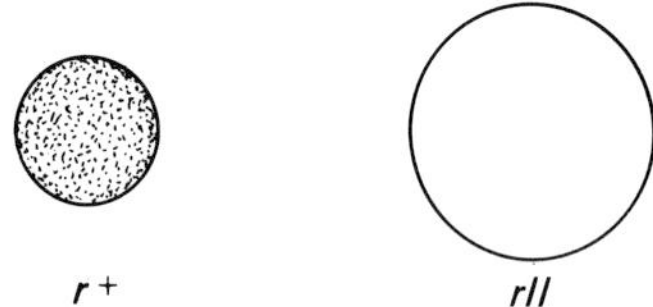

Fig. 9.1. Diagrammatic representations of the plaque morphologies of r^+ and *rII* strains of phage T4 growing on *E. coli.*

Regarding host-range properties, wild type can grow on both the *B* and *K12*(λ) strains of *E. coli*, whereas *rII* strains only grow on *B*. These two properties are summarized in Table 9.1.

Table 9.1. Host range properties of r^+ and *rII* strains of phage T4.

	Plaque morphology on *E. coli*	
Phage strain	Strain *B*	Strain *K12*(λ)
r^+	r^+ type (turbid)	r^+ type
rII	*r* type (clear)	no growth

Crick and coworkers started with an *rII* mutant that had been induced by proflavin treatment, which has the effect of causing either the addition or deletion of a base pair in the DNA. They treated the *rII* mutant with proflavin and isolated a number of r^+ revertants, which could be detected by their ability to grow on *K12*(λ). Genetic analysis showed that several of the revertants resulted from a second mutation within the *rII* gene so that the combination of the two mutations gave an almost wild-type (pseudo-wild) phenotype. (A mutation that reverses the phenotypic effects of another mutation at a different site within the same gene or in a different gene is called a **suppressor mutation.**) The second mutation alone also resulted in a *rII* phenotype, and indeed by treating a strain carrying only that mutation with proflavin, a new series of revertants were obtained, many of which carried two mutations. The explanation for these revertants is straightforward. Suppose that the wild-type DNA sequence is transcribed to produce a mRNA with the nucleotide sequence shown in Fig. 9.2. If read in groups of three, a seven-amino acid polypeptide will be produced, with each amino acid being the same. If the original proflavin-induced *rII* mutation involved a deletion (−) of a nucleotide pair in the DNA, the mRNA might be as shown in Fig. 9.3. Here only the first amino acid will be the same as the wild type and the rest will be changed. All pseudo-wild revertants of this strain presumably involved an additional (+) mutation close to the first (−) mutation. This restores the reading frame with only a few amino acids in-between being erroneous ones. The resulting protein is likely to be almost as functional as the wild-type protein (Fig. 9.4).

mRNA:	CAG	CAG	CAG	CAG	CAG	CAG	CAG
Amino acid:	1	1	1	1	1	1	1

Fig. 9.2. An hypothetical mRNA with a repeating triplet coding for a polypeptide chain made up of one amino acid type. (Assumes translation begins at same point.)

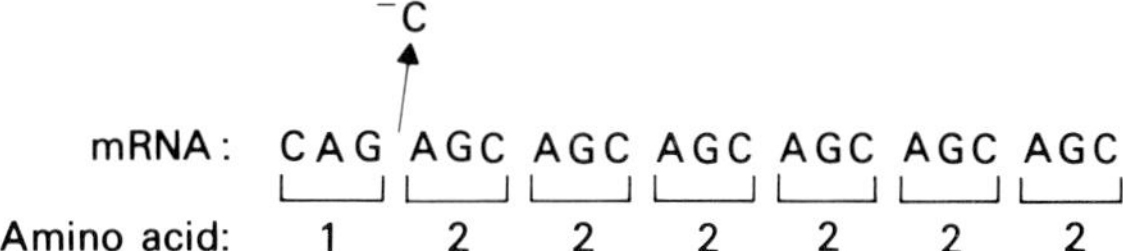

Fig 9.3. Consequences of the deletion of a nucleotide pair on the codon and amino acid sequence of Fig. 9.2.

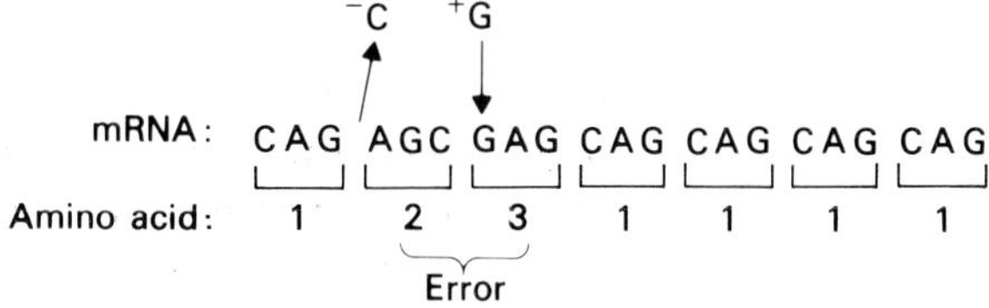

Fig. 9.4. Restoration of reading frame in the message by the addition of a nucleotide pair near the original deletion.

Similarly, all pseudo-wild revertants of the secondary (+) mutation must be deletion (−) mutations close by. In this way, then, a collection of (+) and (−) strains can be isolated. (Note that the [+] and [−] designations are arbitrary as from their experiments it was not possible to ascertain whether a particular mutation involved an addition or a deletion.) Then it was possible to combine a number of (+) mutations into one strain by genetic crosses. If the (+) mutations were close enough in the DNA (testable by genetic analysis), recombinants containing three (+) mutations sometimes gave functional products as evidenced by the strains' abilities to grow in *K12*(λ). A similar result was obtained with certain sets of three (−) mutations, whereas combinations of two (−) or two (+) did not produce pseudo-wilds. The conclusion was drawn, therefore, that the code is a *triplet code*, since the triple mutation strains restored the reading frame (Fig. 9.5).

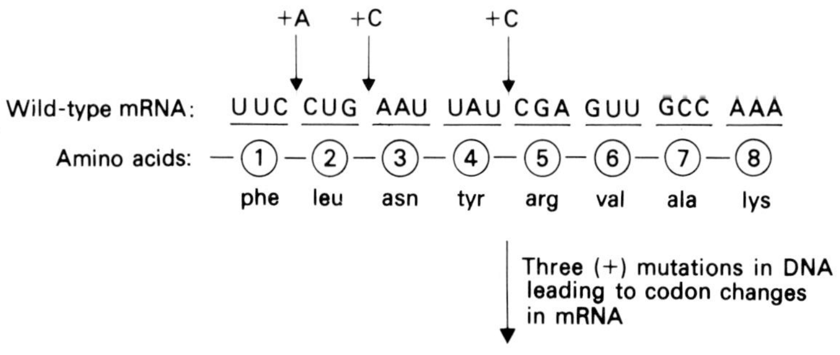

Fig. 9.5. Illustrations of how the reading frame of an mRNA can be restored by three separate, closely linked nucleotide pair additions in the DNA. The result is a net increase of one amino acid in the polypeptide.

Elucidation of the genetic code

Elucidation of which triplets code for which amino acids was largely based on the use of a cell-free, protein-synthesizing system. M. Nirenberg in 1961 found that an extract of *E. coli* containing ribosomes, tRNAs, aminoacyl synthetases, mRNA, amino acids, and other ingredients was able to incorporate amino acid into protein. This reaction only goes on for a few minutes but protein synthesis does occur again if fresh mRNA is added (Fig. 9.6).

This was an exceptionally important discovery, since once the natural mRNA was used up, an artificial, enzymatically synthesized mRNA could be introduced into the system. Such mRNAs were found to function poorly. In these experiments, the ionic conditions were altered from the normal physiological state so that translation began randomly without the need for a

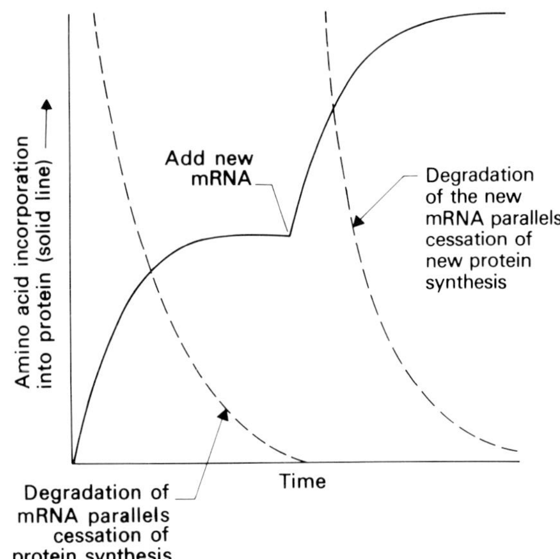

Fig. 9.6. Diagrammatic representation of results showing the dependence of an in vitro protein-synthesizing system on the presence of mRNA. Protein synthesis decreases as mRNA degrades and resumes when fresh mRNA is added. (After the work of M. Nirenberg and J. Matthaei.)

specific start codon. In the first experiments, Nirenberg and J. H. Matthaei set up a series of reaction mixtures each containing the 20 amino acids but with a different amino acid radioactively labeled in each. Each mixture also contained all other ingredients for in vitro protein synthesis, except natural mRNA. The assay mixtures were then primed with a synthetic mRNA, for example, a polynucleotide containing only one base, such as poly(U). These synthetic messengers were made using the activity of the enzyme polynucleotide phosphorylase, which catalyzes the reaction shown in Fig. 9.7. Normally the reaction goes from left to right, but in the presence of high concentrations of the diphosphate, the reaction is forced toward the left and an RNA is formed.

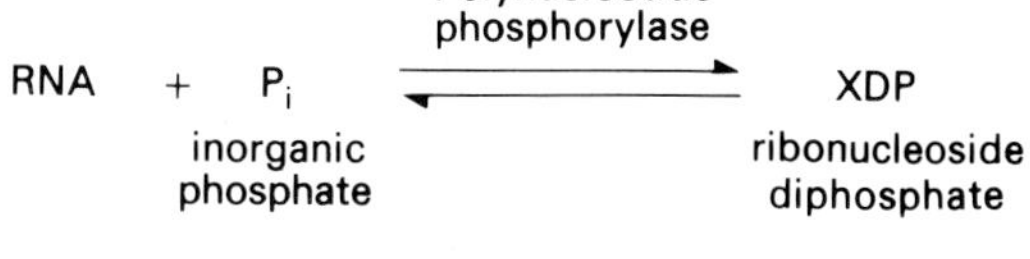

$$\underset{}{\text{RNA}} + \underset{\text{inorganic phosphate}}{P_i} \underset{}{\overset{\text{Polynucleotide phosphorylase}}{\rightleftharpoons}} \underset{\text{ribonucleoside diphosphate}}{\text{XDP}}$$

Fig 9.7. Reaction catalyzed by polynucleotide phosphorylase.

After incubation of the reaction mixtures with the mRNA, the proteins in each tube were precipitated by treatment with trichloroacetic acid (TCA) and the precipitates were then assayed for the presence of radioactivity, which would indicate the incorporation of a radioactive amino acid into the protein. Using this procedure they found that poly(U) caused the incorporation of labeled phenylalanine into the TCA-insoluble protein. Thus UUU must be a codon in the mRNA for the amino acid phenylalanine. Similar experiments showed that AAA is a codon for lysine and CCC is a codon for proline.

The rest of the codon assignments were determined by a number of in vitro experimental approaches, including the use of mRNAs made randomly from mixtures of two nucleotides and the use of mRNAs with alternating bases, such as UCUCUCUCU---, or a repeating series of three bases such as AAGAAGAAGAAG---. In the latter, three polypeptides will be produced by reading the three different possible repeating codons: polylysine (AAG), polyarginine (AGA), and polyglutamic acid (GAA).

In an elegant approach, Nirenberg and P. Leder developed the tRNA-binding technique in which the addition of synthetic trinucleotides of known sequence to a reaction mixture of ribosomes, tRNA, aminoacyl synthetases, amino acids, GTP, etc., caused the formation of a complex between an aminoacyl-tRNA, ribosome, and trinucleotide. More specifically the trinucleotide binds to the 30S subunit as does a mRNA, thus facilitating the binding of the aminoacyl-tRNA with the complementary anticodon. Thus, for example, if the trinucleotide is 5′-GAG-3′, a glutamic acid-tRNA with the anticodon 5′-CUC-3′ will become bound to the ribosome. This complex can be separated from the uncomplexed aminoacyl-tRNAs by using a filter with pores large enough to let them through but small enough to retain the complex. Thus if 20 assay mixtures are set up with a different radioactive amino acid in each, in only one tube will radioactivity be retained on the filter and that will indicate the amino acid coded by the trinucleotide being tested. For example, radioactive glutamic acid will be retained on the filter when GAG is the triplet.

None of the procedures alone produced completely unambiguous results. However, by a combination of these techniques, 61 of the 64 possible codons were assigned to specific amino acids. The three other codons we now know to be chain-terminating triplets, which do not code for any amino acid. The codon assignments of the amino acids and for chain termination in protein synthesis are called the genetic code, and this is presented in Fig. 9.8.

Characteristics of the genetic code

1. The code is *comma-free*; that is, the message is read three nucleotides at a time in a nonoverlapping way.
2. The code is *universal*; that is, all organisms use the same language. This means that a message from an animal cell will produce the same protein whether it is translated with animal cell or *E. coli* protein-synthesis machinery.
3. The code is *degenerate*; that is, with two exceptions (AUG and UGG),

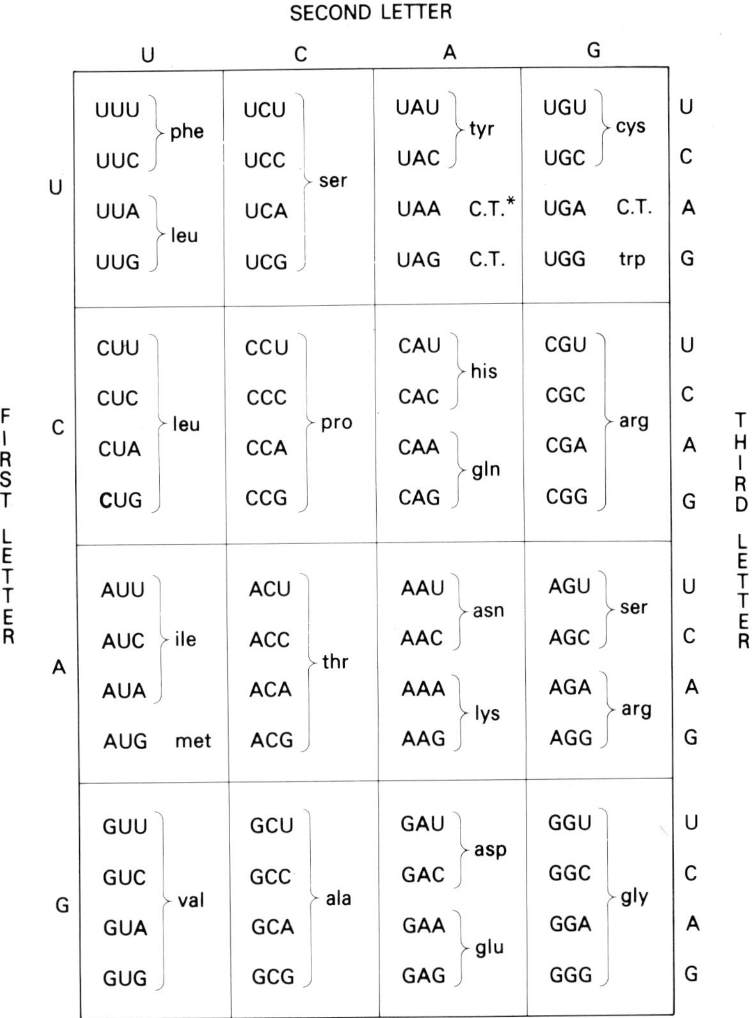

Fig. 9.8. The genetic code.

there is more than one codon for each amino acid. Some patterns exist. For example, when the first two nucleotides are identical, the third nucleotide can be either C or U and the codon may still code for the same amino acid. A similar situation prevails for A and G in the third position.

4. The code uses specific start and stop codons. AUG is the start signal and its absence from the 5′ end of the synthetic messengers explains their inefficient translation. The chain termination codons are UAG, UAA, and UGA, and in many natural mRNAs there are adjacent stop codons presumably to ensure a stop in polypeptide synthesis.

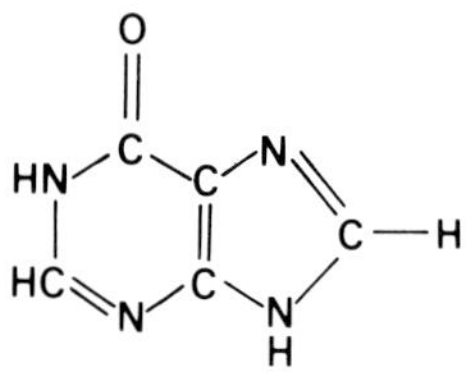

"Wobble"

Inosine

Fig. 9.9. Structure of inosine, a base found in the anticodon of some tRNAs.

It was first thought that 61 different tRNA molecules would be needed with specific anticodons for each of the amino acid–coding triplets. There is now evidence that purified tRNA species can recognize several different codons. From sequence analysis of tRNA molecules, it was found that several tRNAs have inosine as one of the anticodon bases. The structure of inosine is shown in Fig. 9.9. This has a number of base-pairing possibilities as will become apparent.

Sequence analysis also showed that the base at the 5′ end of the anticodon (complementary to the third letter of the codon) is not as sterically confined as the other two, thus allowing it potentially to pair with more than one base at the 3′ end of the codon. This is called base-pairing "wobble", and only certain pairings are possible in this regard (Table 9.2). As can be seen from the table, wobble does not allow any single tRNA molecule to recognize four different codons. As an example, Fig. 9.10 shows how a single tRNA can pair

Table 9.2. A summary of base-pairing wobble.

Base at 5′ end of anticodon	pairs with	Base at 3′ end of codon
G		U or C
C		G
A		U
U		A or G
I (inosine)		A, U, or C

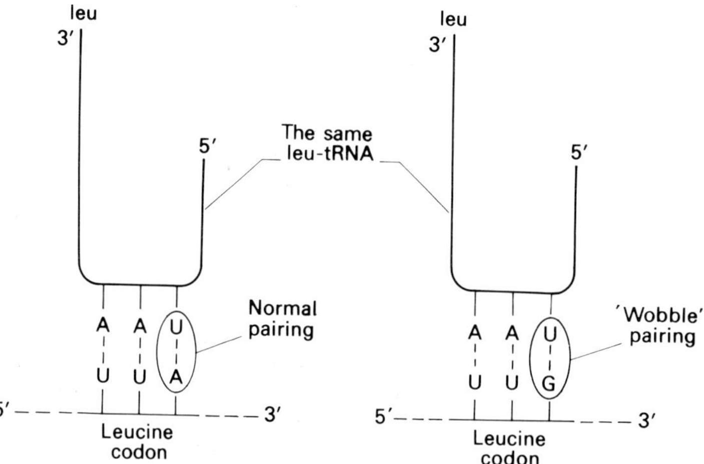

Fig. 9.10. An example of base-pairing wobble. Here a single leu-tRNA recognizes two different leucine codons.

with different codons. Here the anticodon 5′-UAA-3′ (conventionally written in the 5′-to-3′ direction) can recognize both leucine codons UUA and UUG.

Mutations and the genetic code

A previous chapter discussed the mechanism of action of a variety of chemical mutagens. In brief they bring about base-pair transitions, deletions, or additions, and the effect of these changes depends on whether an amino acid change is produced in the polypeptide that diminishes or abolishes its function. Three basic types of mutations can be defined with this in mind. All three produce a nonfunctional protein that results in a detectable mutant phenotype:

Missense mutation. Here a single base-pair change in the DNA leads to a different codon in the mRNA that may code for a different amino acid in the polypeptide. If an altered amino acid results, the effect on the function of the polypeptide will depend on the location of the amino acid in the chain.

Nonsense mutation. It is possible that a single base-pair change in the DNA will lead to the generation of a premature chain termination codon in the mRNA. The result will be a shorter than normal polypeptide chain that is likely to be nonfunctional.

Frameshift mutation. A base-pair deletion or addition in the DNA will produce a mRNA in which the reading frame is shifted one space. Past this point in the mRNA, the codons will code for a completely different set of amino acids, thus most likely resulting in a nonfunctional protein (unless the change is very close to the C-terminal end of the polypeptide).

Questions and problems

9.1 What are the characteristics of the genetic code?

9.2 Base-pairing wobble occurs in the interaction between the anticodon of the tRNAs and the codons. On the theoretical level, determine the minimum number of tRNAs needed to read the 61 sense codons.

9.3 Random copolymers were used in some of the experiments that revealed the characteristics of the genetic code. For each of the following ribonucleotide mixtures, give the expected codons and their frequencies, and give the expected proportions of the amino acids that would be found in a polypeptide directed by the copolymer in a cell-free protein–synthesizing system:

(a) 4 A:6 C
(b) 4 C:1 C

(c) 1 A:3 U:1 C
(d) 1 A:1 U:1 G:1 C

9.4 Three of the codons in the genetic code are chain-terminating codons for which no naturally occurring tRNAs exist. Just like any other sites in the DNA, though, these codons can change as a result of base-pair changes in the DNA. Confining yourself to single base-pair changes at a time, determine which amino acids could be inserted in a polypeptide by mutation of the chain-terminating codons:
(a) UAG
(b) UAA
(c) UGA
(The genetic code is in Fig. 9.8.)

9.5 The amino acid substitutions shown in the accompanying figure occur in the α and β chains of human hemoglobin. Those connected by lines are related by a proposed single nucleotide changes. Propose the most likely codon or codons for each of the numbered amino acids. (Refer to the genetic code listed in Fig. 9.8.)

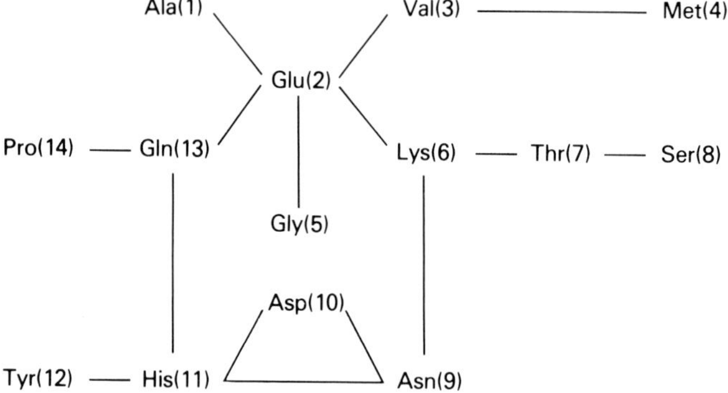

9.6 Yanofsky studied the tryptophan synthetase of *E. coli* in an attempt to identify the base sequence specifying this protein. The wild type gave a protein with a glycine in position 38. Yanofsky isolated two *trp* mutants, A23 and A46. Mutant A23 had Arg instead of Gly at position 38, and mutant A46 had a Gly at position 38.

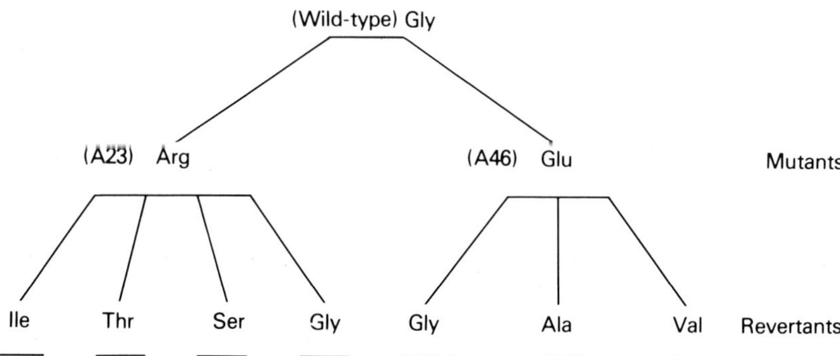

Mutant A23 was plated on minimal medium, and four spontaneous revertants to prototrophy were obtained. The tryptophan synthetase from each of four revertants was isolated, and the amino acids at position 38 were identified. Revertant 1 had Ile, revertant 2 had Thr, revertant 3 had Ser, and revertant 4 had Gly. In a similar fashion three revertants from A46 were recovered and the tryptophan synthetase from each isolated and studied. At position 38 revertant number 1 had Gly, number 2 had Ala, and number 3 had Val. A summary of these data appear in the accompanying figure. Using the genetic code shown in Fig. 9.8, deduce the codons for the wild type, the mutants A23 and A46, and the revertants, and place each designation in the space provided in the figure.

9.7 Consider an enzyme, chewase, from a theoretical microorganism. In the wild-type cell, the chewase has the following sequence of amino acids at positions 39 to 47 (reading from the amino end) in the polypeptide chain:

- Met	- Phe	- Ala	- Asn	- His	- Lys	- Ser	- Val	- Gly
39	40	41	42	43	44	45	46	47

A mutant of the organism was obtained that lacks chewase activity. The mutant was induced by a mutagen known to cause single base-pair insertions or deletions. Instead of making the complete chewase chain, the mutant makes a short polypeptide chain only 45 amino acids long. The first 38 amino acids are in the same sequence as the first 38 of the normal chewase, but the last 7 amino acids are:

- Met	- Leu	- Leu	- Thr	- Ile	- Arg	- Val
39	40	41	42	43	44	45

A "partial revertant" of the mutant was induced by treating it with the same mutagen. The revertant makes a partly active chewase, which differs from the wild-type enzyme only in the region:

- Met	- Leu	- Leu	- Thr	- Ile	- Arg	- Gly	- Val	- Gly
39	40	41	42	43	44	45	46	47

Using the genetic code given in Fig. 9.8, deduce the nucleotide sequences for the mRNA molecules that specify this region of the protein in each of the three strains.

References

Cold Spring Harbor Symposia for Quantitative Biology, vol. 31, 1966. *The Genetic Code.* Cold Spring Harbor Laboratory, New York.

Crick, F.H.C., 1966. Codon-anticodon pairing: the wobble hypothesis. *J. Mol. Biol.* **19:**548–555.

Crick, F.H.C., L. Barnett, S. Brenner and R.J. Watts-Tobin, 1961. General nature of the genetic code for proteins. *Nature* **192:**1227–1232.

Garen, A., 1968. Sense and nonsense in the genetic code. *Science* **160:**149–159.

Khorana, H.G., 1966–67. Polynucleotide synthesis and the genetic code. *Harvey Lectures* **62:**79–105.

Khorana, H.G., H. Buchi, H. Ghosh, N. Gupta, T.M. Jacob, H. Kossel, R. Morgan, S.A. Narang, E. Ohtsuka and R.D. Wells, 1966. Polynucleotide synthesis and the genetic code. *Cold Spring Harbor Symp. Quant. Biol.* **31:**39–49.

Morgan, A.R., R.D. Wells and H.G. Khorana, 1966. Studies on polynucleotides. LIX. Further codon assignments from amino acid incorporation directed by ribopolynucleotides containing repeating trinucleotide sequence. *Proc. Natl. Acad. Sci. USA* **56:**1899–1906.

Nirenberg, M. and P. Leder, 1964. RNA code words and protein synthesis. *Science* **145:**1399–1407.

Nirenberg, M. and J.H. Matthaei, 1961. The dependence of cell-free protein synthesis in *E. coli* upon naturally occurring or synthetic polyribonucleotides. *Proc. Natl. Acad. Sci. USA* **47**:1588–1602.

Nirenberg, M., T. Caskey, R. Marshall, R. Brimacombe, D. Kellog, B. Doctor, D. Hartfield, J. Levin, F. Rottman, S. Pestka, M. Wilcox and F. Anderson, 1966. The RNA code and protein synthesis. *Cold Spring Harbor Symp. Quant. Biol.* **31**:11–24.

Streisinger, G., Y. Odada, J. Emrich, J. Newton, A. Tsugita, E. Terzaghi and M. Inouye, 1966. Frameshift mutations and the genetic code. *Cold Spring Harbor Symp. Quant. Biol.* **31**:77–84.

Chapter 10 Phage Genetics

Outline

We now move from the area of molecular genetics to more classical areas of genetics, centering on transmission genetics. With the background already presented in molecular genetics, is should be possible to consider all of the material of this and future chapters in molecular terms.

Phage T4

A previous chapter discussed phages, their chromosomes, and how they infect and lyse bacterial cells. In this chapter we will concentrate our attention on phages and in particular the phage T4, which is one of the virulent phages, meaning that it will always enter the lytic cycle when it infects its host, *E. coli* (assuming that its DNA is not degraded by enzymes of the host). The structure of T4 was shown in Fig. 2.1 and the T4 life cycle was shown in Fig. 2.2.

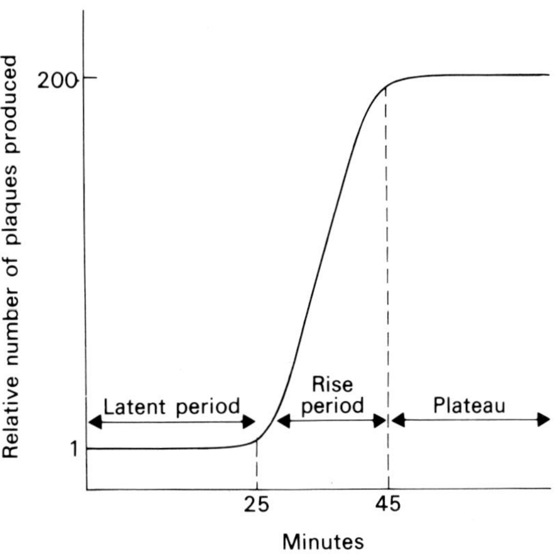

Fig. 10.1. One-step growth curve for phage T4 infection of *E. coli*.

Life cycle

In outline, a T4 phage particle makes contact with the host bacterium, and, after adsorbing to the wall or membrane by means of the tail spikes and tail fibers, the DNA is injected into the host. The DNA replicates and is tran-

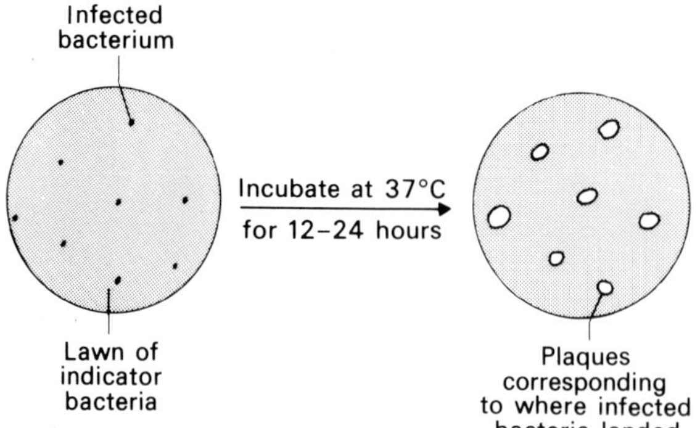

Fig. 10.2. Diagrammatic representation of the production of phage plaques on a bacterial lawn when phage-infected bacteria are spread on it.

scribed and the resulting phage-specific proteins and DNA assemble into progeny phage. At the same time a bacterial lysing enzyme (lysozyme) is made, and eventually this causes the cells to break open, releasing the progeny phages. About 200 progeny phages are produced per infected bacterium.

The life cycle of phage T4 was studied in detail by M. Delbruck in 1940. He reasoned that kinetic analysis of the steps in the life cycle required large numbers of cells in which phage infection occurred synchronously, so Delbruck allowed the phages to infect bacteria for a short time and then removed the unadsorbed phages by treatment with antibodies made against the phage particles. The infected bacterial population was incubated at 37°C, and at various times samples were taken and added to a lawn of bacteria on a petri dish. The plaques that were produced were counted and the number was graphed as a function of time, resulting in the *one-step growth curve* shown in Fig. 10.1.

The curve shows three distinct regions:

1. *Latent period.* During the latent period (which lasts about 25 min at 37°C) the bacteria have been infected and the biosynthesis of phage components is proceeding. No progeny phages are released (by definition) during this time. Thus, when samples from this period are plated on the lawn of indicator bacteria, it is actually infected bacteria that are dispersed over the plate. Each infected bacterium will eventually lyse, releasing progeny phages that will infect surrounding bacteria on the plate. After a number of rounds of infection and bacterial lysis, a plaque will be apparent on the indicator lawn (Fig. 10.2).
2. *Rise period.* At the beginning of the rise period, some infected cells have lysed, releasing progeny phages. For the next 20 minutes at 37°C the remainder of the infected cells lyse synchronously until all have lysed. Platings during this period included a mixture of infected bacteria and progeny phages.
3. *Plateau period.* All of the infected bacteria have lysed, and thus plating of the culture involves entirely progeny phages.

From one-step growth experiments, it was concluded that 25–45 minutes at 37°C are required for one life cycle of phage T4. Further experiments by A. Doermann, in which bacterial cells were broken open at various times during the latent period, showed that about 12 minutes (about half way through the latent period) are required before any mature phage particles are found within the bacterial cells. That 12-minute period involves the synthesis of the phage-specific molecules. From about 12–25 minutes, mature phages are accumulating in the cells. Cell lysis then occurs when the lytic enzyme reaches a high enough concentration.

Genetic recombination

Phage T4 contains one linear chromosome. The useful mutations of T4 fall into two main classes:

1. Those that result in visible phenotypes, usually altered plaque morphology. The disadvantage here is that relatively few of the 80 or so genes of T4 mutate to give plaque-type mutants.

2. Those that are conditionally lethal, that is, heat-sensitive or cold-sensitive mutants in which no mature progeny phage particles are produced at the respective nonpermissive temperatures (i.e. at a higher-than-normal temperature in the case of heat-sensitive mutants and at a lower-than-normal temperature in the case of cold-sensitive mutants).

These two groups of mutants have led to the establishment of a fairly complete genetic map of T4 by genetic recombination techniques.

Genetic recombination in T4 was discovered by M. Delbruck and A. Hershey in 1946. They did genetic crosses by infecting *E. coli* cells simultaneously with two types of visible mutants: one, a host-range (*h*) mutant, and the other, a rapid lysis (*r*) mutant. (Since two genes are involved in the cross, the cross is called a *two-factor cross*.) Four types of progeny phage particles were released: the two parental types (*h* and *r*), and two recombinant types (the double mutant *hr* and a wild type). In these genetic studies, it became apparent that recombination occurs between a large pool of phage DNA molecules. This and the fact that recombination occurs throughout the DNA replication phase of the life cycle make the genetic analysis more in the realm of statistics than genetics. Nevertheless, we can at least consider this process at a simple level here (Fig. 10.3).

In the example, genetic recombination will occur at random along the length of the T4 genome. Any recombination event that occurs between the two gene sites will generate two recombinant phage types, *ab* and ++ in this case. If recombination does not occur between the genes, two parental progeny phage types, *a*+ and +*b*, will be produced. Since the recombination event is random, the probability of it occurring between two genes under

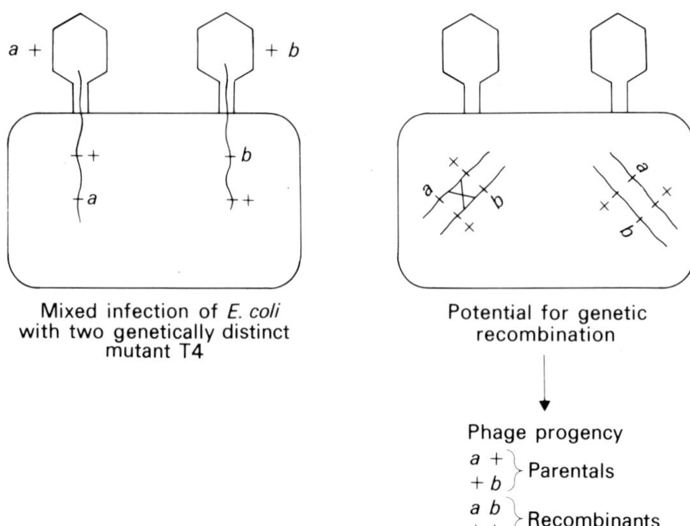

Fig 10.3. Scheme for mapping genes in bacteriophages.

investigation is a function of how far apart the two genes are on the DNA. Thus if 2% of the phages are recombinant types, we would say that the two genes are relatively close to one another (i.e. two map units). By doing a number of two-factor crosses such as that described or by doing three-factor crosses (e.g. *abc*×+++), the genetic map was constructed. This map is circular, although the chromosome is known to be linear. This is a result of circular permutation of the genome as was discussed in Chapter 2.

The genetic analysis procedures described for phage T4 are also applicable to other bacteriophages such as lambda, T2, T7, etc. All that is needed are genetic mutants with phenotypes that are distinguishable alone and in combinations so that parental and recombinant types can be counted; then two- and three-factor crosses may be carried out. Particularly useful are temperature-sensitive, host-range, morphological, and nonsense mutants. Such genetic analyses have led to the identification of most of the genes of the respective phage genomes and the construction of detailed genetic maps.

Genetic fine structure

The classical view of a gene was that it is the unit of genetic material that can mutate to alternative forms, recombine with other genes, and function in the organism. Genes were originally viewed as "indivisible beads on a string" with, for example, recombination occurring between the beads. This classical view was drastically modified in the 1950s when S. Benzer did a series of elegant experiments designed to elucidate that fine structure of the gene. Benzer set out to define the *unit of mutation*, the *unit of recombination*, and the *unit of function* at the molecular level. For his studies, Benzer chose the phage T4 because of the large number of progeny produced in a short time. He worked with *rII* mutants, discussed earlier, since they produced plaques with a morphology easily distinguishable from that of the wild type and because of the host-range properties; namely, the inability of *rII* mutants to grow in *E. coli K12*(λ^1) (Table 10.1).

Table 10.1. Host range and plaque morphology properties of r^+ and *rII* strains of T4.

Strain	Plaque morphology on *E. coli*	
	Strain *B*	Strain *K12*(λ)
r^+ (+)	r^+ type (turbid)	r^+ type
rII	*r* type (clear)	no plaques

Benzer realized that the growth defect of *rII* on *K12*(λ) was a powerful selective tool for detecting the presence of a very small proportion of r^+ genotypes within a large population of *rII* mutants. In particular it is possible to score selectively for the very rare r^+ recombinants that arise in genetic crosses between two very closely linked *rII* mutants.

Initially 60 *rII* mutants were isolated and these were crossed in all pairwise combinations in *E. coli B* (Fig. 10.4). The progeny produced from each cross were analyzed for the frequency of recombination to construct a genetic map.

To do this, the progeny phages were plated on both *B* and *K12*(λ). Appropriate dilutions of the phage suspension were made to get a reasonable number of plaques on the plates so that mathematical accuracy could be ensured. The dilution factor is obviously taken into consideration in the calculations. All of the progeny phages will produce plaques on *B*, and thus the number of plaques on *B* gives information about the total number of progeny phages produced from the cross, usually expressed as plaque-forming units per milliliter (pfu/ml) of suspension. On the other hand, the plaques growing on *K12*(λ) can only be the wild-type recombinants since *rII* mutants cannot grow on *K12*(λ).

A hypothetical cross is presented in Fig. 10.5. Four types of progeny phages will be produced:

r7 } Parentals—clear plaques on *B*
r12 }

r7, r12 } Recombinants—*r7, r12* gives clear plaque on *B*
r^+ } r^+ grows on *K12*(λ) and on *B*

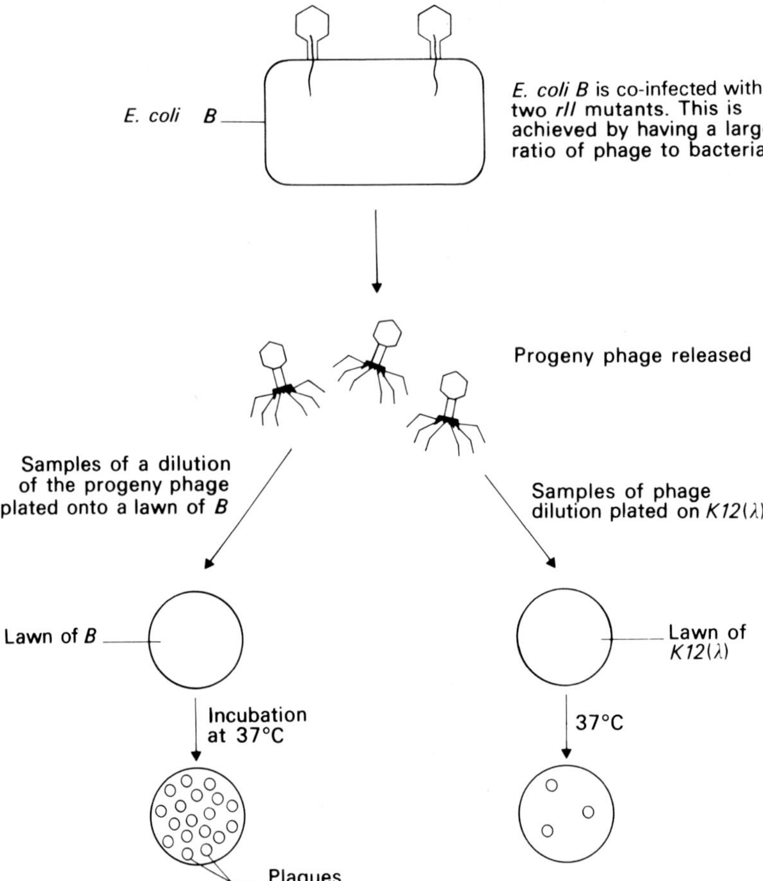

Fig. 10.4. Protocol for determining the number of wild-type recombinants (= one-half the total number of recombinants) produced in a cross of two *rII* mutants of T4.

Thus the total number of progeny per milliliter of suspension is calculated from the number of plaques produced on *B*, and the number of wild-type recombinants is determined from the number of plaques on *K12*(λ). Each cross-over event that produces an r^+ recombinant also generates an *r7*, *r12* double mutant, which produces clear plaques on *B* but does not grow on *K12*(λ). Map distance between two mutations is given by the percentage of recombinant progeny among all progeny from crosses of two mutants. In the case of these *rII* crosses, the number of recombinants is twice the number of r^+ plaques on *K12*(λ). In our example, the map distance between *r7* and *r12* is given by:

$$\frac{2 \times \text{number of } r^{+} \text{plaques on } K12(\lambda)}{\text{Total plaques on } B} \times 100\%$$

The number obtained is in map units.

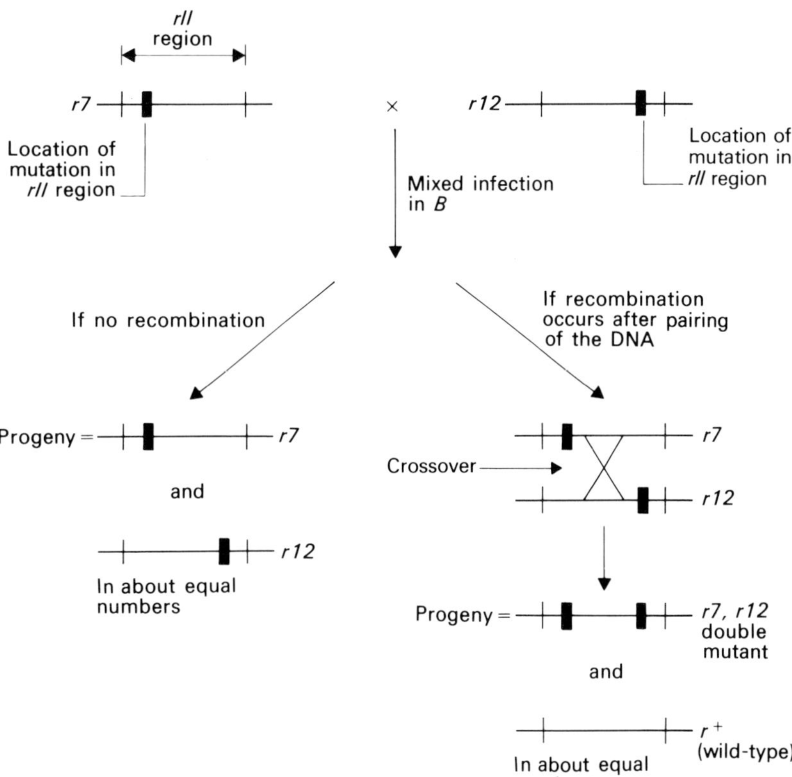

Fig. 10.5. Chromosomal diagram of the production of recombinants in a cross of two *rII* mutants, *r7* and *r12*, of phage T4.

From the recombination data accumulated from the crosses of the *rII* mutants, Benzer constructed a preliminary fine structure genetic map in which the mutations were arranged in a linear order. Among the first set of 60 mutants, crosses between some pairs of mutants produced no r^+ recombinants. This was interpreted to mean that the two mutations involved were extremely close or perhaps altered the same nucleotide pair. (The latter conclusion is a retrospective one since the work was done to define the lower limit for recombination.) Mutations within a gene that do not recombine are *homoalleles*. Those pairs of mutants that did produce r^+ recombinants were considered to carry *heteroallelic* mutations. From the initial map constructed, it was ascertained that the lowest frequency with which r^+ recombinants were found among progeny of crosses of two *rII* mutants was 0.01%, which was much higher than the 0.0001% limit of resolution in the experiment. Given this figure it is possible to make a rough calculation of the physical distance between the two mutations on the DNA. From other mapping experiments, the T4 genome is known to be about 1500 map units. If two *rII* mutants produce 0.01% r^+ recombinants, the mutations involved are separated by 0.02 map units, or $0.02/1500 = 1.3 \times 10^{-5}$ of the genome. The T4 genome consists of about 2×10^5 base pairs, so the minimum recombination distance is $(1.3\times10^{-5})\times(2\times10^5)$, or about three base pairs. This estimate is slightly too high, since we now know that the nucleotide pair is the unit of recombination.

Unit of mutation. The smallest mutational unit is the nucleotide pair. A change of a single nucleotide pair is called a **point mutation** and each should occupy only a single site on the genetic map. Each point mutation in the *rII* region should be able to yield r^+ recombinants in crosses with other point mutations, unless both are changed in the same nucleotide pair. Point mutations should also be revertible to the wild-type state.

Some *rII* mutants do not behave as point mutants as they do not yield r^+ recombinants in crosses with two or more *rII* mutants previously identified as nonallelic point mutants. These *rII* mutants also do not revert. M. Nomura and S. Benzer discovered that these new types of *rII* mutants were *deletion mutants*, which involved deletions of several nucleotide pairs of the DNA. Genetic experiments were used to prove that these nonrevertible *rII* mutants were deletion mutants; an example follows. The mutant *r1695* was proposed to be a deletion since it produced no wild-type recombinants in crosses with known point mutations and because it did not revert. In the experiment two point mutations (*r168* and *r924*) were used that mapped on opposite sides of *r1695* but were not covered by it (i.e. no r^+ recombinants were produced in crosses of either with *r1695*). The frequencies of r^+ recombinants were as shown in the map presented in Fig. 10.6.

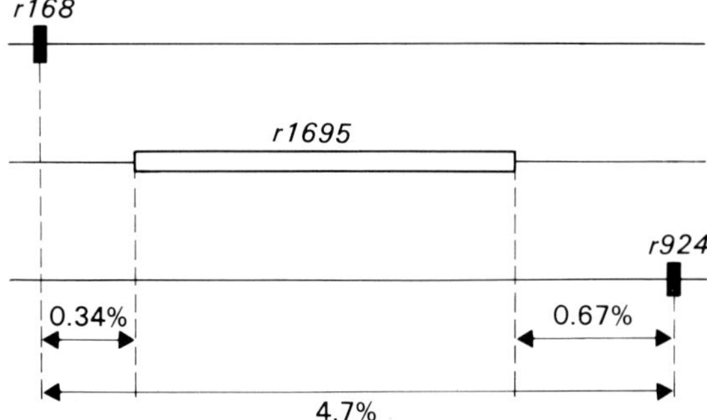

Fig. 10.6. Genetic structure of three *rII* mutants. *r168* and *r924* are two point mutants 4.7 map units apart, and *r1695* has a deletion of most of the DNA between the two.

Then, by genetic crosses, the double mutants *r168, r1695* and *r1695, r924* were constructed. These two were crossed in *E. coli B* and the progeny were isolated. It was reasoned that if *r1695* is a deletion (as we have assumed), the distance between *r168* and *r924* should be shorter than the 4.7 map units determined from *r168*×*r924* crosses. The results showed that the distance between the two was now 1 map unit, thus proving that *r1695* is a deletion. Indeed 800 nucleotides have been lost in the deletion strain (Fig. 10.7).

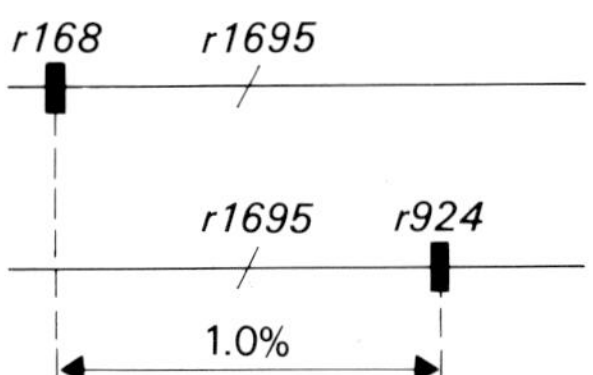

Fig. 10.7. Map distance between *r168* and *r924* is 1.0 map units in a cross of *r168, r1695*×*r1695, r924*, providing evidence that *r1695* is a deletion of 3.7 map units of DNA (see Fig. 10.6 for comparison).

Note that in the experiment just described it is not possible to generate r^+ recombinants since every progeny type has in common the *r1695* deletion, and thus they all have the *rII* phenotype. Therefore all progeny from the cross had to be cloned and backcrossed to the three grandfathers *r168, r1695,* and *r924* to determine the genotypes so that parental and recombinant types could be sorted out and map distance computed (Table 10.2).

Table 10.2. Results of crossing progeny of *r168, r1695*×*r1695, r924* with grandparents. − = no r^+ recombinants; + = r^+ recombinants formed.

	Parentals		Recombinants	
Grandparents	*r168, r1695*	*r1695, r924*	*r1695*	*r168, r1695, r924*
r168	−	+	+	−
r1695	−	−	−	−
r924	−	−	+	−

The patterns of r^+ recombinants formed or not formed in the crosses unambiguously indicated the genotype of the progeny phage involved.

In summary, mutation in the phage genome can involve a change in a single nucleotide pair or the deletion of up to many hundreds of nucleotide pairs.

Mapping using deletion mutants. Deletion mutants made it possible to map mutations within the *rII* region (or in any other gene) in a way that made it unnecessary to cross every mutant with every other mutant. Benzer used a series of overlapping deletions whose ends were mapped by crosses with point mutants of known locations. In the early experiments, the *rII* region was divided into a number of segments, each of which was defined by the length of the map covered by one particular deletion but not by another (segments A1 to A6 and B in Fig. 10.8).

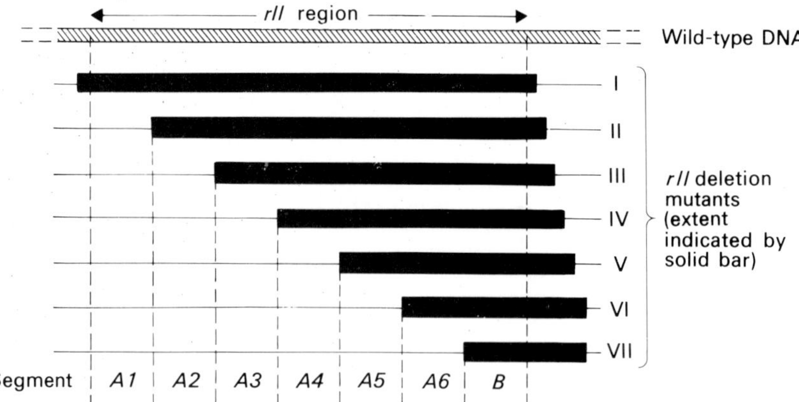

Fig. 10.8. Division of *rII* region into segments by overlapping deletion mutants.

Given a set of overlapping deletions, it is relatively simple to place each new *rII* point mutant into its appropriate segment by crossing the mutant to each deletion mutant and establishing the deletions with which the mutant does and does not produce wild-type (r^+) recombinants. For example, a point mutation in region A4 would give r^+ recombinants with deletions V, VI, and VII but not with the rest. Once mutations were localized to the seven regions, Benzer used a number of other deletions, which ended at various points within the seven segments discussed. In two further rounds of crosses with deletions involving increasing levels of subdivision, then, a point mutation could be localized to one of 47 distinct segments of the *rII* region. The final step in constructing a fine structure map was then pairwise crosses of all mutants within each of these segments to establish mutation order and map distance. In this way, Benzer showed that the 3000 *rII* mutants defined more than 300 sites in the *rII* region. The distribution of the mutants was not random in that certain hot spots were represented by a large number of independent point mutations.

In summary the recombination test for *rII* mutants is:

1. Cross two *rII* point mutants in *B*.

2. Test sample of progeny on *B* to determine total number of progeny.

3. Test sample of progeny on *K12*(λ) to determine number of r^+ recombinants.

4. Map distance = 2×frequency of r^+ recombinants.

Unit of function

The three basic aspects of genes are mutation, recombination, and function. Here we shall discuss the role of a gene as the unit of function. As we know, many genes code for polypeptides, which may have a structural, regulatory, or functional role in the cell. The properties of a particular polypeptide are derived from the three-dimensional shape and the amino acid sequence of the molecule. A mutation at a particular genetic site can therefore result in a mutant phenotype because it can result in an amino acid change at a corresponding site in the polypeptide, thereby altering the latter's function either completely or partially. Deletion mutants usually result in a nonfunctional polypeptide while point mutations may lead to either a complete or partial loss of function.

Benzer designed experiments to elucidate the unit of function of the *rII* region. The fact that he had a number of *r* mutants that produced the same phenotype and mapped close to one another did not mean necessarily that the mutations they carried were all in the same functional unit (gene). To determine the number of functional units in the *rII* region, Benzer adapted to his system the **cis-trans** or **complementation test** previously developed by Lewis in his work with *Drosophila*. For the purposes of the discussion, we will start out by stating that the *rII* region consists of two units of function, A and B, and a mutation in either will produce the *rII* phenotypes. Obviously this conclusion followed the experiments that will now be described.

The initial observation of importance for this test was that if a *K12*(λ) cell is infected with an *rII* mutant and an r^+ phage at the same time, both genomes will replicate and both phage types will be found among the progeny (Fig. 10.9). The explanation here is that the normal *rII* gene (*A* in the example) of the wild type is able to supply the function necessary for growth in *K12*(λ) of both phages. This result is obtained for *rII* mutations either in *A* or in *B*.

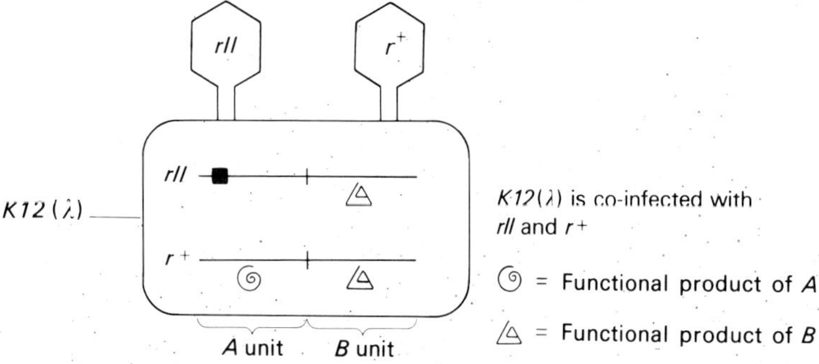

Fig. 10.9. Control experiment for testing for complementation of *rII* mutants. Here the nonpermissive host, *K12*(λ), is co-infected with r^+ and a *rII* mutant. Phages of both types are produced showing that r^+ products permit the replication of both genomes.

The next step is to infect *K12*(λ) with pairs of *rII* mutants to see if any progeny phage can be produced. If progeny phages are produced, the two mutants are said to have *complemented*, and the two mutations must be in different functional units (Fig. 10.10a). If no progeny phages are produced, the mutants do not complement, and the mutations must be in the same functional unit (Fig. 10.10b).

In the example shown in Fig. 10.10a, complementation occurs since the

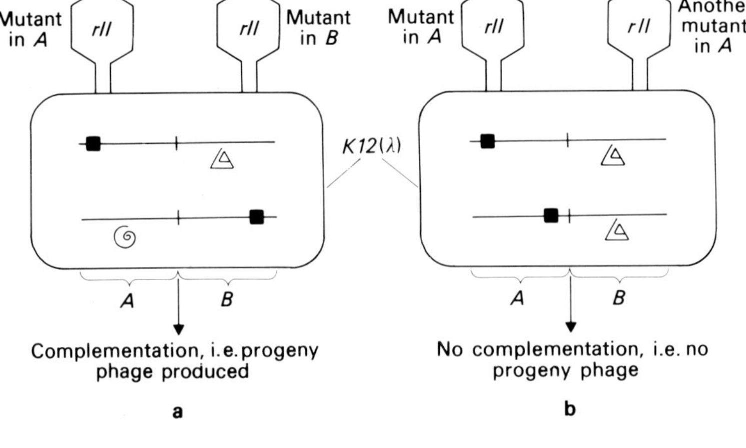

Fig. 10.10. Complementation tests for determining the units of function of *rII* mutants. (**a**) If the nonpermissive host *K12*(λ) is co-infected with an *rII* phage carrying a mutation in the *A* cistron and with an *rII* phage with a mutation in the *B* cistron, the two strains complement each other and progeny are produced. (**b**) Co-infection with two different *rIIA* or with two different *rIIB* phages results in no progeny; that is, complementation does not occur.

strain with a mutation in *A* makes a functional *B* product and the *B* mutant makes a functional *A* product. Between the two mutants, then, both products necessary for growth in *K12*(λ) are produced and thus progeny phages can be assembled and released. Each mutant makes up for the other's defect. In the example shown in Fig. 10.10b, no complementation occurs since both strains have in common the lack of the *A* function, thus preventing growth in *K12*(λ).

On the basis of this sort of test, Benzer found the *rII* mutants fall into the two functional groups, *A* and *B*. That is, all *rIIA* mutants complement all *rIIB* mutants, but *rIIA* mutants do not complement *rIIA* mutants and *rIIB* mutants do not complement other *rIIB* mutants. These two groups of mutations can be assigned definite positions on the genetic map of the *rII* region. No *A* mutants are found in the *B* area and vice versa (Fig. 10.11). *A* and *B* are called *complementation groups*. Since the cis-trans test was used in the studies, Benzer called each a *cistron*. A cistron is a pseudonym of a gene in some senses and is considered to code for a polypeptide. The *rIIA* cistron is 6 map units and about 800 nucleotide pairs long, and the *rIIB* cistron is 4 map units and about 500 nucleotide pairs long.

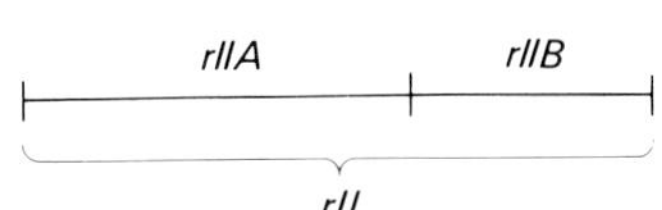

Fig. 10.11. Organization of the *rII* region into two units of function (i.e. *A* and *B* cistrons).

It is remarkable that throughout this work, Benzer did not know the nature of the polypeptides coded by the *A* and *B* cistrons.

Phage ΦX174

We conclude this chapter by discussing briefly the genetic organization of bacteriophage ΦX174 since it has some rather unusual features.

ΦX174 is an icosahedral bacteriophage and thus roughly resembles the head of phage T4. The genetic material of ΦX174 is circular, single-stranded DNA, and there is no redundancy of genes as in phages T2 and T4, for example.

Life cycle (Fig. 10.12)

After infecting *E. coli* the DNA of the phage attaches to the host cell membrane and a double-stranded circular DNA is produced by the production of the complementary DNA strand in polymerization reactions catalyzed by host cell enzymes. The infecting DNA strand is called the + strand since it is of the same "sense" as the phage-coded mRNA molecules. Thus the complementary strand that is made is the − strand, and the membrane-attached double-stranded DNA is called *replicative form 1* (RF1).

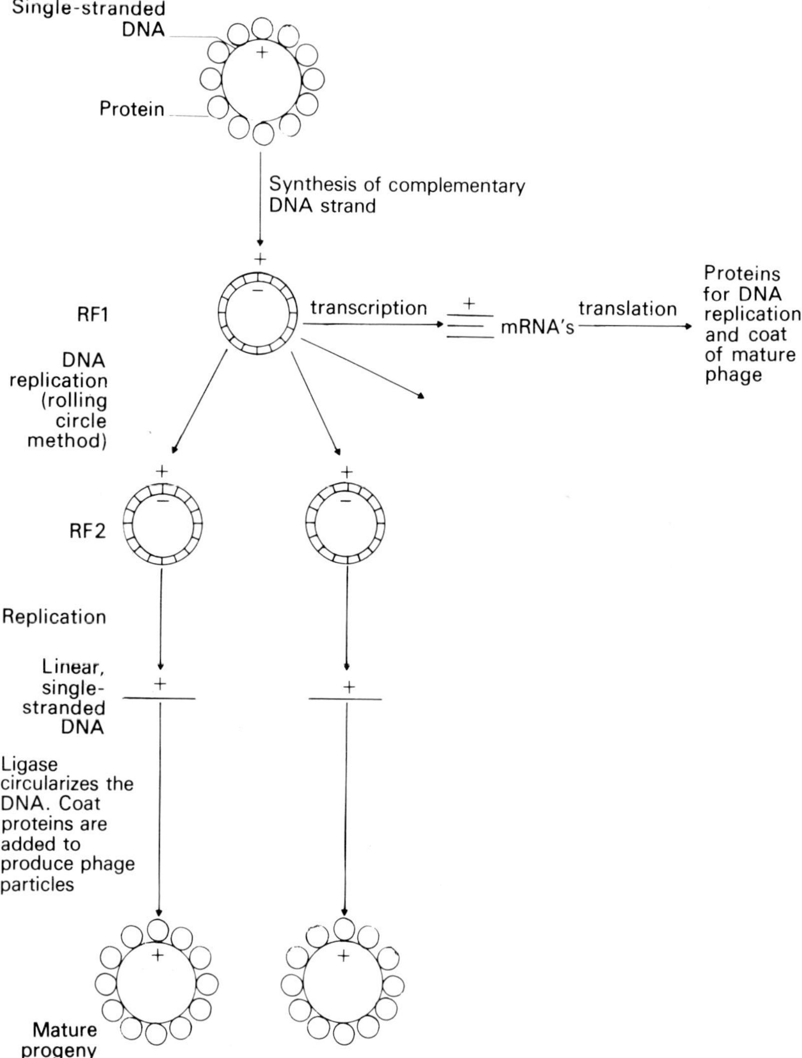

Fig. 10.12. Life cycle of phage ΦX174. The details are given in the text. The + and − designate the sense of the strand, the original infecting strand being the same sense as the mRNAs produced (+).

Once the RF1 has been produced, mRNAs are transcribed from it and these have the + sense like the infecting DNA strand. The mRNAs code for the proteins required for replication of phage DNA and the components needed for the assembly of progeny bacteriophage particles. In addition, the parental RF1 repeatedly replicates by the rolling-circle method to produce many copies of nonmembrane-bound progeny RF molecules called RF2s. Then, again by the rolling-circle method of replication the RF2s generate linear viral DNA, which are + as was the original infecting strand. These linear progeny viral strands are circularized by DNA ligase and then become associated with proteins to form the mature phage particle that are released from the cell when the host lyses.

Genetic organization

Phage ΦX174 is a relatively small bacteriophage. Its single-stranded DNA genome is 5386 nucleotides long and contains nine known genes (as defined by genetic mutations), which code for the nine phage-specific proteins (A through H) that have been identified in *E. coli* following infection by ΦX174. Protein A functions in double-stranded DNA replication; B, C, and D function in the production of single-stranded progeny; proteins F, G, H, and J are components of the phage particle (called the *capsid*); and protein E is responsible for the lysis of the host cell. The order of the genes on the genetic map is *A B C D E J F G H* , and the origin for DNA replication is located within gene *A*.

When the number of nucleotides needed to code for the known amino acid sequences of the nine proteins was determined, it was realized that the entire ΦX174 genome had approximately 700 too few nucleotides. To resolve this apparent paradox, Sanger and colleagues determined the entire nucleotide sequence of ΦX174 through the use of DNA-sequencing procedures (see Chapter 12). From the DNA sequence and amino acid sequence data, they made the following conclusions:

1. The protein coding sequence of gene *B* is totally contained within gene *A*, with the two reading frames of the mRNAs they code for staggered by one nucleotide. The *A* gene itself extends 85 nucleotides (28 amino acids) beyond the end of the *B* gene. The genes *A* and *B*, then, are examples of *overlapping genes*.

2. Gene *C* is located between genes *B* and *D* and illustrates a second type of overlap. Specifically, the sequence of gene C for the initiation codon of the mRNA overlaps by one nucleotide the sequence coding for the termination codon of gene *A* as shown in Fig. 10.13a. A similar one nucleotide overlap is shown for genes *D* and *J* (Fig. 10.13b).

3. Genes *D* and *E* provide a second example of complete gene overlap, with

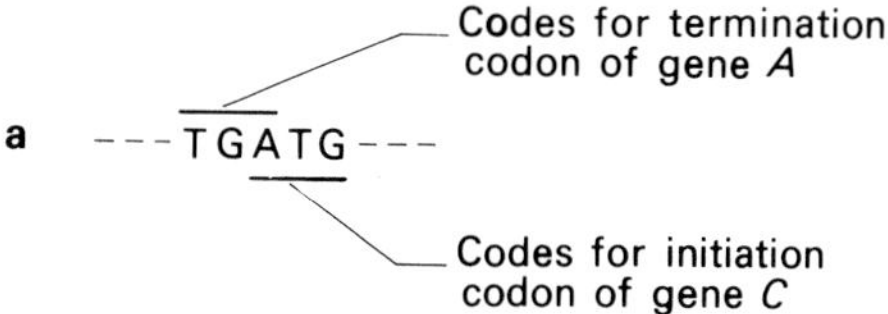

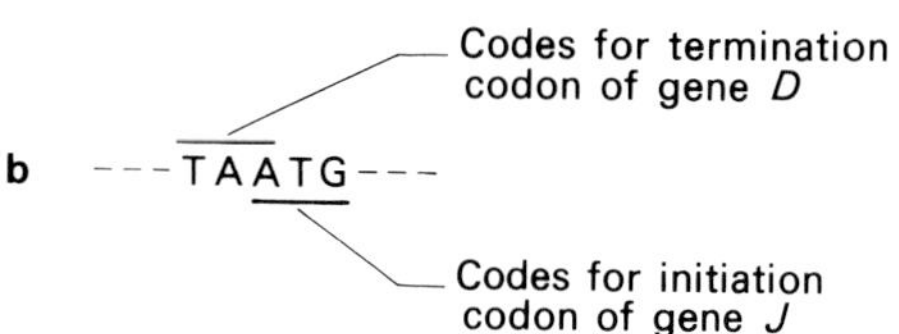

Fig. 10.13. Examples of overlapping genes in phage ΦX174. (**a**) The DNA sequence shown contains the sequence for the termination codon of gene *A*, overlapping by one nucleotide the sequence for the initiation codon of gene *C*. (**b**) Overlap by one nucleotide of the sequences for the termination codon of gene *D* and for the initiation codon of gene *J*.

the sequence of the gene *E* completely contained within the sequence of gene *D*. The two genes differ in reading frame, and their products have different functions within the cell; *D* codes for a protein needed for the production of single-stranded viral DNA, and *E* codes for a lysing protein. The two genes behave completely independently by genetic tests. (Indeed, this was part of the evidence that two distinct reading frames are involved.) Thus nonsense mutants in *E* usually result only in a missense change in *D* or, because of code degeneracy, perhaps no change at all. The same situation has been shown also for nonsense mutations in gene *E*.

4. After the termination sequence of gene *J*, there is a 39-nucleotide gap before the initiation of gene *F*. A 111-nucleotide gap is present between *F* and *G*, and a 66-nucleotide gap is between *H* and *A*. The functions, if any, of these gaps are unknown. Genes *F*, *G*, and *H* do not share nucleotides with any other genes.

In summary, ΦX174 provides very good examples of overlapping genes, both where one gene is wholly within another and where only one nucleotide is shared. This situation is not unique among organisms. For example, Platt and Yanofsky have found an example of the latter type of overlap in the mRNA of the tryptophan operon of *E. coli*. In addition, Reddy and colleagues have analyzed the DNA sequence of simian virus 40 (SV40), a double-stranded DNA virus that infects mammalian cells, and have shown that this virus also has examples of limited overlap of genes.

Questions and problems

10.1 The following two-factor crosses were done to analyze the genetic linkage between certain genes in phage λ:

Parents	Progeny			
$c + \times + mi$	1213 $c+$,	1205 $+mi$,	84 $++$,	75 $c\,mi$
$c + \times + s$	566 $c+$,	808 $+s$,	19 $++$,	20 $c\,s$
$co + \times + mi$	5162 $co+$,	6510 $+mi$,	311 $++$,	341 $co\,mi$
$mi + \times + s$	502 $mi+$,	647 $+s$,	65 $++$,	56 $mi\,s$

Construct a genetic map of the four genes.

10.2 Three gene loci in T4 that affect plaque morphology in easily distinguishable ways are *r* (rapid lysis), *m* (minute), and *tu* (turbid). A culture of *E. coli* is mixedly infected with two types of phage, $r\ m\ tu$ and $r^+\ m^+\ tu^+$. Progeny phages are collected and the following genotype classes are found:

Genotype	Number
$r^+\ m^+\ tu^+$	3729
$r^+\ m^+\ tu$	965
$r^+\ m\ \ tu^+$	520
$r\ \ m^+\ tu^+$	172
$r^+\ m\ \ tu$	162
$r\ \ m^+\ tu$	474
$r\ \ m\ \ tu^+$	853
$r\ \ m\ \ tu$	3467
Total	10,342

Construct a map of the three genes. What is the coefficient of coincidence, and what does the value suggest?

10.3 The *rII* mutants of bacteriophage T4 grow in *E. coli B*, but not in *E. coli K12*(λ). *E. coli* strain *B* is doubly infected with two *rII* mutants. A 6×10^7 dilution of the lysate is plated on *E. coli B*. A 2×10^5 dilution is plated on *E. coli K12*(λ). Twelve plaques appeared on strain *K12*(λ) and 16 on strain *B*. Calculate the amount of recombination between these two mutants.

10.4 Wild-type (r^+) T4 grows on both *E. coli B* and *E. coli K12*(λ), producing turbid plaques. The *rII* mutants of T4 produce clear plaques on *E. coli B* but do not grow on *E. coli K12*(λ). This "host-range" property permits the detection of a very low number of r^+ phage among a large number of *rII* phages. With this sensitive system, it is possible to determine the genetic distance between two mutations within the same gene, in this case, the *rII* locus. Suppose *E. coli B* is mixedly infected with *rIIx* and *rIIy*, two separate mutants in the *rII* locus. Suitable dilutions of progeny phage are plated on *E coli B* and *E. coli K12*(λ). A 0.1 ml sample of a 1000-fold dilution plated on *E. coli B* showed 672 plaques. A 0.2 ml sample of undiluted phage plated on *E. coli K12*(λ) showed 470 turbid plaques. What is the genetic distance between the two *rII* mutations?

10.5 Wild-type (r^+) strains of T4 produce turbid plaques, whereas *rII* mutant strains produce larger clearer plaques. Five *rII* mutations (*a–e*) in the A cistron of the *rII* region of T4 give the following percentages of wild-type recombinants in two-point crosses:

$a \times b$ 0.2%
$a \times c$ 0.9%
$a \times d$ 0.4%
$b \times c$ 0.7%
$e \times a$ 0.3%

$e \times d$ 0.7%
$e \times c$ 1.2%
$e \times b$ 0.5%
$b \times d$ 0.2%
$d \times c$ 0.5%

What is the order of the mutational sites?

10.6 A set of seven different *rII* deletion mutants of bacteriophage T4, 1 through 7, were mapped with the results shown in the following figure:

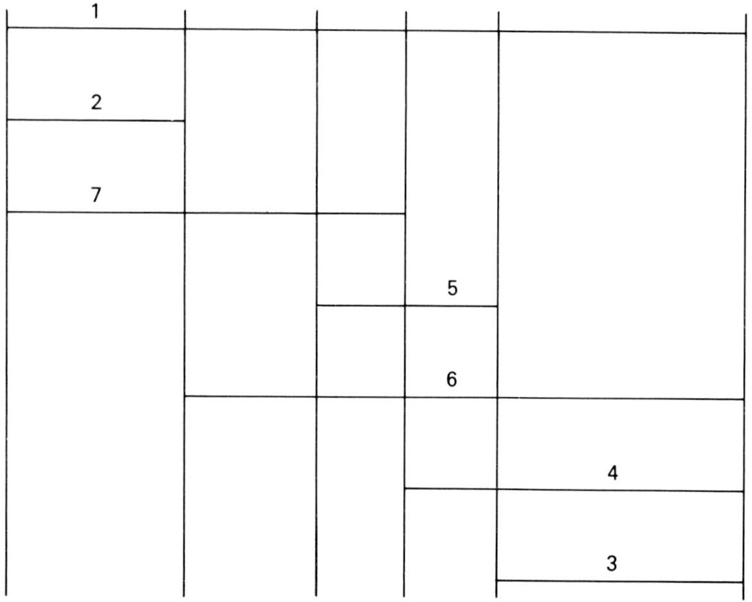

Five *rII* point mutants were crossed with each of the deletions. The following table shows the results, where – = r^+ recombinants were obtained, 0 = no r^+ recombinants were obtained:

Point mutants	Deletion mutants						
	1	2	3	4	5	6	7
a	0	+	+	+	0	0	0
b	0	0	+	+	+	+	0
c	0	+	+	0	0	0	+
d	0	+	0	0	+	0	+
e	0	+	+	+	+	0	0

Map the locations of the point mutants.

References

Barell, B.G., G.M. Air and C.A. Hutchison, 1976. Overlapping genes in bacteriophage ΦX174. *Nature* **264**:34–41.

Benbow, R.M., C.A. Hutchison, J.D. Fabricant and R.L. Sinsheimer, 1971. Genetic map of bacteriophage ΦX174. *J. Virol.* **7**:549–558.

Benzer, S., 1959. On the topology of the genetic fine structure. *Proc. Natl. Acad. Sci. USA* **45**:1607–1620.

Benzer, S., 1961. On the topography of the genetic fine structure. *Proc. Natl. Acad. Sci. USA* **47**:403–415.

Delbruck, M., 1940. The growth of bacteriophage and lysis of the host. *J. Gen. Physiol.* **23**:643–660.

Doermann, A.H., 1952. The intracellular growth of bacteriophages. I. Liberation of intracellular bacteriophage T4 by premature lysis with another phage or with cyanide. *J. Gen. Physiol.* **35**:645–656.

Eisenberg, S., J.F. Scott and A. Kornberg, 1976. Enzymatic replication of viral and complementary strands of duplex DNA of phage ΦX174 proceeds by separate mechanisms. *Proc. Natl. Acad. Sci. USA* **73**:3151–3155.

Ellis, E.L. and M. Delbruck, 1939. The growth of bacteriophage. *J. Gen. Physiol.* **22**:365–384.

Hayes, W., 1968. *The Genetics of Bacteria and Their Viruses*, 2nd ed. Wiley, New York.

Hershey, A.D. and R. Rotman, 1949. Genetic recombination between host-range and plaque-type mutants of bacteriophage in single bacterial cells. *Genetics* **34**:44–71.

Reddy, V.B., B. Thimmappaya, R. Dhar, K.N. Subramanian, B.S. Zain, J. Pan, P.K. Ghosh, M.L. Celma and S.M. Weissmann, 1978. The genome of simian virus 40. *Science* **200**:494–502.

Sanger, F., G.M. Air, B.G. Barrell, N.L. Brown, A.R. Coulson, J.C. Fiddes, C.A. Hutchison, P.M. Slocombe and M. Smith, 1977. Nucleotide sequence of bacteriophage ΦX174. *Nature* **265**:687–695.

Smith, M., N.L. Brown, G.M. Air, B.G. Barrell, A.R. Coulson, C.A. Hutchison and F. Sanger, 1977. DNA sequence at the C termini of the overlapping genes *A* and *B* in bacteriophage ΦX174. *Nature* **265**:702–705.

Chapter 11 Bacterial Genetics

Outline

Conjugation in *E. coli*
- The *F* factor
- *Hfr* strains
- Transfer of DNA from donor to recipient
- Mapping by conjugation

Transduction
- Generalized transduction
- Specialized transduction

Transformation
- Mechanism of transformation
- Determination of gene linkage by transformation

We now turn our attention from bacteriophages to bacteria, which are much more complex in many ways. In this chapter we will consider three ways that genetic material can exchange between different bacteria: **conjugation**, **transduction**, and **transformation**. Before we discuss these, it is appropriate to consider the ways of growing a bacterium such as *E. coli*.

Bacteria can be cultured either on a solidified surface (usually containing agar) or in a liquid medium. In both cases nutrients and necessary salts and minerals must be present. In liquid culture, the bacteria multiply exponentially until the nutrients are exhausted or until toxic products accumulate. The number of bacteria present at any time in a liquid culture can be determined simply. A small sample can be pipetted onto a petri dish containing a solid nutrient medium and spread over the surface using a spreader. Thus individual bacteria will be distributed evenly over the plate, and as each cell divides during the incubation period, a clone is produced that remains together in a clump called a *colony*. Each cell will give rise to a discrete colony (Fig. 11.1).

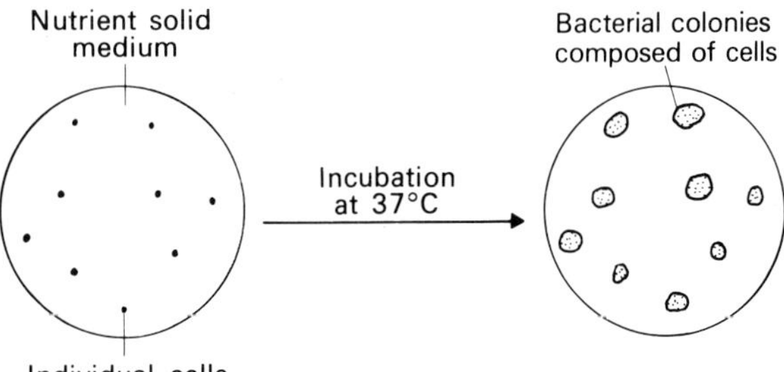

Fig. 11.1. Diagrammatic representation of the production of bacterial colonies on solid culture medium.

Conjugation in *E. coli*

Sexual conjugation in *E. coli* was discovered by J. Lederberg and E. Tatum and hinged on the appearance of prototrophs, $a^+ b^+ c^+ d^+ e^+ f^+$ (all wild-type alleles for genes *a* through *f*), from a mixed culture of two kinds of auxotrophs: $a\ b\ c\ d^+ e^+ f^+$ (requires substances a, b, and c for growth) and $a^+ b^+ c^+ d\ e\ f$ (requires d, e, and f for growth). Since three different auxotrophic mutations were present in each strain, it was extremely unlikely that the prototrophs arose by reversion of one or other of the strains.

The necessity for contact for prototroph formation was shown in an experiment using a U-tube apparatus (Fig. 11.2). In this experiment nutrient medium is present in both sides of the U-shaped tube, and the two types of bacterial cells are separated by a filter through which medium, but not bacteria, can pass. By alternately sucking and blowing on the side tube, the medium can be moved between compartments. As long as the filter is present, no prototrophs are formed. They occur only in the absence of the filter, indicating that cell contact is required for this phenomenon. The interpretation was that prototrophs are formed by the transfer of genes between bacteria in a process involving cell contact. This process is called **conjugation.**

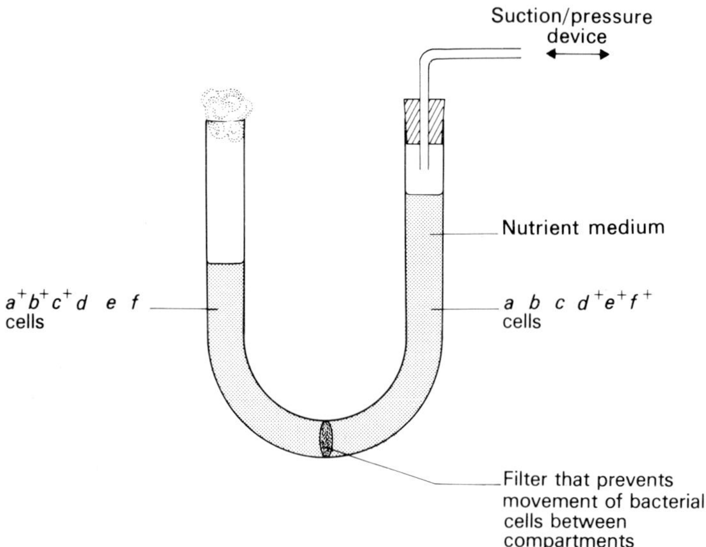

Fig. 11.2. Use of U-tube apparatus to demonstrate that cell contact between genetically distinct *E. coli* strains is needed for prototroph formation.

The *F* factor

Hayes proposed the hypothesis that the genetic transfer described was mediated by some kind of infectious vector that is found in so-called donor (D) cells. The vector facilitates the transfer of genes from the donor cells to recipient (R) cells under certain conditions. The D and R cells differ by the presence (F^+) or absence (F^-), respectively, of an extrachromosomal factor called an *F* factor or a *sex factor*. By conjugation, the donors transmit the *F* factor in high frequency to recipient cells, which then become donors (Fig. 11.3). The transmission of the *F* factor between conjugating D and R cells occurs concurrently with the replication of the *F* factor. The transmission of *F* does not involve the transmission of chromosomal genes of the host.

The *F* factor itself is circular, double-stranded DNA consisting of about 100,000 nucleotide pairs (about 1/40th that of the host chromosome). The

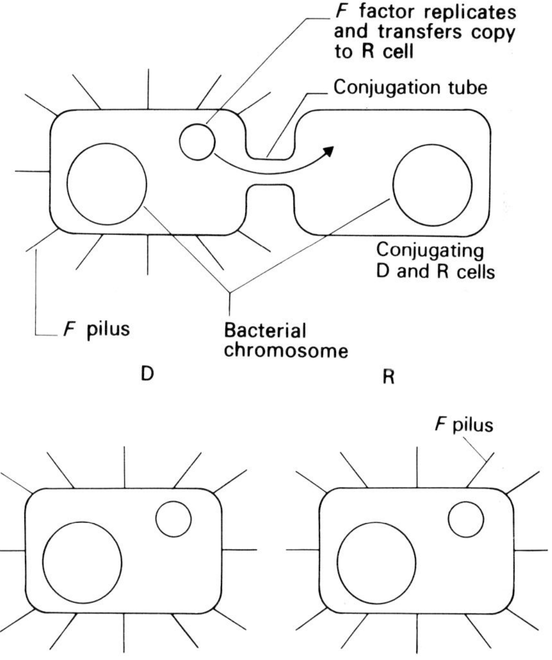

Fig. 11.3. Conjugation between F^+ (donor) and F^- (recipient) cells of *E. coli* results in transfer of a copy of the *F* factor and the conversion of the F^- to the F^+ state.

F^+ cell (or donor) contains one copy of the *F* factor per cell, and this replicates in concert with replication of the host cell chromosome. For its replication, the *F* factor attaches to a unique site on the bacterial membrane.

The *F* factor contains genes that result in the formation of hairlike, cell surface components called *F*-pili on cells containing *F*. The pilus is thin and flexible and may be very long. Other genes are responsible for forming the conjugation tubes between D and R cells that are necessary for transfer of the *F* factor. Once that has occurred, replication of the *F* factor independently of the host chromosome, accompanied by the movement of a copy of the *F* factor through the conjugation tube, results in the conversion of the F^- recipient to a F^+ donor.

Hfr strains

L. Cavalli-Sforza and W. Hayes discovered that some F^+ populations gave rise to donor cells that could transmit chromosomal genes at high frequency. These were called **high-frequency recombinant** or Hfr strains for short. These strains are physically similar to F^+ cells in that they have *F*-pili. In crosses of *Hfr* strains with F^- strains, there is a gradation in frequencies in which various donor loci appear among recombinant cells. For example, in the cross

Hfr $a^+ b^+ c^+ d^+ \ldots z^+ \times F^- a\,b\,c\,d \ldots z$, the most frequent class of recombinants was a^+, the next most frequent class was b^+, and the least frequent class of recombinants was z^+ (Fig. 11.4). In addition, the recombinant progeny remain F^- (Fig. 11.4). From this it was concluded that *Hfr* gene a^+ entered into the F^- first and therefore had the greatest chance of recombination. The remainder of the genes b^+ to z^+ entered sequentially in that order, and their frequency of recombination was determined by the order of entry of the genes.

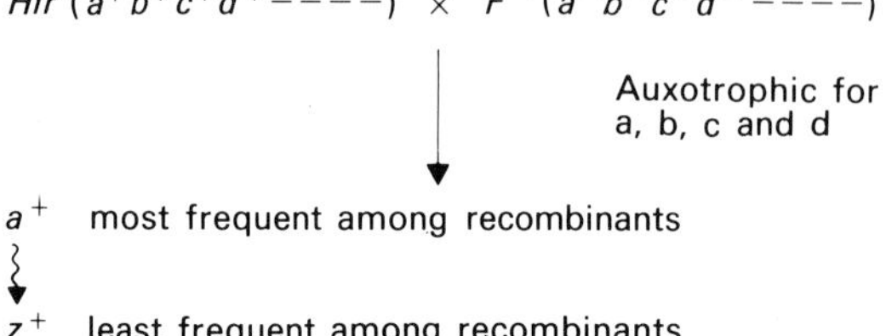

Fig. 11.4. Hypothetical cross between an *Hfr* strain and a F^- strain to show the sequence in which host genes are transferred from the donor to the recipient.

The difference between F^+ and *Hfr* cells is that in *Hfr* cells the *F* factor has become integrated into the bacterial chromosome. Since the *F* genes are still in the cell, the *F* functions are all present, including the capacity to form conjugation tubes with F^- cells and the ability to transfer the *F* factor to the F^- cell. Since *F* is integrated into the chromosome, this transfer process is accompanied by transfer of the host chromosome across the conjugation tube.

The integration of the *F* factor occurs by a genetic recombination event involving a single crossover (Fig. 11.5). A given *Hfr* can revert to the F^+ state with an extrachromosomal *F* factor by the reverse of this process.

The *F* factor is able to integrate at a number of sites over the host chromosome, and thus the order of genes transferred to an F^- cell varies between *Hfr* strains. Since the *F* factor has polarity in terms of the transfer properties, different *Hfr*s transfer genes in different orientations. In Fig. 11.6, the arrowheads indicate a number of the known *Hfr* cells, and the direction of transfer of host genes. For example, *HfrH* (*H* here stands for Hayes) transfers genes in the order a^+ b^+ c^+ d^+ e^+ . . ., whereas *HfrJ4* transfers genes in the order u^+ t^+ s^+ r^+ q^+ . . ., and so on.

For a given *Hfr* strain, not all of the host genes can be transferred to the F^- since the conjugation tube is fragile and there is a good chance the cells will break apart before the 90 minutes at 37°C that would be required for transfer of the whole chromosome. Thus, to build the complete map of the *E. coli* chromosome, a number of *Hfr* strains with different start points and orientations were used. In the start of the process of gene transfer, the *F* factor breaks along its length so that part of the *F* factor is transferred first, whereas

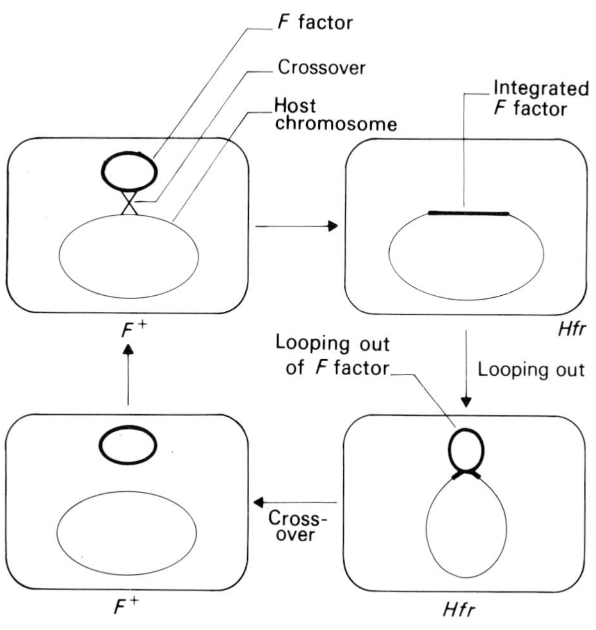

Fig. 11.5. Diagram of the formation of an *Hfr* strain by integration of the *F* factor, and reversion of the *Hfr* to the F^- state by the reverse process.

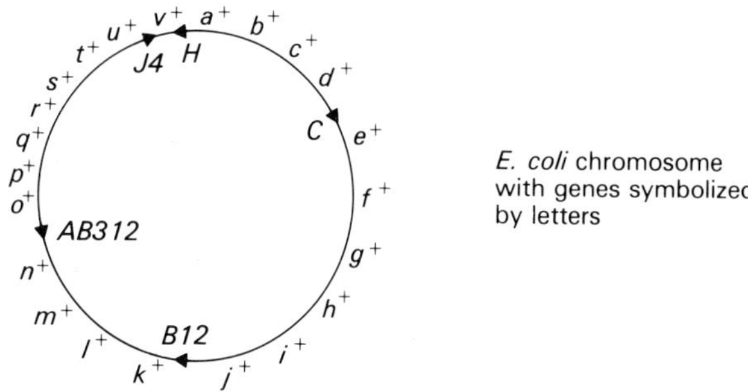

Fig. 11.6. Stylized representation of the *E. coli* chromosome with letters designating genes and arrowheads signifying the points of integration and orientation of the *F* factor to produce various *Hfr* strains. Note that an *F* factor is present at only one site in any given *Hfr* strain.

the rest is at the tail end of the *E. coli* chromosome. It is for this reason that the recipients usually remain F^- in phenotype since only after transfer of the entire chromosome would a complete *F* factor be found in the recipient cell.

Transfer of DNA from donor to recipient

$F^+ \times F^-$ matings. The two cells form a conjugation tube and a copy of the *F* factor is transferred from the F^+ to the F^-. During this event, a single strand of the *F* factor passes through the conjugation bridge and the complementary strand is synthesized in the recipient, eventually resulting in a circularized *F* factor and a conversion of the F^- to an F^+. Since the *F* factor is small (relative to the host chromosome), there is a likelihood of transfer of the complete *F* factor copy before the F^+ and F^- cells pull apart.

Hfr×F⁻ matings. Essentially the same sequence of events occurs as in the $F^+ \times F^-$ matings. A conjugation tube is formed between an *Hfr* cell and an F^- cell, and the integrated *F* factor in the *Hfr* breaks into two, one end passing through the conjugation tube followed by chromosomal genes (Fig. 11.7).

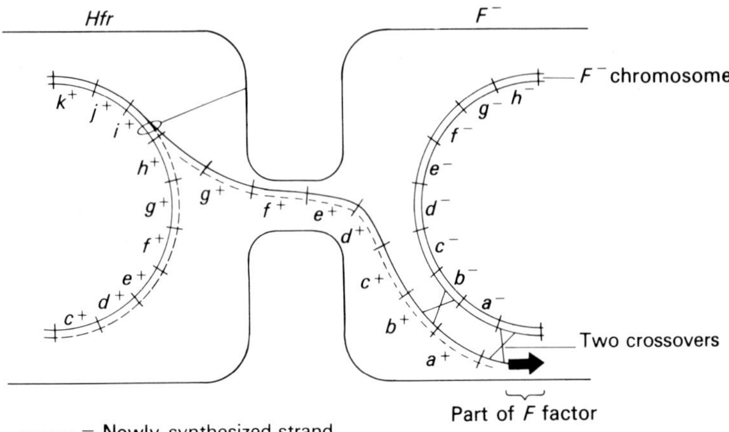

Fig. 11.7. Transfer of host genes to the recipient in an $Hfr \times F^-$ mating.

Only one strand is transferred, the complementary one being synthesized in the F^- recipient cell. If the recipient has different alleles of the genes being transferred, recombinant cells resulting from crossover can be detected and studied. Since the rest of the *F* factor is not transferred until all the chromosome has moved across the bridge (which is very unlikely), the F^- rarely becomes an *Hfr*.

In Fig. 11.7, the crossovers shown would generate recombinants with a^+ phenotype and all descendants of the F^- would have that phenotype. During the transfer, the donor chromosome can break anywhere, so only partial transfer occurs. The chance of breakage is random and thus the genetic recombination between donor and recipient chromosomes transferred during the "zygote" phase.

Even when a donor chromosome is present in the recipient, it does not necessarily mean that it is going to recombine with the recipient chromosome. Fortunately for genetic studies this *postzygotic coefficient of integration* is about the same for most chromosomal markers. When a bacterium contains its own chromosome plus a fragment transferred from a donor cell, it is called a *merodiploid* or *merozygote*.

Mapping by conjugation

F. Jacob and E. Wollman realized that the random breakage of conjugation tubes afforded a means of mapping gene sequences by timing the entry of different genes into a recipient. They used an *interrupted mating technique* in which *Hfr* and F^- cells were mated and then at various intervals a sample of the culture was taken and blended in a Waring blender to separate the cells. Thus the length of donor chromosome that entered the F^- was controlled by timing the interval between the onset of conjugation and the blending treatments. The experiments were set up so that recombinant progeny for a number of genes could be scored at the various times. The relationships between genes and their positions on the chromosome could be mapped, therefore, in terms of time units measured in minutes. Note that one must take into account that part of the *F* factor enters first and that this takes time. One must therefore only compare distance in time units for chromosomal genes. Here is an example of an experiment done by Jacob and Wollman involving the following cross:

HfrH $str^S \times F^-$ *thr leu azi tonA lac gal* str^R

The F^- cell carries a number of mutant genes: it is auxotrophic for the threonine (*thr*) and leucine (*leu*), it is sensitive to inhibition by azide (*azi*), it is sensitive to phage T1 infection (*tonA*), it is unable to ferment (and hence utilize as a sole carbon source) lactose (*lac*) and galactose (*gal*), and it is resistant to streptomycin, an antibiotic that will kill sensitive cells. The *Hfr* cell, on the other hand, has the wild-type alleles of all these genes and it is sensitive to streptomycin.

The two cell types are mixed together in nutrient medium to encourage mating. After a few minutes the culture is diluted so that no new mating pairs can form, thus ensuring some synchrony in the chromosome transfer for those mating pairs already established. Then, at various times, samples of the culture are taken, blended, and plated on a selective medium that is designed to allow particular recombinant types to grow while counterselecting against (killing) the *Hfr* and F^- parental cells. In our example, the thr^+ and leu^+ genes are transferred first in the cross, and thus the selective medium omitted both threonine and leucine, but other essential nutrients were included. On such a medium thr^+ leu^+ recombinants could grow but the F^- cell could not. The medium also would contain streptomycin in this case so that the *Hfr* parental would be killed. The F^- recombinants, of course, would be unaffected by the antibiotic. In this way a selection of colonies, each representing a clone of a thr^+ leu^+ recombinant, can be isolated. Each must have arisen by recombination between a transferred donor chromosome and the recipient chromosome. If one selects for a gene transferred early, there is a good chance of obtaining a large number of recombinants.

One can now test the thr^+ leu^+ recombinants for whether they are also recombinant for the other marker genes of the F^- by using other selective platings or tests. The results of doing this for the cross we are discussing are shown in Fig. 11.8.

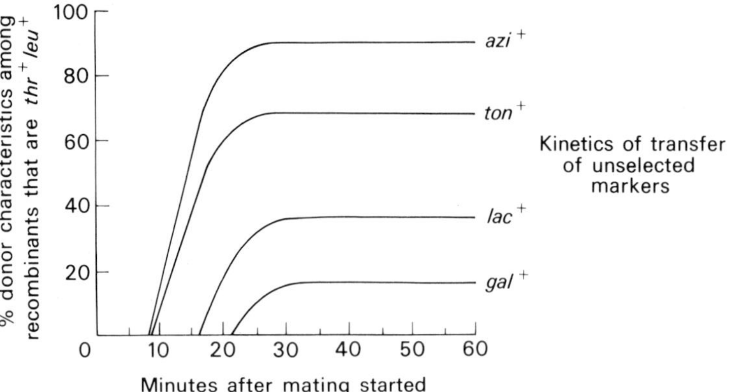

Fig. 11.8. The kinetics of transfer of unselected donor markers among selected *thr*$^+$*leu*$^+$ recombinants as a function of time of conjugation. (After F. Jacob and E. L. Wollman, 1961. *Sexuality and the Genetics of Bacteria.* Copyright © Academic Press, New York.)

After 8 minutes, *azi*$^+$ recombinants have appeared and thus that gene on the donor chromosome must have entered the recipient cell. Similarly, the *ton*$^+$ gene entered at 10 minutes, the *lac*$^+$ gene at 17 minutes and the *gal*$^+$ gene at 23 minutes. It can be concluded, therefore, that the order of genes on the transferred donor chromosome is origin (part of *F* factor)-*thr-leu-azi-ton-lac-gal*. The distance between the genes can be computed from the times of entry, that is, the times at which recombinant progeny for each gene is observed. Thus, for example, the *ton* and *lac* genes are 7 minutes apart on the map.

As we mentioned before, a number of *Hfr* strains are necessary to map the entire circular chromosome of *E. coli* since the mating pairs generally break apart before all the chromosome has been transferred. There is a very good agreement for the distance in time units between two particular genes when different *Hfr* strains are used, thus lending credibility to this method of mapping. The entire map is 90 minutes long in *E. coli*, and conventionally, the threonine gene is placed at 0 minutes on the map.

Transduction

The most widely used approach to mapping closely linked bacterial genes is **transduction**, a process by which bacteriophages mediate the transfer of genetic material from one bacterium (the donor) to another (the recipient). We shall consider two types of transduction, generalized and specialized, which differ in the mechanisms involved. Before discussing these, we need to distinguish between two types of phage. The "virulent" phages such as T2 and T4 always result in a lytic response (that is, the production of progeny particles) when they infect their host. Other phages are "temperate", meaning that they can either elicit a lytic response or they can infect and remain dormant in bacterial cell. This latter phenomenon is called the **lysogenic** response, and the dormant phage in that state is called a **prophage**.

At any time there can be a transition from the prophage to the vegetative state, resulting in the lytic cycle and release of progeny phage particles. **Lysogeny** is an advantage to the bacterium because it also prevents superinfection with a second particle of the same phage type.

Generalized transduction

N. Zinder and J. Lederberg in 1952 showed that temperate phages may act as carriers for genes between one bacterial cell and another. The motivation of these researchers was to see if genetic exchange previously shown in *E. coli* (conjugation) also existed in the mouse typhus bacterium, *Salmonella typhimurium*. They mixed together two double amino acid auxotrophic cells, *phe trp met*$^+$ *his*$^+$ (strain LA22) and *phe*$^+$ *trp*$^+$ *met his* (strain LA2), on minimal medium and found wild types. The wild types did not occur unless the strains were mixed and were apparently the result of genetic exchange. The exchange did not occur with the same frequency for each strain examined, and the combination of strains LA22 and LA2 resulted in the highest frequency of wild types for many gene combinations.

Zinder and Lederberg used the U-tube apparatus discussed earlier to test the mode of genetic exchange (Fig. 11.9). The left hand side of the U-tube contained the *phe*$^+$ *trp*$^+$ *met his* LA2 strain and the right hand side contained the *phe trp met*$^+$ *his*$^+$ LA22 strain. Wild-type bacteria appeared on the right side but not on the left side, suggesting that a filterable agent was produced by LA2 that could produce wild-type recombinants in LA22. This filterable agent only arose when the two strains shared the same growth medium. The

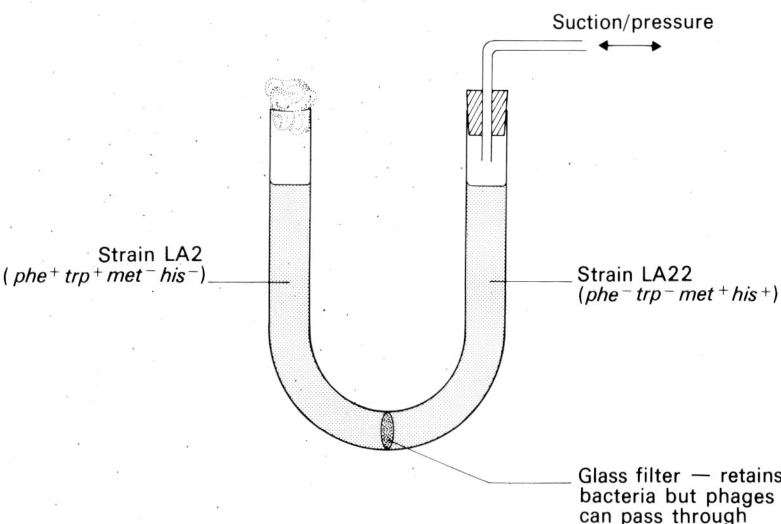

Fig. 11.9. Use of a U-tube apparatus to demonstrate transfer of genetic material between strains of *Salmonella typhimurium* does not require cell contact.

agent was shown not to be naked DNA or RNA since the appearance of prototrophs was not abolished either by DNase or RNase treatment. The formal explanation is that the filterable agent is the temperate phage P22, which can lysogenize the LA22 strain. In the experiment, progeny P22 particles were produced from some LA22 bacteria in which the prophage was converted to the vegetative state. These phages moved through a filter as the medium was moved between compartments by alternate suction and pressure. The phages then infected the nonlysogenic LA2 strain, and new phages were produced by the lytic cycle. During this process, occasionally a piece of host DNA can become wrapped up in a phage coat. The phage particles move back to the right side and the phages lysogenize the LA22 cells. This time some phage particles are associated with genetic material from LA2, some of which is wild type for the mutant genes of LA22. Thus wild-type recombinants are found among the LA22 cells as a result of recombination between the LA2 material carried by P22 and the LA22 chromosome. In other words the LA22 was *transduced*, and the phenomenon is called **transduction**. Since any donor gene can be transduced to a recipient cell by a phage, this is *generalized transduction*. Bacterial cells with new genotypes resulting from transduction are called *transductants*. No wild-type transductants were found in the LA2 side since that strain is nonlysogenic.

The frequency of transduction is very low. The relative efficiency of transduction of any P22 phage lysate is the ratio of the number of transductions produced to the number of P22 phage particles with which the recipient bacteria had been infected. This efficiency is around 10^{-5}–10^{-7}.

Transduction is not confined to *Salmonella typhimurium*. Genes can be transduced in *E. coli* with phage P1 and *Bacillus subtilis* with phage SP10 for example.

How can generalized transduction be used in genetic analysis? The amount of DNA in P22 is about 1/100 that of *Salmonella typhimurium* and thus the transducing particles can only carry a very small part of the host chromosome. Thus transduction can provide information about whether two mutations are closely linked and it can also help ascertain gene order if three genes are being examined. An example using the P1 transducing phage of *E. coli* illustrates this. Two strains of *E. coli* were used, the donor was *leu*$^+$ *thr*$^+$ *azi*^{r} and the recipient was *leu thr azi*^{s}. Phage P1 was grown on the donor cells and the progeny were used to transduce the recipient. In such an experiment, one can select for any of the donor markers in the recipient and then, as was done in the conjugation experiment, one can look for the presence of the other unselected markers among the transductants (with the results shown in Table 11.1).

In the experiment selecting for *leu*$^+$ the results show that the *leu* and *azi* genes are close together, and both are distant from the *thr* gene. The results of

Table 11.1. Cotransduction frequencies for markers in an experiment involving P1 phage, a *leu*$^+$ *thr*$^+$ *azi*^{r} donor, and the *leu thr azi*^{s} recipient.

Selected marker	Unselected marker
leu$^+$	50% = *azi*^{r}
	2% = *thr*$^+$
thr$^+$	3% = *leu*$^+$
	0% = *azi*^{r}

the experiment selecting the *thr*$^+$ marker show that the *leu* gene is closer to the *thr* gene than is the *azi* gene. Thus the order of genes is:

thr leu azi

In an experiment such as this, the recombinants result by recombination between the linear DNA fragment brought in from the donor by the phage particle and the circular DNA of the recipient. A double crossover is necessary for each genetic exchange (Fig. 11.10).

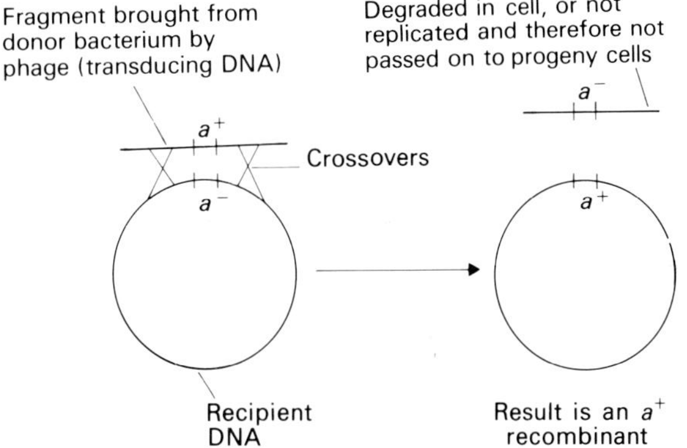

Fig. 11.10. Exchange of gene on transducing DNA with homologous gene on the host genome by a double crossover event.

Given a donor and recipient with two or more closely linked genes, different classes of transductants can result from different crossover positions. Since map distance is calculated from the frequency of crossing over between two genes, it is usual to determine the frequencies of the various transductant classes. Then map distance can be computed exactly as described for the similar example of mapping by transformation presented later in the chapter.

Specialized transduction

E. Lederberg discovered that the *E. coli K12* is a lysogenic strain in that it can carry a temperate phage (λ) (see Fig. 2.5). The lysogeny of *K12* (thus *K12*(λ), c.f. Benzer's *rII* work) was found after nonlysogenic derivatives were accidentally isolated. Phage λ is a DNA-containing phage whose genetic material consists of about 50,000 nucleotide pairs, which is about one-fourth that of T-even phage DNA. The DNA of phage λ is for the most part double-stranded and complementary.

When *E. coli* is infected with λ, the phage DNA circularizes and can either replicate and go through the lytic cycle, or it can integrate into the host chromosome to produce the prophage state. The integration step is similar to the *F*-factor integration and involves a specific λ attachment site (*att*-λ locus) on the host DNA that is homologous with a site on the phage DNA (*b2*). The integration then results from a recombination event using both phage-specific and host enzymes (Fig. 11.11), and this places the λ genome between the *gal* (galactose) and *bio* (biotin) genes of the *E. coli* chromosome.

Steps in λ integration

a Infection of *E. coli*

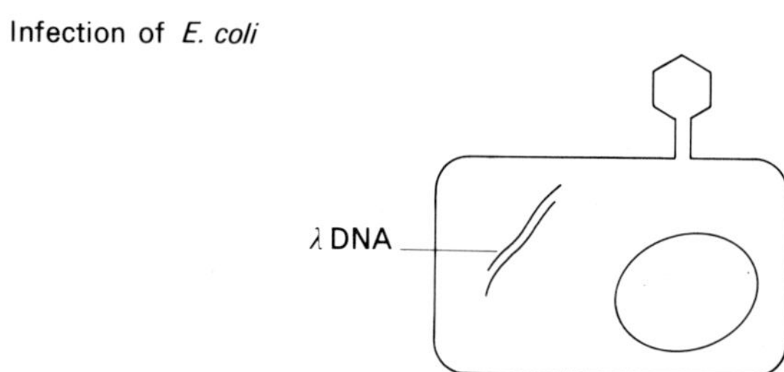

b Circularization of λ genome

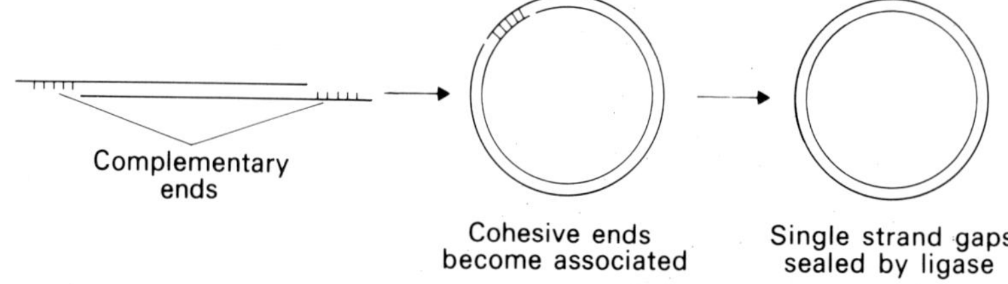

c Integration

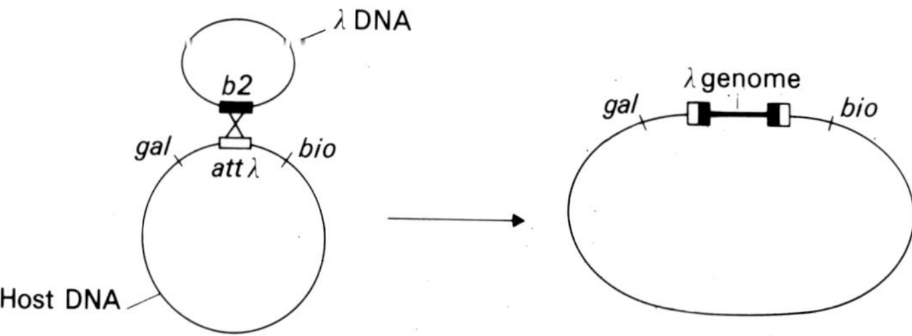

Fig. 11.11. Steps of integration of phage λ genome into the *E. coli* chromosome at the normal recognition site, *att*-λ.

The lysogeny of the λ phage is brought about by a repressor that is coded by one of the genes of the phage itself. The repressor is a protein molecule consisting of four identical subunits, each with a molecular weight of 38,000. The repressor acts by blocking the transcription of phage genes.

In 1956, J. Lederberg tested whether λ can tranfer *E. coli* genes from donor to recipient cells. He took a lysogenic wild-type *K12*(λ) strain and induced the λ prophage with ultraviolet light. This destroys the repressor and thus the phage goes through the lytic cycle generating a lysate of λ phages. He then infected a variety of genetically marked nonlysogenic cultures of *K12* with the phages and plated them on selective media to see if any of the wild-type genes of the *K12*(λ) donor cells had been transferred to the now-lysogenized mutant recipients. The results were mostly negative, the exception being that about 1 in 10^6 of the λ-infected gal^- bacteria (unable to ferment galactose) had acquired the gal^+ phenotype of the donor. Thus λ is capable of transduction, but it is restricted to the *gal* genes in the vicinity of the *att*-λ region: as a result, this phenomenon is called *specialized transduction*.

Most of the transductants are genetically unstable in that each gal^+ colony contains about 1–10% of gal^- cells. The gal^+ transductants, then, are actually gal^+/gal^- partial heterozygotes; that is, the gal^+ donor fragment brought in by the transducing phage has been added to the recipient genome rather than exchanged for the gal^- gene (in this case). Thus there is a potential for losing the gal^+ gene.

All of the events involved in specialized transduction can be summarized diagrammatically. The first step involves erroneous looping out of the *lambda* genome when the prophage is induced (Fig. 11.12). A crossover generates a circular DNA containing most, but not all, of the λ genome as well as some host genome—here including the gal^+ genes. Enzyme action converts this circular DNA to a linear molecule, which is then assembled into a phage particle. The result of these events is the production of a defective phage particle called λ*dg* (λ-defective-galactose), which carries bacterial gal^+ genes. This incorrect looping out occurs only very rarely among prophage excisions.

The λ*dg* phage is a transducing phage since it can transfer the gal^+ genes to a nonlysogenic recipient cell. When it infects such a cell, the λ*dg* DNA can integrate into the recipient chromosome by crossing-over at the homologous *gal* regions (Fig. 11.3).

Here the crossover generates a continuous DNA containing a defective prophage (λ*def*) between two bacterial *gal* genes. In the example, the dominance of the donor gal^+ gene over the recipient gal^- gene results in a gal^+ phenotype. Reversal of the above event in the transductant will produce a gal^- segregant.

The number of λ*dg* phages present in a phage lysate is very small.

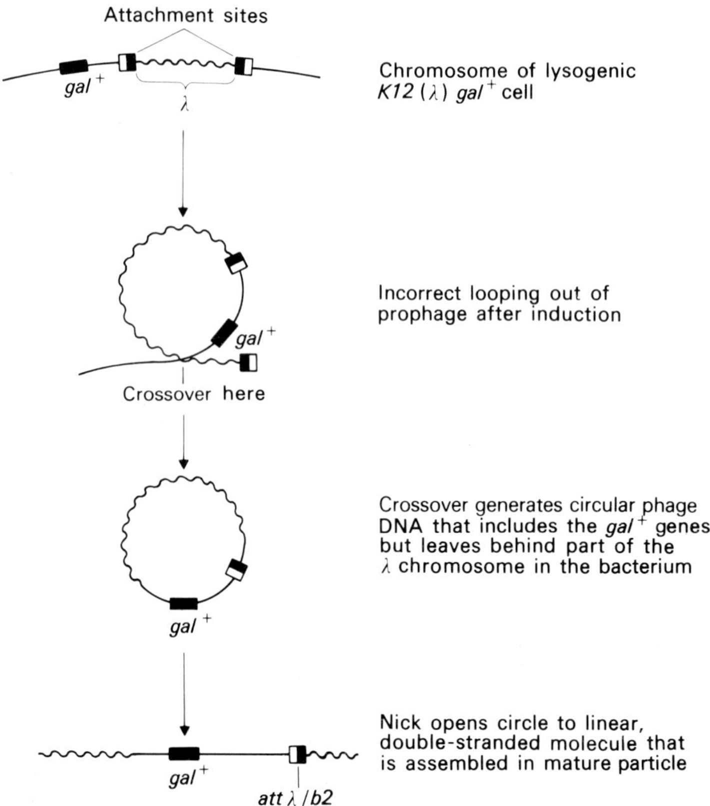

Fig. 11.12. Generation of λ transducing phage (λ*dg*) carrying host *gal*$^+$ genes by incorrect looping out of the prophage from the *E. coli* chromosome.

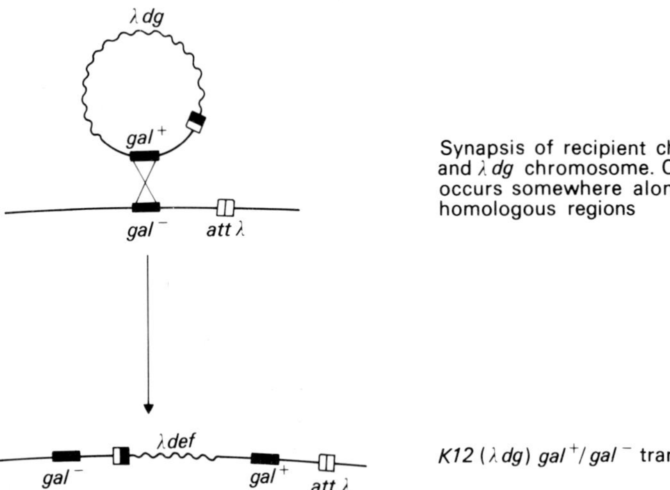

Fig. 11.13. Integration of λ*dg* transducing phage into the *E. coli K12* chromosome by crossing-over in the *gal* region. This results in a *K12*(λ*dg*)*gal*$^+$/*gal*$^-$ transductant.

Therefore if the bacterial infection is done with a relatively large number of phages compared with bacterial cells, it is possible that a λ^+ phage will co-infect with the λdg phage. In this case the λ^+ can integrate at the normal *att*-λ site to produce a double lysogenic *K12*(λ)(λdg) (Fig. 11.14).

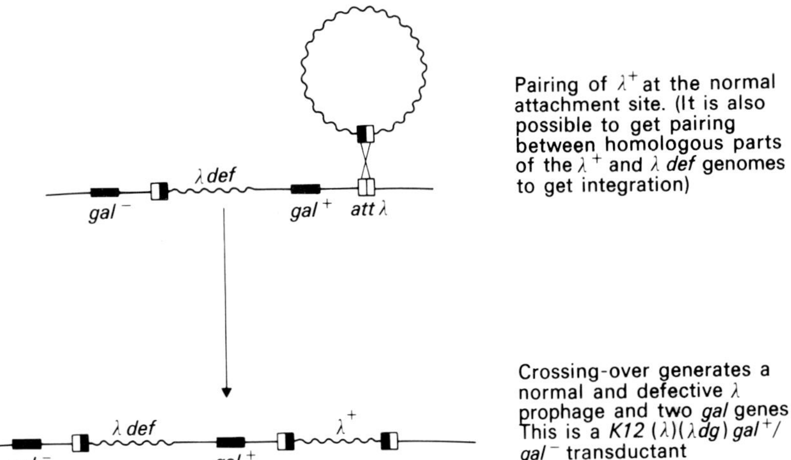

Fig. 11.14. Production of a *K12*(λ)(λdg)gal^+/gal^- transductant by simultaneous integration of λ and λdg DNA into the *E. coli* chromosome. The λdg integrates as in Fig. 11.13 and the λ DNA integrates at the *att*-λ site.

If the *K12*(λ)(λdg)gal^+/gal^- transductant (which has a gal^+ phenotype) is induced with ultraviolet light, then, by the reversal of the processes described, about equal proportions of the λ^+ and λdg phages will be produced. The resulting lysate will have a very high capacity for transducing gal^- recipient cells, and hence it is called an HFT (high-frequency transducing) lysate to contrast it with the LFT (low-frequency transducing) lysate described earlier where only 1 in 10^6 phages carried donor *gal* genes.

How can specialized transduction be used? One example is that complementation tests can be performed for mutations in the *gal* region to define the number of cistrons. (Indeed the *gal* region is an operon consisting of three cistrons.) Thus one can have lysogenic donor cells that carry one type of gal^- mutation and generate transducing phages that are used to infect nonlysogenic recipient cells carrying a different gal^- mutation. If the two mutations are in different cistrons (complementation groups), complementation will occur and the recipient will become gal^+. If the two mutations are in the same cistron, no gal^+ transductants will be found.

Transformation

The previous sections in this chapter have demonstrated how gene maps can be constructed in bacteria that conjugate and/or that have transducing phages. There is a third method of DNA transfer between bacteria, **transformation**, a process involving the transfer of genetic material between donor and recipient cells by means of extracellular pieces of DNA; no cell contact or phage vectors are involved. Bacteria with new genotypes resulting from transformation are called *transformants*. Transformation can be used for gene mapping in certain bacteria, including some that are not amendable to conjugation or transduction. One example of transformation in Chapter 1 was described when evidence that DNA is the genetic material was discussed.

Mechanism of transformation

The basic mechanism of transformation involves bacteria taking up fragments of DNA that may then exchange with the homologous DNA of the cells by crossing-over. Thus, as in the discussions of conjugation and transduction, it is appropriate to consider donor and recipient cells. For purposes of mapping, the two strains are usually manipulated to have different genotypes. Experimentally, DNA is extracted from donor cells and the fragmented molecules are added to the recipient cell population. The recipient, then, takes up random pieces of DNA.

Not all bacterial species have the ability to take up DNA, and even those that do may need to be in a particular phase of the growth cycle or in a particular growth medium for them to be competent to take in the DNA. In this regard the bacterial species *Diplococcus pneumoniae* (see Chapter 1) and *Bacillus subtilis* are relatively easy to render *competent* (able to take up DNA), whereas *E. coli* must be genetically deficient for two exonucleases and must be incubated in the presence of a high concentration of calcium chloride to make the membrane permeable to DNA. The transformation of *E. coli*, then, has not been utilized for gene mapping; conjugation or transduction are preferable procedures to use. However, as will be discussed in the next chapter, transformation of *E. coli* is an integral part of recombinant DNA technology.

For efficient transformation of a bacterium such as *Bacillus subtilis*, the DNA must be double-stranded and of relatively high molecular weight (1×10^6 to 8×10^6 daltons). As the DNA crosses the membrane of a competent bacterium, one of the strands of the DNA is broken down and provides energy for the transfer of the DNA. The resulting single-stranded DNA then may exchange with the homologous region of the recipient's chromosome, and this event can be detected given appropriate genetic differences between donor and recipient cell.

Determination of gene linkage by transformation

Suppose we have a donor bacterial strain that is $a^+ b^+$ and a recipient strain that is $a\ b$ and we wish to determine whether the two genes are linked. The first step is to use donor DNA to transform the recipient cell under selective conditions whereby the frequency of transformation (the proportion of recipients that are transformed to the wild-type state) can be calculated for the *a* and *b* genes separately. In such an experiment the frequency of transformation of a single gene will be rare, ranging from 1 in 10^6 to 1 in 10^3 depending on how much DNA was used. (Therefore, it is important to use the same amount of DNA when comparing frequencies of transformation.) In the next step, the experiment is repeated, but this time the frequency of a^+ b^+ transformants (or, in other words, the cotransformation frequency) is determined. If the two genes a^+ and b^+ are not closely linked in the donor cell, then the cotransformants that are formed can do so only by taking up at least two separate pieces of DNA, with one carrying the a^+ gene and one carrying the b^+ gene. The probability of such cotransformation in this case, then, is the product of the probabilities of transformation for the two genes separately. So, if the observed cotransformation frequency is significantly greater than this calculated frequency, the two genes must be closely linked on the chromosome. Indeed if the observed frequency is very close to the transformation frequency for a single gene, the genes must be very closely linked on the chromosome such that there is a high likelihood that a piece of transforming DNA will carry both genes into the recipient.

Given that two genes are closely linked, it is possible to determine the map distance between them by transformation. The assumption is that the two genes are close enough together such that they are located on the average fragment size of transforming DNA. Then, with an a^+ b^+ donor and an *a b* recipient, three classes of transformants are produced with the respect to the allelic pairs. These are a^+ *b*, *a* b^+, and a^+ b^+. In each case the transformant is produced by a double crossover that exchanges part of the donor DNA with the recipient. If one crossover is to the left of the *a* gene and the second between the *a* and *b* genes as shown here:

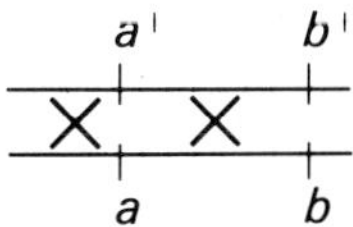

the a^+ *b* transformants are produced. Or, if one crossover occurs between the *a* and a crossover to the right of *b* as follows:

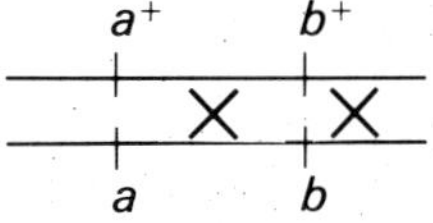

the *a* b^+ transformants are generated. Finally, a crossover to the left of *a* and a crossover to the right of *b* as shown here:

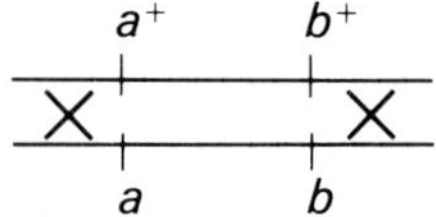

gives the $a^+ b^+$ transformants.

Thus the map distance between a and b is given by the number of transformants that are produced in which there is a crossover between a and b expressed as a percentage of the total number of transformants. In the example, the a-b map distance is given by:

$$\frac{\text{Number of } (a^+ b + a b^+) \text{ transformants}}{\text{total number of transformants}} = 100\%$$

Transformation analysis is also often used in three-factor experiments.

In summary the linkage of genes can be established by determining the frequency of cotransformation as compared with frequencies of transformations of single markers alone. With the appropriate number of gene differences between the donor and recipients, a gene order and hence a gene map can be constructed by these sorts of experiments.

Questions and problems

11.1 Distinguish between F^-, F^+ and *Hfr* strains of *E. coli*.

11.2 In $F^+ \times F^-$ crosses, the F^- recipient is converted to a donor with very high frequency. However, it is rare for a recipient to become a donor in $Hfr \times F^-$ crosses. Explain why.

11.3 Distinguish between the lysogenic and lytic cycles.

11.4 Distinguish between generalized and specialized transduction.

11.5 Using the technique of interrupted mating, four *Hfr* strains were tested for the sequence in which they transmitted a number of different genes to an F^- strain. Each *Hfr* strain was found to transmit its genes in a unique sequence, as shown in the accompanying table (only the first six genes transmitted were scored for each strain).

Order of transmission	*Hfr* strain 1	2	3	4
First	O	R	E	O
	F	H	M	G
	B	M	H	X
	A	E	R	C
	E	A	C	R
Last	M	B	X	H

What is the gene sequence in the original strain from which these *Hfr* strains derive? Indicate on your diagram the origin and polarity of each of the four *Hfr*s.

11.6 When *Hfr* donors conjugate with F^- recipients that are lysogenic for phage λ, the recipients (the exconjugant zygotes) usually survive. However, when *Hfr* donors that are lysogenic for λ conjugate with F^- cells that are nonlysogens, the exconjugant zygotes produced from matings that have lasted at least 100 minutes usually lyse, releasing mature λ phage particles. This is called zygotic induction of λ.

(a) Explain zygotic induction.

(b) Explain how the locus of the integrated λ prophage can be determined.

11.7 Cotransduction of genes leu^+ and trp^+ to recipient *leu trp* cells produced the following transductants:

369 leu^+ trp^+
31 leu trp^+
46 leu^+ trp

What is the map distance between *leu* and *trp*?

11.8 If an *E. coli* auxotroph *A* could only grow on a medium containing thymine, and an auxotroph *B* could only grow on a medium containing leucine, how would you test whether DNA from *A* could transform *B*?

11.9 In a transformation experiment, DNA was isolated from donors of genotype p^+ m^+ and was used to transform recipient cells of genotype *p m*. The resulting transformants were as follows:

Class I	$p^+ m$	228
Class II	$p\ m^+$	141
Class III	$p^+ m^+$	631
		1000 total transformants

Assuming that the transforming DNA fragments were all the same size, determine the map distance between *p* and *m*.

References

Conjugation

Campbell, A., 1969. *Episomes.* Harper and Row, New York.
Curtiss, R., 1969. Bacterial conjugation. *Annu. Rev. Microbiol.* **23**:69–136.
Gilbert, W. and D. Dressler, 1969. DNA replication: the rolling circle model. *Cold Spring Harbor Symp. Quant. Biol.* **33**:473–484.
Susman, M., 1970. General bacterial genetics. *Annu. Rev. Genetics* **4**:135–176.
Vielmetter, W., F. Bonhoeffer and A. Schutte, 1968. Genetic evidence for transfer of a single DNA strand during bacterial conjugation. *J. Mol. Biol.* **37**:81–86.
Wollman, E.L., F. Jacob and W. Hayes, 1962. Conjugation and genetic recombination in *E. coli K-12. Cold Spring Harbor Symp. Quant. Biol.* **21**:141–162

Transduction

Campbell, A.M., 1962. Episomes. *Adv. Genetics* **11**:101–145.
Jacob, F., 1955. Transduction of lysogeny in *Escherichia coli. Virology* **1**:207–220.

Jacob, F. and E.L. Wollman, 1961. *Sexuality and the Genetics of Bacteria*. Academic Press, New York.

Morse, M.L., E.M. Lederberg and J. Lederberg, 1956. Transduction in *Escherichia coli K-12*. *Genetics* **41**:142–156.

Ozeki, H. and H. Ikeda, 1968. Transduction mechanisms. *Annu. Rev. Genetics* **2**:245–278.

Zinder, N.D. and J.L. Lederberg, 1952. Genetic exchange in *Salmonella*. *J. Bacteriol.* **64**:679–699.

Transformation

Archer, L.J., 1973. *Bacterial Transformation*. Academic Press, New York.

Dubnau, D., D. Goldthwaite, I. Smith and J. Marmur. 1967. Genetic mapping in *Bacillus subtilis*. *J. Mol. Biol.* **27**:163–185.

Goodgal, S.H., 1961. Studies on transformation of *Hemophilus influenzae*. IV. Linked and unlinked transformations. *J. Gen. Physiol.* **45**:205–228.

Hotchkiss, R.D. and M. Gabor, 1970. Bacterial transformation, with special reference to recombination processes. *Annu. Rev. Genetics* **4**:193–224.

Hotchkiss, R.D. and J. Marmur, 1954. Double marker transformations as evidence of linked factors in deoxyribonucleate transforming agents. *Proc. Natl. Acad. Sci. USA* **40**:55–60.

Lacks, S., B. Greenberg and M. Neuberger, 1974. Role of a deoxyribonuclease in the genetic transformation of *Diplococcus pneumoniae*. *Proc. Natl. Acad. Sci. USA* **71**:2305–2309.

Ravin, A.W., 1961. The genetics of transformation. *Adv. Genetics* **10**:61–163.

Tomaz, A., 1969. Some aspects of the competent state in genetic transformation. *Annu. Rev. Genetics* **3**:217–232.

Chapter 12 Recombinant DNA

Outline

In recent years, experimental procedures have been developed that have allowed researchers to construct **recombinant DNA molecules** in the test tube; each DNA molecule is composed of genetic material from two different sources. This has opened the way for new and exciting research possibilities and affirms the plausibility of *genetic engineering*. In this chapter we will discuss the techniques by which recombinant DNA can be made and we will present some examples of how **recombinant DNA technology** is furthering knowledge of the structure and function of prokaryotic and eukaryotic genomes.

In outline, recombinant DNA is made in the following way. A piece of DNA from the organism of interest is sliced into either a plasmid or a lambda phage (called the **cloning vehicle** or **cloning vector**) and the resulting chimeric molecule is used to transform or inject, respectively, a host bacterial cell. The latter is often a special strain of *E. coli* that is unable to reproduce without special culture conditions. Reproduction of the *E. coli* results in the replication ("cloning") of the recombinant DNA molecule (the process is thus also called *molecular cloning*), thus producing many copies for analysis.

Restriction endonucleases

The restriction-modification phenomenon

One reason for the rapid development of recombinant DNA technology was the discovery of a variety of enzymes that catalyze the cleavage of DNA at a small number of reproducible sites. These enzymes are called *restriction endonucleases* or *restriction enzymes*. (Recall from previous discussions that an endonuclease results in a cut within a nucleic acid chain.)

In the 1950s, S. Luria and colleagues found that the progeny phage produced after infection of a particular strain of *E. coli* (*B/4*) by phage T2 could no longer reproduce in the normal host strain of *E. coli*. They reasoned that the phage released from *B/4* had become modified in some way such that the normal host was no longer permissive for growth of T2. This phenomenon was called *host-controlled modification and restriction*.

In the 1960s W. Arber and D. Dussoix shed some light on the restriction phenomenon. They grew phage λ on *E. coli K* and used the progeny to infect *E. coli B*, with the result that most of the DNA was broken into pieces; that is, the DNA was *restricted*. Apparently a few DNA strands remained intact since some progeny phages were produced from the injection of strain *B*. These could now infect *E. coli B* normally and about 2% of them could also infect *E. coli K*.

To investigate this in more detail, Arber's group grew phage λ in *E. coli K* that was growing in a culture medium containing the heavy isotopes ^{15}N and ^{2}H (deuterium) so that the DNA of the resulting progeny phage was more dense than normal phage DNA. The heavy phages were then used to infect *E. coli B* growing in normal light medium and the densities of the progeny phage's DNA were determined by cesium chloride density gradient centrifugation. Most of the progeny phages contained normal light DNA, which had been entirely synthesized from new material. These phages could not grow in *E. coli K*. A few of the progeny phage were heavy, and these retained the ability to grow in *E. coli K*. They reasoned, therefore, that phages grown in particular bacterial strains become modified in a way specific for that particular strain. Further, the modification occurred on the DNA itself and involved a covalent bond since it was not lost by passage through the bacterial host. It now appears that in many bacteria the modification involves the addition of methyl groups on certain bases in particular DNA sequences in the genome.

In 1970 a major breakthrough occurred. This was the identification of restriction enzymes, which play a role in the restriction phenomenon. These enzymes recognize a specific nucleotide sequence in DNA and this may then result in cleavage. The way this relates to the modification-restriction phenomenon is as follows. A bacterial cell has two enzymes (or sets of two enzymes, depending on the number of modifications of which it is capable) both of which have a recognition site for a particular nucleotide-pair sequence. One of them, the *modification enzyme*, catalyzes the addition of a modifying group (often a methyl group) to one or more nucleotides in the sequence. The other enzyme, the *restriction enzyme*, can only recognize the unmodified nucleotide sequence and hence serves to digest invading DNA unless it too is modified like the host DNA. To date a number of restriction-modification systems have been identified in bacateria. However, not all of the enzymes that are being used in recombinant DNA experiments have been shown to have a role in a restriction-modification system in the bacterium from which they have been isolated, even though they are called restriction enzymes.

Properties of restriction enzymes

Over 100 restriction enzymes have been isolated so far. All of these enzymes have been found in prokaryotes; no similar enzymes have been identified in the few eukaryotic organisms that have been examined.

The restriction enzymes fall into two classes with regard to the way they cleave DNA. Class I enzymes recognize a specific nucleotide-pair sequence and then cleave the DNA at a nonspecific site away from that recognition site. The enzymes involved in the *E. coli K* and *B* restriction system belong to this class. Class II enzymes, on the other hand, cleave the DNA at the specific recognition site. For that reason, and the fact that the recognition sites have twofold rotational symmetry (Fig. 12.1), these enzymes are valuable for constructing recombinant DNA molecules, as we shall see.

Class II restriction endonucleases have been isolated from a large number of microorganisms. All of the enzymes are sequence specific, and thus the number of cuts they make in a particular DNA molecule or population of molecules is dependent upon the number of times the particular sequence is present in the DNA. The cleavage sites for a number of restriction enzymes are shown in Table 12.1 along with the number of cuts each enzyme makes in the commonly tested DNAs from phage λ, adenovirus-2 (Ad2), and simian

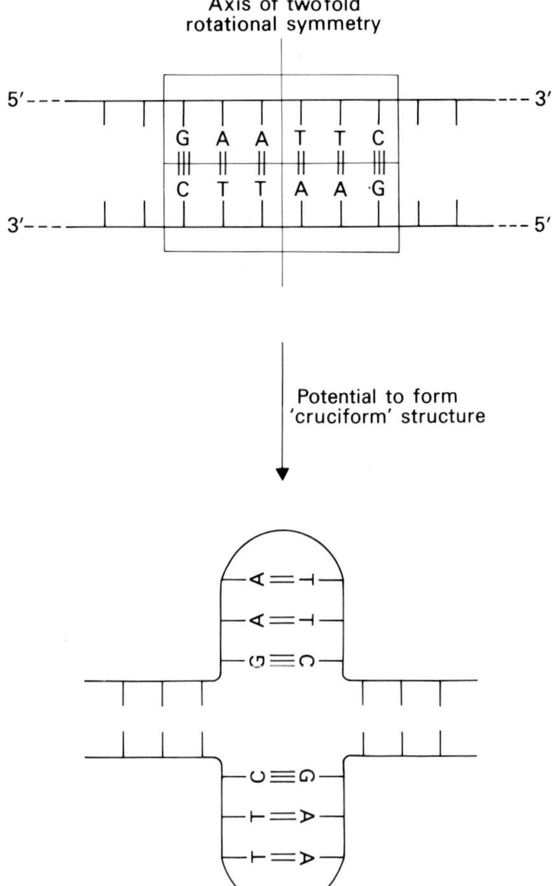

Fig. 12.1. Region of DNA showing twofold rotational symmetry of nucleotide-pair sequence. The sequence shown is actually the recognition site for the restriction endonuclease EcoRI.

Table 12.1. Characteristics of some restriction endonuclease. (After R. J. Roberts, 1976. *Crit. Rev. Biochem.* **4:**123.)

Enzyme name	Organism from which enzyme is isolated	Recognition sequence and position of cut	Number of cleavage sites in DNA from:		
			λ	Ad2	SV40
BamHI	*Bacillus amyloliquefaciens H*	5′ G↓G A T C C 3′* 3′ C C T A G↑G 5′	5	3	1
BglIII	*Bacillus globigi*	A↓G A T C T T C T A G↑A	5	12	0
EcoRI	*E. coli RY13*	G↓A A T T C C T T A A↑G	5	5	1
HaeIII	*Haemophilus egyptius*	G G↓C C C C↑G G	50	50	18
HhaI	*Haemophilus hemolyticus*	G C G↓C C↑G C G	50	50	2
HindIII	*Haemophilus influenzae* R_d	A↓A G C T T T T C G A↑A	6	11	6
PstI	*Providencia stuartii*	C T G C A↓G G↑A C G T C	18	25	3
SmaI	*Serratia marcescens*	C C C↓G G G G G G↑C C C	3	12	0

*In this column the two strands of DNA are shown with the sites of cleavage indicated by arrows. Since there is an axis of twofold rotational symmetry in each recognition sequence, the DNA molecules resulting from the cleavage are symmetrical. In some cases, the two cuts are staggered and hence the DNA molecules produced have complementary, single-stranded ends (e.g. BamHI), while in other cases the point of cleavage is at the axis of symmetry and the DNA molecules produced are double-stranded with no single-stranded ends; these are called "blunt" ends (e.g. HaeIII).

virus 40 (SV 40). This information can be used as a guideline for choosing an enzyme for a particular application. As can be seen, some enzymes cut both strands of DNA between adjacent nucleotide pairs, whereas others make staggered cuts in the symmetrical nucleotide pair sequence.

Cloning vehicles

To clone a piece of DNA for study, it must first be attached to a **cloning vehicle (vector)**. One type of cloning vehicle used for these experiments is **plasmids**. Plasmids are extrachromosomal genetic elements that replicate autonomously within bacterial cells. Their DNA is circular and double-stranded and they carry the genes required for replication of the plasmid as well as for the other functions that plasmids have. The plasmids most extensively used in recombinant DNA experiments carry antibiotic-resistance genes that play an important role in the identification and selection of recombinant DNA molecules. In general the plasmid vehicles have a different buoyant density than that of the host DNA and thus they can easily be purified. Some plasmids have the ability to integrate into the host's chromosome, and these are called *episomes*. The *F* factor that is involved with conjugation of *E. coli* is an example of an episome. Further, only some of the plasmids are able to promote conjugation between bacteria. Thus to clone a piece of DNA the DNA is spliced into the plasmid DNA, and the chimeric molecule is used to transform a host bacterium such as *E. coli*. The plasmid then replicates within the host as the host grows and divides, and the piece of DNA of interest becomes cloned.

A plasmid that has been used extensively for molecular cloning is pBR322 (Fig. 12.2). This plasmid is of the nonconjugal type; it will not promote conjugation, and each *E. coli* cell transformed with it will have six to eight copies per host chromosome.

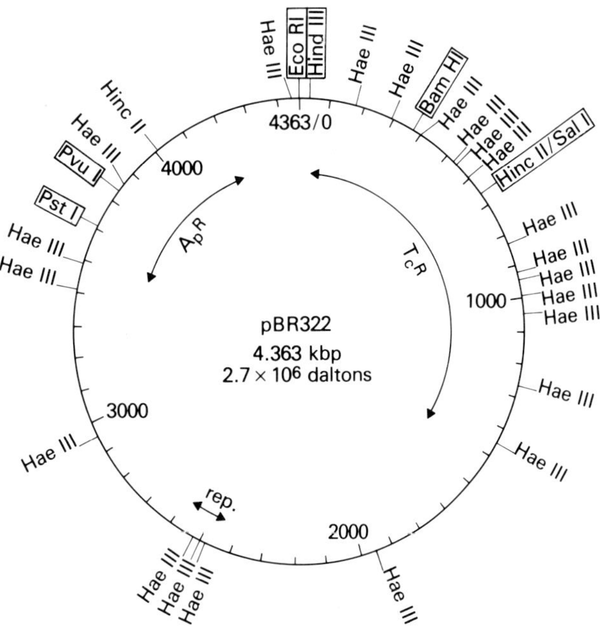

Fig. 12.2. Physical map of the pBR322 cloning vehicle. The plasmid consists of 4363 base pairs and contains genes that confer ampicillin resistance (Ap^R) and tetracycline resistance (Tc^R) upon *E. coli* cells in which the plasmid replicates. *rep* is the origin of replication sequence. Cleavage sites are shown for a few of the restriction enzymes that cut this plasmid. The boxed enzymes cut within specific six-nucleotide-pair sequences, whereas HaeIII cuts within a specific four-nucleotide-pair sequence.

pBR322 is a plasmid that has been "engineered" in the laboratory from natural plasmids so that it has features useful for molecular cloning experiments. Its replication in the *E. coli* cell is dependent upon the presence of *rep*, the origin of replication sequence (about "7 o'clock" in Fig. 12.2). pBR322 is 4363 base pairs (4.363 kilobase pairs [kbp]) long, weighing 2.7×10^6 daltons. This compares with 3600 kbp (2500×10^6 daltons) for the *E. coli* chromosome. pBR322 is cleaved once by any one of the following enzymes: EcoRI, HindIII, BamHI, SalI, PstI, PvuI, PvuII, AvaI, and ClaI. The sites for HindIII,

BamHI, and SalI all lie within the Tc^R gene (which confers tetracycline resistance upon cells with the plasmid), and the sites for PstI and PvuI are within the Ap^R gene (which confers ampicillin resistance upon cells with the plasmid). Such cleavage produces linear plasmid molecules that are suitable for use in the construction of recombinant DNA molecules as will be described later. None of the aforementioned enzymes cut within the *rep* sequence so that any recombinant DNA molecules generated are capable of replication within the bacterial host. A number of other restriction enzymes cut the plasmid but at two or more sites. These enzymes are not useful in cloning experiments, however, since the plasmid is cleaved into two or more pieces by their action and hence it is not possible simply to insert a piece of "foreign" DNA into the plasmid sequence.

Bacteriophages or their derivatives are also used as cloning vehicles. For example, engineered mutants of λ can be used. The wild-type (linear) λ genome consists of 46.5 kbp of DNA, and within this are five EcoRI cleavage sites. Derivatives have been made in the laboratory that, for example, have only two EcoRI sites. Other derivatives have been made similarly with two BamHI sites or two HindIII sites, or two of a variety of other enzyme sites. In these special strains, the two cleavage sites are located on either side of an approximately 25 kbp segment of the λ genome that is not essential for phage replication in *E. coli*; this segment can be replaced with a piece of DNA to be cloned. The normal λ genome contains 49 kbp of DNA, but the phage head can hold between 38 and 53 kbp of DNA without affecting normal function. Thus, large pieces of DNA can be cloned using a λ vector.

Construction and cloning of recombinant DNA molecules

We have already discussed plasmid vehicles and restriction enzymes. Here we describe how these two can be employed to prepare recombinant DNA molecules, that is, the insertion of foreign DNA into the plasmid vehicle.

Insertion of DNA into the plasmid vehicle

In Table 12.1, it can be seen that a number of restriction endonucleases make *staggered cuts* at specific recognition sites. Thus, for example, foreign DNA can be cut into segments by a restriction enzyme that also makes one cut in the plasmid vehicle (for example, pBR322), converting the latter to a linear molecule. Fig. 12.3 illustrates this for EcoRI. By the very fact that EcoRI cleaves at a specific site in this way, the single-stranded ends of the linear plasmid vehicle are complementary to the single-stranded ends of the EcoRI-generated segments of foreign DNA. In solution the two DNAs can come together to produce a larger circular DNA held together by hydrogen

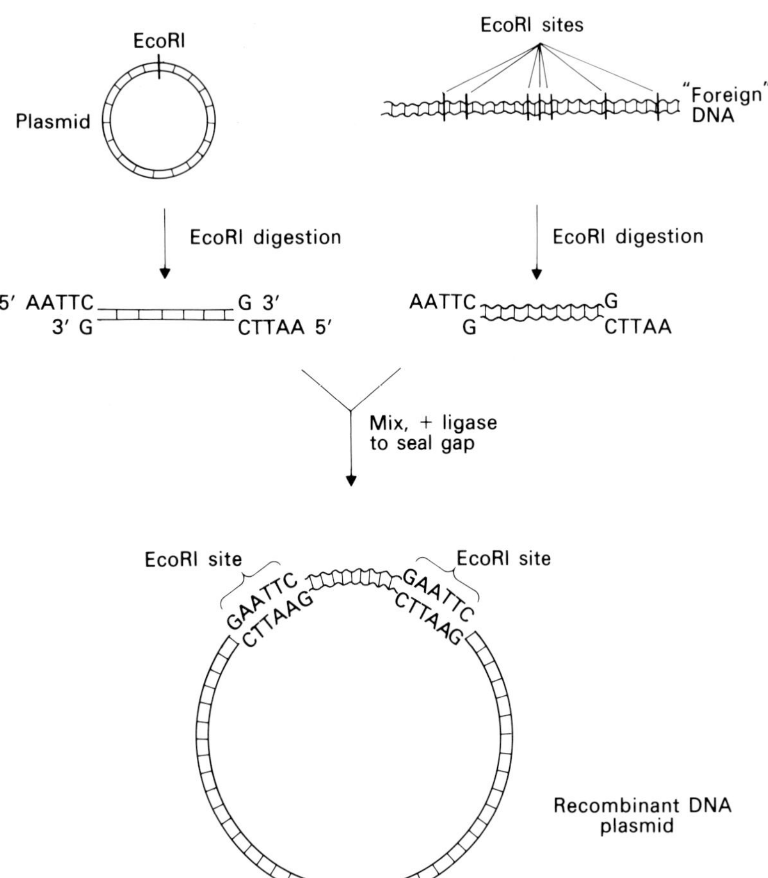

Fig. 12.3. Construction of a recombinant DNA plasmid through the use of the restriction enzyme EcoRI. Details of the procedure are given in the text.

bonding of the complementary ends. In the presence of the enzyme polynucleotide ligase, the single-stranded gaps in the sugar-phosphate backbones are sealed and the structure is stabilized. The result is a recombinant DNA molecule. Since the ends of the DNA pieces produced by EcoRI digestion are all identical, the foreign DNA can insert into the plasmid vehicle in two orientations, and this will occur in a random way. This orientation may have effects on transcription of the genes or gene fragments on the foreign DNA since the initiation of transcription is likely to depend on the location of promoter and other controlling sites on the vehicle.

Practically speaking, the number of single-stranded nucleotides on the DNA after digestion with a restriction enzyme that makes a staggered cut is small, and thus the probability of the complementary sequences finding one another in solution is relatively small. Further, some restriction enzymes do not make staggered cuts, and some methods for producing the DNA fragment to be cloned result in completely double-stranded (blunt-ended) DNA. In these cases it is possible to synthesize a single-stranded polynucleotide chain on the DNA molecules using the enzyme terminal deoxynucleotidyl transferase. Thus, for example, in the presence of dATP, this enzyme will catalyze the production of a poly(dA) tail on each 3′ end of the DNA. To apply this to the insertion of a DNA fragment into the plasmid vehicle, poly(dA) tails (approximately 100 nucleotides long) can be polymerized on the linearized plasmid, and poly(dT) tails of the same length can be polymerized onto the foreign DNA (Fig. 12.4). Then, by mixing the DNAs in solution and adding polynucleotide ligase, a recombinant DNA molecule can be produced. Note that in this procedure the only circular DNA that can result is one in which the foreign DNA has been inserted into the plasmid.

Cloning of recombinant DNA

Once a recombinant DNA molecule has been formed, the next step is to transform an *E. coli* (or other bacterial) strain with it. As mentioned in the previous chapter, transformation of *E. coli* with relatively large pieces of DNA can be facilitated by treating the cells with $CaCl_2$ to make them permeable. With this procedure, the transforming DNA enters the cell intact and the host bacterium remains viable. Then, during growth of the *E. coli* cell, the plasmid replicates under the control of its genes as discussed previously.

Ideally one needs to have a way of determining whether the host cell has been transformed with the plasmid. Further, since it is the foreign DNA that one is interested in studying, it would be useful if one could isolate just those transformed cells that carry recombinant plasmids. The presence of a plasmid in an *E. coli* cell can be shown by the fact that it will confer antibiotic resistance on the bacterium as a result of the plasmid-specific genes it carries. For example, if the plasmid pBR322 transforms *E. coli*, the bacterium will become resistant to both ampicillin and tetracycline. When recombinant DNA plasmids are constructed using the method described, that is, using the complementarity of the restriction enzyme-generated single-stranded sequences, then the population of DNA molecules used to transform the cells will contain a large number of nonrecombinant plasmids (the plasmid vehicles alone that have spontaneously recircularized). In such cases it may not be as easy to determine whether a clone of the transformed *E. coli* cell carries a recombinant DNA molecule or just the plasmid vehicle. However, if a restriction enzyme is used that makes a cut in one of the antibiotic-resistant genes, the insertion of a foreign DNA fragment will split that gene apart and the recombinant plasmid is no longer able to confer resistance to that antibiotic on the bacterial cell that it transforms. Thus, for example, if PstI is used to generate DNA fragments and to open the pBR322 plasmid, any recom-

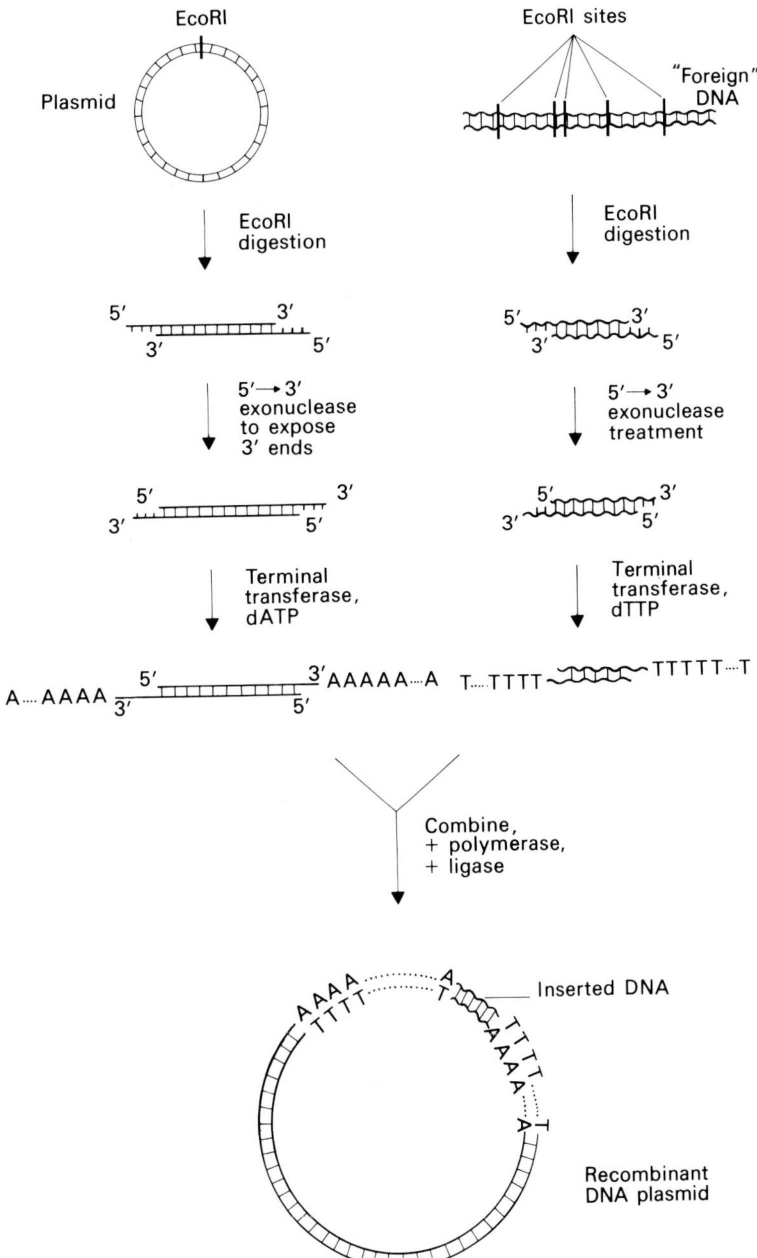

Fig. 12.4. Construction of a recombinant DNA plasmid using the enzyme terminal transferase to synthesize complementary ends on the linearized plasmid and a restriction enzyme generated fragment of foreign DNA. Details of the procedure are given in the text.

binant plasmid that is produced will make the cell it transforms tetracycline resistant but *not* ampicillin resistant. Therefore, if the transformed cells are cloned by plating them and allowing them to form colonies on solid culture medium, this can easily be tested by using the replica-plating technique described in Chapter 6. The cells transformed with nonrecombinant plasmids, then, can be detected since the bacteria will be resistant to both antibiotics and these clones can be removed from further consideration. However, if a piece of DNA is inserted into the EcoRI site of pBR322, as in Fig. 12.2, other means must be used to identify transformants carrying recombinant DNA molecules.

Advances are being made in this area all the time so it is now possible to clone recombinant DNA molecules in other host cells, including yeast and some mammalian tissue culture lines.

Genomic libraries

Researchers typically want to study a small segment of an organism's genome, such as a gene. One approach to obtaining a particular cloned segment of the genome is to isolate it from a *genomic library* through the use of a specific probe. A genomic library is a collection of clones that together contain at least one copy of every DNA sequence in the genome.

Genomic libraries of eukaryotic DNA may be prepared in two ways:

1. Genomic DNA is digested to completion with a restriction enzyme and the fragments are inserted into a suitable vector, usually λ (see earlier). One drawback of this method is that if the sequence of interest contains recognition site(s) for the restriction enzyme used, the sequence will be cloned in two or more pieces. Another drawback is that the average size of the fragment produced by digestion of eukaryotic DNA with restriction enzymes that have six base-pair recognition sequences is relatively small (about 4 kbp). Thus, an entire library would of necessity contain a very large number of recombinant phages, and screening by hybridization would be laborious.

2. Both of the problems of the first method can be avoided by cloning large (about 20 kbp) DNA fragments that are generated by random shearing of eukaryotic DNA. This method ensures that sequences are not excluded from the cloned library simply because of the distribution of restriction sites. The procedure is done as follows: High-molecular-weight eukaryotic DNA is fragmented randomly so that a population of molecules is produced with an average size of 20 kbp. Random fragmentation can only be achieved by mechanical shearing, but DNA prepared in this way must be subjected to several additional enzymatic manipulations (e.g. the addition of restriction enzyme sites) so that the fragments can be cloned. So, instead, the method typically used is partial digestion of the DNA with restriction enzymes that recognize frequently occurring four base-pair recognition sequences. Sucrose density gradient centrifugation or agarose gel electrophoresis can then be used to collect fragments of the size desired. The result is a population of overlapping fragments that is close to random and that can be cloned directly, since the ends of the fragments were produced by restriction enzyme digestion. For example, if the DNA is digested with the enzyme Sau3A (which produces fragments with ends that are complementary to ends produced by BamHI digestion), the fragments generated can be cloned into a particular λ vector that has a central BamHI fragment which is not essential for phage replication in *E. coli* (see earlier in this chapter). This is done by digesting the λ DNA with BamHI, thereby producing a λ left arm, a λ right arm, and the disposable central segment. If the Sau3A digested eukaryotic DNA is mixed with the cut λ DNA, recombinant DNA molecules can be produced that can replicate in *E. coli*; these contain a left arm and a right arm on either side of a partially digested Sau3A fragment.

The aim of the method just described is to produce a library of recombinant molecules that is as complete as possible. However, not all sequences of the eukaryotic genome are equally represented in such a library. For example, if the restriction sites are very far apart or extremely close together in a particular region, the chances of obtaining a fragment of clonable size are small. Additionally, some regions of eukaryotic chromosomes may contain sequences that affect the ability of λ clones containing them to replicate in *E. coli*; these sequences would then be lost from the library. There is also evidence that the presence of tandemly repeated sequences in the cloned eukaryotic DNA segment can lead to deletion of some of the sequence by recombination during phage propagation in *E. coli.*

Finally, the probability of having any DNA sequence represented in the genomic library can be calculated from the formula:

$$N = \frac{\ln(1-P)}{\ln(1-f)}$$

where N is the necessary number of recombinants, P is the probability desired, and f is the fractional proportion of the genome in a single recombinant (i.e. f is the average size, in base pairs, of the fragments used to make the library divided by the size of the genome, in base pairs).

Selection of specific recombinant DNA clones

In many instances the genome of an organism is digested with a restriction enzyme and the fragments produced are then cloned by the procedures described. By using different restriction enzymes, the genome can be cut up in a number of ways, and this results in the formation of an extremely large number of recombinant DNA clones. However, in general an investigator wishes to study a clone carrying a particular DNA segment. In some cases the clone or clones of interest can be identified in a relatively easy way.

As an example, we shall discuss the cloning of a piece of the ribosomal DNA repeat unit of *Neurospora crassa*, which contains the 17S rRNA gene.

Neurospora rDNA. As discussed in Chapter 7, eukaryotic ribosomes consist of two dissimilar subunits, each containing RNA and proteins. There are four different RNA species in the ribosome, which are named according to their sedimentation in density gradients: 17S, 25S, 5S, and 5.8S for the molecules from *Neurospora* ribosomes. The genes for these RNAs are found in the moderately repetitive DNA since they are present in many copies in the chromosomes. In the case of *Neurospora* there are about 200 copies each of the 17S, 25S, 5.8S rRNA genes and 100 copies of the 5S gene. The genes for 17, 5.8 and 25S rRNAs are found in a tandem array with the general organization shown in Fig. 12.5 (refer also to Fig. 7.16).

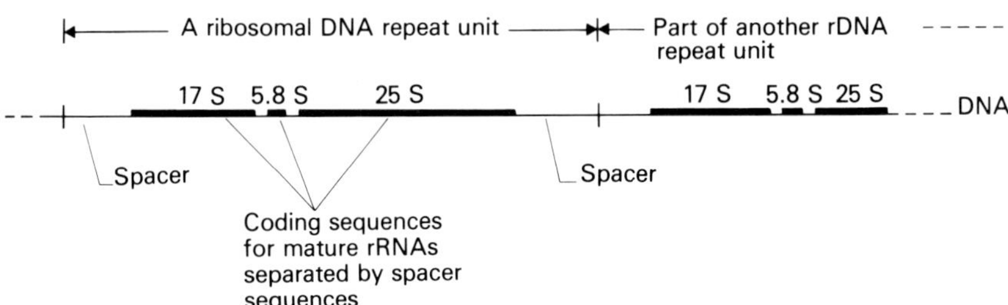

Fig 12.5. Organization of the ribosomal DNA repeat unit of *Neurospora crassa*.

In *Neurospora*, the repeat units are homogeneous; that is, there is no variation in the length or organization of the rDNA repeat units. Fig. 12.6 shows the locations of the known restriction enzyme cleavage sites within the rDNA repeat units.

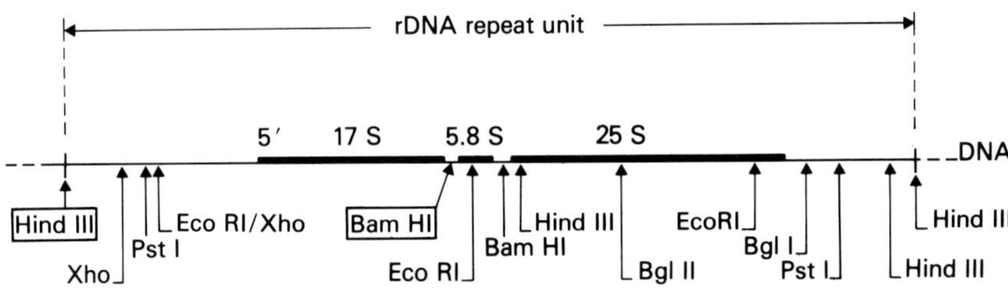

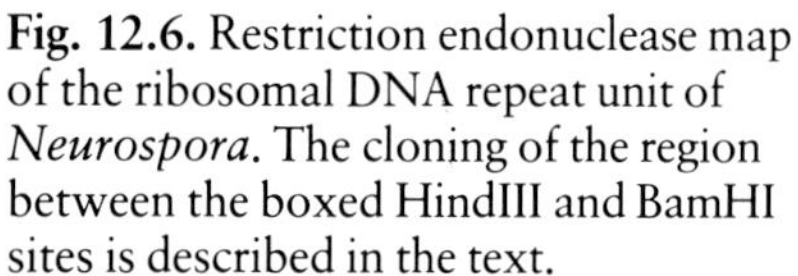

Fig. 12.6. Restriction endonuclease map of the ribosomal DNA repeat unit of *Neurospora*. The cloning of the region between the boxed HindIII and BamHI sites is described in the text.

Preparation and analysis of the DNA fragment for cloning. The experiment discussed in this section has been reported in the literature and involves the cloning of approximately the left half of the ribosomal DNA repeat unit.

Suppose that the part of the DNA we wish to clone extends from the leftmost HindIII site to the BamHI site that is near the 3′ end of the 17S rRNA coding sequence; these enzyme sites are boxed in Fig. 12.6. If total genomic DNA isolated from *Neurospora* is treated with both HindIII and BamHI, a large number of DNA fragments will be produced, and these will have a broad range of size owing to the distribution of the restriction sites over the genome. Because of the relative order of the HindIII and BamHI restriction sites along the DNA, various fragment types will be produced (Fig. 12.7): some fragments will have HindIII sites at each end, some will have BamHI sites at each end, and some will have a HindIII site at one end and a BamHI site at the other end. Among this last class will be the HindIII → BamHI fragment from the rDNA repeat unit that we want to clone. Since there are 200 copies of the repeat unit in the genome, there are 200 times as many of these restriction fragments as any other HindIII → BamHI fragment cut from unique sequence DNA.

Fig. 12.7. The result of cleaving a theoretical section of DNA with both HindIII and BamHI. This produces a number of fragments which have one or other of the enzyme sites at each end, reflecting the arrangement of restriction sites in the DNA.

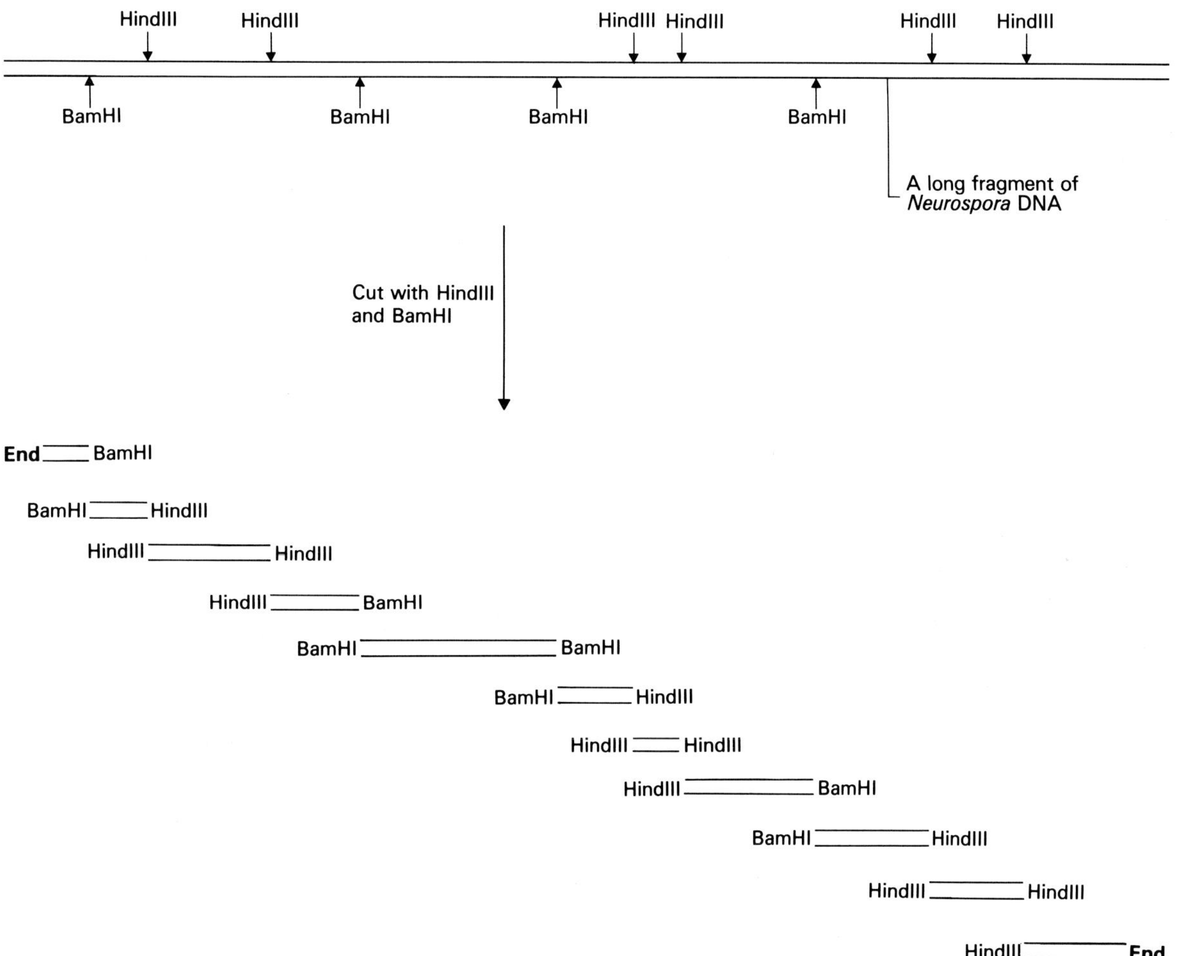

The results of the HindIII+BamHI restriction enzyme digest can be analyzed by subjecting a sample of the cut DNA to *agarose gel electrophoresis*. In this procedure, the DNA sample is placed into a well of a horizontal slab of agarose. Both the gel and the electrophoresis buffer contain ethidium bromide, which binds to DNA, causing the DNA to fluoresce under ultraviolet light illumination. A potential difference is generated along the length of the gel, and the negatively charged DNA fragments migrate toward the anode. They do so as an approximately linear function of the logarithm of their length in base pairs, with the smallest fragments moving fastest and the largest fragments moving slowest. It is usual to have DNA size standards alongside the DNA sample so that the sizes of the DNA fragments in the sample can be calculated.

After electrophoresis, the gel is examined under ultraviolet light to determine the distribution of the DNA fragments in the cut sample. The expected result for the HindIII+BamHI digest would be a smear of fluorescence down the gel lane. The reason for this is as follows: the genome of *Neurospora* contains a large number of restriction sites for the two enzymes used. Treatment with the two enzymes therefore produces an extremely heterogeneous collection of fragments with respect to size and this is reflected in the smear on the gel.

The smear on the gel is not wasted information. We can *probe* the smear in a very powerful way and identify the fragment of DNA that we want to clone. (Or we can probe for any other piece of DNA if we have the right probes.) To do this, the DNA fragments are first transferred from the gel to a sheet of nitrocellulose filter as they were in the gel. This is done by what has become known as the *Southern blot technique* (Fig. 12.8), named after its developer, E. M. Southern. In brief, the gel is treated with alkali to denature the DNA to single strands. The gel is neutralized with buffer and is then placed on filter paper that acts as a wick to pull buffer from a reservoir. A piece of nitrocellulose filter is placed on top of the gel, and on top of the filter is placed a stack of absorbent filter paper. The latter wicks up buffer through the gel, then through the nitrocellulose filter, and onto the stack. By the movement of the buffer, DNA fragments are transferred out of the gel and onto the nitrocellulose to which they stick owing to the properties of the nitrocellulose itself.

Once the Southern transfer is completed, the nitrocellulose filter is dried and then it can be probed, meaning that a (usually) radioactive nucleic acid is added. At this stage single-stranded DNA fragments are fixed on the nitrocellulose filter, and thus any radioactive nucleic acid (probe) added will bind to any DNA fragment to which it is complementary. In this example, we want to find the DNA fragment that contains the 17S rRNA coding sequence.

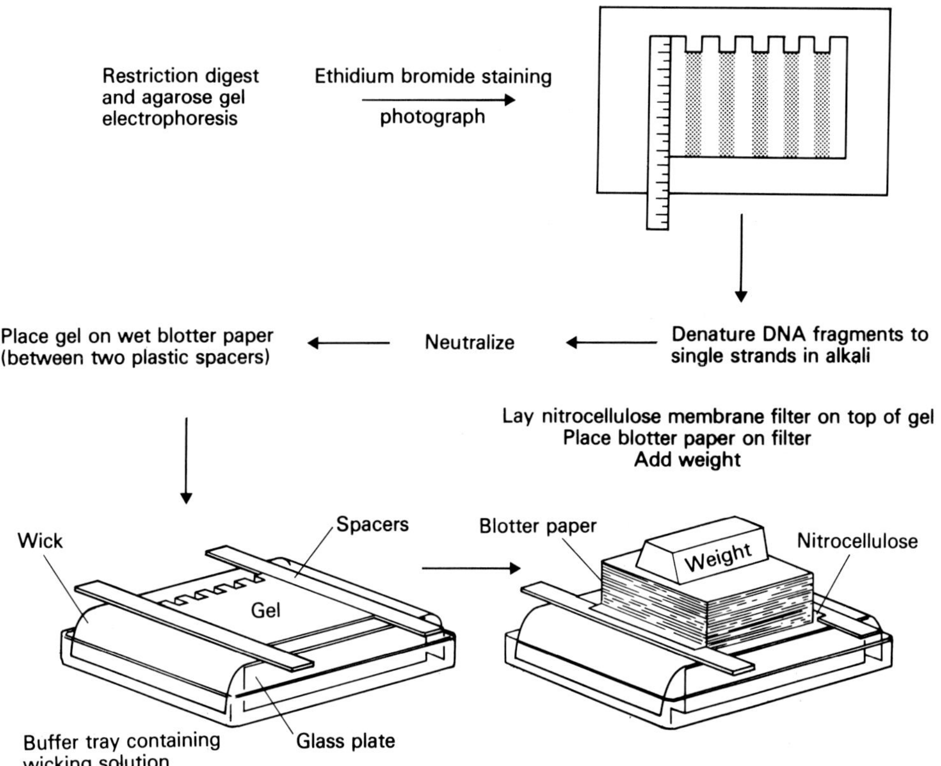

Fig. 12.8. Diagram of the Southern blot technique for the transfer of DNA fragments from agarose gels onto nitrocellulose filters that can be used in radioactive probe experiments. The technique is described in the text.

Therefore the most appropriate probe to use is ^{32}P-labeled 17S rRNA. (Note: In general RNA probes are inferior to DNA probes.) If we hybridize with that probe, the 17S rRNA will hydrogen bond (hybridize) to the HindIII → BamHI fragment in which we are interested. If the filter is now washed free of any remaining unbound radioactive molecules and an autoradiograph prepared of the filter, the result shown in Fig. 12.9 is seen. As can be seen, there is only one band of silver grains on the autoradiograph, and this corresponds to a DNA fragment of about 4.4 kbp. We were able to pick out this fragment from all the rest by using the appropriate radioactive probe. This procedure is limited, therefore, only by the availability of the appropriate probe for doing the hybridization. Thus, as another example, we could probe for the DNA fragment or fragments for any enzyme, if we could isolate and purify the mRNA for that enzyme.

Constructing the recombinant DNA molecule containing a HindIII → BamHI DNA fragment. From the previous discussion, we know that the

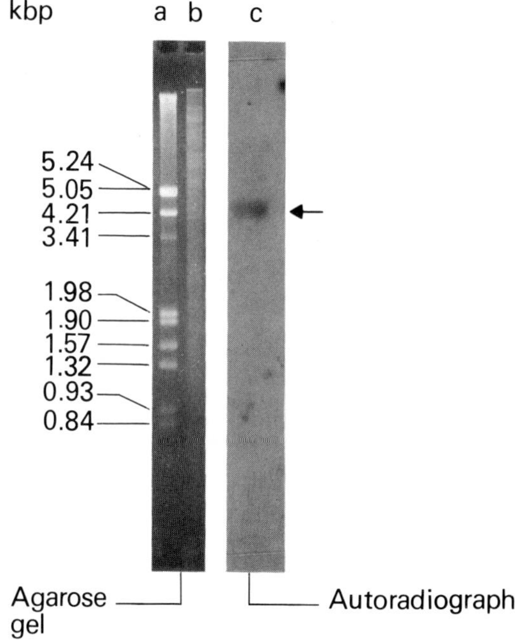

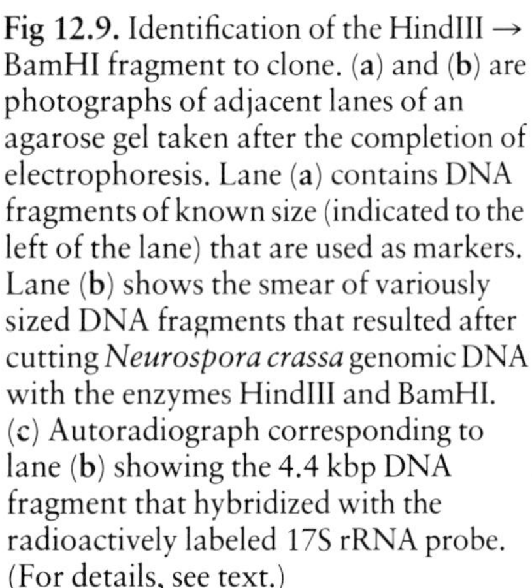

Fig 12.9. Identification of the HindIII → BamHI fragment to clone. (**a**) and (**b**) are photographs of adjacent lanes of an agarose gel taken after the completion of electrophoresis. Lane (**a**) contains DNA fragments of known size (indicated to the left of the lane) that are used as markers. Lane (**b**) shows the smear of variously sized DNA fragments that resulted after cutting *Neurospora crassa* genomic DNA with the enzymes HindIII and BamHI. (**c**) Autoradiograph corresponding to lane (**b**) showing the 4.4 kbp DNA fragment that hybridized with the radioactively labeled 17S rRNA probe. (For details, see text.)

DNA fragment we are interested in is about 4.4 kbp. We cannot purify enough of the fragment out of the smear on the gel to study it. Instead, we have to clone the fragment and again make use of the specific probe to check for the presence of the fragment along the way. In other words, we need to make a recombinant DNA molecule containing the fragments, and then we need to clone it. Fig. 12.10 shows the scheme for cloning HindIII → BamHI cut *Neurospora* DNA, which contains a number of fragments with one end resulting from a HindIII cut and the other end resulting from the BamHI cut. The other ingredient is the appropriate cloning vehicle, in this case the bacterial plasmid pBR322 (see Fig. 12.2). pBR322 has one HindIII restriction site and one BamHI site so that if this plasmid is cut with these two enzymes, two pieces will be produced, one large and one small. Only the larger of the two contains the sequence that is required for replication of the plasmid in *E. coli*. If the cut pBR322 is mixed with the cut *Neurospora* DNA in the presence of the enzyme, DNA ligase, recombinant DNA molecules will be produced (see Fig. 12.10). However, any one of a number of HindIII → BamHI fragments can be inserted into the cut plasmid, including the plasmid piece that was originally cut back. It remains, then, to find the plasmid carrying the piece of *Neurospora* rDNA that we want. To do that, the recombinant plasmids are first cloned.

Cloning the recombinant DNA plasmids. At this point we have in the test

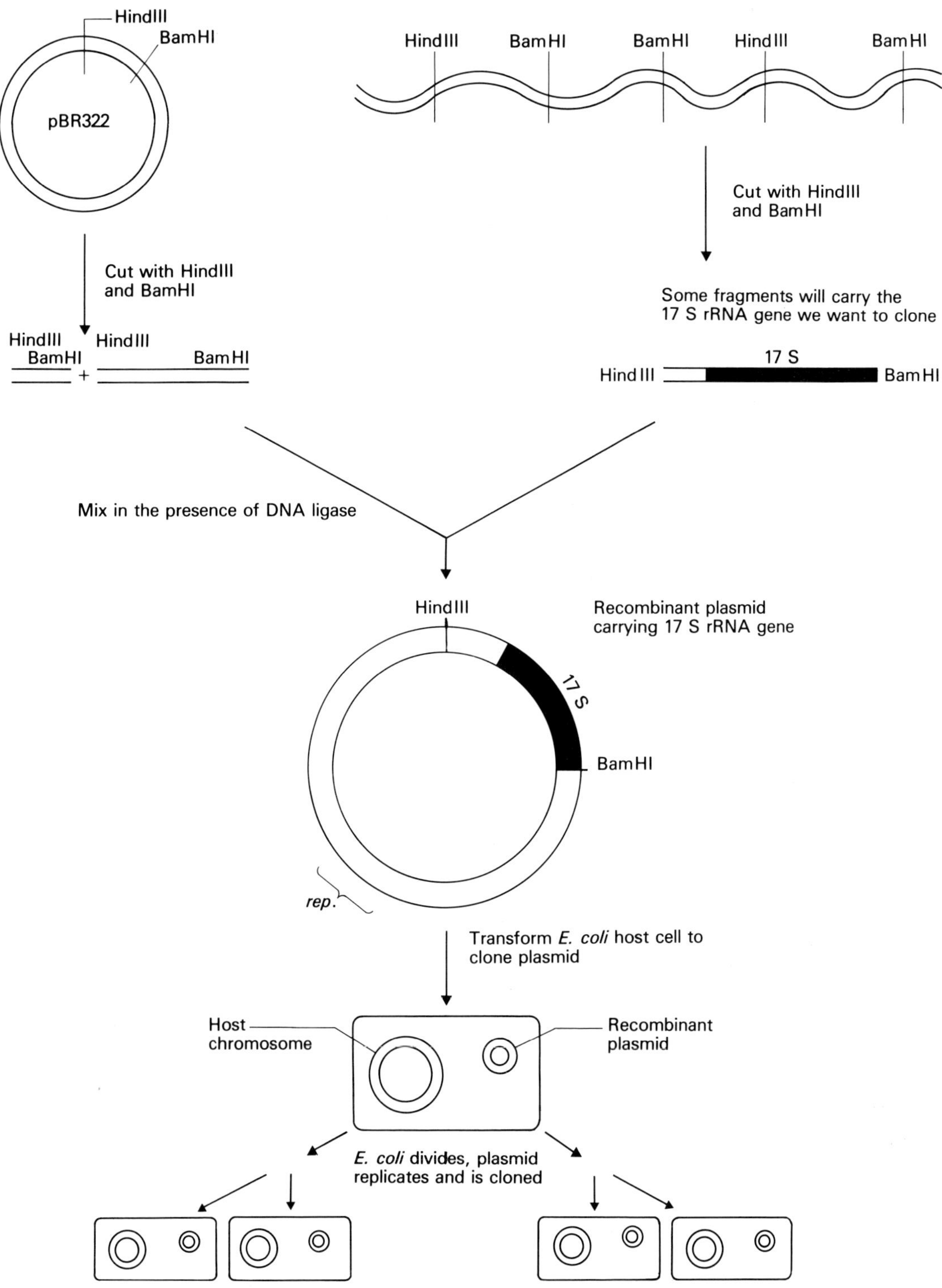

Fig. 12.10. Scheme for the cloning of HindIII → BamHI fragments of *Neurospora* using the pBR322 plasmid cloning vehicle.

tube a mix of plasmid+*Neurospora* recombinant DNA molecules and some reconstructed pBR322 plasmids. Any circular DNA containing the large plasmid piece is able to replicate within the *E. coli* host cell; thus the next step is to get the molecules into the bacterial cell. This is done by a procedure called *transformation*. The *E. coli* strain used for cloning experiments is not wild-type strain, but a mutated strain that has little chance of surviving if it is released into the environment.

To transform *E. coli*, cells are treated with calcium chloride to make the membrane permeable to DNA. The permeabilized cells are then mixed with the recombinant DNA molecules, and in many cases the bacteria will be transformed; that is, they will take up the DNA. The circular DNAs that include the plasmid *rep* sequence then replicate as the *E. coli* grows and divides. This is cloning the plasmid.

Identification of the clones containing the DNA of interest. After the plasmid has been cloned within the bacterial host, we need to identify those clones containing the DNA in which we are interested, in this case the rDNA piece. To make things simpler, there are some selection procedures that can be used.

First, we need to look only at those bacteria that were transformed, that is, those which had taken up circular plasmids capable of replication. This is where we can exploit the drug-resistance genes carried by the plasmid. Recall that pBR322 carries both ampicillin-resistance and tetracycline-resistance genes. Recombinant DNA plasmids containing HindIII → BamHI fragments have an intact ampicillin-resistance gene, but the small HindIII → BamHI fragment that has been replaced in pBR322 contains part of the tetracycline-resistance genes. Thus the recombinant plasmids are resistant to ampicillin and sensitive to tetracycline. The host *E. coli* that is transformed is sensitive to both ampicillin and tetracycline. Therefore transformed bacteria can be selected by growing the cells in a culture medium containing ampicillin to kill the untransformed bacteria.

Once the transformed cells have grown, they are spread onto plates of solid medium containing ampicillin. On the plate, each bacterium will give rise to a colony, and each colony can be picked to a microtiter dish containing wells of ampicillin-containing medium (see Fig. 12.11). This produces a series of clones in liquid culture. Replicas of each culture are printed onto a nitrocellulose filter that is on a plate of ampicillin containing medium. Incubation of the plate gives rise to a matrix of colonies on the filter. The filter is peeled from the plate, and it is treated to lyse the cells, to denature the DNA, and to cause the DNA to stick to the filter. The filter is then processed more or less as for the Southern blot. It is dried, and an appropriate radioactive probe is added. In our case, the probe is ^{32}P-labeled 17S rRNA. Each colony that contained a recombinant DNA plasmid carrying the HindIII → BamHI

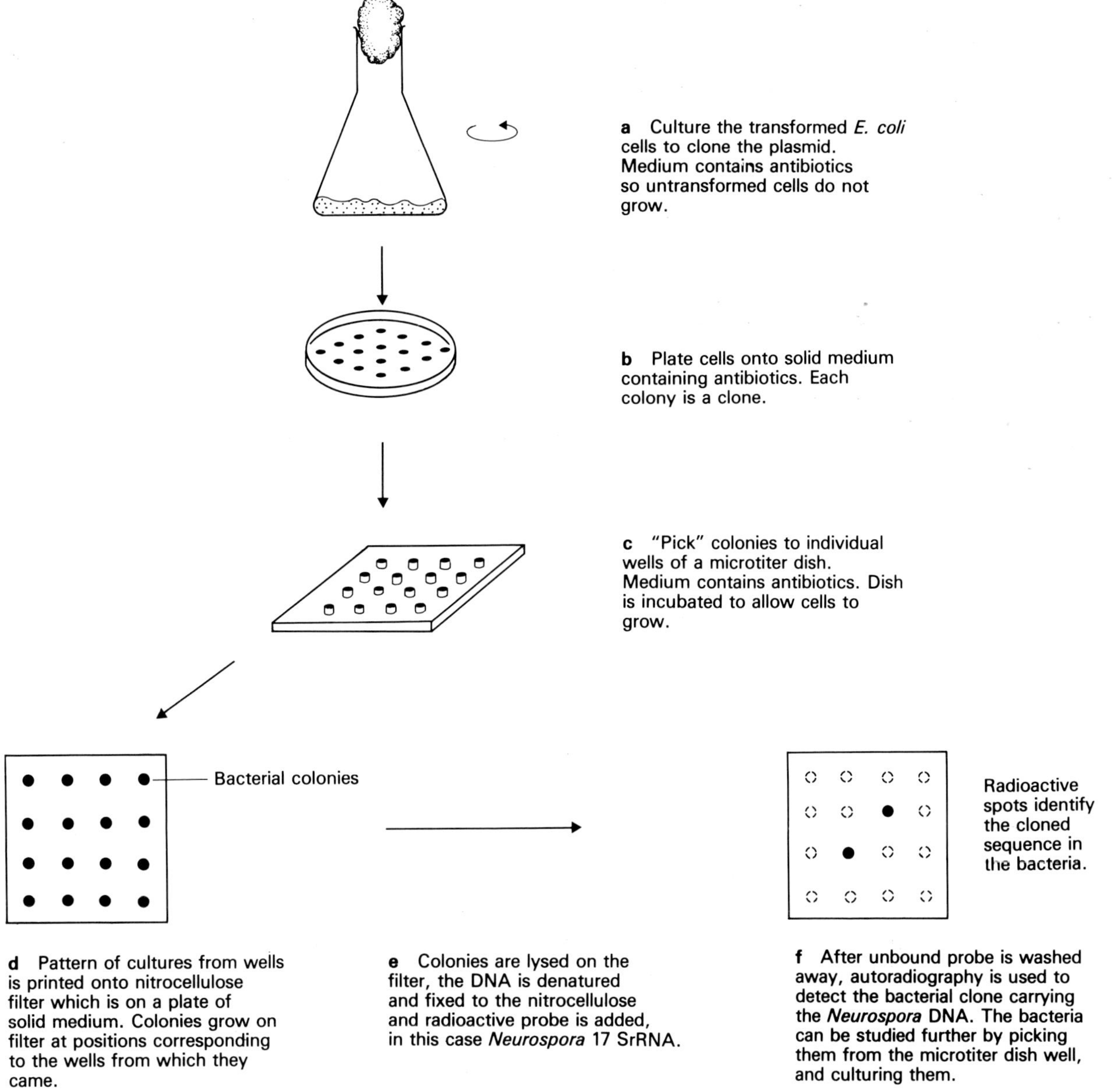

Fig. 12.11. Scheme for the detection of the HindIII → BamHI fragment of the ribosomal repeat unit from among a number of cloned HindIII → BamHI fragments by using a specific radioactive probe. For details, see text.

fragment with the 17S rRNA coding sequence is identified as a dark spot on the resulting autoradiograph. We can then go back to the microtiter well and grow large quantities of that clone for whatever purposes we have.

The radioactive probing here is highly selective. It will not pick up reconstructed pBR322 plasmids. Thus, in the few steps we have described, we can obtain the desired recombinant DNA clone.

Genes coding for proteins have often been cloned using a different approach. This approach involves the isolation of the mRNA for a protein. This approach is simplest for those mRNAs that are produced in large quantities by a cell (e.g. globin mRNA from rabbit reticulocytes). Once the mRNA has been isolated, a complementary DNA (cDNA) copy of it can be made in reactions involving, first, RNA-dependent DNA polymerase and, then, *E. coli* DNA polymerase I. The former enzyme was first identified separately by D. Baltimore and H. Temin in 1970 as a component of the RNA tumor viruses, Rous' sarcoma virus (RSV) and mouse leukemia virus (MLV), respectively. The genetic material of these viruses is RNA, and they are able to transform a cell that they infect to the tumor state. This process apparently involves the production of a DNA copy of the viral genome by the RNA-dependent DNA polymerase (also called *reverse transcriptase*, since the process it catalyzes, *reverse transcription*, is the opposite of transcription).

The use of reverse transcriptase in molecular cloning is illustrated in Fig. 12.12. The starting point is a polyadenylated mRNA, such as globin mRNA. The steps are as follows:

1. An oligo (dT) primer (a short, synthetic sequence of Ts) is annealed to the poly(A) tail of the mRNA.

2. The primer is extended in a reaction catalyzed by reverse transcriptase and using the four deoxyribonucleoside triphosphates as precursors. The result is a DNA-mRNA double-stranded molecule.

3. The RNA strand is destroyed by alkaline hydrolysis, and continued DNA synthesis by DNA polymerase I forms a short double-stranded, hairpin loop structure.

4. The 3′ end of the hairpin loop acts as a primer for DNA polymerase I to catalyze the synthesis of the second DNA strand.

5. S1 nuclease cuts the loop, resulting in a linear, double-stranded cDNA copy of the original poly(A)-mRNA. The cDNA can be inserted into a linearized plasmid and cloned.

The cloned cDNA may then be purified and used as a probe in any one of several interesting applications such as the following: (a) to obtain sequence information about the mRNA (its coding and noncoding sequences); (b) to identify precursors to the mRNA (such as heterogeneous nuclear RNA); (c) to use as a hybridization probe to identify those recombinant DNA clones made from shared total DNA that contains part or all of the DNA region coding for

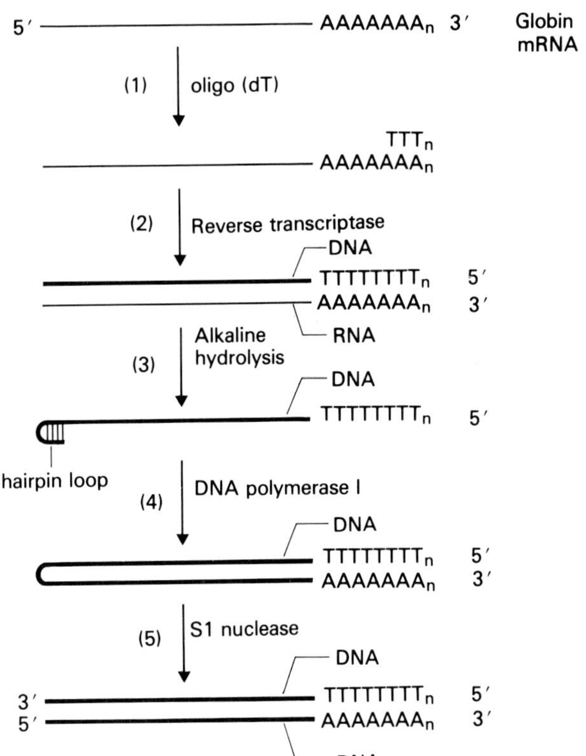

Fig. 12.12. The synthesis of double-stranded cDNA from a polyadenylated mRNA using reverse transcriptase.

globin mRNA; and (d) to attempt to get the gene produced synthesized in a bacterial or other simple organism host. In view of the discovery of intervening sequences, this can allow regions of interest, such as controlled sites or the intervening sequences themselves, to be analyzed by DNA sequencing.

Rapid DNA sequencing

Two techniques for the rapid sequencing of DNA molecules were developed by F. Sanger and A. R. Coulson in 1975 and by A. M. Maxam and W. Gilbert in 1977. Sanger and Coulson's technique is an enzymatic method, and Maxam and Gilbert's is a chemical method (described subsequently). Coupled with the technology for cloning specific DNA fragments, these methods have allowed a rapid advance of our knowledge of gene structure and function. For example, a number of genes and their controlling sequences have been completely sequenced, and detailed information about eukaryotic introns has been accumulated.

Maxam and Gilbert DNA-sequencing technique

This technique uses chemical reactions to break the DNA at specific nucleotides. The starting point is a homogeneous population of DNA molecules, for example, specific purified restriction fragments of a cloned DNA molecule. In brief, the steps for DNA sequencing are:

1. *Labeling the DNA.* One end (either 5′ or 3′) needs to be labeled with ^{32}P. This is accomplished in an enzyme-catalyzed reaction that is specific for the 5′ or 3′ end of a chain. The result is a double-stranded DNA with a ^{32}P molecule at each end but on different strands. Then, either the DNA is denatured into single strands, each of which is labeled at only one terminus, or the DNA is cut with a restriction enzyme to generate two double-stranded fragments, each of which has only one strand with a terminal ^{32}P molecule. In the former case, one strand is sequenced first and then the other is sequenced to verify the DNA sequence obtained.

2. *Chemical modification and cleavage of the DNA.* Consider a homogeneous population of single-stranded DNA molecules all labeled at the 5′ end with ^{32}P. After dividing the sample into four fractions, each fraction is chemically treated in a different way to modify and remove a base from its sugar. Subsequently the DNA backbone is broken at that point. The conditions for chemically modifying the DNA are adjusted so the reaction is a limited one, for example, affecting one group for every 100 bases or so in the DNA fragment. The reaction for cleaving the DNA is adjusted so that it will go to completion.

One of the four fractions is treated to indicate the position of every G in the strand. This involves a methylation reaction, then heating at neutral pH to release the base, and then heating in alkali to break the DNA of that point. The specific reaction used also works with about five times lower specificity on As.

The second fraction is treated initially in the same way as the first, but then incubation with cold dilute acid release As preferentially over Gs from the DNA. Alkali treatment is again used to break the DNA.

The third fraction is treated to modify both Ts and Cs and then to cleave at those points, and the last fraction is treated so cleavage only occurs at Cs in the chain.

Example of DNA sequencing analysis

Consider a 10-base DNA fragment with the sequence: 5′ A-G-C-C-T-A-G-A-C-T. The DNA is labeled at the 5′end with ^{32}P, and four samples are subjected to the reactions just described. The cleavage chains from each reaction are separated according to their length by electrophoresis in an acrylamide gel. In this, the smallest fragment migrates the fastest. After electrophoresis, the so-called ladder gel is subjected to autoradiography (it is placed against an x-ray film) to determine the distribution of fragments. The autoradiogram may be read to determine the base sequence of the fragment. Fig. 12.13

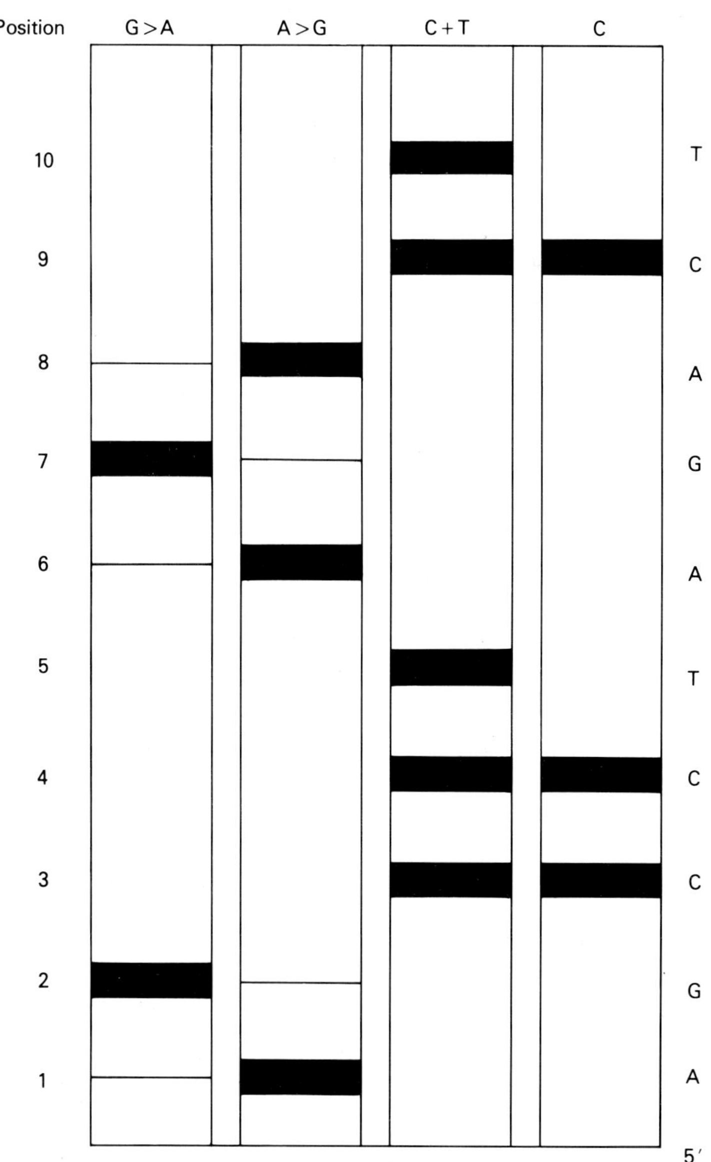

Fig. 12.13. Diagram of autoradiogram expected after the Maxam and Gilbert rapid DNA-sequencing technique is applied to the DNA fragment: 5′[32]P-A-G-C-C-T-A-G-A-C-T.

diagrams the autoradiogram expected of the theoretical 10-base DNA fragment.

The four lanes in the autoradiogram represent the four different reactions in the order they were discussed. Consider first the G > A lane. The reaction conditions were arranged so that each chain was cleaved on the average only once. Therefore the sample applied to the gel contained the following ^{32}P-labeled fragments, each ending at the base before where there was a G in the chain and less frequently where there was an A:

^{32}P
^{32}P-A
^{32}P-A-G
^{32}P-A-G-C-C-T
^{32}P-A-G-C-C-T-A
^{32}P-A-G-C-C-T-A-G

Note that, since the ^{32}P is only at one end of the molecule, and that only ^{32}P molecules are seen (as bands) on the autoradiogram, any other piece or pieces or each chain produced after cleavage is not seen. Also, since the first reaction shows preference for G and A, the fragment generated by cleavage at Gs appear as thicker bands than those generated by cleavage at As. The actual results may be interpreted to mean that the second and seventh bases were Gs, and that the first, sixth, and eighth were As.

The second lane shows the A > G results so the band densities are exactly the opposite of those in the first lane. The last two lanes define the C and T bases in the chain. Lane three shows the C+T positions and lane four shows the C positions. Comparison of the two lanes therefore enables the C and T positions to be deduced.

Therefore it is relatively easy to read the DNA sequence from the band positions in the gel. This is done by starting at the bottom and finding the band or bands at each step of the ladder and deducing the base from the banding patterns. In the example, position 1 has two bands, the one in the A > G lane being denser than the one in the G > A, thereby indicating an A. Position 2 has a dense band in G > A and a faint band in A > G, indicating a G. Positions 3 and 4 have bands in both the C+T and C lanes, indicating Cs. Position 5 has a band in the C+T lane but no band in the C lane so this is a T. Following similar logic, positions 6 through 10 are A, G, A, C, and T, respectively. In actual practice the rapid DNA-sequencing technique can be used on DNA fragments consisting of 500 bases or more. Rapid RNA-sequencing techniques have also been developed that use enzymatic means to break an end-labeled RNA chain at specific bases.

Applications of recombinant DNA technology

In the years since recombinant DNA technology has been developed, it has been used in a large number of studies of both prokaryotic and eukaryotic genomes, and it is impossible to deal with them all adequately in this brief discussion. To give but three examples, recombinant DNA cloning has made it possible to obtain large amounts of particular genes, to determine the functions of segments of nuclear and organellar DNA, and to map genomes (e.g ΦX174 and SV40 were mapped this way). Particularly useful applications of the techniques are studies in which the controlling regions for prokaryotic and eukaryotic genes are cloned with an eye towards understanding the regulation of gene expression at a very basic level.

The application of recombinant DNA cloning and the associated techniques has the potential to benefit mankind. For example, the genes for a number of human genes have been cloned in appropriately engineered vehicles so that the host bacterial or yeast cell synthesize the gene product. Thus, human insulin is being made using recombinant DNA methodologies and the hormone can be used to treat diabetics, particularly those who are sensitive to the porcine insulin now used. Similarly, to name but a few other examples, human growth hormone is being synthesized to treat pituitary dwarfism, human interferon is being made from cloned genes to test as antiviral and antitumor agents, and bovine growth hormone is being made and used to increase milk production. Another application is to "construct" organisms to carry out specific functions. Thus, organisms have been engineered to "eat" oil spills, and experiments are underway to produce other organisms that can deal with toxic wastes or spill.

In medical diagnosis, progress is being made to use recombinant DNA methodologies to detect genetic diseases. For example, at least some genetic mutations affect the sites for restriction enzymes. It then becomes possible to distinguish wild type and mutant genes by digesting a sample of genomic DNA with the appropriate restriction enzyme(s), separating the fragments on a gel, transferring the fragments to a filter, and hybridizing with an appropriate labeled probe. Since fetal cells can be obtained by taking a sample of the amniotic fluid through a syringe, this approach can be used to screen fetuses for any genetic diseases that show a restriction enzyme digestion pattern difference.

Finally, there is a great potential for the use of recombinant DNA methodologies in plant genetics, particularly for improving crop yields. One exciting area of research concerns nitrogen fixation. This is the process whereby leguminous (e.g. peas, soybeans) capture atmospheric nitrogen and reduce it to a form that can be utilized directly by the plant. Special bacteria in the root nodules called rhizobia are responsible for this nitrogen conversion.

The many genes involved are called *nif* genes, and these are found in plasmids in the rhizobia. If these can be manipulated, it should be possible to increase the amount of nitrogen fixation in the plants. Experiments are also being done with cloned *nif* genes to try to induce nonleguminous plants to fix atmospheric nitrogen. As of this writing, a corn strain has been produced that get 1% of its nitrogen from nitrogen fixation.

In summary, great strides are being made in the application of recombinant DNA methodologies to projects related to human welfare. There is no doubt that significant achievements will continue to occur.

Questions and problems

12.1 A new restriction endonuclease is isolated from a bacterium. This enzyme cuts DNA into fragments that are, on the average, 4096 base pairs long. Like all other known restriction enzymes, the new one recognizes a sequence in DNA that has twofold rotational symmetry. From the information given, how many base pairs of DNA constitutes the recognition sequence for the new enzyme.

12.2 Restriction endonucleases are used to construct restriction maps of linear or circular pieces of DNA. The DNA is usually produced in large amounts by recombinant DNA techniques. The generation of restriction maps is similar to the process of putting the pieces of jigsaw together. Suppose we have a circular piece of double-stranded DNA that is 5000 base pairs long. If this DNA is digested completely with restriction enzyme I, four DNA fragments are generated: fragment *a* is 2000 base pairs long; fragment *b* is 1400 base pairs long; *c* is 900 base pairs long; and *d* is 700 base pairs long. If, instead, the DNA is incubated with the enzyme for a short time, the result is incomplete digestion of the DNA; not every restriction enzyme site in every DNA molecule will be cut by the enzyme, and all possible combinations of adjacent fragments can be produced. From an incomplete digestion experiment of this type, fragments of DNA were produced from the circular piece of DNA, which contained the following combinations of the above fragments: *a-d-b, d-a-c, c-b-d, a-c, d-a, d-b,* and *b-c*. Lastly, after digesting the original circular DNA to completion with restriction enzyme I, the DNA fragments were treated with restriction enzyme II under conditions conducive to complete digestion. The resulting fragments were: 1400, 1200, 900, 800, 400, and 300. Analyze all of the data to locate the restriction enzyme sites as accurately as possible.

12.3 Draw the banding pattern you would expect to see on a DNA-sequencing gel if you applied the Maxam and Gilbert DNA-sequencing method to the following single-stranded DNA fragment (which is labeled at the 5′ end with ^{32}P): 5′^{32}P-A-A-G-T-C-T-A-C-G-T-A-T-A-G-G-C-C-3′.

References

Aaij, C. and P. Borst, 1972. The gel electrophoresis of DNA. *Biochim. Biophys. Acta* **296:**192–200.

Arber, W., 1965. Host-controlled modification of bacteriophage. *Annu. Rev. Microbiol.* **19:**365–378.

Arber, W., 1974. DNA modification and restriction. *Prog. Nucl. Acid Res. Mol. Biol.* **14**:1–37.

Arber, W. and D. Dussoix, 1962. Host specificity of DNA produced by *Escherichia coli*. I. Host controlled modification of bacteriophage lambda. *J. Mol. Biol.* **5**:18–36.

Arber, W. and S. Linn, 1969. DNA modification and restriction. *Annu. Rev. Biochem.* **38**:467–500.

Baltimore, D., 1970. Viral RNA-dependent DNA polymerase. *Nature* **226**:1209–1211.

Blattner, F.R., B.G. Williams, A.E. Blechl, D. Denniston-Thompson, D.O. Kiefer, D.D. Moore, J.W. Schumm, E.L. Sheldon and O. Smithies, 1977. Charon phages: safer derivatives of bacteriophage lambda for DNA cloning. *Science* **196**:161–169.

Boyer, H.W., 1971. DNA restriction and modification mechanisms in bacteria. *Annu. Rev. Microbiol.* **25**:153–176.

Chan, H.W., M.A. Israel, C.F. Garon, W.P. Rowe and M.A. Martin, 1979. Molecular cloning of polyoma virus DNA in *Escherichia coli*: plasmid vector system. *Science* **203**:883–892.

Chang, L.M.S. and F.J. Bollum, 1971. Enzymatic synthesis of oligodeoxynucleotides. *Biochemistry* **10**:536–542.

Curtiss, R. III, 1976. Genetic manipulation of microorganisms: potential benefits and biohazards. *Annu. Rev. Microbiol.* **30**:507–533.

Danna, K. and D. Nathans, 1971. Specific cleavage of simian virus 40 DNA by restriction endonuclease of *Haemophilus influenzae*. *Proc. Natl. Acad. Sci. USA* **68**:2913–2917.

Danna, K.J., G.H. Sack and D. Nathans, 1973. Studies of simian virus 40 DNA. VII. A cleavage map of the SV40 genome. *J. Mol. Biol.* **78**:363–376.

Dussoix D. and W. Arber, 1962. Host specificity of DNA produced by *Escherichia coli*. II. Control over acceptance of DNA from infecting phage lambda. *J. Mol. Biol.* **5**:37–49.

Freifelder, D., 1978. *Recombinant DNA: Readings From Scientific American.* W.H. Freeman and Co., San Francisco.

Kelley, T.J. and H.O. Smith, 1970. A restriction enzyme from *Haemophilus influenzae*. II. Base sequence of the recognition site. *J. Mol. Biol.* **51**:393–409.

Kessler, C., P.S. Neumaier and W. Wolf, 1985. Recognition sequences of restriction endonucleases and methylases—a review. *Gene* **33**:1–102.

Lee, A.S. and R.L. Sinsheimer, 1974. A cleavage map of bacteriophage ΦX174. *Proc. Natl. Acad. Sci. USA* **71**:2882–2886.

Lobban, P.E. and A.D. Kaiser, 1973. Enzymatic end-to-end joining of DNA molecules. *J. Mol. Biol.* **78**:453–471.

Luria, S.E., 1953. Host induced modification of viruses. *Cold Spring Harbor Symp. Quant. Biol.* **18**:237–244.

Maniatis, T., E.F. Fritsch and J. Sambrook, 1982. *Molecular Cloning. A Laboratory Manual.* Cold Spring Harbor Laboratory, Cold Spring Harbor, NY.

Maxam, A.M. and W. Gilbert, 1977. A new method for sequencing DNA. *Proc. Natl. Acad. Sci. USA* **74**:560–564.

Maxam, A.M., R. Tizard, K.G. Skryabin and W. Gilbert, 1977. Promoter region for yeast 5S ribosomal RNA. *Nature* **267**:643–645.

Meselson, M., R. Yuan and J. Heywood, 1972. Restriction and modification of DNA. *Annu. Rev. Biochem.* **41**:447–466.

Mulder, C., J.R. Arrand, H. Delius, W. Keller, U. Pettersson, R.J. Roberts and P.A. Sharp, 1974. Cleavage maps of DNA from adenovirus types 2 and 5 by restriction endonucleases EcoRI and HpaI. *Cold Spring Harbor Symp. Quant. Biol.* **39**:397–400.

Sanger, F. and A.R. Coulson, 1975. A rapid method for determining sequences in DNA by primed synthesis with DNA polymerase. *J. Mol. Biol.* **94**:441–448.

Sharp, P.A., B. Sugden and J. Sambrook, 1973. Detection of two restriction endonuclease activities in *Haemophilus parainfluenzae* using analytical agarose-ethidium bromide electrophoresis. *Biochemistry* **12**:3055–3063.

Sinsheimer, R.L., 1977. Recombinant DNA. *Annu. Rev. Biochem.* **46**:415–438.

Smith, H.O. and K.W. Wilcox, 1970. A restriction enzyme from *Haemophilus influenzae*. I. Purification and general properties. *J. Mol. Biol.* **51**:379–391.

Southern, E.M., 1975. Detection of specific sequences among DNA fragments separated by gel electrophoresis. *J. Mol. Biol.* **98**:503–517.

Struhl, K., J.R. Cameron and R.W. Davis, 1976. Functional genetic expression of eukaryotic DNA in *Escherichia coli. Proc. Natl. Acad. Sci. USA* **73**:1471–1475.

Temin, H.M. and S. Mizutani, 1970. RNA-dependent DNA polymerase in virions of Rous sarcoma virus. *Nature* **226**:1211–1213.

Wu, R., 1978. DNA sequence analysis. *Annu. Rev. Biochem.* **47**:607–634.

Chapter 13

Eukaryotic Genetics: Mendel and his Laws

Outline

Chapter 5 discussed the behavior of chromosomes in meiosis. We can now relate that behavior to genetics by discussing the segregation of genes as it was first studied by Gregor Mendel. Mendel's experiments provided the first contributions to our knowledge of chromosomal genetics, and ironically the significance of his work was not realized until almost 30 years after his death.

Mendel and the history of classical genetics

1822 Gregor Mendel was born in Austria.

1843 Mendel was admitted as a novice at a monastery in Brunn.

1847 Mendel became a priest.

1850 Mendel failed an examination for a teaching certificate in natural science.

1854–1855 Mendel obtained 34 strains of peas and checked them for consistency of characteristics.

1856–1863 Mendel conducted his famous pea experiments concerning gene segregation.

1865 Mendel read a paper on his results to the Brunn Society of Natural History.

1866 A paper on his work was published in the Proceedings of the Brunn Society of Natural History. His work was largely ignored.

1875 O. Hertwig showed that the nucleus of the cell was required for fertilization and cell division, and hence presumably contained information necessary for those activities.

1882–1885 E. Strasburger and W. Flemming showed that nuclei contained chromosomes, and A. Weissman proposed a theory of heredity and development in which chromosomes contained the hereditary material.

1884 Gregor Mendel died after having held an administrative position for many years.

1900 Around 1900, three different workers produced results that confirmed Mendel's work in the segregation of factors responsible for characters in organisms. H. de Vries found "Mendelian" segregation ratios for crosses of a number of different plant species and published a paper mentioning Mendel's results. C. Correns also published data for maize and peas confirming Mendel's conclusions. Finally, E. von Tschermark, working with peas obtained ratios paralleling those found by Mendel.

1902 W. Bateson showed that Mendelian principles apply to animals. In addition, he coined the terms "genetics", "zygote", and "allelomorph" (now shortened to "allele") and proposed F1 and F2 as symbols for the two generations of progeny commonly followed in genetic crosses. In 1909 W. Johannsen proposed the term "gene" to replace Mendelian "factor".

1902–1903 As a result of his work with grasshoppers, W. S. Sutton proposed that different pairs of chromosomes become oriented at random in meiosis and that this was responsible for the independent segregation of separate pairs of genes. This was a very important hypothesis since it related chromosomes to Mendelian factors.

1905 W. Bateson and R. C. Punnett showed that not all genes segregated independently; that is, two genes in sweet pea showed incomplete linkage. Also in this year, N. M. Stevens and E. B. Wilson separately showed that there were different chromosomes in the two sexes of many insects: XX in females and XY in males.

1909 F.A. Janssens investigated exchanges between chromosome strand and proposed that they followed the formation of chiasmata (singular = chiasma).

1910 The fruit fly, *Drosophila melanogaster* made its appearance in genetics laboratories. In this year T. H. Morgan found the first sex-linked gene, *white*, and soon many others were discovered. This was the beginning of a long period of work with *Drosophila* by T. H. Morgan and many colleagues, which established a number of important genetic principles. Only some of their contributions to genetics will be considered in the following.

1911 Linkage between two sex-linked genes, *yellow* and *white*, was shown. Morgan proposed that linkage was the result of the genes involved being located in the same chromosome pair. Breaking up of the linkage to produce recombinant flies was proposed to be the result of crossing-over between the genes. Further, Morgan reasoned that close linkage implied infrequent crossing-over between the genes. This was a major breakthrough since it tied together data following the inheritance of genes and data concerning chromosomes and meiosis. In addition, the conclusion fitted Janssens' chiasma hypothesis of 1909.

1913 Once the relationship of crossing-over and linkage was established, A. H. Sturtevant took the next step in constructing the first linear chromosome map of five sex-linked genes.

1919 T. H. Morgan and C. B. Bridges discovered autosomal linkage in *Drosophila melanogaster*.

1927 H.J. Muller used X-rays to generate mutations in *Drosophila* as a basis for further genetic studies.

1931 C. Stern working with *Drosophila* and H. B. Creighton and B. McClintock working with maize showed that genetic recombination was accompanied by a physical exchange of homologous chromosomes.

After that, genetics ballooned. Many other organisms became the subject of investigation, biochemical pathways were dissected using genetic mutants, DNA was confirmed as the genetic material, and molecular genetics studies of cell function were begun. Mendel's work and some of the classical genetics experiments will now be discussed.

Mendel's first law: the principle of segregation

Mendel studied the pea, *Pisum sativum*. The pea normally reproduces by **self-fertilization** (**selfing**). That is, pollen produced from the stamen will land on the pistil within the same flower and fertilize the plant. Selfing can be prevented by removing the stamens from a developing flower before they produce mature pollen. The pistils of that emasculated flower can then be pollinated with pollen taken from the stamen of another flower: this is called **cross-fertilization** or a **cross**.

Mendel allowed each strain to self-fertilize for many generations so that he could be sure that he worked only with those strains in which the genetic character remained unchanged from parent to offspring for many generations. Such strains are called **true-breeding strains**; they are homozygous for the genes under study. He looked at only one character at a time in the early experiments, and he counted everything, probably because of his training in the physical science.

In his pea strains, Mendel noticed seven alternate phenotypes—tall v. short stems, smooth v. wrinkled seeds, yellow v. green seeds, grey v. white seed coats, green v. yellow pods, inflated v. pinched pods, and axial v. terminal flowers—that differed in a single character. He made crosses between true-breeding strains and examined the progeny for the parental phenotypes. Such crosses are called monohybrid crosses. An example is shown in Fig. 13.1.

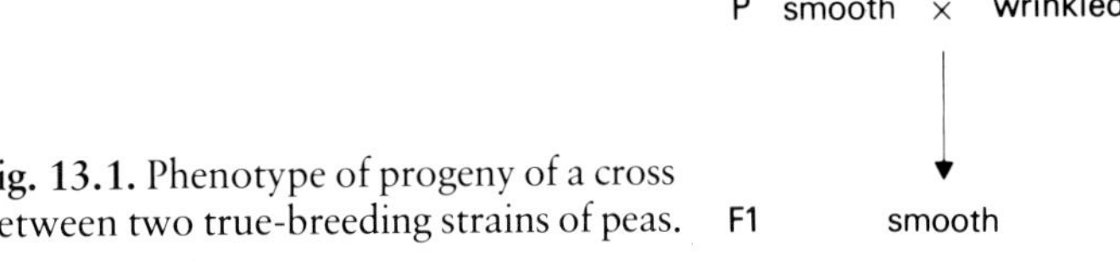

Fig. 13.1. Phenotype of progeny of a cross between two true-breeding strains of peas.

Since the F1 seeds were smooth, we can describe the smooth character as being dominant over the wrinkled character. Conversely, wrinkled is recessive to smooth. Next, Mendel planted the seeds and allowed the F1 plants to

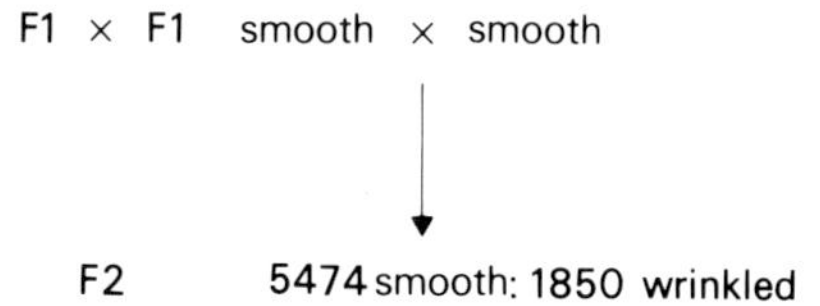

Fig 13.2. F2 phenotypes and ratios resulting from selfing the F1 of the cross shown in Fig. 13.1.

produce the F2 seed by self-fertilization so that he could determine the seed types. The results are shown in Fig. 13.2.

The ratio of F2 phenotypes is 2.96 : 1, or roughly a 3 : 1 ratio of smooth to wrinkled. To obtain more information about the F2 progeny seeds, Mendel planted the seeds, allowed 565 of the resulting plants to self, and examined the progeny for the two characters. He found that 193 of the F2s gave only smooth progeny and 372 gave both smooth and wrinkled progeny. This is approximately a 1 : 2 ratio.

Mendel drew some interesting conclusions from the data. He recognized that to get true-breeding lines as his parentals were, both egg and pollen must be of the same type. When the F1 plant were selfed, the F2 progeny seeds showed *both* parental characters whereas the F1 seeds had exhibited only one parental phenotype. This showed that the F1 must be carrying one copy of each character and the F2s are produced in the ratios found if it is assumed that the gamete types occur in equal frequency and the uniting of egg and pollen occurs by chance. We can relate this now to chromosome behavior. That is, somatic cells are diploid and contain one chromosome from each parent. Gametes are haploid and are produced by meiosis. The smooth× wrinkled cross can then be illustrated using genetic symbols (Fig. 13.3). (Note that these symbols are in common usage for plant systems and that alternative symbols are used in animal and other systems. These will be introduced as

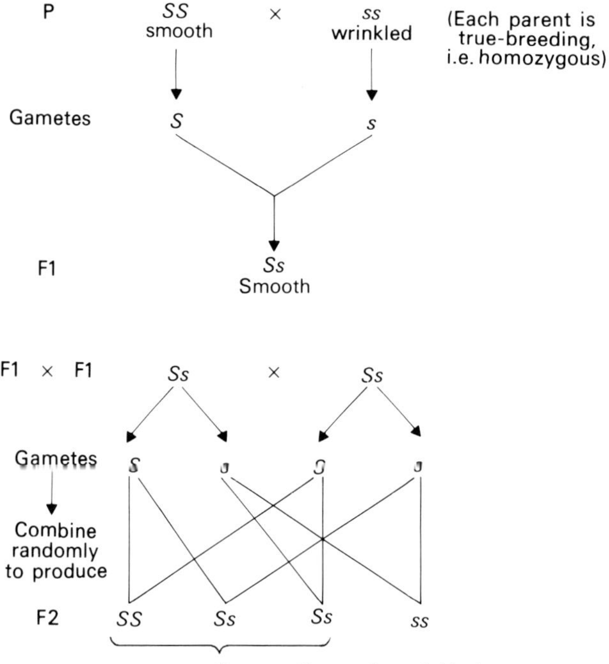

Fig. 13.3. Principle of segregation: genotypic representation of the parental, F1 and F2 generations of the crosses shown in Figs. 13.1 and 13.2.

examples are presented.) The F1 is heterozygous and has a smooth phenotype owing to the dominance of the *S* allele over the *s* allele, and the F2 derived by selfing the F1 has a 3:1 phenotypic ratio of smooth to wrinkled.

An alternative, branching diagram, way of representing the production of F2 by random association of gametes is to express the proportion of each gamete as a fraction of the total number of gamete types. The F1×F1 then becomes as shown in Fig. 13.4. The data also illustrate the fact that the genotypic ratio in the F2 is 1*SS*:2*Ss*:1*ss*. As was indicated before, this was tested by selfing the F2 strains with the dominant phenotypes and thus shows a 1:2 ratio of *SS*:*ss* (Fig. 13.5).

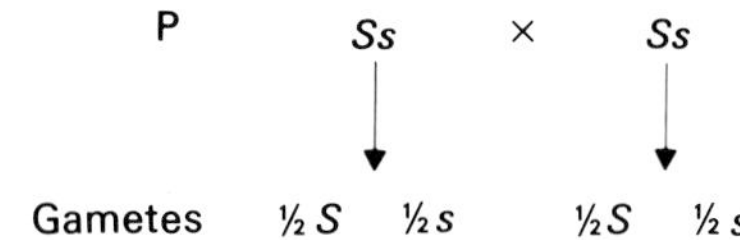

Random combination of gametes results in

One parent	Other parent	Progeny genotype		Progeny phenotype
½*S*	½*S*	¼*SS*		¾ *S*_ (shortened form of *SS* or *Ss* indicating dominant smooth phenotype)
	½*s*	¼*Ss*	½*Ss*	
½*s*	½*S*	¼*Ss*		
	½*s*	¼*ss*		¼*ss* (wrinkled, recessive phenotype)

Fig. 13.4. Calculation of the ratios of phenotypes in the F2 of the cross of Fig. 13.3 using a branching diagram approach.

F2 selfs: *SS* × *SS* → all *SS* (smooth) progeny

Ss × *Ss* → ¾ *S*_ (smooth): ¼ *ss* (wrinkled) i.e. both kinds of progeny

Fig. 13.5. Determination of the genotypes of the F2 smooth progeny of Fig. 13.4 by selfing.

A more common way of testing for this, and one that overcomes the problem that most organisms do not self, is using the **testcross**, which is a cross of an unknown with a homozygous recessive parent. We can illustrate this for the same *SS*, *Ss* question (Fig. 13.6). Here one-third of the F2 smooth

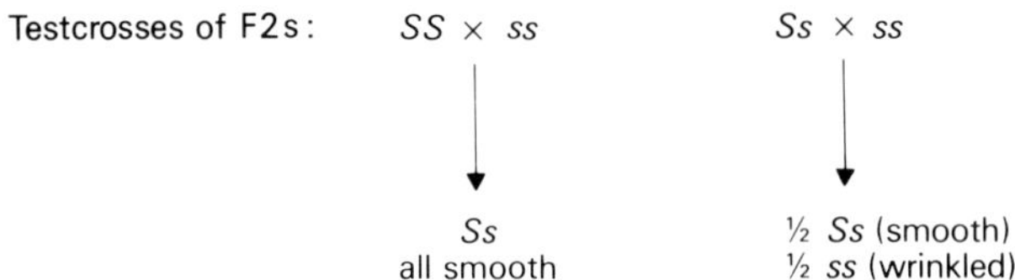

Fig. 13.6. Determination of the genotypes of the F2 smooth progeny of Fig. 13.4 by testcrossing with a homozygous recessive strain.

progeny will give all smooth progeny in a testcross and two-thirds of the F2 smooth will give one-half smooth and one-half wrinkled progeny.

From all of this, Mendel proposed his first law, the **principle of segregation**: (a) the gametes are pure in that they can carry only one allele of each gene (i.e. they are haploid), and (b) the gametes segregate from each other and new progeny are produced by the random combination of gametes from the two parents. Thus, in proposing the law Mendel had distinguished between the factor that caused the traits and the traits themselves.

Incomplete dominance

In the examples discussed up to now, there was complete dominance of one character over the alternative character such that heterozygotes are phenotypically indistinguishable from the homozygous dominant individuals. This is not always the case. Some genes exhibit **incomplete dominance**, and therefore the three genotypes *AA*, *Aa*, and *aa* have three distinct phenotypes. One example of this is flower color in snapdragons (*Antirrhinum*) where *AA* flowers are red, *aa* flowers are white, and *Aa* flowers are pink. A cross of two pink-flowered plants will therefore produce plants with red, pink, and white flowers in a ratio of 1:2:1, respectively. The sample explanation here is that *AA* individuals make red pigment, *aa* plants make white pigment, and *Aa* plants make both red and white pigments, which blend to produce pink flowers.

Another example of incomplete dominance involves plumage color in Andalusian fowls. Crosses between a true-breeding black variety and a true-breeding white strain give F1 birds that are intermediate in plumage color, or a gray. Chicken breeders call the gray birds Andalusian blues, which are not true-breeding since they are heterozygous.

Molecular model of genetic dominance

In most cases the mutant allele of a gene is completely recessive to the wild-type allele so that the heterozygote carrying the two alleles results in a wild-type phenotype. One explanation for this is the following. If the gene codes for an enzyme, the mutant allele may produce a protein that lacks or has little enzyme activity (e.g. a misense mutation) or it may produce no protein or a protein fragment (e.g. a nonsense mutation). Thus, barring regulation of transcription or translation rate, the heterozygote will produce approximately half the amount of active enzyme as the homozygous wild type. If that amount is all that is necessary for the cell or organism to carry out normal biochemical functions, a wild-type phenotype will result and by definition the wild-type allele is completely dominant over the mutant allele. Presumably this situation has evolved through natural selection.

There are many examples of mutant alleles that are dominant over the wild-type allele. One explanation for this is that the locus codes for an enzyme that, in the mutant, has a greater affinity for the substrate of the reaction it catalyzes than does the wild-type enzyme. However, the mutant enzyme is unable to catalyze the reaction or it does so with very low efficiency. Thus, a mutant phenotype results either when the mutant allele is homozygous or when it is heterozygous with the wild-type allele. An example of this is the dominant *Stubble* (*Sb*) mutation of *Drosophila melanogaster*, which results in short bristles. This is one of the genes used in an example of genetic mapping in the next chapter.

Mendel's second law: the principle of independent assortment

After having shown that each of the pairs of alternate characters behaved in the same way as the smooth/wrinkled pair, Mendel turned his attention to analyzing data for the segregation of two gene pairs in the same cross. Fortunately he chose only those gene pairs that were unlinked genetically so genetic crossing-over was not a factor in the experiments. We can consider an example involving smooth/wrinkled and yellow/green characters, where smooth and yellow are the dominant alleles, respectively. Mendel made crosses between true-breeding smooth-yellow plants and wrinkled-green plants with the results shown in Fig. 13.7.

Each F1 is a doubly heterozygote and produces gametes of four types: *SY*, *Sy*, *sY*, *sy*, each occurring with equal frequency. In the F1 selfing these four gamete types pair randomly in all possible combinations to give rise to the diploid progeny. Thus, in the above so-called Punnett square, 16 gamete combinations are produced. Owing to genetic dominance, only four distinct phenotype classes are found: smooth, yellow; smooth, green; wrinkled, yellow; and wrinkled, green. These are predicted to occur with a relative proportion of 9:3:3:1, respectively, as a result of the random fusion of the gametes and the independent assortment of the two gene pairs into the gamete as a result of chromosome segregation during meiosis.

Generally, geneticists concern themselves with ratios of phenotypic classes. It is cumbersome to construct a Punnett square of gamete combinations and then count up the numbers of each phenotypic class. Actually it

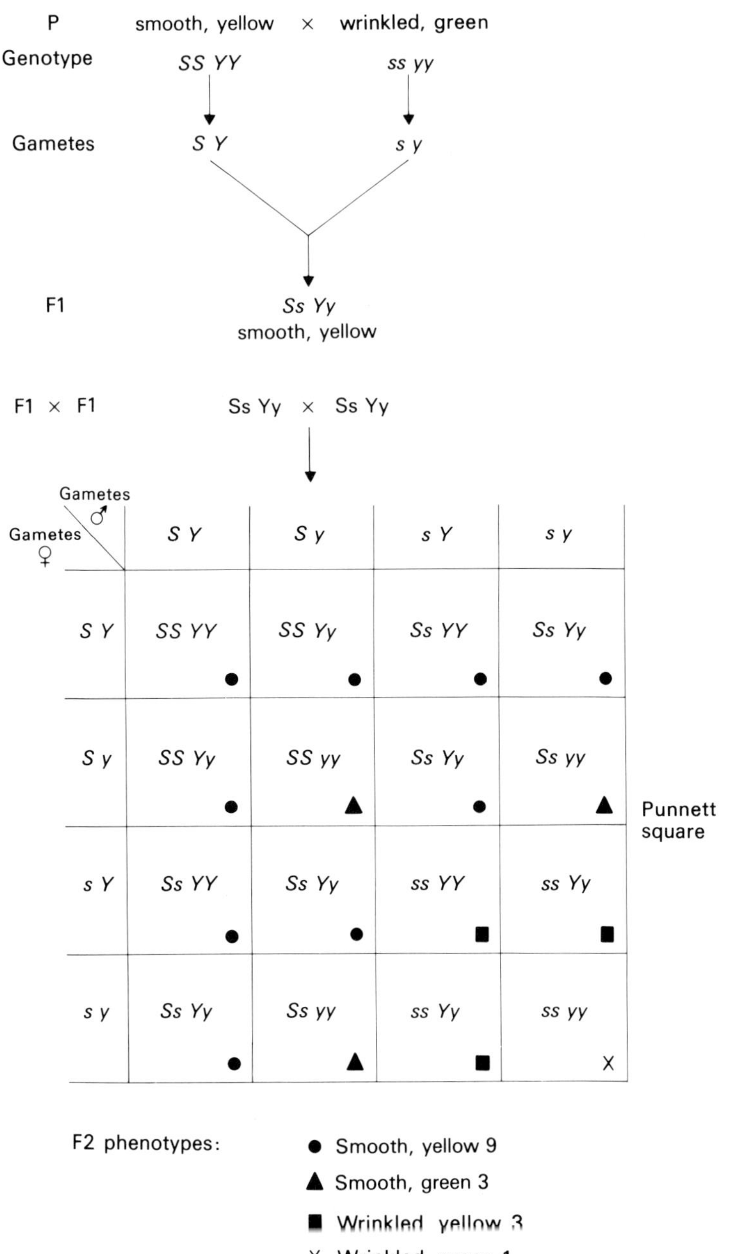

Fig. 13.7. Principle of independent assortment: demonstration of F2 9:3:3:1 phenotypic ratio for two unlinked gene pairs by using the Punnett square.

is not difficult when two gene pairs are being considered, but any more than that and it becomes complex. Therefore it is easier to deal directly with the expected ratios of phenotypic classes. If we consider the same example where the two gene pairs assort independently into the gametes, we can think about each pair in turn. We discovered earlier that an F1 self of a smooth/wrinkled (*Ss*) heterozygote gave 3/4 *S*_ (i.e. genotypically *SS* or *Ss*: smooth) and 1/4 *ss* (wrinkled) progeny. This is also the case with an F1 self of a *Yy* heterozygote. Since these two events occur independently, we can analyze the expected F2s with a branching diagram (Fig. 13.8). The **independent assortment** of the gene pairs gives the 9/16:3/16:3/16:1/16 or 9:3:3:1 ratio of F2 phenotypes. This ratio will be obtained from double heterozygote selfs (*Aa Bb*×*Aa Bb*) whenever gene *A* is unlinked to gene *B*. In sum, **Mendel's second law**, the **principle of independent assortment** states that genes in different nonhomologous chromosomes behave independently in the production of gametes.

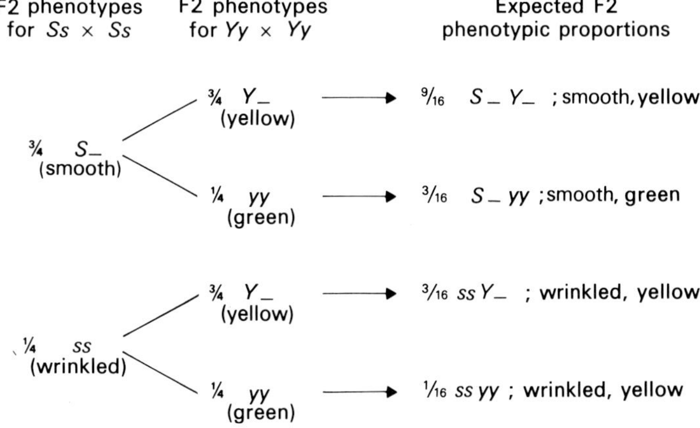

Fig. 13.8. Same as Fig. 13.7, but using the branching-diagram approach.

Finally, the testcross can be used to analyze the genotypes of the F1 or F2 progeny of a cross involving two gene pairs. The expected ratios of phenotypes among the progeny of such crosses are shown in Table 13.1. All of the patterns of gene segregation shown in the table are directly related to chromosome behavior during meiosis.

Table 13.1. Proportions of phenotypic classes expected from testcrosses of strains with various genotypes for two gene pairs.

	Proportion of phenotypic classes			
Testcrosses	*A*_ *B*_	*A*_ *bb*	*aa B*_	*aa bb*
AA BB×*aa bb*	1	0	0	0
Aa BB×*aa bb*	1/2	0	1/2	0
AA Bb×*aa bb*	1/2	1/2	0	0
Aa Bb×*aa bb*	1/4	1/4	1/4	1/4
AA bb×*aa bb*	0	1	0	0
aa BB×*aa bb*	0	0	1	0
aa Bb×*aa bb*	0	0	1/2	1/2

Extensions of Mendelian principles

Multiple alleles

In a population of individuals, there can be not just two but **multiple alleles** of genes. Any given diploid individual can possess only two different alleles. Multiple alleles obey the same rule of transmission as alleles of which there are only two kinds. Dominance relationships among multiple alleles vary in that for some groups of alleles, every homozygous and heterozygous genotype produces a different phenotype, whereas in others the alleles may be arranged in a descending series in which every allele is dominant over all alleles below it. An example is the human ABO blood group system in which three alleles, I^A, I^B, and i, determine the blood group. That is, I^AI^B specifies AB blood group, I^AI^A or I^Ai specifies A, I^BI^B or I^Bi specifies B and ii specifies O.

Codominance

Codominance is different from incomplete dominance (see earlier). In the case of codominance, the heterozygote exhibits the phenotypes of both homozygotes rather than exhibiting an intermediate phenotype as in incomplete dominance. For example, the I^A and I^B alleles (see above) are codominant since I^AI^B individuals have a phenotype that is essentially a combination of those shown by individuals with A and B blood groups, at least in terms of the red blood cell surface antigens specified by the alleles involved.

Gene interaction

Nonallelic genes may not function independently in determining the phenotypic characteristics in an organism. Interaction between gene products may occur to produce new phenotypes without modifying typical Mendelian ratios. Or, interaction between gene products may cause modifications of Mendelian ratios by one gene product interfering with the phenotypic expression of another nonallelic gene or genes. Here, the phenotype is controlled mostly by the former gene and not the latter when both genes occur together in the genotype. This type of interaction is called **epistasis**.

Lethal alleles

A **lethal allele** is one which, when expressed, is lethal to the individual. Both recessive lethal and dominant lethal alleles are known. A recessive lethal allele may or may not have an effect in the heterozygous condition. A dominant lethal allele, to be detected, must exert its lethal effects at some time during development or during the lifetime of the organism. The human genetic disease Huntington's chorea, for example, is caused by a dominant lethal gene that exerts its effects generally in the adult.

Penetrance

In some cases not all individuals with a given genotype exhibit the phenotype specified by that genotype. The frequency with which a dominant or homozygous recessive gene manifests itself in the phenotype of the individuals is called the **penetrance** of the gene. For example, if 65% of the individuals with a particular gene show the corresponding phenotype, there is 65% penetrance.

Expressivity

Expressivity refers to the kind of phenotypic expression of a penetrant gene or genotype. Expressivity may be slight, intermediate, or severe.

Quantitative traits

All of the traits discussed so far have been *discontinuous traits* in that there are clearly distinguished alternative characters. All seven traits studied by Mendel were discontinuous traits. However, many traits show a large continuum over a range, for example, height, eye color or hair color in humans, and height and yield in crop plants. These traits show a quantitative variation and are called **quantitative** or **continuous traits**. The properties of such traits are as follows:

1. The F1 from a cross between two phenotypically distinct, true-breeding parents has a phenotype intermediate between the parental phenotypes. The F2 shows much more variability than the F1, with the extreme values overlapping well into the ranges of the two parental values.

For example, consider the height character in a theoretical organism. We shall assume that all of the organisms have a common set of genes that determine a basic height of 20 cm. Let us assume also that there are two independently assorting gene pairs *A*/*a* and *B*/*b* that control height beyond the basic level such that each dominant allele adds 4 cm. Therefore *aa bb* will be 20 cm, and *AA BB* will be 36 cm. The F1 from *AA BB*×*aa bb* is *Aa Bb*, which with two dominant alleles is 28 cm tall, or intermediate between the two parentals. (In nature, variation between organisms of the same genotype due to the environment would result in a narrow distribution of heights around a mean value.) The F2 from an F1×F1 cross will have nine different genotypes in the following distribution: 1 *AA BB*:2 *AA Bb*:1 *AA bb*:2 *Aa BB*:4 *Aa Bb*:2 *Aa bb*:1 *aa BB*:2 *aa Bb*:1 *aa bb* (refer to Fig. 13.7). In terms of the number of dominant alleles, this is a 1:4:6:4:1 ratio of 4, 3, 2, 1, and 0 dominant alleles. This means that the F2 would consists of one-sixteenth 36 cm, four-sixteenths 32 cm, six-sixteenths 28 cm, four-sixteenths 24 cm, and one-sixteenth 20 cm tall organisms.

2. As is already implied in (**1**), quantitative traits have as their basis a number of genes called a *multiple-gene* or *polygenic series*. A simple model is that there are contributing and noncontributing alleles in the series so that, as the number of contributing alleles increases, there is an additive (or sometimes a multiplicative) effect on the phenotype. Depending upon the number of genes in the multiple-gene series, there may be discrete phenotypic classes in the range or a continuum of phenotypic variation

Questions and problems

13.1 In tomatoes, red fruit color is dominant to yellow. Suppose a tomato plant homozygous for red is crossed with one homozygous for yellow. Determine the appearance of the following:

(a) the F1
(b) the F2
(c) the offspring of a cross of the F1 back to the red parent
(d) the offspring of a cross of the F1 back to the yellow parent

13.2 A red-fruited tomato plant when crossed with a yellow-fruited one produces progeny about half of which are red-fruited and half of which are yellow-fruited. What are the genotypes of the parents?

13.3 In guinea pigs, rough coat (*R*) is dominant over smooth coat (*r*). A rough-coated guinea pig is bred to a smooth one, giving eight rough and seven smooth progeny in the F1.

(a) What are the genotypes of the parents and their offspring?
(b) If one of the F1 animals is mated to its rough parent, what progeny would you expect?

13.4 In cattle, the polled (hornless) condition (*P*) is dominant over the horned (*p*) phenotype. A particular polled bull is bred to three cows. With cow A, which is horned, a horned calf is produced; with a polled cow B, a horned calf is produced; and with horned cow C, a polled calf is produced. What are the genotypes of the bull and the three cows, and what phenotypic ratios do you expect in the offspring of these three matings?

13.5 In the Jimsonweed, purple flowers are dominant to white. When a particular purple-flowered Jimsonweed is self-fertilized, there are 28 purple-flowered and 10 white-flowered progeny. What proportion of the purple-flowered progeny will breed true?

13.6 Two black female mice are crossed with the same brown male. In a number of litters, female X produced nine blacks and seven browns and female Y produced

14 blacks. What is the mechanism of inheritance of the black and brown coat color in mice? What are the genotypes of the parents?

13.7 In Jimsonweed, purple flower (*P*) is dominant to white (*p*), and spiny pods (*S*) is dominant to smooth (*s*). In a cross between a Jimsonweed homozygous for white flowers and spiny pods and one homozygous for purple flowers and spiny pods, determine the phenotype of the following:

(a) the F1
(b) the F2
(c) the progeny of a cross of the F1 back to the white, spiny parent
(d) the progeny of a cross of the F1 back to the purple, smooth parent

13.8 What progeny would you expect from the following Jimsonweed crosses?

(a) *PP ss*×*pp SS*	(d) *Pp Ss*×*Pp Ss*
(b) *Pp SS*×*pp ss*	(c) *Pp Ss*×*Pp ss*
(c) *Pp Ss*×*Pp SS*	(f) *Pp Ss*×*pp ss*

13.9 In summer squash, white fruit (*W*) is dominant over yellow (*w*), and disc-shaped fruit (*D*) is dominant over sphere-shaped fruit (*d*). In the following, the appearances of the parents and their progeny are given. Determine the genotypes of the parents in each case.

(a) White, disc×yellow, sphere gives 1/2 white, disc and 1/2 white, sphere.

(b) White, sphere×white, sphere gives 3/4 white, sphere and 1/4 yellow, sphere.

(c) Yellow, disc×white, sphere gives all white disc progeny.

(d) White, disc×yellow, sphere gives 1/4 white disc, 1/4 white, sphere, 1/4 yellow, disc, and 1/4 yellow, sphere.

(e) White, disc×white, sphere give 3/8 white, disc, 3/8 white, sphere, 1/8 yellow, disc, and 1/8 yellow, sphere.

13.10 In garden peas, tall stem (*T*) is dominant over short stem (*t*); green pods (*G*) over yellow pods (*g*); and smooth seeds (*S*) over wrinkled seeds (*s*). Suppose a homozygous short, green wrinkled pea plant is crossed with homozygous tall, yellow smooth one.

(a) What will be the appearance of the F1?

(b) What will be the appearance of the F2?

(c) What will be the appearance of the offspring of a cross of the F1 back to its short, green, wrinkled parent?

(d) What will be the appearance of the offspring of a cross of the F1 back to its tall, yellow, smooth parent?

13.11 The coat color of mice is controlled by several genes. The agouti pattern, characterized by a yellow band of pigment near the tip of the hairs, is produced by the dominant allele *A*; homozygous *aa* mice do not have the band and are nonagouti. The dominant allele *B* determines black hairs, and the recessive allele, *b*, brown. Homozygous c^hc^h individuals allows pigments to be deposited only at the extremities (e.g. feet, nose, and ears), in a pattern called Himalayan. The genotype C_ allows pigment to be distributed over the entire body.

(a) If a true-breeding black mouse is crossed with a true-breeding brown agouti Himalayan mouse, what will be the phenotypes of the F1 and F2?

(b) What proportion of the black agouti F2 will be genotype *Aa BB* Cc^h?

(c) What proportion of the Himalayan mice in the F2 are expected to show brown pigment?

(d) What proportion of all agoutis in the F2 are expected to show black pigment?

13.12 In rabbits, C = agouti coat color, c^{ch} = chinchilla, c^{h} = Himalayan, and c = albino. The four alleles constitute a multiple allelic series. The agouti C is dominant to the three other alleles, c is recessive to all three other alleles, and chinchilla is dominant to Himalayan. Determine the phenotypes of progeny from the following crosses:

(a) $CC \times cc$	(d) $Cc^{h} \times c^{h}c$	(g) $c^{h}c \times cc$
(b) $Cc^{ch} \times Cc$	(e) $Cc^{h} \times cc$	(h) $Cc^{h} \times Cc$
(c) $Cc \times Cc$	(f) $c^{ch}c^{h} \times c^{h}c$	(i) $Cc^{h} \times Cc^{ch}$

13.13 In humans, the three alleles I^{A}, I^{B}, and i constitute a multiple allelic series that determine the ABO blood group system described in this chapter. For the following problems, state whether the child mentioned can actually be produced from the marriage. Explain your answer.

(a) An O child from the marriage to two A individuals
(b) An O child from the marriage of an A and B
(c) An AB child from the marriage of an A to O
(d) An O child from the marriage of an AB to A
(e) An A child from the marriage of an AB to B

13.14 In snapdragons, red-flower color (R) is incompletely dominant to white (r); the Rr heterozygotes are pink. A red-flowered snapdragon is crossed with a white-flowered one. Determine the flower color of the following:

(a) the F1
(b) the F2
(c) the progeny of a cross of the F1 to the red parent
(d) the progeny of a cross of the F1 to the white parent.

13.15 In shorthorn cattle, the heterozygous condition of the alleles for red coat color (R) and white coat color (r) is roan coat color. If two roan cattle are mated, what proportion of the progeny will resemble their parents in coat color?

13.16 What progeny will a roan shorthorn have if bred to:

(a) red
(b) roan
(c) white

13.17 In peaches, fuzzy skin (F) is completely dominant over smooth (nectarine) skin (f), and the heterozygous conditions of oval glands at the base of the leaves (O) and no glands (o) gives round glands. A homozygous fuzzy, no-gland peach variety is bred to a smooth, oval-gland variety.

(a) What will be the appearance of the F1?
(b) What will be the appearance of the F2?
(c) What will be the appearance of the offspring of a cross of the F1 back to the smooth, oval-glanded parent?

13.18 In guinea pigs, short hair (L) is dominant over long hair (l), and the heterozygous conditions of yellow coat (W) and white coat (w) gives cream coat. A short-haired, cream guinea pig is bred to a long-haired, white guinea pig and a long-haired, cream baby guinea pig is produced. When the baby grows up, it is bred back to the short-haired, cream parent. What phenotypic classes are expected among the offspring and in what proportions will they be found?

13.19 The shape of radishes may be long (LL), oval (Ll) or round (ll), and the color of radishes may be red (*RR*), purple (*Rr*), or white (*rr*). If a long, red radish plant is crossed with a round, white plant, what will be the appearance of the F1 and F2?

13.20 In poultry, the genes for rose comb (*R*) and the pea comb (*P*), if present together, give walnut comb. The recessive alleles of each gene, when present together in a homozygous state, give single comb. What will be the comb characters of the offspring of the following crosses?

(a) *RR Pp*×*rr Pp* (c) *Rr pp*×*rr Pp* (e) *Rr pp*×*Rr pp*
(b) *rr PP*×*Rr Pp* (d) *Rr Pp*×*Rr Pp*

13.21 For the following crosses involving the comb character in poultry, determine the genotypes of the two parents:

(a) A walnut crossed with a single produces offspring, 1/4 of which are walnut, 1/4 rose, 1/4 pea, and 1/4 single.

(b) A rose crossed with a walnut produces offspring, 3/8 of which are walnut, 3/8 rose, 1/8 pea, and 1/8 single.

(c) A rose crossed with a pea produces five walnut and six rose offspring.

(d) A walnut crossed with a walnut produces one rose, two walnut, and one single offspring.

13.22 A locus in mice is involved with pigment production; when parents heterozygous for this locus are mated together, 3/4 of the progeny are colored and 1/4 are albino. Another phenotype concerns the coat color produced in the mice; when two yellow mice are mated together, 2/3 of the progeny are yellow, and 1/3 are agouti. The albino mice cannot express whatever alleles they may have at the independently assorting agouti locus.

(a) When yellow mice are crossed with albinos, they produce an F1 consisting of 1/2 albino, 1/3 yellow, and 1/6 agouti. What are the probable genotypes of the parents?

(b) If yellow F1 mice are crossed among themselves, what phenotypic ratio would you expect among the progeny? What proportion of the yellow progeny produced here would be expected to be true-breeding?

13.23 In *Drosophila melanogaster*, a recessive autosomal gene, ebony (*e*), produces an ebony (i.e. black) body color when homozygous, and an independently assorting autosomal gene, black (*bl*), has a similar effect. Flies with genotypes *e e* bl^+–, e^+– *bl bl*, and *e e bl bl* are phenotypically identical with respect to body color; they are all black. If true-breeding ebony flies are crossed with true-breeding black flies,

(a) What will be the phenotype of the F1s?

(b) What phenotypes and what proportions would occur in the F2 generation?

(c) What phenotypic ratios would you expect to find in the progeny of these backcrosses: (i) F1×true-breeding ebony? (ii) F1×true-breeding black?

13.24 In four o'clock plants, two genes, *Y* and *R*, affect flower color. Neither is completely dominant, and the two interact on each other to produce seven different flower colors:

YY RR = crimson   *Yy RR* = magenta
YY Rr = orange-red   *Yy Rr* = magenta-rose
YY rr = yellow   *Yy rr* = pale yellow
yy RR, *yy Rr*, and *yy rr* = white

(a) In a cross of a crimson-flowered plant with a white one (*yy rr*), what will be the appearances of the F1, the F2 and the offspring of the F1 backcrossed to the crimson parent? Give the ratio of each phenotypic class in each case.

(b) What will be the flower colors in the offspring of a cross of orange red × pale yellow, and in what proportions will they be found?

(c) What will be the flower colors in the offspring of a cross of a yellow with a *yy Rr* white, and in what proportions will they be found?

13.25 Two four o'clock plants were crossed and gave the following: 1/8 crimson; 1/8 orange-red; 1/4 magenta; 1/4 magenta rose; and 1/4 white. Unfortunately, the person who made the crosses was color-blind and could not record the flower colors of the parents. From the results of the cross, deduce the genotypes and flower colors of the two parents.

13.26 In *Drosophila*, there is a mutant strain that has plum-colored eyes. A cross between a plum-eyed male and a plum-eyed female gives 2/3 plum-eyed and 1/3 red-eyed (wild type) progeny flies. A second mutant strain of *Drosophila*, called stubble, has short bristles instead of the normal long bristles. A cross between a stubble female and a stubble male gives 2/3 stubble, and 1/3 normal-bristle flies in the offspring. Assuming that the *plum* gene assorts independently from the *stubble* gene, what will be the phenotypes and their relative proportions of the progeny of a cross between two plum-eyed, stubble-bristled flies? (Both genes are autosomal.)

13.27 Two pairs of genes with two alleles each, *A/a* and *B/b*, determine plant height additively in a population. The homozygote *AA BB* is 50 cm tall; the homozygote *aa bb* is 30 cm tall.

(a) What is the F1 height in a cross between the two homozygous stocks?

(b) What genotypes in the F2 will show a height of 40 cm after a F1×F1 cross?

(c) What will be the F2 frequency of the 40 cm plants?

13.28 Three independently segregating genes (*A, B, C*), each with two alleles, determine height in a plant. Each capital letter allele adds 2 cm to a base height of 2 cm.

(a) What are the heights expected in the F1 progeny of a cross between homozygous strains *AA BB CC* (14 cm)×*aa bb cc* (2 cm)?

(b) What is the distribution of heights (frequency and phenotype) expected in a F1×F1 cross?

(c) What proportion of F2 plants will have heights equal to the original two parental strains, *AA BB CC* and *aa bb cc*?

(d) What proportion of the F2 would breed true for the height shown by the F1?

13.29 Assume that the difference between a race of oats yielding about 4 g per plant and one yielding 10 g is the result of three equal and cumulative multiple gene pairs *AA BB CC*. If you cross the type yielding 4 g with the type yielding 10 g, what will be the phenotypes and their relative distribution of the F1 and the F2?

References

Bateson, W., 1909. *Mendel's Principles of Heredity*. Cambridge University Press, Cambridge.

Brewbaker, J.L., 1964. *Agricultural Genetics*. Prentice-Hall, Englewood Cliffs, New Jersey.

Davenport, C.B., 1913. Heredity of skin color in negro-white crosses. *Carnegie Inst. Wash. Pub. 554*. Washington, D.C.

East, E.M., 1919. A Mendelian interpretation of variation that is apparently continuous. *Amer. Naturalist* **44**:65–82.

Falconer, D.S., 1961. *Introduction to Quantitative Genetics*. Ronald Press, New York.

Mather, K. and J.L. Jinks, 1971. *Biometrical Genetics. The Study of Continuous Variation*. Cornell University Press, Ithaca, New York.

Mendel, G., 1866. Experiments in plant hybridization (translation). In *Classic Papers in Genetics*, J.A. Peters (ed.). Prentice-Hall, Englewood Cliff, New Jersey.

Sturtevant, A.H., 1965. *A History of Genetics*. Harper and Row, New York.

Sutton, W.S., 1903. The chromosomes in heredity. *Biol. Bull.* **4**:213–251.

Tschermark-Seysenegg, E. von, 1951. The rediscovery of Mendel's work. *J. Heredity* **42**:163–171.

Chapter 14

Eukaryotic Genetics: Meiotic Genetic Analysis in Diploids

Outline

One of the fundamental principles of genetics is the independent assortment of genes during meiosis. Certain genes, however, are inherited together because they are on the same chromosome; such genes are called **linked genes** and are said to belong to a *linkage group*. The linkage relationships of genes can be determined through the use of testcrosses, and this type of analysis and construction of genetic maps is the subject of this chapter.

The testcross and genetic linkage

The principle of independent assortment states that, if an individual that is doubly heterozygous for two recessive, unlinked genes is crossed with another individual of the same genotype (i.e. *Aa Bb*×*Aa Bb*), the *phenotypic ratio* in the resulting progeny will be 9 AB:3 Ab:3 aB:1 ab. In a testcross of *Aa Bb*×*aa bb*, the phenotypic ratio of the progeny would be 1 AB:1 Ab:1 aB:1 ab. The testcross is a very useful cross for determining whether two genes are linked or unlinked since, if there is significant deviation from the 1:1:1:1 ratio such that too many parental genotypes and too few recombinant genotypes are found, then the two genes in question are considered linked (Fig. 14.1). The typical statistical test used to determine if the deviation is significant is the **chi-square test**, which is described in Appendix I.

In the example, the recombinant types can only be produced by a physical exchange (crossing-over) of homologous chromosome parts during meiosis in the doubly heterozygous F1 individual. If we assume that crossing-over is a random event along the chromosome, the proportion of the total progeny that are recombinant types will be directly related to how far apart the two genes are on the chromosome. Genes very close together, for example, will be separated by crossing-over rarely as compared with genes far apart. As was discussed for prokaryotic systems, we can use the frequency of recombinant types (the recombination frequency) to generate genetic maps. Note that the recombination frequency calculated for two genes from testcross data reflects physical exchanges of the chromosomes and is not related to the allelic forms of the genes. Therefore, the same percentage of recombinant types should be found either when both wild-type alleles are carried on the same chromosome in the F1 ($a^+ b^+/a\ b$) (a situation known as **coupling**) or when each chromosome carries one of wild-type alleles ($a^+ b/a\ b^+$) (a situation known as **repulsion**).

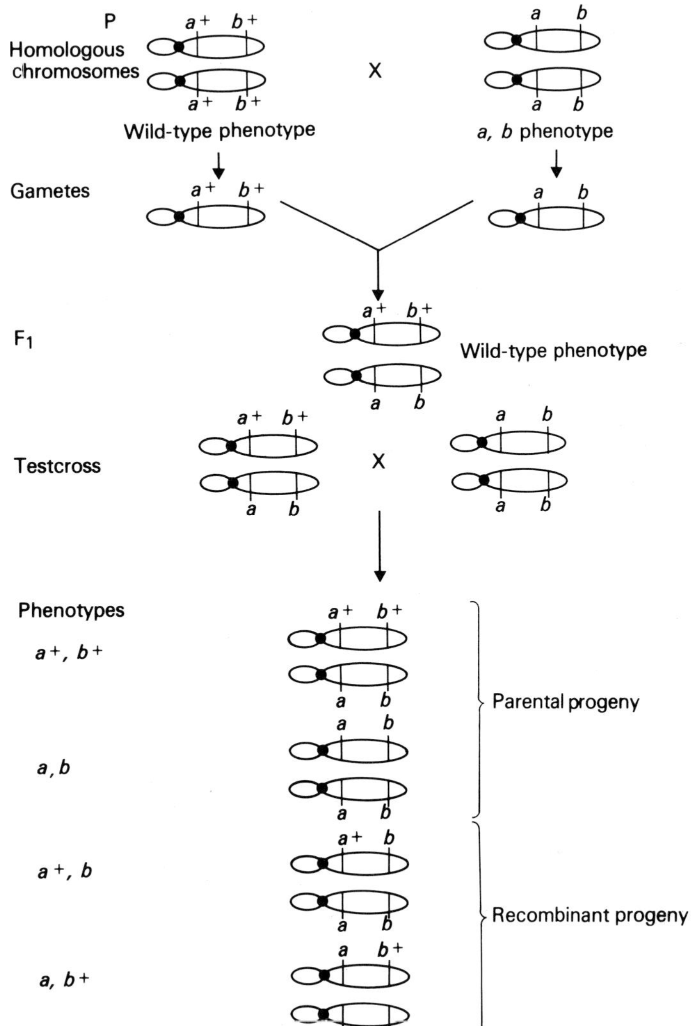

Fig. 14.1. Representation of a testcross when two genes are linked. Here *a* and *b* are linked genes on a chromosome. The alleles *a* and *b* are recessive to the wild-type alleles a^+ and b^+. An excess of parental progeny over recombinant progeny indicates that the two genes are linked.

Historically, T. H. Morgan working with the fruit fly *Drosophila melanogaster* found that many genes did not segregate independently of other genes during meiosis. The results of testcrosses were that the parental phenotypic classes were the most frequent while the recombinant phenotypic classes occurred much less frequently. Morgan concluded that during meiosis, certain genes tend to remain together because they lie near each other on the same chromosomes. Genes on the same chromosome are said to belong to the same *linkage group*, the number of linkage groups in all organisms being equal to the haploid number of chromosomes. In *D. melanogaster*, for example, there are four linkage groups.

Sex chromosomes and sex linkage

The chromosome complement of animal cells consists of the autosomes and the sex chromosomes. In most cases the male of the species is XY and the female is XX with respect to the sex chromosomes. One of the distinguishing features about the sex chromosomes is that the X and Y are not homologous and very few genes have been found on the Y chromosome that also occur on the X (the "bobbed" gene of *Drosophila* is an example). For our purposes we can consider the Y chromosome to be "silent" in terms of discussing the dominance or recessiveness of genes on the X chromosome. The Y chromosome does contain genes and they play a very important role in, for example, development, but they are different genes from those on the X. In an XY individual the genes carried on the X chromosome are described as being **hemizygous** (literally "half zygote") and constitute the X-linkage group. Characters resulting from genes on the X chromosome are referred to as being **sex-linked**, although **X-linked** is a better term.

Sex-linked genes exhibit different segregation ratios from autosomal genes. In *Drosophila* there is a recessive mutation called white (w) that results in individuals with white eyes instead of the red eyes characteristic of the wild type. When reciprocal crosses are done between a true-breeding wild type and a true-breeding white strain, the two F1s differ phenotypically (Fig. 14.2). This is a characteristic of sex-linked genes.

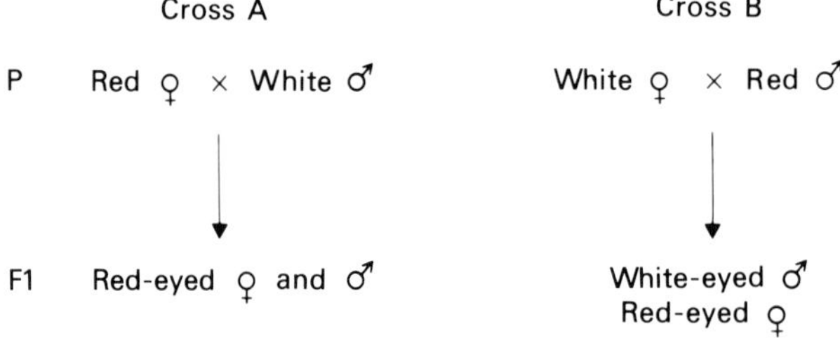

Fig. 14.2. Phenotypic results of reciprocal crosses of a strain carrying sex-linked, recessive mutation (white eyes) with the wild type (red eyes).

If the gene in question was located on an autosome, both sexes of both F1s would be expected to show the wild-type trait. The results may be explained by the fact that the white gene is located on the X chromosome and the two crosses in Fig. 14.2 can be written using genotypic symbols. The letter Y, or ↾, indicates the Y chromosome, and || indicates the two homologous X chromosomes (Fig. 14.3). F1 females from each reciprocal cross receive one X chromosome from each parent and therefore are heterozygous and express the wild-type phenotype. F1 males receive their sole X chromosome from their mother and a Y chromosome from their father. The phenotypes of the F1 males therefore depend upon the genes on the inherited X chromosome.

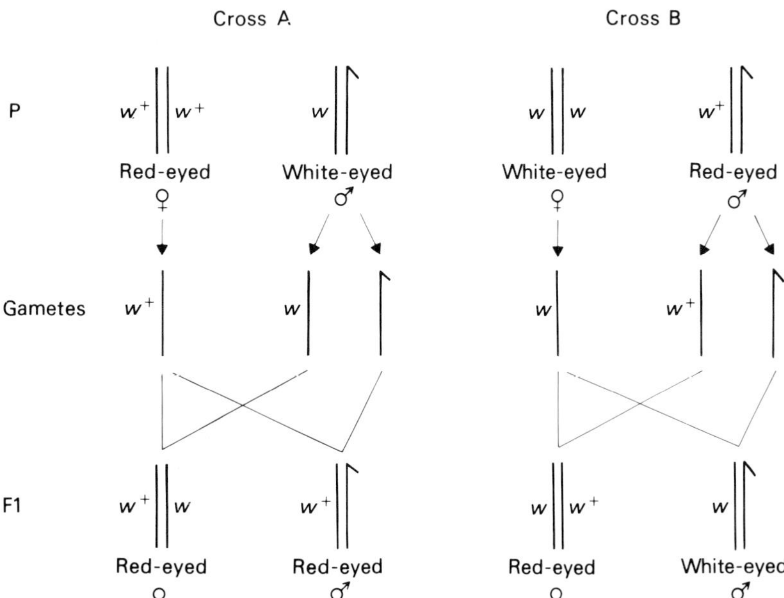

Fig. 14.3. Genotypic diagrams of the parental and F1 generations of the reciprocal crosses of Fig. 14.2.

For cross A, then, the F1 male is w^+Y and the red-eyed, and for cross B the F1 male is w Y and white-eyed.

We showed earlier that a 9:3:3:1 ratio will result in crosses of strains carrying unlinked autosomal genes when doubly heterozygous F1s are crossed together. When the two genes are autosomal, the 9:3:3;1 ratio will be found for both male and female F2 progeny. When a sex-linked gene and an autosomal gene are involved in a cross, the F2 ratios are different from the case of two autosomal genes. In the following examples, the gene symbolism will be abbreviated further and the two mutant genes are recessive to the wild type: gene x is on the X chromosome and gene a is on an autosome. We will refer to the *phenotypes* controlled by the genes as follows: (x^+) = phenotype of the dominant allele of the X chromosome locus (i.e. genotypically x^+x^+ or x^+x in females or x^+Y in males); (x) = phenotype of the recessive allele of the X chromosome locus (i.e. genotypically xx in females, or xY in males); (a^+) = phenotype of dominant allele of the autosomal locus (i.e. genotypically a^+a^+ or a^+a in females or males); and (a) = phenotype of the autosomal locus (i.e. genotypically aa in females or males). The examples apply to any organism with sex chromosomes. Again we will follow the reciprocal crosses A and B, and calculate the F2 ratios using branching diagrams.

Fig. 14.4 shows cross A, or $x^+x^+ \; a^+a^+ \times xY \; aa$. The F1s from this cross are doubly heterozygous, $x^+x \; a^+a$ (wild-type) females and x^+Y a^+a (wild-type) males. If we consider only the X-linked gene, the F1×F1 is $x^+x \times x^+$Y. By random segregation of the resulting gametes, this will result in four F2

Cross A P $x^+x^+ \; a^+a^+$ ♀ × $x\upharpoonleft aa$ ♂

wild-type x, a phenotypes

(abbreviation of

$x^+ \| x^+ \; a^+ \| a^+$)

F1 $x^+x \; a^+a$ ♀ and $x^+\upharpoonleft \; a^+a$ ♂

wild-type wild-type

Combining this with the autosomal ratio, we have:

F2 phenotypic ratios

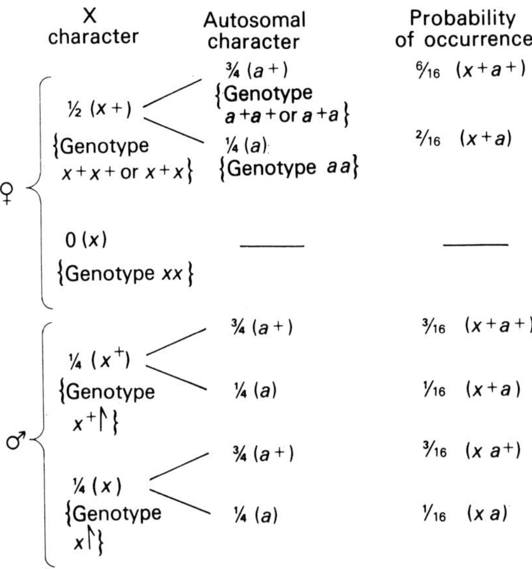

Summarizing, we have:

Phenotype ratios

	(x^+a^+)		(x^+a)		$(x\,a^+)$		$(x\,a)$
♀	6	:	2	:	0	:	0
♂	3	:	1	:	3	:	1
♀ and ♂	9	:	3	:	3	:	1

Fig. 14.4. Genotypes and phenotypes of the parental F1 and F2 generations of a cross of a wild-type female with a male carrying a sex-linked recessive mutation and homozygous for an autosomal recessive mutation.

genotypes occurring with equal frequencies: x^+x^+ (wild-type female), x^+x (wild-type female), x^+Y (wild-type male),and x^+Y (mutant male). Phenotypically all females are wild type, one-half of the males are wild-type and one-half are mutant. Considering the autosomal gene, the $a^{+}a \times a^{+}a$ cross will generate a F2 progeny with a phenotypic ratio of 3/4 (a^{+}) and 1/4 (a). Since the X chromosome and the autosome assort independently at meiosis, the overall *phenotypic* ratio of the F2 can be calculated by multiplying the probabilities of occurrence of the X autosomal characters together in all possible ways. In this the females and males are considered separately. For the four different phenotypes, the ratios for males and females separately and for the two combined are shown in Fig. 14.4.

As can be seen, a 9:3:3:1 phenotypic ratio is the case when females and males are combined, but quite different ratios are apparent for the two sexes considered separately: 6:2:0:0 for females and 3:1:3:1 for males. This results from the inheritance characteristics of sex-linked genes stemming from the lack of homologous genes on the Y chromosome.

The reciprocal cross B is $xx\ aa \times x^{+}\mathrm{Y}\ a^{+}a^{+}$; it can be analyzed in the same way as for cross A (Fig. 14.5). In the F2 of cross B, neither sex shows a 9:3:3:1 ratio, but instead each sex exhibits a 3:1:3:1 ratio. For males and females combined, the F2 phenotypic ratio is 6:2:6:2.

Crossing-over and genetic recombination

Genetic crosses can be used to show recombination of genes from generation to generation. This genetic recombination results from a physical exchange of parts of homologous chromosomes by a process called crossing-over. Crossing-over occurs in the first division of meiosis during the time when four chromatids are present for each pair of chromosomes (the four-strand stage). Proof for this will be given when meiotic genetic analysis of fungi is discussed.

That genetic recombination does result from a physical exchange of parts of homologous chromosomes was demonstrated separately in 1931 by H. B. Creighton and B. McClintock working with *Zea mays* (corn) and by C. Stern working with *Drosophila melanogaster*. We will discuss Stern's experiments.

Stern overcame the difficulties of distinguishing the members of a pair of homologous chromosomes by using "abnormal" chromosomes that could be distinguished cytologically, that is, by using cytological markers. The *Drosophila* cross he used is shown in Fig. 14.6. The male parent carries normal X and Y chromosomes and on the X is the recessive mutant allele *car* (carnation) and the wild-type allele of the *B* (bar) gene. The male therefore has carnation-colored eyes. The female parent has two abnormal and cytologically distinguishable X chromosomes. One carries the wild-type alleles of both the *car* and *B* genes and in addition has a part of the Y chromosome attached to it. The other X carries the recessive mutation *car* and the dominant mutation *B*. This X chromosome is shorter than normal because part of it has broken off and is attached to chromosome Y. Since the *car*

Cross B P $xx\,aa$ ♀ × $x^{+}\upharpoonleft\,a^{+}a^{+}$ ♂

↓

F1 $x^{+}x\,a^{+}a$ ♀ × $x\upharpoonleft\,a^{+}a$ ♂

wild-type (x, a^{+}) phenotype

Considering the X-linked gene alone:

F1 × F1 $x^{+}x$ ♀ × $x\upharpoonleft$ ♂

↓

F2 ¼ $x^{+}x$, ¼ xx , ¼ $x^{+}\upharpoonleft$, ¼ $x\upharpoonleft$

wild ♀ x ♀ wild ♂ x ♂

Combining this with the autosomal ratio in the same way as for Cross A, we have:

F2 phenotypic ratios

	X character	Autosomal character	Probability of occurrence
♀	¼ x^{+} $(x^{+}x)$ {Genotype $x+x+$ or $x+x$}	¾ $(a+)$ {Genotype $a^{+}a^{+}$ or $a^{+}a$}	$^{3}/_{16}$ $(x+a+)$
		¼ (a) {*Genotype aa*}	$^{1}/_{16}$ $(x+a)$
	¼ x $(x\,x)$ {Genotype xx}	¾ $(a+)$	$^{3}/_{16}$ $(x\,a+)$
		¼ (a)	$^{1}/_{16}$ $(x\,a)$
♂	¼ x^{+} $(x^{+}\upharpoonleft)$ {Genotype $x+\upharpoonleft$}	¾ $(a+)$	$^{3}/_{16}$ $(x+a+)$
		¼ (a)	$^{1}/_{16}$ $(x+a)$
	¼ x $(x\upharpoonleft)$ {Genotype $x\upharpoonleft$}	¾ $(a+)$	$^{3}/_{16}$ $(x\,a+)$
		¼ (a)	$^{1}/_{16}$ $(x\,a)$

Summarizing, we have:

Phenotype ratios

	$(x+a+)$		$(x^{+}a)$		$(x\,a^{+})$		$(x\,a)$
♀	3	:	1	:	3	:	1
♂	3	:	1	:	3	:	1
♀ and ♂	6	:	2	:	6	:	2

Fig. 14.5. Genotypes and phenotypes of the parental F1 and F2 generations of a cross of a female homozygous for a sex-linked recessive and an autosomal recessive mutation, with a wild-type male.

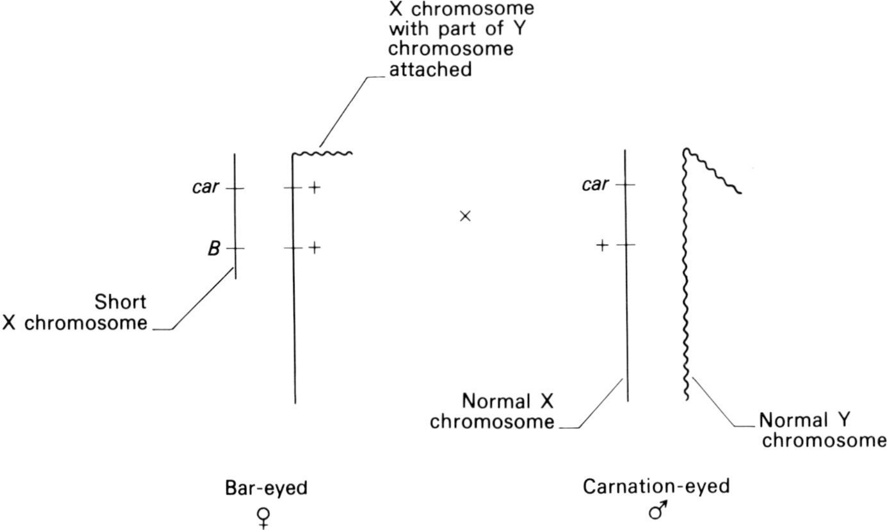

Fig. 14.6. Diagrammatic representation of the chromosomes in the male and female *Drosophila* that were crossed to show that genetic recombination results from a physical exchange of chromosomes.

mutation is heterozygous, the females have wild-type eye color. For dominant mutations, the heterozygote will show the mutant rather than the wild-type phenotype and thus the *B*/+ females have bar-shaped eyes rather than round eyes.

In gamete formation, the male produces only two classes of sperm: the Y-bearing and the X-bearing, which carries the *car* allele and the wild-type allele of *B*. The female produces four gamete classes: two of these result from meiosis in which no crossovers occurred between the *car* and *B* loci, and the other two are recombinant gametes produced by such a crossover. The gamete types and the diploid progeny are shown in Fig. 14.7.

Examination of the chromosomes of the progeny whose phenotypes indicated that genetic recombination had occurred showed that in every case the recombination event was accompanied by an exchange of identifiable chromosome segments.

Gene mapping in diploids

T. H. Morgan pioneered gene mapping with his experiments using *Drosophila*. He crossed a female homozygous for the sex-linked genes white-eyed (*w*) and miniature-winged (*m*) with a wild-type male and analyzed the F1 and F2 progeny (Fig. 14.8). (Note that, in this case, the F1×F1 cross is also a testcross.)

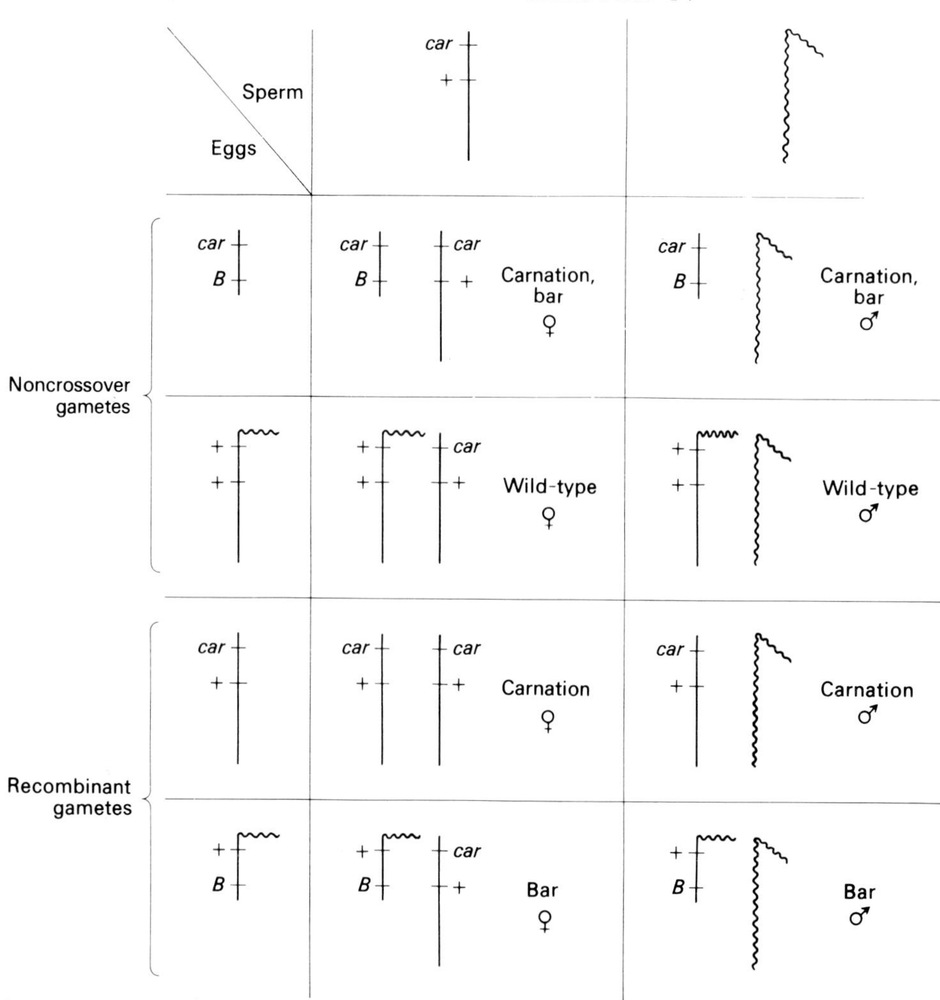

Fig. 14.7. Progeny of cross shown in Fig. 14.6 showing that genetic recombination results from crossing over.

The results showed that 36.9% of the progeny had recombinant (i.e. nonparental combinations of phenotypes). The recombinant flies were distributed approximately equally between females and males. The recombinant types resulting from crossing-over between the *w* and *m* loci; and the same frequency of recombination would be found if the genes were in repulsion.

The recombination frequency between particular pairs of linked genes occurs at stable and characteristic frequencies in all organisms. The frequency found depends on the two loci involved, and it can be used as an indicator of the *genetic distance* between the two loci on the genetic map. The value for the percentage of recombinants is usually converted into **map units**, where 1 crossover percent = 1 map unit. In our example, this means that *w* and *m* are 36.9 map units apart, meaning that 36.9% of gametes arose as a result of crossovers.

The case described was an example of two-point mapping analysis where the distance between two genes is determined. By doing a series of two-point

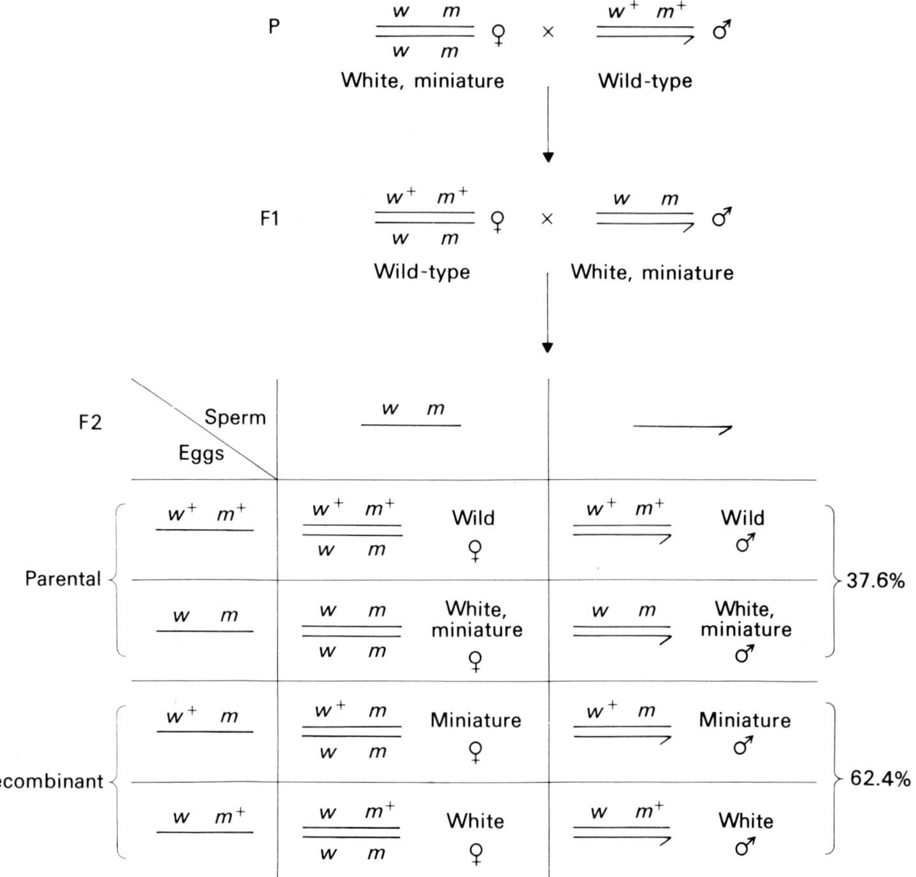

Fig. 14.8. Experimental crosses of *Drosophila* performed by T. H. Morgan to determine the map distances between the sex-linked genes, white and miniature.

crosses, the linkage relationships of a number of sex-linked genes of *Drosophila* were determined. In 1913, A. Sturtevant, a student of Morgan, devised the testcross method to analyze the linkage relationships between genes. He was the scientist who first proposed that the percentage of recombinants could be used as a quantitative measure of the distance between two gene pairs of a genetic map: this distance is measured in map units (mu). The map unit was later named a **centiMorgan** (cM) in honor of Morgan.

Sturtevant's concept of genetic map distance led him to the realization that the distances between a series of linked genes are additive. That is, the genes on a chromosome can be represented by a linear genetic map that depicts the genes in a linear order as they are arranged along the chromosome. A gene's position on a genetic map is called its **locus**.

Crossover and recombination values are used to show a linear order of the genes on a chromosome and to calculate the relative distance between any

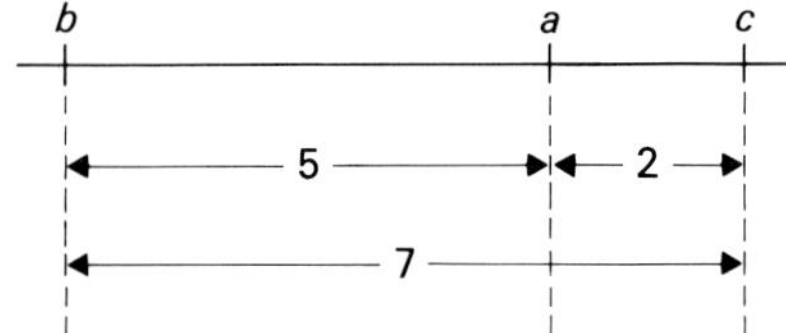

Fig. 14.9. Linear relationship of genes *a*, *b*, *c* as determined by all possible pairwise crosses and recombination analysis.

two genes. The crossover frequency is directly proportional to the genetic distance in map units. The logic for constructing a genetic map of linked genes is then straightforward. For example, if *a* and *b* are 5 map units apart, *b* and *c* are 7 map units apart, and *a* and *c* are 2 map units apart, the linear relationship of the three genes can be determined (Fig. 14.9).

In genetic mapping analysis, it is important to set up the most appropriate cross or crosses. For mapping recessive sex-linked genes, the crosses are set up so the F1 female is heterozygous for the two genes and the F1 male is hemizygous for both mutant alleles (see Fig. 14.8). Since the Y carries no homologous genes to the one on the X, the F1×F1 cross is essentially a testcross, which is the ideal cross for mapping analysis as we have discussed. A testcross is used also for mapping recessive autosomal genes, and the usual sequence of crosses (here with genes in coupling) for a diploid organism would be as shown in Fig. 14.10.

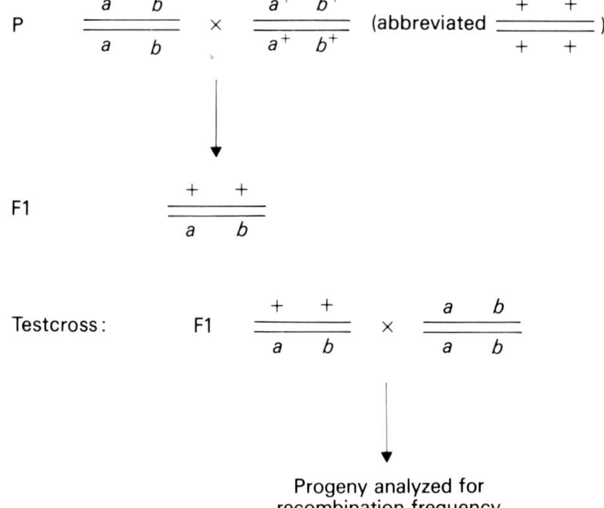

Fig. 14.10. Testcross procedure for mapping autosomal recessive mutant genes.

For mapping dominant mutations, the testcross is also used and here that means crossing the double heterozygote to a strain carrying the wild-type alleles of the mutant genes, which are, of course, recessive (Fig. 14.11). By convention the dominant mutation is indicated by a capital letter (e.g. *Cy* for the curly wing mutation in *Drosophila*). For this example, the wild-type allele would be signified by Cy^+.

Three-point testcross and gene mapping

Genetic recombinants result from crossing-over, the interchange of segments between homologous chromosomes. Testcross data provide recombination

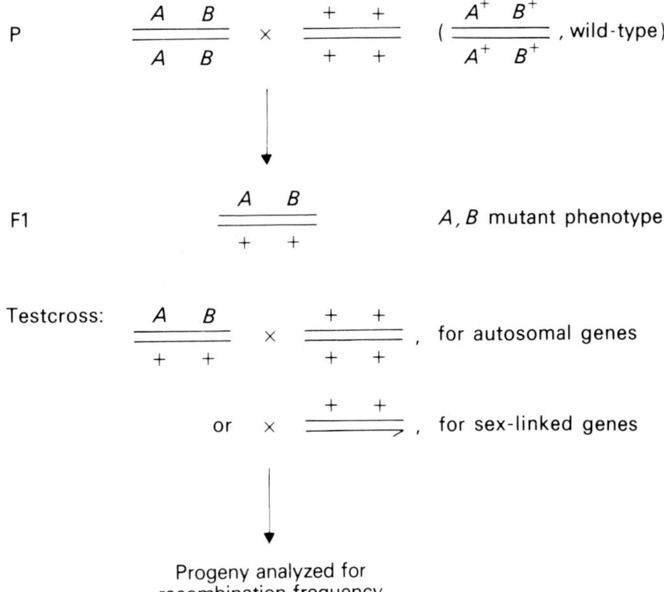

Fig. 14.11. Testcross procedure for mapping autosomal (or sex-linked) dominant mutant genes.

frequencies as well as giving an estimate of crossing-over frequency. The latter may not be accurate owing to multiple crossovers occurring between genes, and this can complicate the computation of map distance. (Note that *recombination frequency* is the fraction of haploid cells that are **recombinant** for two genetic loci, whereas *map distance* is the average number of exchanges occurring between the two loci.) If we assume that crossing-over occurs randomly along the length of a chromosome, the crossover frequency reflects the distance apart of any two genes. However, the only ways that crossovers can be counted is by looking at recombinant progeny that must have resulted from a crossover. Therefore, especially if genes are a reasonable distance apart, double- and other even-numbered crossovers can occur between them, but they will not be counted since the resulting progeny will not be recombinant (Fig. 14.12). Thus, since only odd numbers of exchanges between marked loci recombine those loci (if the same two chromatid strands are involved), recombination frequency and map distance are not in general equivalent.

This phenomenon is most significant for double crossovers as the frequency of higher number crossovers between two genes is relatively small. They do contribute to the map distance calculations since odd-numbered crossovers generate recombinant gametes and even-numbered crossovers generate parental gametes. For map distances under 5 map units, double crossing-over occurs only infrequently. Thus an accurate linkage map can best be constructed by a series of two-point testcrosses where each pair of

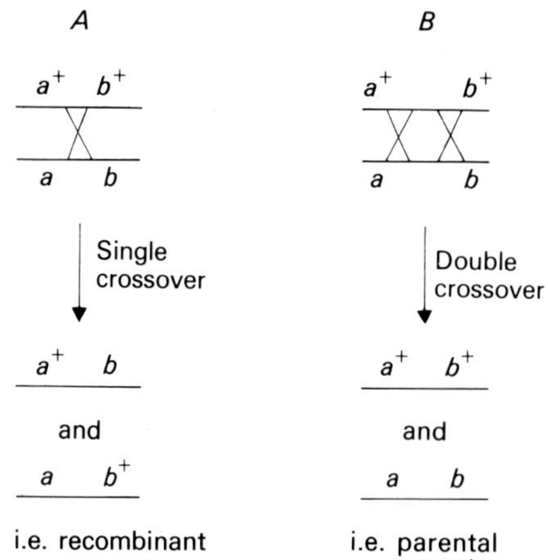

Fig. 14.12. Crossing over between linked genes *a* and *B*: (*A*) A single crossover generates recombinant gametes. (*B*) A double crossover generates parental gametes.

genes is closely linked, so that recombination frequencies give an accurate measure of crossover frequencies and hence of map distances.

One way to obtain good recombination data is to use three-point testcrosses involving three genes located within a relatively short segment of a chromosome. Three-point testcrosses are preferable to two-point testcrosses for mapping because they aid in the detection of double crossovers, and because the relative order of the genes on the chromosome can be established from a single cross. If we put a gene *c* between *a* and *b* in the previous example, one of the advantages of such a testcross is that the double crossover would be recognizable since it generates recombinant progeny (Fig. 14.13). Obviously double crossovers in the *a-c* and *b-c* region would not be detected among the progeny.

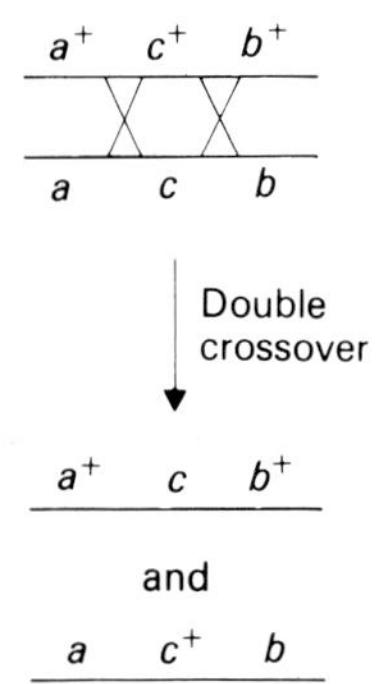

Fig. 14.13. The presence of a third gene, *c*, between genes *a* and *b* allows certain double crossovers between *a* and *b* to be detected as recombinants for *c*.

To illustrate how a three-point testcross can be used to map genes, we shall discuss theoretical data such as might be obtained in laboratory experiments with the fruit fly *Drosophila melanogaster*. The data presented are testcross values involving the segregation of three mutant alleles. These are: r = red, a recessive mutation whose wild-type allele, r^+, specifies a black phenotype; m = miniature, a recessive mutation whose wild-type allele, m^+, specifies a large phenotype; and o = oval, a recessive mutation whose wild-type allele, o^+, specifies a spherical phenotype. We will take as given that the three genes are linked on one chromosome. The parental cross was $r^+m^+\ o/r^+m^+\ o \times r\ m\ o^+/r\ m\ o^+$ to give a triply heterozygous F1 with the genotype $r^+m^+\ o/r\ m\ o^+$ in which genes r and m are in coupling and both r and m are in replusion with gene o. Fig. 14.14 shows the testcross $r^+m^+\ o/r\ m\ o^+ \times r\ m\ o/r\ m\ o$ that was done to obtain data for map distance calculations. In this, the order of the genes is arbitrarily given at the moment.

Testcross $\frac{r^+ \; m^+ \; o}{r \; m \; o^+}$ X $\frac{r \; m \; o}{r \; m \; o}$

(black, large, spherical) (red, miniature oval)

Progeny

Class	Phenotypes of testcross progeny[a]	Number of individuals	Genotypes of gamete from heterozygous parent responsible for phenotype
1	black, large, *oval*	433	$r^+ \; m^+ \; o$
2	red, *miniature*, spherical	456	$r \; m \; o^+$
3	black, *miniature*, spherical	29	$r^+ \; m \; o^+$
4	red, large, *oval*	38	$r \; m^+ \; o$
5	black, large, spherical	2	$r^+ \; m^+ \; o^+$
6	red, *miniature*, oval	3	$r \; m \; o$
7	black, *miniature*, *oval*	48	$r^+ \; m \; o$
8	*red*, large, spherical	61	$r \; m^+ \; o^+$
Total progeny: 1070			

[a] Mutant phenotypes are in italic

Fig. 14.14. Three-point mapping analysis showing the testcross used and the resultant progeny.

Each progeny type has the phenotype dictated by the genotype of the gamete produced by the heterozygous parent since that always pairs with a gamete from the other parent that carries all recessive alleles. Thus class 1 is $r^+m^+ \; o/r \; m \; o$ in genotype, and this has a black, large, oval phenotype. The gametes produced by meiosis with no crossovers in the triply heterozygous parent are $r^+m^+ \; o$ and $r \; m \; o^+$. By pairing with the $r \; m \; o$ gamete from the other parent, these gametes give rise to classes 1 and 2 of the progeny. These are the parental-type progeny, and these classes have the largest numbers of individuals, with about equal numbers in each class as would be expected from such a reciprocal event. In general, even if the parental genotype is not known, the parental-type progeny can be identified by the fact that they have the largest numbers.

The double crossover progeny can also be picked out by inspection. A double crossover involves the simultaneous occurrence of two events, each of which alone has a relatively low probability of occurrence. Thus the double crossover gametes are the least frequent reciprocal pair, in this case classes 5 and 6, $r^+m^+ \; o^+$ and $r \; m \; o$, respectively.

Once the parental and double crossover progeny have been established, the order of the genes in the chromosome can be determined. As Fig. 14.13 shows, a double crossover changes the arrangement of the central gene with respect to the two other genes. To aid the analysis of the data, the parental and double crossover gamete types are illustrated in Fig. 14.15. The only arrangement of the three genes that is compatible with the data is $r^+o^+m^+$ and the heterozygous parent is $r^+o \; m^+/r \; o^+ \; m$. Fig. 14.16 illustrates the two crossovers and the resulting gametes.

Parental gametes	Double crossover gametes
$r^+ \; m^+ \; o$	$r^+ \; m^+ \; o^+$
and	and
$r \; m \; o^+$	$r \; m \; o$

Fig. 14.15. Comparison of the parental and double crossover gamete types from the testcross progeny data of Fig. 14.14.

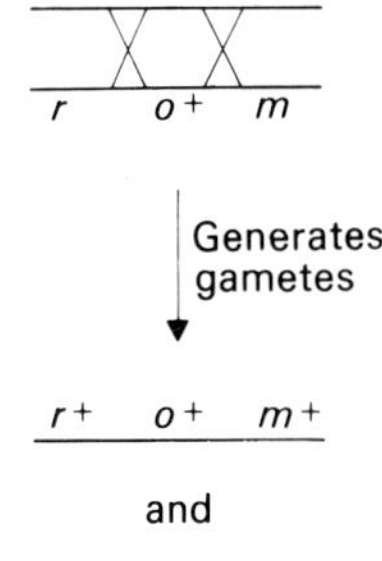

Fig. 14.16. Arrangement of the three genes on the chromosome is *r o m*. This figure shows how the two crossovers in the parent generates the double crossover recombinant types.

The data can now be rewritten, and for convenience we will call the region between *r* and *o*, region I and the region between *o* and *m*, region II (Fig. 14.17).

Map distances are calculated as before, that is, by computing the frequency of crossing-over between two genes. In the example, the crossing-over between the *r* and *o* loci (in region I) produces classes 3 and 4 (single crossover, region I) and classes 7 and 8 (double crossover, regions I and II) above. Out of the 1070 progeny produced, 72 represent crossovers between

Testcross: $\frac{r^+ \quad o \quad m^+}{r \quad o^+ \quad m} \times \frac{r \quad o \quad m}{r \quad o \quad m}$

Region I (between *r* and *o*); Region II (between *o* and *m*)

Progeny:

Class	Gametic type	Number of individuals	Type
1	*r* + *o* *m* +	433	Parental types; no crossover
2	*r* *o* + *m*	456	
3	*r* + *o* + *m*	29	Recombinant — single crossover (sco) region I
4	*r* *o* *m* +	38	
5	*r* + *o* *m*	48	Recombinant sco region II
6	*r* *o* + *m* +	61	
7	*r* + *o* + *m* +	2	Recombinant — double crossover regions I and II
8	*r* *o* *m*	3	

Total progeny = 1070

Fig. 14.17. A rewritten form of the cross of Fig. 14.14 based on the newly determined gene order.

the two loci. This is 6.7% of the progeny, and thus *r* and *o* are 6.7 map units apart. In other words the *r*-*o* map distance is:

$$= \frac{\text{Frequency of sco, region I} + \text{frequency of dco}}{\text{Total}} \times 100$$

$$= \frac{67+5}{1070} \times 100$$

$$= 6.7\%$$

Similarly, the *o*-*m* map distance involves crossovers in region II and is as follows:

$$= \frac{\text{Frequency of sco, region II} + \text{frequency of dco}}{\text{Total}} \times 100$$

$$= \frac{(48+61)+(2+3)}{1070} \times 100$$

$$= \frac{114}{1070} \times 100$$

$$= 10.7\%$$

Therefore the *o*-*m* distance is 10.7 map units. From these data, the segment of the chromosome involving the three genes is as depicted in Fig. 14.18.

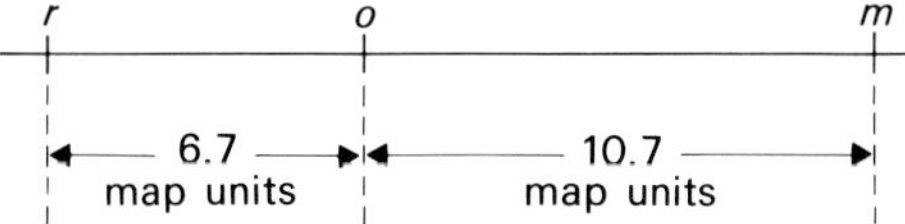

Fig. 14.18. Genetic map of the *r*-*o*-*m* region of the chromosome determined from the data presented in Fig. 14.17.

Note that in calculating map distance from three-point testcross data, the double crossover figure must be added to each of the single crossover figures since in each case a double crossover represents single crossing-over in *both* regions I and II.

Interference and coincidence

The map distances obtained from crosses such as those described enable the investigator to gauge whether the expected number of double crossovers occur or whether the occurrence of one crossover diminishes the probability of a second crossover occurring close by. In our example, the *r-o* distance of 6.7 map units means that 6.7% of the gametes are expected to reflect crossing-over between those two loci. Similar arguments can be made for the *o-m* gene loci. If we assume for the moment that a crossover event in region I is independent of a crossover event in region II, then the probability of crossovers occurring simultaneously in both regions in a meiosis is equal to the product of the probabilities of the two crossovers occurring independently. In this case, this would be $0.067 \times 0.107 = 0.0072$, or in other words, 0.72% double crossover progeny would be expected in the cross we analyzed. Actually, only 5/1070 or 0.47% double crossovers occurred in the cross. Indeed, it is characteristic of three-point testcross data that the observed number of double crossover progeny is lower than the expected number of double crossover progeny. This signifies that one crossover, perhaps by physical perturbations of the paired homologs, reduces the probability of a second crossover occurring nearby. The phenomenon is called **chromosome** or **chiasma interference** and the magnitude of the interference may vary throughout the genome. H. Muller called the ratio of observed double crossover frequency/expected double crossover frequency the **coefficient of coincidence.** *Interference* is (1−coefficient of coincidence). For the *r-m* region of the example that we have been discussing, the coincidence is 0.47/0.72 = 0.65. Coincidence values normally vary from 0 to 1 and they vary inversely with interference values. Thus a coincidence of 0 means complete interference; two crossovers did not occur simultaneously at all in that case. Conversely, a coincidence of 1 would indicate absolutely no interference. In our example, the interference value is 0.35, or in other words, only 65% of the expected double crossovers took place in the region being studied.

Summary

1. The most accurate mapping analysis is done when genes involved are closely linked. owing to multiple crossovers, and particularly double crossovers, the distances obtained for genes far apart are usually underestimated. Note that the highest recombination frequency that can be obtained for any two genes located far apart on the same chromosome is 50% since the number of even-numbered crossovers (producing parental progeny) between them will equal the number of odd-numbered crossovers (producing recombinant progeny).

2. The map distances obtained from mapping experiments are genetic distances and reflect the probability of occurrence of crossing-over for the region being mapped. Although it is common to assume that crossing-over is random throughout the genome, it is almost certain that this is not true. For example, there is good evidence that crossing-over occurs infrequently near centromeres. Genes located in those areas, then, would appear to be closely linked genetically, whereas physically they could be far apart. Therefore, while the order of the genes is equivalent on the genetic map and on the chromosome, the genetic distance may or may not be an accurate reflection of physical distance, depending on what part of the genome one is studying.

3. The genetic crosses described allow one to construct linkage maps for genes in any organism, which provides useful information about the distribution of genes with related function throughout the genome.

Questions and problems

14.1 In *Drosophila*, white eyes is a sex-linked character. The mutant allele for white eyes (w) is recessive to the wild-type allele for brick red eye color (w^+).

(a) A white-eyed female is crossed with a red-eyed male, and an F1 female from this cross is mated with her father and an F1 male is mated with his mother. What will be the appearance of the offspring of these last two crosses with respect to eye color?

(b) A white-eyed female is crossed with a red-eyed male, and the F2 from this cross is interbred. What will be the appearance of the F3 with respect to eye color?

14.2 One form of color-blindness (c) in humans is caused by a sex-linked recessive mutant gene. A woman with normal color vision (c^+) whose father was color-blind marries a man of normal vision whose father was also color-blind. What proportion of their offspring will be color-blind. (Give your answer separately for males and females.)

14.3 In *Drosophila*, vestigial wings (vg) are recessive to normal long wings (vg^+), and the gene for this trait is autosomal. The gene for the recessive white eye trait (w) is on the X chromosome. Suppose a homozygous white, long-winged female fly is crossed with a true-breeding red, vestigial-winged male.

(a) What will be the appearance of the F1?

(b) What will be the appearance of the F2?

(c) What will be the appearance of the offspring of a cross of the F1 back to each parent?

14.4 In *Drosophila*, two red-eyed, long-winged flies are bred together and produce the following offspring: Females are 3/4 red, long and 1/4 red, vestigial, and males are 3/8 red, long; 3/8 white, long; 1/8 red, vestigial; and 1/8 white, vestigial. What are the genotypes of the parents?

14.5 Suppose gene A is on the X chromosome, and genes B, C, and D are on three different autosomes. Thus, $A\_$ signifies the dominant phenotype in the male or female. An equivalent situation holds for $B\_$, $C\_$, and $D\_$. The cross $AA\ BB\ CC\ DD$ females $\times$ aY $bb\ cc\ dd$ males is made.

(a) What is the probability of obtaining an $A\_$ individual in the F1?

(b) What is the probability of obtaining an a male in the F1?

(c) What is the probability of obtaining an $A\_\ B\_\ C\_\ D\_$ female in the F1?

(d) How many different F2 genotypes will there be?

(e) What proportion of F2s will be heterozygous for the four genes?

(f) Determine the probabilities of obtaining each of the following types in the

F2: (i) $A\_\ bb\ CC\ dd$ (female); (ii) $aY\ BB\ Cc\ Dd$ (male); (iii) $AY\ bb\ CC\ dd$ (male); (iv) $aa\ bb\ Cc\ Dd$ (female).

14.6 A cross $AA\ BB \times aa\ bb$ results in an F1 of phenotype $A\ B$; the following numbers are obtained in F2 (phenotypes):

$A\ B$: 110
$A\ b$: 16
$a\ B$: 19
$a\ b$: 15

Total 160

Are genes at the a and b loci linked or independent? What F2 numbers would otherwise be expected?

14.7 In corn, a dihybrid for the recessives a and b is testcrossed. The distribution of the phenotypes was as follows:

$A\_\ B\_$	122
$A\_\ bb$	118
$aa\ B\_$	81
$aa\ bb$	79

Are the genes assorting independently? Test the hypothesis with a (χ^2) test (see Appendix I). Explain tentatively any deviation from expectation, and tell how you would test your explanation.

14.8 The F1 from a cross of $A\ B/A\ B \times a\ b/a\ b$ is testcrossed, resulting in the following phenotypic ratios:

$A\ B$	308
$A\ b$	190
$a\ b$	292
$a\ B$	210

What is the frequency of recombination between genes a and b?

14.9 In rabbits, the English type of coat (white-spotted) is dominant over non-English (unspotted), and short hair is dominant over long hair (Angora). When homozygous English, short-haired rabbits were crossed with non-English Angoras and the F1 crossed back to non-English Angoras, the following offspring were obtained: 72 English, short-haired; 69 non-English, Angora; 11 English, Angora; and 6 non-English, short-haired. What is the map distance between the genes for coat color and hair length?

14.10 In *Drosophila* the mutant black (b) has a black body in contrast to the wild-type, which has a grey body; the mutant vestigial (vg) has wings that are much shortened and crumpled compared with the long wings of the wild-type. In the following cross, the true-breeding parents are given, together with the counts of offspring of F1 females × black, vestigial males:

P black, normal × grey vestigial
F1 females × black, vestigial males give:

grey, normal	283
grey, vestigial	1294
black, normal	1418
black, vestigial	241

From these data, calculate the map distance between the black and vestigial genes.

14.11 Use the following two-point recombination data to map the genes concerned. (Show order and length of shortest intervals.)

Gene loci	% recombination	Gene loci	% recombination
a, b	50	*b, d*	13
a, c	15	*b, e*	50
a, d	38	*c, d*	50
a, e	8	*c, e*	7
b, c	50	*d, e*	45

14.12 The following data are from C. Bridges and T. H. Morgan's work on the recombination between the genes black, curved, purple, speck, and vestigial in chromosome II of *Drosophila*. On the basis of the data, map the chromosome for these five genes as accurately as possible. Remember, that determinations for short distances are more accurate than those for long ones.

Genes in cross	Total progeny	Number of recombinants
black, curved	62,679	14,237
black, purple	48,931	3,026
black, speck	685	326
black, vestigial	20,153	3,578
curved, purple	51,136	10,205
curved, speck	10,042	3,037
curved, vestigial	1,720	141
purple, speck	11,985	5,474
purple, vestigial	13,601	1,609
speck, vestigial	2,054	738

14.13 Genes *a* and *b* are linked, with 10% recombination. What would be the phenotypes and the probability of each among progeny of the following cross.

$$\frac{a\,b^+}{a^+\,b} \times \frac{a\,b}{a\,b}$$

14.14 Genes *A* and *B* are sex-linked and located 7 map units apart in the X chromosome of *Drosophila*. A female of genotype *A b*/*a B* is mated with a wild-type (*A B*) male.

(a) What is the probability that one of her sons will be either *A B* or *a B* in phenotype?

(b) What is the probability that one of her daughters will be *A B* in phenotype?

14.15 In maize, the dominant genes *A* and *C* are both necessary for colored seeds. Homozygous recessive plants give colorless seeds regardless of the genes at the second locus. *A* and *C* show independent segregation. At a third locus, the dominant gene *Wx* gives starchy endosperm while the recessive mutant allele *wx*, when homozygous, gives waxy endosperm. The *Wx* gene is linked with *C*, showing 20% recombination with it.

(a) What phenotypic ratios would be expected when a plant of constitution *c Wx/C wx; A/A* is testcrossed?

(b) When a plant of constitution *c Wx/C wx*; *A/a* is testcrossed?

14.16 Assume that genes *A* and *B* are linked and show 20% crossing-over.

(a)If a homozygous *A B/A B* individual is crossed with one which is *a b/a b*, what will be the genotype of the F1? What gametes will the F1 produce and in what proportions and genotypes of the offspring?

(b) If, instead, the original cross is *A b/A b*×*a B/a B*, what will be the genotype of the F1? What gametes will the F1 produce and in what proportions? If the F1 is testcrossed with a double homozygous recessive, what will be the proportions and genotypes of the offspring?

14.17 In tomatoes, tall vine is dominant over dwarf, and spherical fruit shape is dominant over pear shape. Vine height and fruit shape are linked, with a crossover percentage of 20. A certain tall, spherical fruited tomato plant crossed with a dwarf, pear-fruited one produces 81 tall, spherical; 79 dwarf, pear; 22 tall, pear; and 17 dwarf, spherical. Another tall and spherical plant crossed with a dwarf and pear plant produces 21 tall, pear; 18 dwarf, spherical; 5 tall, spherical; and 4 dwarf, pear. What are the genotypes of the two tall and spherical plans? If they were crossed, what would their offspring be?

14.18 Genes *A* and *B* are in one chromosome, 20 map units apart; *C* and *D* are in another chromosome, 10 map units apart. *E* and *F* are in yet another chromosome and are 30 map units apart. Cross a homozygous *ABCDEF* individual with an *abcdef* one, and cross the F1 back to an *abcdef* individual. What are the chances of getting individuals of the following phenotypes in the progeny?

(a) *ABCDEF*
(b) *ABCdeF*
(c) *AbcDEf*
(d) *aBCdef*
(e) *abcDeF*

14.19 In the Maltese bippy, amiable (*A*) is dominant to nasty (*a*), benign (*B*) is dominant to active (*b*), and crazy (*C*) is dominant to sane (*c*). A true-breeding amiable, active, crazy bippy was mated, with some difficulty, to a true-breeding nasty, benign, sane bippy. An F1 individual from this cross was then used in a testcross (to a nasty, active, sane bippy), and produced, in typical prolific bippy fashion 4000 offspring. From an ancient manuscript entitled *The Genetics of the Bippy, Maltese and Other*, you discover that all three genes are autosomal, *A* is linked to *B* but not to *C*, and the map distance between *A* and *B* is 20 map units.

(a) Predict all the expected phenotypes and the numbers of each type from this cross.

(b) Which phenotypic classes would be missing had *A* and *B* shown complete linkage?

(c) Which phenotypic classes would be missing if *A* and *B* were unlinked?

(d) Again, assuming *A* and *B* to be unlinked, predict all the expected phenotypes of nasty bippies, and frequencies of each type, resulting from a self cross of the F1.

14.20 For each of the following tabulations of testcross progeny phenotypes and numbers, state which locus is in the middle, and reconstruct the genotype of the tested triple heterozygotes.

(a)			(b)			(c)		
	ABC	191		*CDE*	9		*FGH*	110
	abc	180		*cde*	11		*fgh*	114
	Abc	5		*Cde*	35		*Fgh*	37
	aBC	5		*cDE*	27		*fGH*	33
	ABc	21		*CDe*	78		*FGh*	202
	abC	31		*cdE*	81		*fgH*	185
	AbC	104		*CdE*	275		*Fgh*	4
	aBc	109		*cDe*	256		*fGH*	0

14.21 Genes of loci *f*, *m*, and *w* are linked, but their order is unknown. The F1 heterozygotes from a cross of *FF MM WW*×*ff mm ww* are testcrossed. The most frequent phenotypes in testcross progeny will be *F M W* and *f m w* regardless of what the gene order turns out to be.

(a) What classes of testcross progeny (phenotypes) would be least frequent if locus *m* is in the middle?

(b) What classes would be least frequent if locus *f* is in the middle?

(c) What classes would be least frequent if locus *w* is in the middle?

14.22 The following numbers were obtained for testcross progeny in *Drosophila* (phenotypes):

+ *m* +	218
w + *f*	236
+ + *f*	168
w m +	178
+ *m f*	95
w + +	101
+ + +	3
w m f	1
Total	1000

Construct a genetic map.

14.23 Three of the many recessive mutations in *Drosophila melanogaster* that affect body color, wing shape, or bristle morphology are black (*b*) body versus grey in wild-type; dumpy (*dp*), obliquely truncated wings versus long wings in wild-types; and hooked (*hk*) bristles at the tip versus not hooked in wild-type. From a cross of a dumpy female with a black, hooked male, all of the F1 were wild-type for all three characters. The testcross of a F1 female with a dumpy, black, hooked male, gave:

wild-type	169
black	19
black, hooked	301
dumpy, hooked	21
hooked	8
hooked, dumpy, black	172
dumpy, black	6
dumpy	305
Total	1001

(a) Construct a genetic map of the linkage group(s) these genes occupy. If applicable, show the order and give the map distances between the genes.

(b) (i) Determine the coefficient of coincidence for the portion of the chromosome involved in the cross. (ii) How much interference is there?

14.24 In Chinese primroses, long style (l) is recessive to short (L), red flower (r) is recessive to magenta (R), and red stigma (rs) is recessive to green (Rs). From a cross of homozygous short, magenta flower, green stigma with long, red flower, red stigma, the F1 was crossed back to long, red flower, red stigma. The following offspring were obtained:

Style	Flower	Stigma	Number
short	magenta	green	1063
long	red	red	1032
short	magenta	red	634
long	red	green	526
short	red	red	156
long	magenta	green	180
short	red	green	39
long	magenta	red	54

Map the genes involved.

14.25 The frequencies of gametes of different genotypes, determined by testcrossing a triple heterozygote, are:

Gamete genotype	%
+ + +	12.9
a b c	13.5
+ + *c*	6.9
a b +	6.5
+ *b c*	26.4
a + +	27.2
a + *c*	3.1
+ *b* +	3.5
Total	100.0

(a) Which gametes are known to have been involved in double crossovers?
(b) Which gamete types have not been involved in any exchanges?
(c) The order shown is not necessarily correct. Which gene locus is in the middle?

14.26 Genes *a*, *b*, and *c* are recessive. Females heterozygous at these three loci are crossed to phenotypically wild-type males. The progeny are phenotypically as follows:

Daughters: all + + +

Sons:	
+ + +	23
a b c	26
+ + *c*	45
a b +	54
+ *b c*	427
a + +	424
a + *c*	1
+ *b* +	0
Total	1000

(a) What is known of the genotype of the female parents with respect to these three loci? (Give gene order and arrangement in homologs.)

(b) What is known of the genotype of the male parents?

(c) Map the three genes.

14.27 Two normal looking *Drosophila* are crossed and yield the following phenotypes among the progeny:

	Phenotype	Number
Females:	+ + +	2000
Males:	+ + +	3
	a b c	1
	+ b c	839
	a + +	825
	a b +	86
	+ + c	90
	a + c	81
	+ b +	75
	Total	4000

Give parental genotypes, gene arrangement in the female parent, map distances, and coefficient of coincidence.

14.28 Three different semidominant mutations affect the tail of mice. They are linked genes and all three are lethal in the embryo when homozygous. Fused-tail (*Fu*) and kinky-tail (*Ki*) mice have kinky-appearing tails, while Brachyury (*T*) mice have short tails. A fourth gene, histocompatibility-2 (*H-2*), is linked to the three tail genes and is concerned with tissue transplantation. Mice that are *H-2*/+ will accept tissue grafts, whereas +/+ mice will not. In the following crosses, the normal allele is represented by a +. The phenotypes of the progeny are given for four crosses.

(1) $\dfrac{Fu\ +}{+\ Ki} \times \dfrac{+\ +}{+\ +}$

Phenotype	Number
Fused tail	106
Kinky tail	92
Normal tail	1
Fused-kinky tail	1

(2) $\dfrac{Fu\ H\text{–}2}{+\ +} \times \dfrac{+\ +}{+\ +}$

Phenotype	Number
Fused tail, accepts graft	88
Normal tail, rejects graft	104
Normal tail, accepts graft	5
Fused tail, rejects graft	3

(3) $\dfrac{T H\text{–}2}{+\ +} \times \dfrac{+\ +}{+\ +}$

Phenotype	Number
Brachy tail, accepts graft	1048
Normal tail, rejects graft	1152
Brachy tail, rejects graft	138
Normal tail, accepts graft	162

(4) $\dfrac{Fu\ +}{+\ T} \times \dfrac{+\ +}{+\ +}$

Phenotype	Number
Fused tail	146
Brachy tail	130
Normal tail	14
Fused-brachy tail	10

14.29 The cross in *Drosophila* of

$$\frac{a^+\ b^+\ c\ \ d\ \ e}{a\ \ b\ \ c^+\ d^+\ e^+} \times \frac{a\,b\,c\,d\,e}{a\,b\,c\,d\,e}$$

gave 1000 progeny of the following 16 phenotypes:

Genotype	Number	Genotype	Number
(1) $a^+\ b^+\ c\ \ d\ \ e$	220	(9) $a\ \ b^+ c^+ d\ \ e^+$	14
(2) $a^+\ b^+\ c\ \ d\ \ e^+$	230	(10) $a\ \ b^+ c^+ d\ \ e$	13
(3) $a\ \ b\ \ c^+\ d^+\ e$	210	(11) $a^+ b\ \ c\ \ d^+ e^+$	18
(4) $a\ \ b\ \ c^+\ d^+\ e^+$	215	(12) $a^+ b\ \ c\ \ d^+ e$	8
(5) $a\ \ b^+\ c^+\ d^+\ e$	12	(13) $a^+ b^+ c^+ d\ \ e^+$	7
(6) $a\ \ b^+\ c^+\ d^+\ e^+$	13	(14) $a^+ b^+ c^+ d\ \ e$	7
(7) $a^+\ b\ \ c\ \ d\ \ e^+$	16	(15) $a\ \ b\ \ c\ \ d^+ e^+$	6
(8) $a^+\ b\ \ c\ \ d\ \ e$	14	(16) $a\ \ b\ \ c\ \ d^+ e$	7

(a) Draw a genetic map of the chromosome, indicating the linkage of the five genes and the number of map units separating each.

(b) From the single crossover frequencies, what would be the expected frequency of $a^+\ b^+\ c^+\ d^+\ e^+$ flies?

References

Belling, J., 1933. Crossing over and gene rearrangement in flowering plants. *Genetics* **18**:388–413.

Bridges, C.B., 1916. Nondisjunction as a proof of the chromosome theory of heredity. *Genetics* **1**:1–52, 107–163.

Creighton, H.S. and B. McClintock, 1931. A correlation of cytological and genetical crossing over in *Zea mays*. *Proc. Natl. Acad. Sci. USA* **17**:492–497.

Gillies, C.B., 1975. Synaptonemal complex and chromosome structure. *Annu. Rev. Genetics* **9**:91–109.

Levine, R.P., 1955. Chromosome structure and the mechanism of crossing over. *Proc. Natl. Acad. Sci. USA* **41**:727–730.

McClung, C.E., 1902. The accessory chromosome—sex determinant? *Biol. Bull.* **3**:43–84.

McKusick, V.A. and F.H. Ruddle, 1977. The status of the gene map of the human chromosomes. *Science* **196**:390–405.

Morgan, T.H., 1910. Sex-limited inheritance in *Drosophila*. *Science* **32**:120–122.

Morgan, T.H., 1911. An attempt to analyze the constitution of the chromosomes on the basis of sex-limited inheritance in *Drosophila*. *J. Exp. Zool.* **11**:365–414.

Muller, H.J., 1916. The mechanism of crossing over. II. *Amer. Nat.* **50**:284–305

Roth, R., 1976. Temperature-sensitive yeast mutants defective in meiotic recombination and replication. *Genetics* **88**:675–686.

Stern, C., 1931. Zytologisch-genetische Untersuchungen als Beweise fur die Morgansche Theorie des Faktorenaustauschs. *Bio. Zentralbl.* **51**:547–587.

Sturtevant, A.H., 1913. The linear arrangement of six sex-linked factors in *Drosophila*, as shown by their mode of association. *J. Exp. Zool.* **14**:43–59.

Sutton, W.S., 1903. The chromosomes in heredity. *Biol. Bull.* **4**:213–251.

Westergaard, M. and D. von Wettstein, 1972. The synaptonemal complex. *Annu. Rev. Genetics* **6**:74–110.

Wilson, E.B., 1905. The chromosomes in relation to the determination of sex in insects. *Science* **22**:500–502.

Chapter 15 Eukaryotic Genetics: Fungal Genetics

Outline

Fungal life cycles

The advantages of fungi for genetic analysis are that many are haploid and have life cycles that enable each of the four products of a single meiosis (the meiotic **tetrad**) to be analyzed. The latter is called **tetrad analysis.** (Tetrad analysis is also possible with certain unicellular algae, for example, *Chlamydomonas reinhardi.*) The life cycles of two fungi—*Saccharomyces cerevisiae* (a budding yeast) and *Neurospora crassa* (a mycelial-form fungus)—that can be used for tetrad analysis (as well as for molecular experiments) will be described.

Yeast (Fig. 15.1)

In yeast there are two mating types, α and a. The haploid vegetative cells propagate mitotically by budding. Fusion of haploid a and α cells produces a diploid cell that is stable and is propagated by budding. The diploid cell can propagate mitotically by budding. However, if the diploid a/α strain is put into conditions of nitrogen starvation, sporulation is induced, which leads to meiosis. The four haploid meiotic products (the *ascospores*) are contained within a spherical structure called an *ascus*. Two of the ascospores are a, and two are α mating type. When the four ascospores are released and germinate, they produce haploid vegetative cells. In this organism the ascospores are organized randomly within the ascus: these tetrads are called **unordered tetrads** because the arrangement of the spores has no relationship to the organization of the four chromatids of each chromosome at metaphase I or meiosis.

Neurospora crassa (Fig. 15.2)

Neurospora crassa is a haploid organism that grows vegetatively by the formation of a weblike, branching growth called a *mycelium.* For asexual propagation of *Neurospora*, the asexual spores (*conidia*) or pieces of mycelia can be used to inoculate a growth medium to produce a new mycelium. The mycelium consists of cellular compartments separated by a septum with a central hole that permits circulation of cell contents, including nuclei, throughout the mycelium.

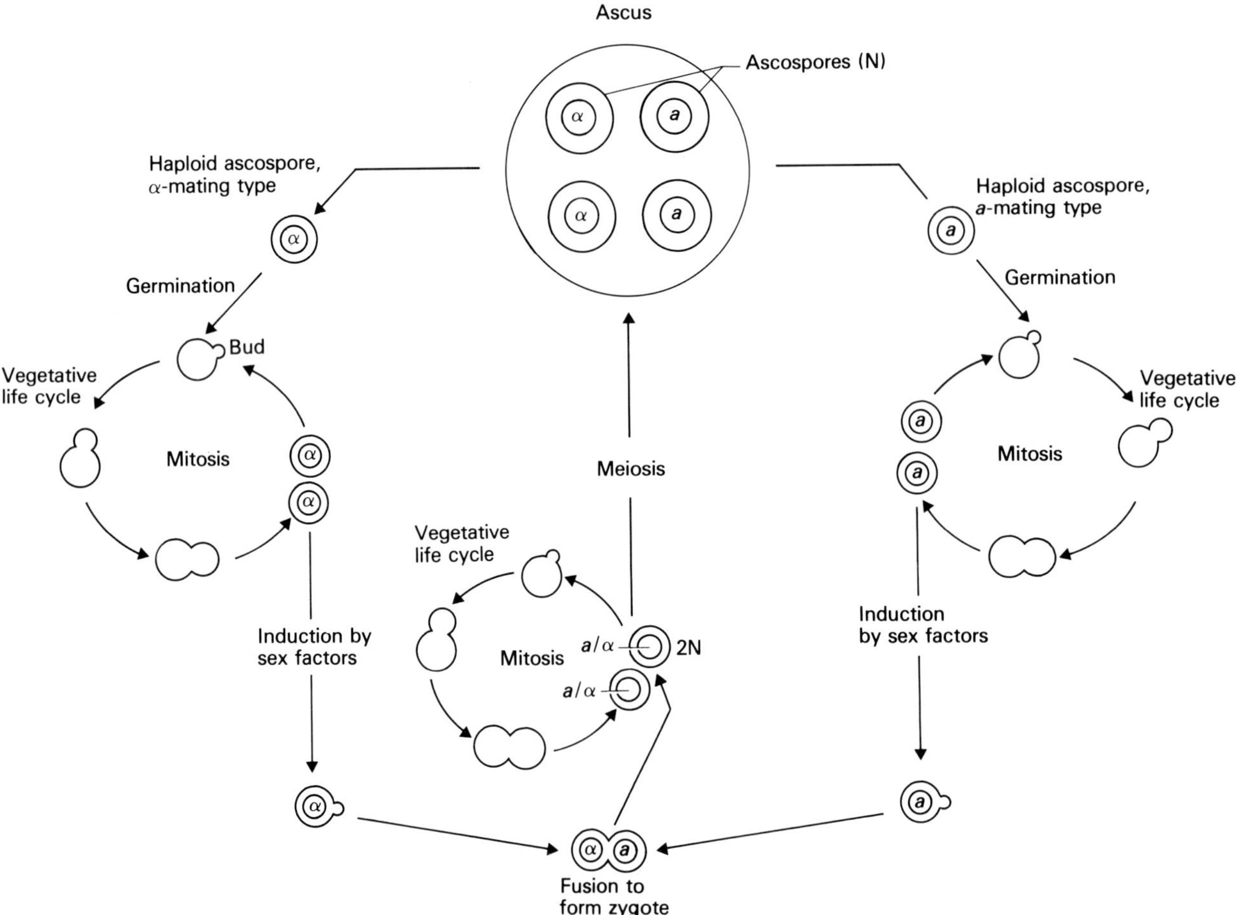

Fig. 15.1. Life cycle of the yeast *Saccharomyces cerevisiae.*

N. crassa has two mating types, *A* and *a*, and these are determined by members of an allelic pair. The two mating types look identical and can be distinguished only by the fact that *A* strains mate with *a* strains but not with *A* strains, and *a* strains mate with *A* strains but not with *a* strains. Sexual reproduction occurs by fusion of nuclei of the opposite mating types to form a diploid zygote that has only a transient existence in the life cycle of this organism. The *A*/*a* zygote undergoes meiosis in an elongating linear tubular structure called an ascus. The two divisions occur in tandem within the developing ascus, producing four haploid nuclei (two *A* and two *a*). Each nucleus divides mitotically, resulting in a linear arrangement of eight haploid nuclei around which spore walls form to produce eight ascospores (four *A* and four *a*). The eight spores of each ascus represent the four meiotic products

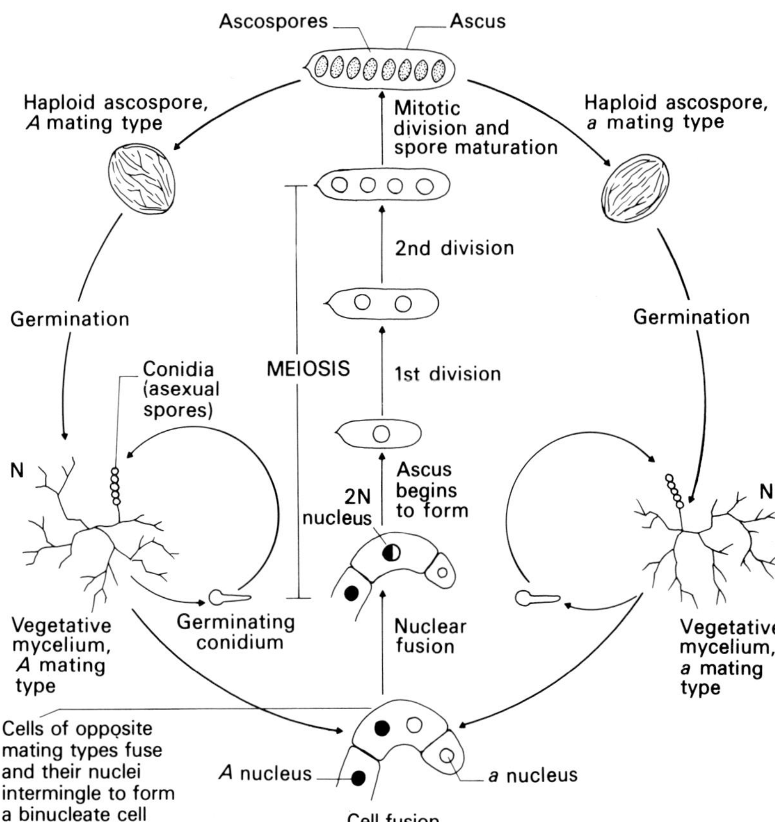

Fig. 15.2. Life cycle of *Neurospora crassa*.

(each doubled) arranged in an order that reflects exactly the orientation of the four chromatids of each tetrad at metaphase I of meiosis: these tetrads are called **ordered tetrads**. When the haploid ascospores germinate, they give rise to the vegetative mycelium.

The asci themselves are formed within a fruiting body called a *perithecium* (Fig. 15.3). When the asci are "ripe" the ascospores are ejected through the neck of the perithecium and may be collected for random spore analysis. If the asci are removed from the perithecium just before they burst, each ascus can be dissected manually for ordered tetrad analysis.

Aspergillus nidulans (Fig. 15.4)

This fungus is a haploid organism and has a colorless, multinucleate

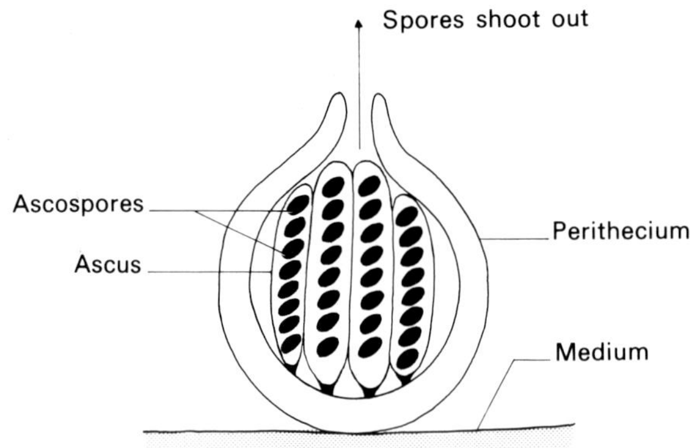

Fig. 15.3. Diagram of a cross-section through the perithecium (fruiting body) of *Neurospora* showing the sets of ascospores arranged linearly within asci.

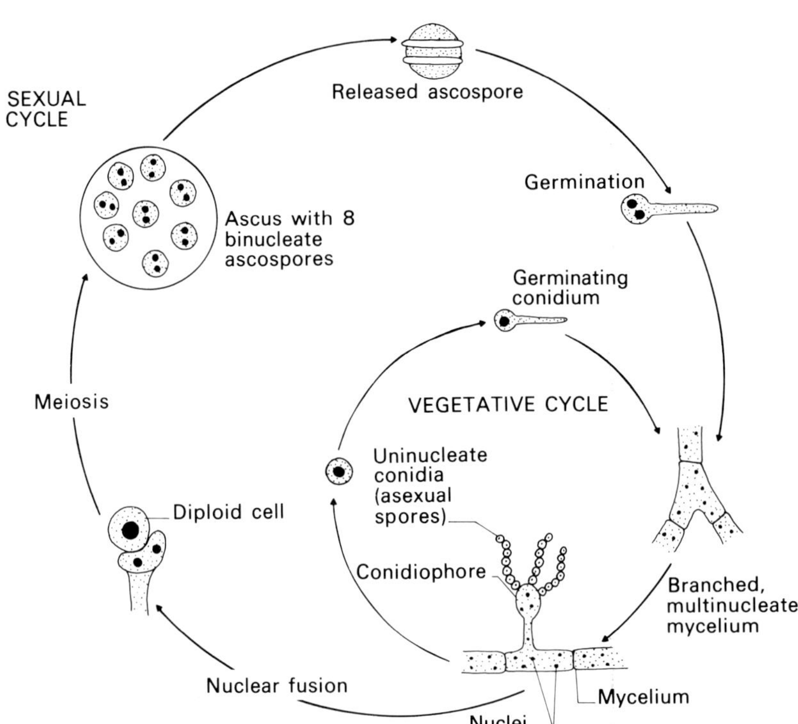

Fig 15.4. Life cycle of *Aspergillus nidulans.*

mycelium. The asexual spores, the conidia, bud off from specialized conidiophores and are uninucleate and dark green in color in the wild type. When these spores germinate, a mycelium is produced and the vegetative cycle is completed. *Aspergillus* has a sexual cycle in which asci are produced with eight ascospores that, when they germinate, give rise to the vegetative mycelium. Unlike *Neurospora*, which requires fusion of nuclei of two different mating types to instigate the sexual cycle, *Aspergillus* is *homothallic*, meaning that it can and does self-cross. Two nuclei from the same culture can fuse to produce a diploid nucleus, which then undergoes meiosis to produce the ascospores. Since this makes the setting up of controlled genetic crosses difficult, meiotic genetic analysis is a problem with this organism. Mitotic genetic analysis, however, is possible in this fungus and this will be described after we have discussed the mechanism of mitotic recombination.

Meiotic genetic analysis of yeast and *Neurospora*

Random spore analysis to determine map distance

In both yeast and *Neurospora*, the ascospores that are released can be collected, induced to germinate, and the resulting cultures analysed for phenotypic characteristics. The ascospores, then, are equivalent to progeny of the crosses described in the chapter on diploid genetic analysis, and two-point and three-point crosses are routinely done in these organisms to map genes. The haploid nature of these fungi simplifies the analysis to some extent.

Tetrad analysis

The ability to isolate ordered tetrads (the four products of meiosis) from certain fungi (such as *Neurospora*) allows the distance between a gene and the centromere to be determined. The determination of gene–centromere distance is usually not possible with unordered tetrads or by random progeny analysis. In addition, the isolation of ordered or unordered tetrads provides a different means of mapping the distance between two or more genes.

Gene–centromere distance determination. This requires the isolation of ordered tetrads such as are produced by *N. crassa*. The example shown in Fig. 15.5 considers the mating type alleles *A* and *a*, which are found in linkage group I. At the zygote stage, each chromosome has doubled and the resulting four-chromatid (tetrad) stage for linkage group I is shown diagrammatically in the figure. Obviously this stage is in a more compact three-dimensional state in the cell than is shown in the diagram, but for our purposes the two-dimensional representation is adequate. In this and the other figures that will be discussed in this example, ● will be used to indicate the centromere derived from the *A* parent and ○ to indicate the centromere of the *a* parent. The centromeres are physically alike, of course, but it will be useful to distinguish the two for the sake of discussion. Note that at this diploid zygote stage the chromosomes have divided but the centromeres have not.

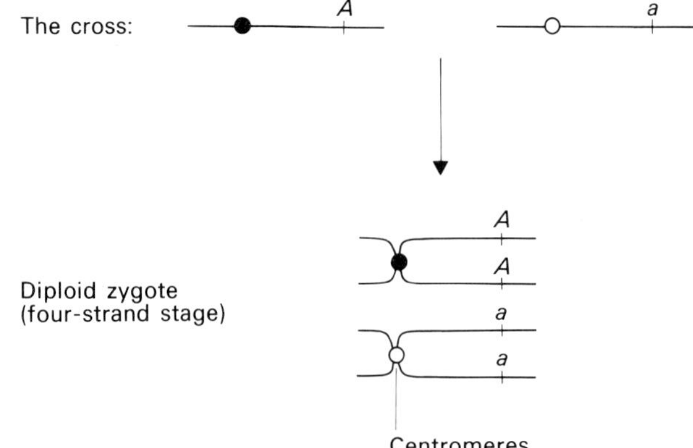

Fig. 15.5. Determination of the gene–centromere distance of the mating-type locus of *Neurospora*: formation of diploid zygote by crossing a strain of mating-type *A* with a mating-type *a* strain.

Fig. 15.6 shows the meiotic divisions and the ascus that results when no crossover occurs between the centromere and the mating-type locus. (Note: for simplification the final mitotic division, which merely replicates each of the four meiotic products, is omitted.) As can be seen the resulting four ascospores directly reflect the arrangement of the four chromatids in the zygote. Since the centromeres do not divide until just before the second meiotic division, there will be a 2:2 segregation (4:4 if we consider the mitotic division that produces eight ascospores) of the centromere from one parent (●) to the centromere of the other parent (o) We say that the centromeres always segregate to different nuclear areas at the first meiotic division; that is, they show *first division segregation*. If no crossover occurs between the gene and its centromere, that gene will also show first division segregation. In Fig. 15.6, after the first division, the two chromatids with the *A*

Fig. 15.6. Determination of the gene–centromere distance of the mating-type locus of *Neurospora*: development of ascus from diploid zygote in which no crossover occurred between the centromere and the mating-type locus. The ascus shows first division segregation for the mating-type alleles.

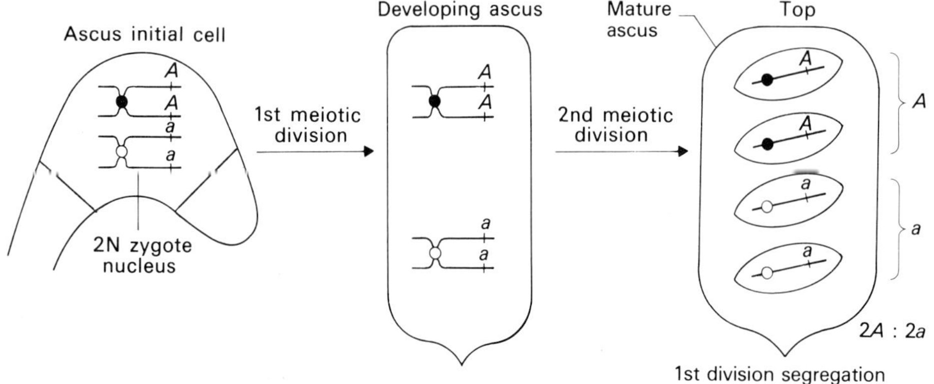

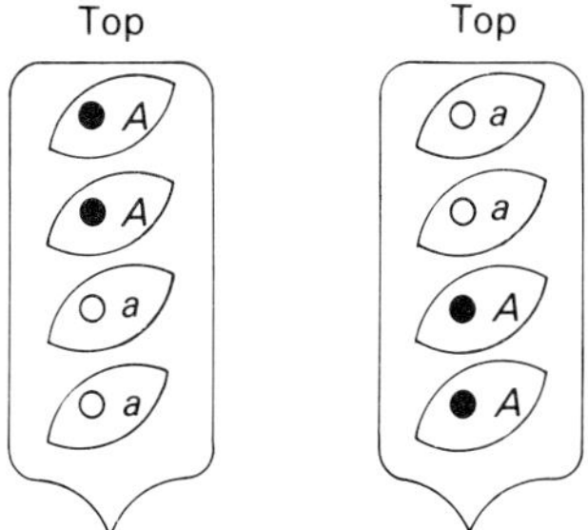

Fig. 15.7. The two possible orientations of mating-type alleles and centromeres (● and ○) in first division segregation asci. The two types occur with equal frequency.

alleles had migrated to a different region of the ascus from the two chromatids carrying the *a* alleles. Also, since it is equally probable that the four chromatids in the zygote are inverted, we would expect to find equal numbers of these two types of asci shown in Fig. 15.7.

If a single crossover occurs between the mating type locus and its centromere, the centromeres will show first division segregation as before (and thus in a sense we can think of them as being "genes" that always segregate at the first division), and the *A/a* alleles will show *second division segregation*. That is, the separation of the allelic genes is delayed until the second meiotic division because of the crossover event. Since the three-dimensional arrangement of the two noncrossover and the two crossover chromatids varies, four second division segregation asci result with equal frequencies (Fig. 15.8). Again the final mitotic division is omitted for the purpose of simplification.

By the analysis of ordered tetrads, the percentage of asci that show second division segregation for a particular allelic pair can be determined. For the mating type locus, this is 14%. As we learned before, the distance between two genes in map units comes directly from the percentage of recombinant progeny from a particular cross. In the *Neurospora* cross we can consider the centromere to be a chromosome marker with parentals of —●——*A*— and —○——*a*—.

A second division ascus can be examined for parental and recombinant configurations of the centromere and mating type alleles (Fig. 15.9). Half of the resulting spores are parental (—●——*A*— and —○——*a*—) and half are recombinant (—●——*a*— and —○——*A*—). Therefore gene–centromere distance is computed by dividing the percentage of second division asci by two. Thus the mating type locus is 14%/2 = 7 map units from the centromere.

Although we will not do so in the following examples, it is possible to combine an analysis of gene–centromere distance with an analysis of map distance between two or more genes if ordered tetrads rather than unordered tetrads are isolated.

Map distance between two genes. With two heterozygous genes in a diploid cell at meiosis there are three possible segregation patterns from a cross. For *a b* × + +, for example, the three types of tetrads are depicted in Fig. 15.10. The **parental ditype** (PD) tetrad has two types of spores, *a b* and + +, which are both parental. The **nonparental ditype** (NPD) tetrad has *a* + and + *b* spores, both of which are recombinant (nonparental). The **tetratype** (T) tetrad has two parental (*a b*, + +) and two recombinant (*a* +, + *b*) spores, hence four types of spores. (Note: the existence of tetratype asci is evidence that crossing-over occurs at the four-strand stage of meiosis.)

By making a cross such as *a b* × + +, we can obtain information about the

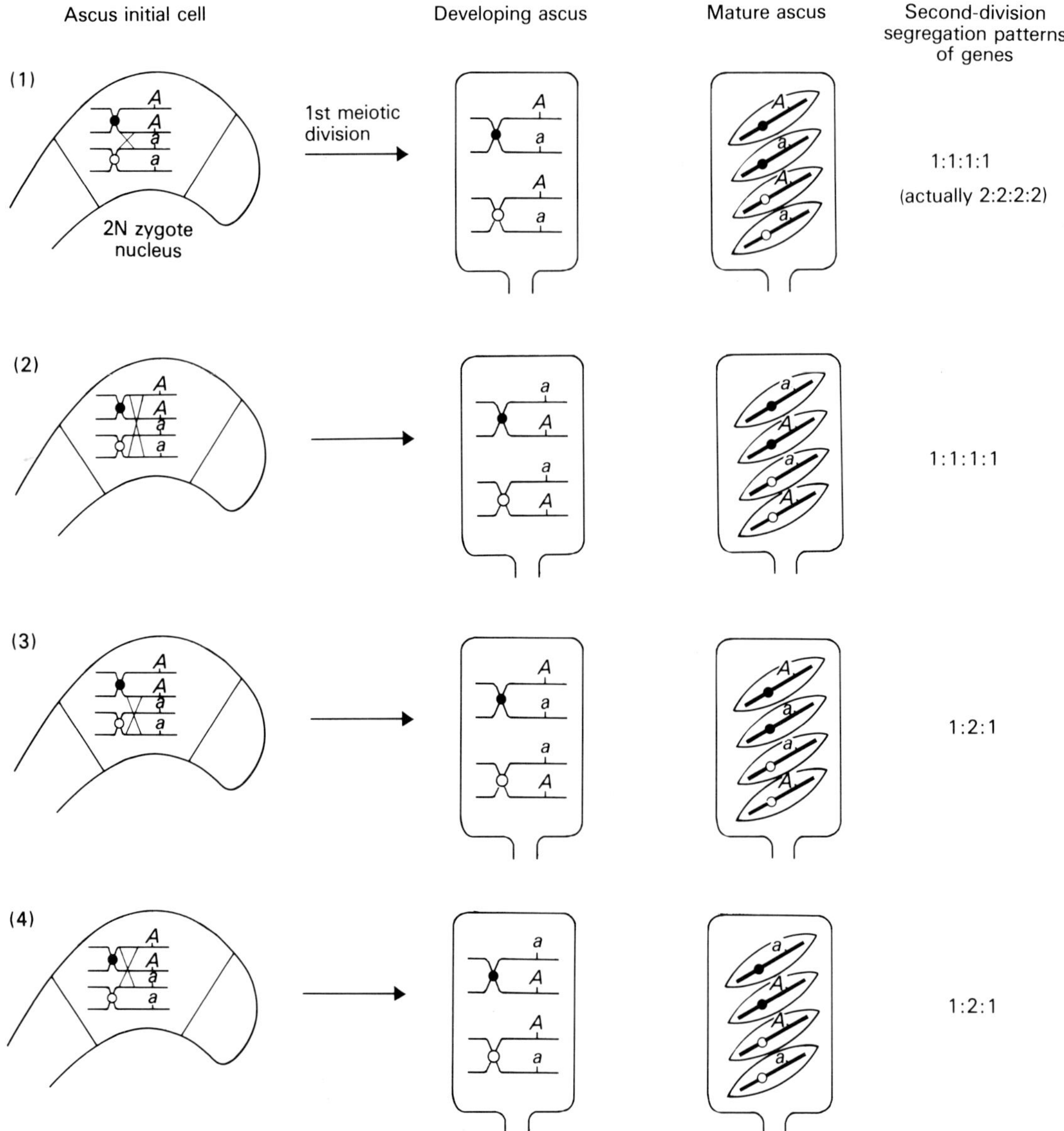

Fig. 15.8. Determination of the gene–centromere distance of the mating-type locus of *Neurospora*. Development of asci from diploid zygotes in which a single crossover has occurred between the centromere and the mating-type locus. The asci produced show second division segregation for the mating-type alleles. The four types of asci are produced in equal proportions.

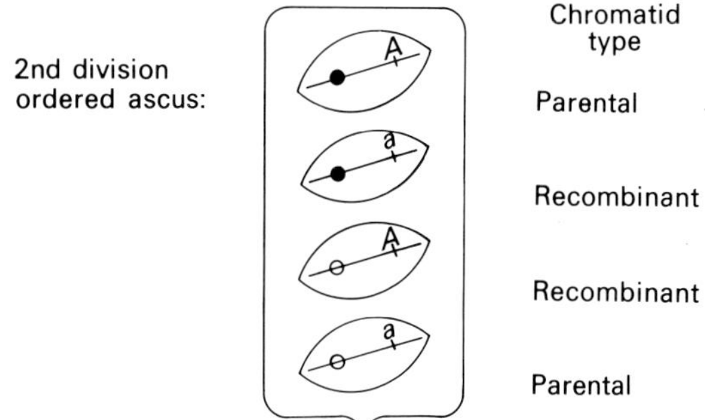

Fig. 15.9. Parental and recombinant configurations of the centromeres and mating-type alleles in a second-division segregation ascus.

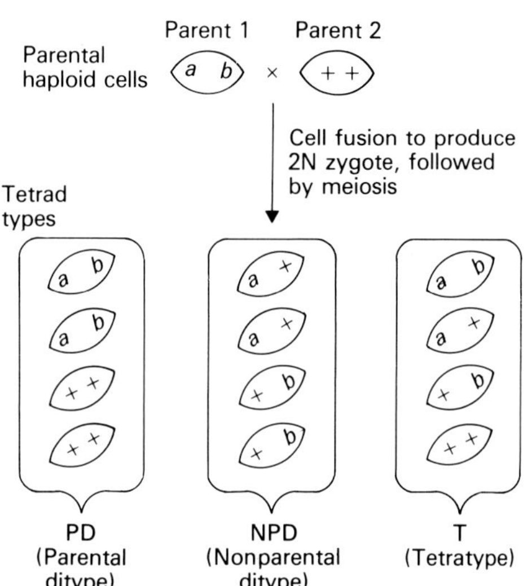

Fig. 15.10. The three types of tetrads that can be produced from a cross *a b*×+ +.

linkage relationships between the two genes by analyzing the progeny derived from unordered or ordered tetrads. There are two situations: the first is genes *a* and *b* on different chromosomes. In this case PDs and NPDs arise with equal frequency as a result of the random alternative alignments of the two sets of four chromatids at metaphase I, and the Ts arise if a single crossover occurs between one or other of the genes and its centromere (Fig. 15.11). The proportion of T asci depends on how far the two genes are from their centromeres.

The second situation is two genes *a* and *b* linked on the same chromosome (Fig. 15.12). Ordered tetrads are drawn in the diagrams, but the analysis is the same with unordered tetrads. The possibilities we shall consider are no (Fig. 15.12a), single (Fig. 15.12b), and double crossovers (Fig. 15.12c).

If no crossover occurs between the two genes (Fig. 15.12a), all of the asci

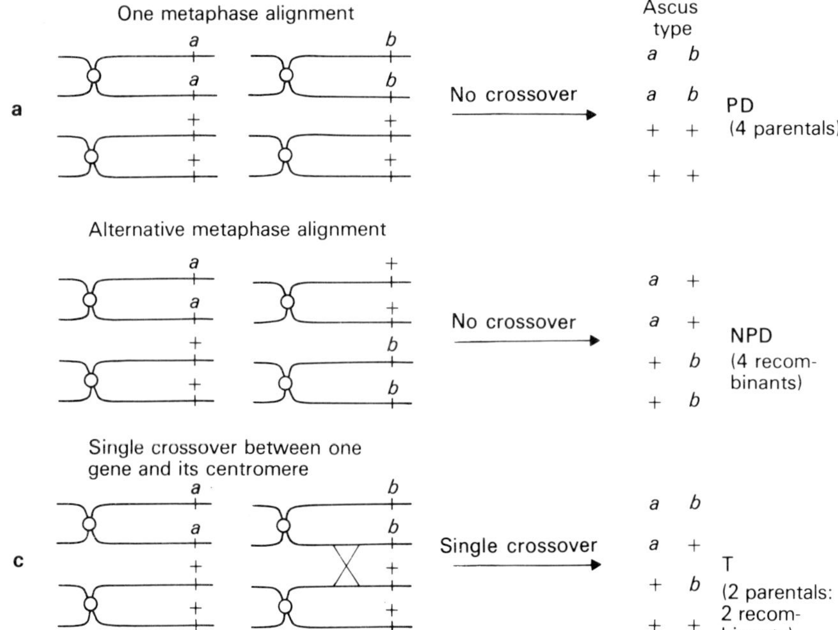

Fig. 15.11. Tetrad analysis for a cross $a\,b \times + +$ where genes a and b are located on different chromosomes (**a**) and (**b**) show the types of tetrads produced by random orientation of the two sets of four chromatids in meiosis when no crossover occurs; (**c**) shows the tetrad type produced when there is a crossover between one gene and its centromere.

are PD asci in which all of the spores are of the parental type. If a single crossover takes place between the two loci (Fig. 15.12b), all of the asci are T asci in which one-half of the spores are parental and one-half are recombinant. In considering double crossovers (Fig. 15.12c), since there are four chromatids in the zygote, it is necessary to consider three types: two-strand, three-strand, and four-strand double crossovers. There are two ways of having three-strand double crossovers, and thus the relative proportion of two-strand: three-strand: four-strand double crossovers is 1:2:1, respectively. Asci resulting after a two-strand double crossover are all PD asci in which all spores are parental (Fig. 15.12c1). Asci from three-strand double crossovers are T asci (one-half parental, one-half recombinant, Fig. 15.12c2), and asci from four-strand double crossovers are NPD asci (all recombinant; Fig. 15.12c3).

As has been discussed previously, the distance between any two genes is computed from the formula:

$$\frac{\text{Number of recombinants}}{\text{Total progeny}} \times 100$$

Examining the tetrads, we see that NPD asci contain four recombinant spores and T asci contain two parental and two recombinant spores. Therefore

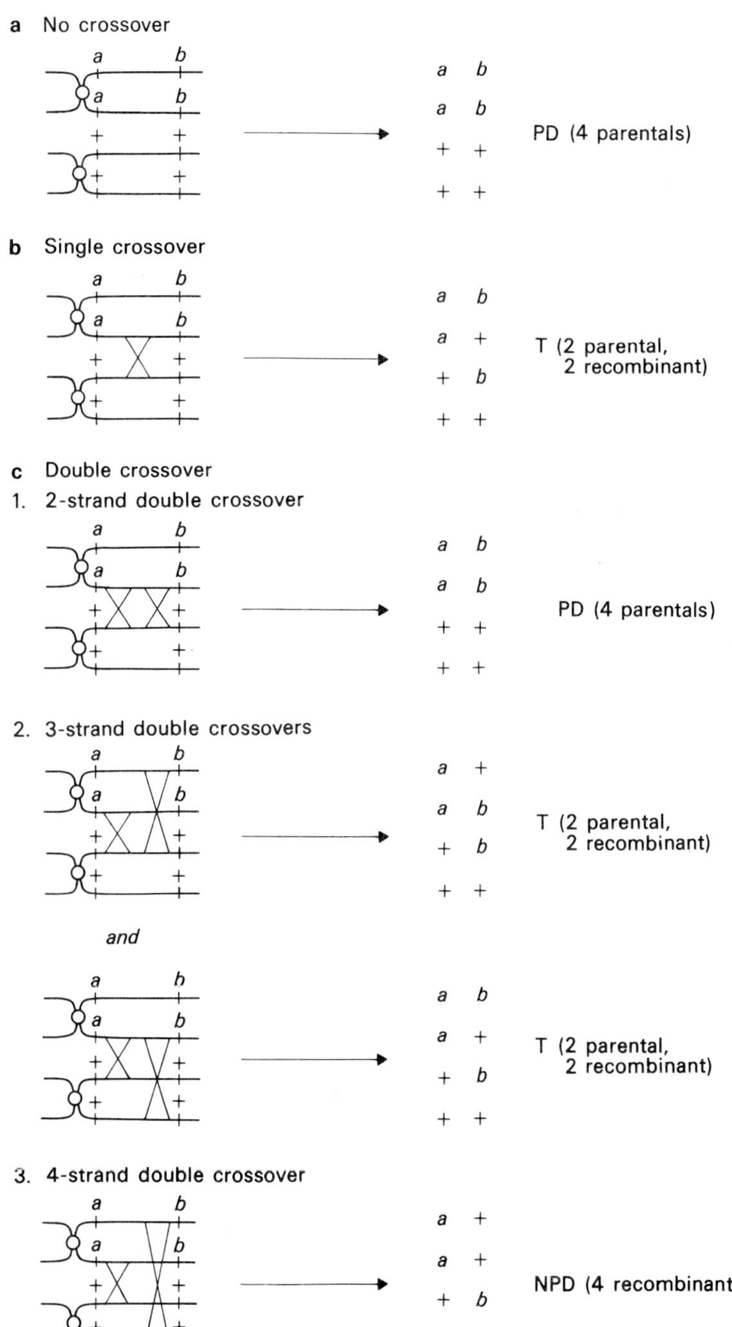

Fig. 15.12. Tetrad analysis for a cross $a\,b \times + +$ where both genes *a* and *b* are located on the same chromosome. (**a**) No crossover results in a PD ascus, (**b**) single crossover between the two genes results in a T ascus, and (**c**) double crossovers between the two genes result in PD, T, or NPD asci, depending on the number of chromatids involved in the exchanges.

converting the general mapping formula into tetrad terms, the *a-b* distance is equal to:

$$\frac{1/2\mathrm{T}+\mathrm{NPD}}{\mathrm{Total}} \times 100$$

Thus if there were 1000 asci, with 900 PD, 96 T and 4 NPD, the *a-b* distance would be:

$$\frac{1/2(96)+4}{1000} \times 100 = 5.2 \text{ map units}$$

However, the formula derived calculates map distances on the basis of recombination frequency rather than on the basis of crossover frequency. A correction can be made for this.

In the theory that was presented, three types of double crossovers were shown: two-strand, three-strand, and four-strand doubles in a ratio of 1:2:1, respectively. The four-strand double crossover frequency is directly determined from the number of NPD asci. In the case of two-strand double crossovers the result is a PD ascus, but here the two crossovers were not counted (see Fig. 15.12c). From the ratio of the three different types of double crossovers, the number of PD asci that result from double crossovers should equal the number of NPD asci.

There are two types of three-strand double crossovers and each gives rise to a T ascus, as does a single crossover. Thus, for each three-strand double crossover, only one of the crossovers is really being considered in the original formula, and between the two three-strand double crossover, the equivalent of two crossovers or one double crossover is not being included in the calculations. As we have seen, the NPD frequency is equivalent to a double crossover frequency. The formula for calculating the distance between two genes from tetrad data can then be modified as in Fig. 15.13.

Recombinants = ½ T + NPD

+ NPD (to add 2-strand double crossover which give PDs with frequency equivalent to NPDs)

+ NPD (to add the equivalent of a double crossover not counted for the two 3-strand double crossovers — again equivalent to NPD frequency)

Total = ½ T + 3 NPD

Therefore, map distance between two genes

$$= \frac{½\mathrm{T} + 3\mathrm{NPD}}{\mathrm{Total}} \times 100$$

Fig. 15.13. Origin of the formula for determining the distance between two genes by tetrad analysis.

Therefore map distances based on crossover frequencies, as manifested by the tetrad types, is given by the formula:

$$\frac{1/2\mathrm{T}+3\mathrm{NPD}}{\mathrm{Total}}\times 100$$

For the numbers given before, this would be:

$$\frac{1/2(96)+3(4)}{1000}\times 100 = 6 \text{ map units}$$

The result is 6 map units vs. 5.2 map units for the "old" recombination frequency formula.

The tetrad formula can be used to calculate the map distance between any two genes. In crosses where more than two genes are segregating, the data should be analyzed by considering two genes at a time and sorting the data in PD, NPD, and T ascus types for those two genes.

Tests for gene linkage using tetrad analysis

Tetrad analysis can be used also to determine whether two genes are linked or unlinked. If two genes are unlinked by being on different chromosomes, the PD frequency will equal the NPD frequency (see Fig. 15.11). PD asci contain only parental-type spores and NPD asci contain only recombinant type spores. T asci contain half parental and half recombinant spores. Therefore, with PD = NPD, whatever the T value, the recombination frequency (RF) is 50%, which signifies no linkage between the two genes.

The frequency of T asci might provide information, however, about whether two unlinked genes are on different chromosomes or far apart on the same chromosome. In the former case, the T asci arise as a result of a single crossover between one or other of the two genes and their respective centromeres, and thus the frequency of T asci will depend on how far away from the centromeres the genes are. In this case the T frequency can vary between 0 (both genes tightly linked to their centromeres) and the limiting frequency of 66.7% (see later). On the other hand, if two genes are very far apart on the same chromosome, there will be large number of crossovers, even-numbered ones occurring with about the same frequency as odd-numbered ones, thus giving rise to equal numbers of PD and NPD asci, respectively. In this situation the T frequency will be 66.7% of all asci, and this will now be explained for a cross $a\ b \times a^+\ b^+$ in which the a and b alleles are initially on chromatids 1 and 2 and the a^+ and b^+ alleles are initially on chromatids 3 and 4 (Table 15.1). If the two loci are far apart, there will be multiple crossovers between them.

If we consider chromatid 1 and the a allele, since multiple crossovers

Table 15.1. Relative frequencies of PD, NPD, and T asci from the cross $a\,b \times a^+\,b^+$ when two genes are far apart on the same chromosome.* (With permission, from *An Introduction to Genetic Analysis*, David T. Suzuki and Anthony J. F. Griffiths. Copyright © 1976 W.H. Freeman and Company.)

Chromatid 1	Chromatid 2	Total probability	Tetrad genotype	Ascus type
$p(b) = 1/2$	$p(b) = 1/3$	1/6	$a\,b\;a\,b\;a^+b^+\;a^+b^+$	PD
	$p(b^+) = 2/3$	2/6	$a\,b\;a\,b^+\;a^+b\;a^+b^+$	T
$p(b^+) = 1/2$	$p(b^+) = 1/3$	1/6	$a\,b^+\;a\,b^+\;a^+b\;a^+b$	NPD
	$p(b) = 2/3$	2/6	$a\,b^+\;a\,b\;a^+b\;a^+b^+$	T

Therefore, frequency of T = 2/6+2/6 = 66.67%
PD = 1/6 = 16.67%
NPD = 1/6 = 16.67%

*The cross is $ab \times a^+b^+$ and, for the purposes of determining tetrad genotype, the two copies of allele a are on chromatids 1 and 2, and the two copies of allele a^+ are therefore on chromatids 3 and 4. Multiple crossovers occur between a and b such that the two genes effectively segregate independently.

occur between the two loci, there is an equal probability that this chromatid will end up being b or b^+. That is, the probability it will be b is $p(b) = 1/2$ and the probability it will be b^+ is $p(b^+) = 1/2$ (Table 15.1). If we now consider chromatid 2 and compute the probabilities of the genotypes, this will then determine the ascus type. Thus, if chromatid 1 carries b, chromatid 2 has a one-third probability to carry b and a two-thirds probability to carry b^+. If the former occurs, chromatid 1 will be $a\,b$ and chromatid 2 will be $a\,b$, and consequently, both chromatids 3 and 4 will be $a^+\,b^+$, thereby giving a PD ascus. If the latter is the case, chromatid 1 will be $a\,b$ and chromatid 2 will be $a\,b^+$, so chromatids 3 and 4 will be $a^+\,b$ and $a^+\,b^+$ (although which will be which cannot be predicted), thereby giving a T ascus.

In summary, when two genes are unlinked and very far apart on the same chromosome, the result will be a ratio of 1 PD : 1 NPD : 4 T asci. For two genes on different chromosomes, PD = NPD, and the relative number of T will depend on the distance the two genes are from their centromeres. A low T frequency compared with PD and NPD would certainly indicate that the two genes in question are on different chromosomes.

If the two genes *are* linked, NPD asci can only arise as a result of a four-strand double crossover, which is a very rare event. Therefore in this case the frequency of PD asci will greatly exceed the frequency of NPD asci (i.e. PD >> NPD). The T asci result from single- and three-strand double crossovers, and these will occur with a frequency intermediate between those for PD and NPD.

Mitotic genetic analysis in *Aspergillus nidulans*

Up to now our discussion of crossing-over has been limited to meiosis. In 1936, C. Stern discovered that crossing-over could take place in the somatic tissue of *Drosophila melanogaster*, during mitosis. Here we will discuss mitotic crossing-over in the mycelial-form fungus *Aspergillus nidulans* and we will show how, in concert with haploidization, mitotic crossing-over analysis can be used to map genes to particular linkage groups and to determine map distances between genes.

Mechanism of mitotic crossing-over

In mitosis each pair of homologous chromosomes replicates. The two pairs of chromatids come independently to the metaphase plate and then segregate to the two daughter cells, which thus have the same genotype as the parental cell. This is shown in Fig. 15.14a for a theoretical cell with all genes heterozygous. Normally the maternally and paternally derived chromatids do not pair together during mitosis. Very rarely, once each chromosome has replicated, the two pairs of chromatids do come together to form a transient tetrad equivalent to the four-strand stage of meiosis. As in meiosis, crossing-over can take place during this tetrad stage after the chromatid pairs separate and align independently in the metaphase plate (Fig. 15.14b). The result of this is that some of the progeny cells will be homozygous for one or more of the genes. When it is a mutant gene that becomes homozygous, the cell will now have a mutant phenotype where the parental cell was heterozygous and wild type.

Like meiotic recombination, mitotic recombination involves only two of the four strands. Because of the randomness with which the two chromatid pairs become oriented at metaphase, in only 50% of the time does a recombinant phenotype appear in one of the two progeny cells. Note that when a mitotic crossover occurs, the recombinant progeny will be homozygous for *all* gene markers distal to the crossover site. In addition, since mitotic recombination is a very rare event, for the purposes of discussion double- and other higher-order crossovers can be ignored.

Formation of diploid strains

The first step in mitotic genetic analysis in *Aspergillus* is the formation of stable diploid strains. This can be accomplished by combining complementary auxotrophic strains on a minimal medium (Fig. 15.15). In the example, each strain carries a conidial color mutation: the w allele is recessive to the wild-type green allele w^+ and results in white spores, and the y allele is

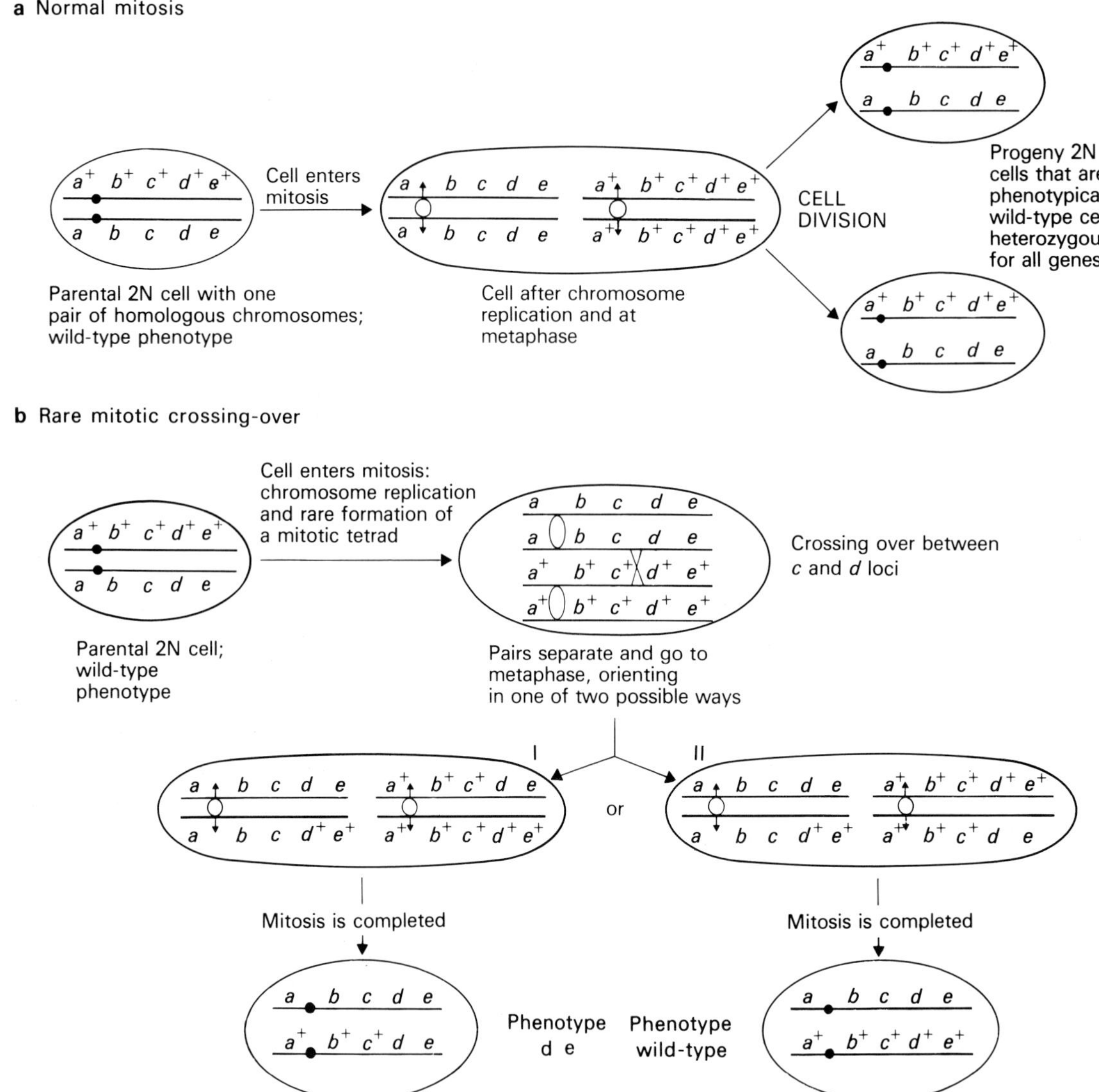

Fig. 15.14. Diagrammatic representation of (**a**) normal mitosis and (**b**) mitosis involving a rare crossing-over event.

recessive to the wild-type green allele y^+ and results in yellow spores. A $w^+ y^+$ strain is green, a $w\, y^+$ strain is white, a $w^+ y$ strain is yellow, and a $w\, y$ strain is white. The two strains also differ in the auxotrophic mutations they carry: the yellow ($y\ w^+$) strain is auxotrophic for adenine (*ad*) and wild type for a thiamine requirement (thi^+), while the white ($y^+ w$) strain is wild type for the adenine requirement (ad^+) and auxotrophic for thiamine (*thi*). After the two strains are mixed together, the only colonies that will grow (barring reversion

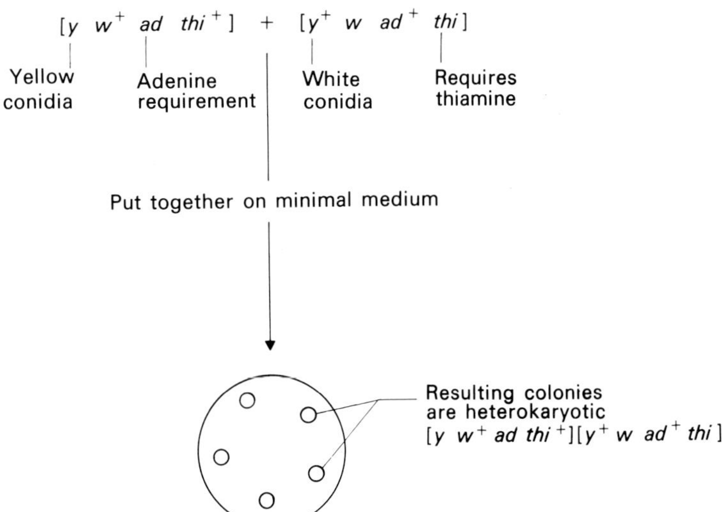

Fig. 15.15. Formation of a diploid strain of *Aspergillus* by combining two strains carrying complementary auxotrophic and conidial color markers.

of a mutation) will be those that result from hyphal fusion of the two strains. The resulting **heterokaryon** contains both types of nuclei in a common cytoplasm, and since the two auxotrophic mutations will complement under these circumstances, growth will occur. When the heterokaryotic strain produces conidia, most of the asexual spores will be uninucleate and haploid, that is, carrying one or other of the parental nuclei. Thus the conidia will either be white or yellow. When plated on minimal medium, these will not survive because of the auxotrophic mutation each contains. Rarely, nuclear fusion will occur in the heterokaryon, and this can lead to uninucleate diploid conidia of the genotype $y\,w^+\,ad\,thi^+/y^+\,w\,ad^+\,thi$. Owing to the dominance of y^+ and w^+, these conidia will be dark green. When the diploid conidia germinate, they are able to grow on minimal medium because of complementation of the auxotrophic mutations, and the resulting diploid colonies can be used for mitotic genetic analysis.

Locating genes to linkage groups by haploidization

The diploid that is formed is relatively stable but has a tendency to break down to haploid segregants in a process called **haploidization**. We can follow this, for example, if we have a diploid that is $+/y$ and $+/w$ for conidial color since some of the haploids that result will have white or yellow conidia and will be detected as different-colored sectors in dark-green colonies. The significant point about haploidization is that which chromosome of a pair will end up in the haploid segregant is a random event. If the chromosomes are genetically marked, we can detect this by segregants having or not having particular sets of genes. Each set, of course, will segregate independently of other sets of genes representing other chromosomes. This will be made apparent by considering an example (Fig. 15.16). At the outset we are giving the conclusion of the experiment by showing that there are three linkage groups.

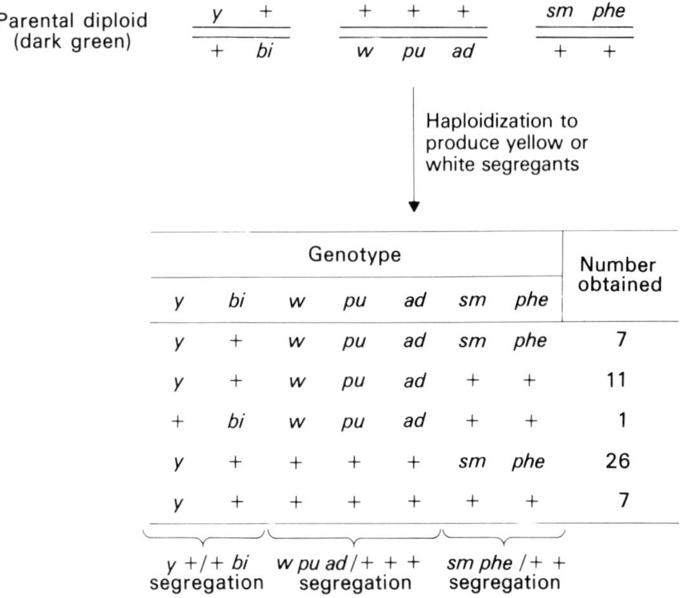

Genotype							Number obtained
y	*bi*	*w*	*pu*	*ad*	*sm*	*phe*	
y	+	*w*	*pu*	*ad*	*sm*	*phe*	7
y	+	*w*	*pu*	*ad*	+	+	11
+	*bi*	*w*	*pu*	*ad*	+	+	1
y	+	+	+	+	*sm*	*phe*	26
y	+	+	+	+	+	+	7

Fig. 15.16. Example of how haploidization of a diploid strain can localize genes to chromosomes. For details, see text.

In the experiment, white or yellow haploid segregants were obtained and analyzed for genotype by further testing. (It should be pointed out that the numbers of each class in the data may not be significant owing to the effects of the markers on viability.) In the haploid segregants, recombination has not occurred between markers on the same chromosome, but independent segregation of the homologous chromosomes into the haploids has taken place. Thus, for example, the segregants are either *y* + or + *bi*, either *w pu ad* or + + +, and either *sm phe* or + +. Thus, without assuming gene order, we can designate which markers are on different chromosomes by analyzing which blocks of genes recombine (actually assort) independently of other blocks. This leads to the assignment of the three linkage groups with particular genotypes shown in Fig. 15.16. Even if a little mitotic recombination occurs in the diploid before haploidization, the same conclusions could be drawn since it would be such a rare event.

Gene mapping by mitotic recombination

Once genes are assigned to linkage groups, their order and map locations can be determined by analysis of segregants that have arisen as a result of mitotic recombination. We have illustrated mitotic crossovers earlier in the chapter; remember that mitotic crossing-over leads to homozygosity for all markers distal to the crossover, and this will be detected in only 50% of the cases because of the randomness with which the homologous chromatid pairs align at metaphase.

The parental diploid and segregant data we will discuss are presented in Fig. 15.17. As in the other cases, the diploid was selected by the use of appropriate markers on this and other chromosomes. The diploid is dark green because it is $+/y$, and this means we can select for yellow segregants. The diploid is also homozygous for an adenine auxotrophic mutation (ad_{20}) and heterozygous for a recessive suppressor of adenine ($su\text{-}ad_{20}$). If the suppressor is homozygous, the ad_{20}/ad_{20} strain would no longer require adenine for growth, and this means that we can also select for adenine independence. The resulting segregants can be analyzed for their phenotypes with respect to the other markers and thus the gene order can be determined.

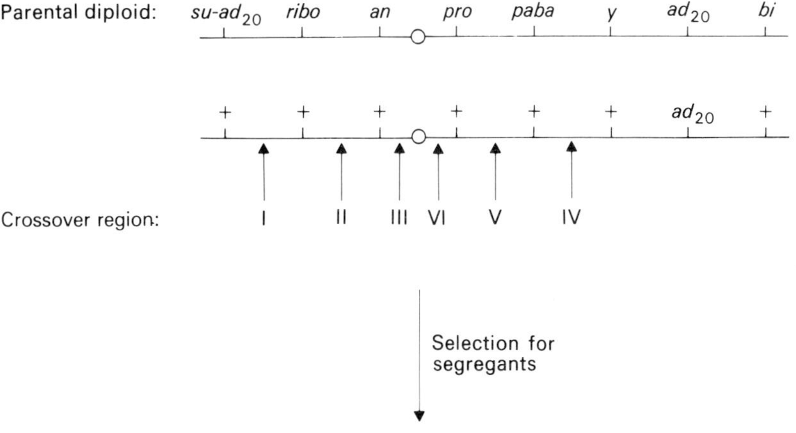

Class	Segregant selected	Phenotype	Number
1	adenine-independence ($su\text{-}ad_{20}$----ad_{20})	+	24
2		*ribo*	9
3		*ribo an*	62
4		*ribo an pro paba y bi*	7
5	Yellow	*y ad bi*	15
6		*paba y ad bi*	42
7		*pro paba y ad bi*	12
8		*ribo an pro paba y bi*	3

Fig. 15.17. Example of how mitotic crossing-over can be used to determine gene order on chromosome arms. For details, see text.

A mitotic crossover will produce homozygosis for markers distal to the crossover point on the same chromosome arm. It is not possible to make the whole chromosome homozygous by a single crossover, and a double crossover (one crossover on each side of a centromere) is so rare that it can be ignored. Thus classes 4 and 8 in Fig. 15.17 must have arisen by haploidization—note that both are adenine-independent as a result of the presence of the recessive suppressor in the haploid with the mutant gene it suppresses. These classes will be ignored in further discussions.

Consider the adenine-independent diploid segregant classes 1, 2, and 3. Homozygosity for *su-sd*$_{20}$ will result following any crossover between the centromere and the *su-ad*$_{20}$ locus, that is, in regions I, II, and III on the parental diploid diagram. A crossover in I will only give homozygosity for *su-ad*$_{20}$, and the *ribo* and *an* genes will remain heterozygous with their respective wild type alleles. A crossover in II will give homozygosity for *ribo* and *su-ad*$_{20}$, but never an *an* (class 2). A crossover in III will give homozygosity for *an*, *ribo*, and *su-ad*$_{20}$ (class 3). Thus, reading from the distal to the proximal, the gene order must be *su-ad*$_{20}$-*ribo*-*an*-centromere.

Applying similar logic to the other chromosome arm, a crossover in IV will give a segregant that is *y*, *ad*$_{20}$, and *bi* in phenotype (class 5). From the data presented, the order of the genes distal to *y* cannot be determined.

Class 6 results from a crossover in region V, and class 7 results from a crossover in region VI. Thus the order must be (*bi*, *ad*$_{20}$)-*y*-paba-pro-centromere, where the order of *bi* and *ad*$_{20}$ relative to *y* is not known. Haploidization showed that all the genes we have discussed are on the same chromosome (see classes 4 and 8) and thus the data allow us to illustrate the gene order as we did in the parental diploid.

We can extend the analysis one step further by calculating the (mitotic) map distances between genes on the chromosome. Here we shall analyze the data for the adenine-independent segregants of Fig. 15.17. These data are present in a different way in Fig. 15.18.

The "other markers segregating" in Fig. 15.18 are the recombinant types resulting from single crossovers in each of the three regions I, II, and III. The frequency of segregants resulting from a crossover in region I, by analogy with meiotic mapping analysis, is a function of the genetic distance between the *ribo* and *su-ad*$_{20}$ loci. The value here is 25.3% of the total, indicating a map distance of 25.3 map units. Similar analyses indicate a *ribo*–*an* distance (crossover in II) of 9.5 map units, and an *an*–centromere distance (crossovers in I) of 65.3 map units. Obviously these are relative distances, and they may or may not tally with meiotic map distances. The gene order would be the same in each case, of course.

In conclusion, fungal genetics is similar, in many respects, to the formal genetics of other eukaryotic organisms. Tetrad analysis in fungi such as yeast

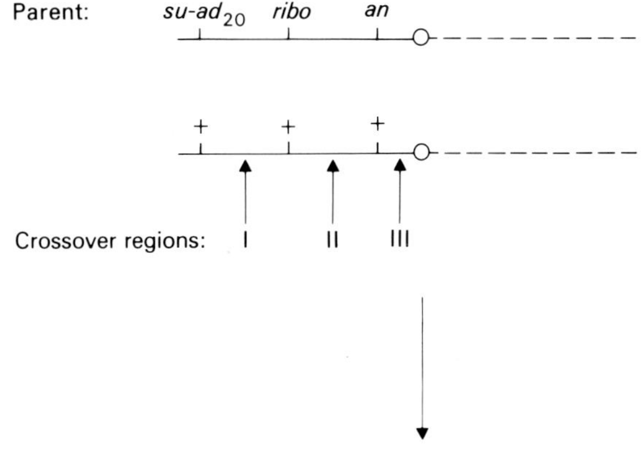

Segregant selected	Other markers segregating	Crossover region	Number	% of total
$\frac{su}{su}$ (adenine-independence)	none	I	24	25.2
	ribo	II	9	9.5
	ribo an	III	62	65.3
		Total	95	

Fig. 15.18. Example of how distances between genes can be determined by mitotic crossing-over. For details, see text.

and *Neurospora* permit particular questions about the recombination mechanisms to be asked that are virtually impossible to ask in other eukaryotes. In addition, *Aspergillus* allows the process of genetic recombination in mitosis to be probed in detail.

Questions and problems

15.1 In *Saccharomyces* and *Chlamydomonas*, what meiotic events give rise to PD (parental ditype), NPD (nonparental ditype), and T (tetratype) tetrads?

15.2 A cross was made between a pantothenate-requiring (*pan*) strain and a lysine-requiring (*lys*) strain of *Neurospora crassa*, and 750 random ascospores were plated on minimal medium (a medium lacking pantothenate and lysine). Thirty colonies subsequently grew. What is the map distance between the *pan* and *lys* loci?

15.3 In *Neurospora crassa*, the following crosses yielded progeny, as shown:

Cross	Number	Progeny	Cross	Number	Progeny
$a^+ b \times a\, b^+ \rightarrow$	981	$a^+ b$	$a^+ c \times a\, c^+ \rightarrow$	850	$a^+ c$
	1000	$a\ b^+$		833	$a\ c^+$
	10	$a^+ b^+$		169	$a^+ c^+$
	9	$a\ b$		148	$a\ c$
Total	2000		Total	2000	

$b^+ c \times b\, c^+ \rightarrow$	850	$b^+ c$
	850	$b\ c^+$
	140	$b^+ c^+$
	160	$b\ c$
Total	2000	

What is the probable gene order, and what are the approximate map distances between adjacent genes?

15.4 Four different albino strains of *Neurospora* were each crossed to wild types. All crosses resulted in 1/2 wild type and 1/2 albino progeny. Crosses were made between the first strain and the other three with the following results:

1×2: 975 albino, 25 wild type
1×3: 1000 albino
1×4: 750 albino, 250 wild type

Which mutations represent different genes and which genes are linked? How did you arrive at your conclusions?

15.5 Genes *met* and *thi* are linked in *Neurospora crassa*; we wish to locate *arg* with respect to *met* and *thi*. From the cross *arg* + + × + *thi met*, the following random ascospore isolates were obtained. Map these three genes.

arg	*thi*	*met*	26	*arg*	+	+	51
arg	*thi*	+	17	+	*thi*	+	4
arg	+	*met*	3	+	+	*met*	14
+	*thi*	*met*	56	+	+	+	29

15.6 Given a *Neurospora* zygote of the following constitution, diagram the significant events producing an ascus where the *A* alleles segregate at first division and the *B* alleles at the second division.

A b
A b
a B
a B

15.7 Double exchanges between two loci can be of several types (called two-strand, three-strand, and four-strand doubles).

(a) Four recombination gametes would be produced from a tetrad in which the two exchanges were as depicted in the following figure. Draw in the second exchange.

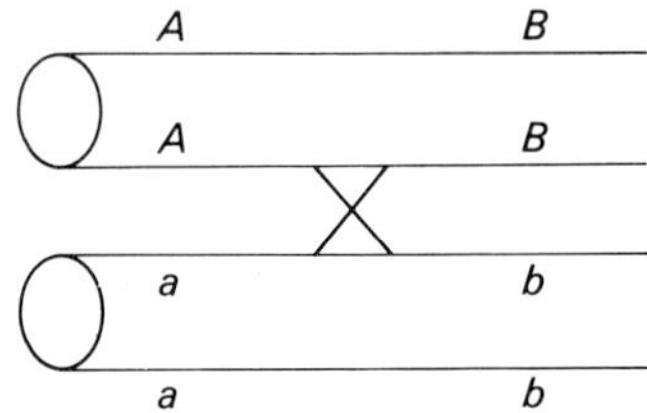

(b) Draw in the second exchange so that four nonrecombination gametes would result from the tetrad shown in the following figure:

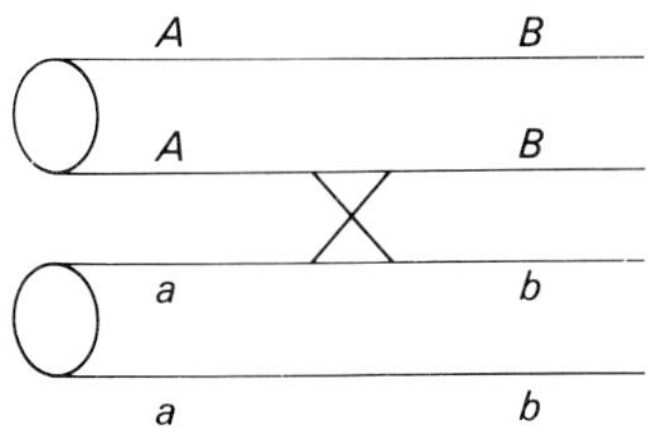

(c) Other possible double-crossover types would result in two recombination and two parental-type gametes. If all possible multiple-crossover types occur at random, the frequency of recombination between genes at two loci will never exceed a certain percentage regardless of how far apart they are on the chromosome. What is that percentage? Explain the reason why the percentage cannot theoretically exceed that value.

15.8 A cross between a pink (p$^-$) a yeast strain of mating type mt^+ and a gray strain (p^+) of mating type mt^- produced the following tetrads:

Number of tetrads	Kind of tetrad
18	$p^+ mt^+, p^+ mt^+, p^- mt^-, p^- mt^-$
8	$p^+ mt^+, p^- mt^+, p^+ mt^-, p^- mt^-$
20	$p^+ mt^-, p^+ mt^-, p^- mt^+, p^- mt^+$

On the basis of these results, are the *p* and *mt* genes on separate chromosomes?

15.9 In *Neurospora* the peach gene (*pe*) is on one chromosome and the colonial gene (*col*) is on another. Disregarding the occurrence of chiasmata, what kinds of tetrads (asci) would you expect, and in what proportions, if these two strains were crossed?

15.10 The following unordered asci were obtained from the cross *leu* +×+ *rib* in yeast. Draw the linkage map and determine the map distance between the genes.

30	45	41	4	39	2	39
leu +	*leu rib*	*leu* +	+ +	*leu* +	*leu rib*	*leu* +
+ *rib*	*leu* +	*leu* +	*leu rib*	+ *rib*	+ +	*leu* +
leu +	+ +	+ *rib*	*leu rib*	+ +	*leu rib*	+ *rib*
+ *rib*	+ *rib*	+ *rib*	+ +	*leu rib*	+ +	+ *rib*

15.11 The genes *a*, *b*,and *c* are in the same chromosome arm in *Neurospora crassa*. The following ordered asci were obtained from the cross *a b* +×+ + *c*.

45	5	146	1	10	20	15	58
a b +	*a b* +	*a b* +	*a b* +	*a b* +	*a b* +	*a b* +	*a b* +
+ *b c*	*a* + +	*a b* +	+ + +	*a* + *c*	+ + *c*	*a b c*	+ *b* +
a + +	+ *b c*	+ + *c*	*a b c*	+ *b* +	*a b* +	+ + +	*a* + *c*
+ + *c*	+ + *c*	+ + *c*	+ + *c*	+ + *c*	+ + *c*	+ + *c*	+ + *c*

(a) Determine the correct gene order.
(b) Determine all of the gene–gene and gene–centromere distances.

15.12 A diploid strain of *Aspergillus nidulans* (forced between a wild-type and a multiple mutant) that was heterozygous for the recessive mutations *y* (yellow), *w* (white), *ad* (adenine), *sm* (small), *phe* (phenylalanine), and *pu* (putrescine) produced haploid segregants. Forty-one haploid yellow segregants were tested and found to have the following genotypes and numbers:

y	*w*	*pu*	*ad*	*sm*	*phe*	7
y	*w*	*pu*	*ad*	+	+	11
y	+	+	+	*sm*	*phe*	16
y	+	+	+	+	+	7

What are the linkage relationships of these genes?

15.13 A (green) diploid of *Aspergillus nidulans* is heterozygous for each of the following mutant genes: *sm, pu, phe, bi, w* (white), *y* (yellow), and *ad*. Analysis of white and yellow haploid segregants from this diploid indicated a number of classes with the following genotypes:

Genotype						
sm	*pu*	*phe*	*bi*	*w*	*y*	*ad*
sm	*pu*	*phe*	+	*w*	*y*	*ad*
+	*pu*	+	+	*w*	*y*	*ad*
+	*pu*	+	*bi*	*w*	+	*ad*
sm	+	*phe*	+	+	*y*	+
+	+	+	+	+	*y*	+
sm	*pu*	*phe*	*bi*	*w*	+	*ad*

How many linkage groups are involved, and which genes are on which linkage group?

15.14 A (green) diploid of *Aspergillus nidulans* is homozygous for the recessive mutant gene *ad* and heterozygous for the following recessive mutant genes: *paba, ribo, y* (yellow), *an, bi, pro*, and *su-ad*. Those recessive alleles that are in the same chromosome are in coupling. The *su-ad* allele is a recessive suppressor of the *ad* allele: the +/*su-ad* genotype does not suppress the adenine requirement of the *ad/ad* diploid, whereas the *su-ad/su-ad* genotype does suppress that requirement. Therefore the parental diploid requires adenine for growth. From this diploid, two classes of segregants were selected: yellow and adenine-independent. The accompanying table lists the types of segregants obtained.

Segregant type selected	Phenotype
Adenine-independent	+
	ribo
	ribo an
	ribo an pro paba y bi
Yellow	*y ad bi*
	paba y ad bi
	pro paba y ad bi
	ribo an pro paba y bi

Analyze these results as completely as possible.

References

Barrett, R.W., D. Newmeyer, D.D. Perkins and L. Garnjobst, 1954. Map construction in *Neurospora crassa*. *Adv. Genetics* **6**:1–93.

Davis, R.H. and F.J. deSeeres, 1970. Genetics and microbiological research techniques for *Neurospora crassa*. In *Methods in Enzymology*, S. Colowick and N.O. Kaplan (eds.), vol. 17A, pp. 80–143. Academic Press, New York.

Emerson, S., 1967. Fungal genetics. *Annu. Rev. Genetics* **1**:201–220.

Esser, K. and R. Kuenen, 1967. *Genetics of Fungi*. Springer Verlag, New York.

Fincham, J.R.S., 1970. Fungal genetics. *Annu. Rev. Genetics* **4**:347–372.

Fincham, J.R.S. and P.R. Day, 1971. *Fungal Genetics*, 3rd ed. Blackwell Scientific Publications, Oxford.

Fink, G.R., 1970. The biochemical genetics of yeast. In *Methods in Enzymology*, S. Colowick and N.O. Kaplan (eds), vol. 17A, pp. 59–78. Academic Press, New York.

Houlahan, M.B., G.W. Beadle and H.G. Calhoun, 1949. Linkage studies with biochemical mutants of *Neurospora crassa*. *Genetics* **34**:493–507.

Kafer, E., 1958. An 8-chromosome map of *Aspergillus nidulans*. *Adv. Genetics* **9**:105–145.

Mortimer, R.K. and D.C. Hawthorne, 1966. Yeast genetics. *Annu. Rev. Microbiol.* **20**:151–168.

Pontecorvo, G., 1956. The parasexual cycle in fungi. *Annu. Rev. Microbiol.* **10**:393–400.

Pontecorvo, G. and E. Kafer, 1958. Genetic analysis by means of mitotic recombination. *Adv. Genetics* **9**:71–104.

Pontecorvo, G., J.A. Roper and E. Forbes, 1953. Genetic recombination without sexual reproduction in *Aspergillus niger*. *J. Gen. Microbiol.* **8**:198–210.

Pritchard, R.H., 1955. The linear arrangement of a series of alleles of *Aspergillus nidulans*. *Heredity* **9**:343–371.

Roper, J.A., 1968. The parasexual cycle. In *The Fungi*, G.C. Ainsworth and A.S. Sussmann (eds.), vol. 2, pp. 589–617. Academic Press, New York.

Ruddle, F.H. and R.P. Creagan, 1975. Parasexual approaches to the genetics of man. *Annu. Rev. Genetics* **9**:407–486.

Stern, C., 1936. Somatic crossing-over and segregation in *Drosophila melanogaster*. *Genetics* **21**:625–730.

Suzuki, D.T. and A.J.F. Griffiths, 1986. *An Introduction to Genetic Analysis*. W.H. Freeman, San Francisco.

Chapter 16 Eukaryotic Genetics: An Overview of Human Genetics

Outline

- Pedigree analysis
  - Principles of pedigree analysis
  - Examples of human pedigree
- Chromosomal aberrancies and human diseases
  - Types of chromosomal aberrancies
  - Down syndrome
  - Anomalies of the sex chromosomes
  - The Lyon hypothesis
- Mapping genes to human chromosomes
  - Mapping using somatic cell hybridization
  - Deletion mapping
  - Mapping using nucleic acid hybridization

The study of human genetics is complicated by the fact that, unlike other species of animals and plants, humans are not bred experimentally. Therefore we cannot apply the types of genetic analysis that have been discussed in previous chapters. Nonetheless we have discovered that many human traits are due to single pairs of segregating genes. In this chapter we will discuss briefly how family studies can be used to discover the genetic basis of human traits. This involves **pedigree analysis** where the phenotypic records of families extending over several generations are compiled so that gene segregation patterns can be hypothesized. Naturally the more complete the pedigree, the more accurate the genetic analysis. As we shall see, pedigree analysis is most useful for traits that are the result of a simple gene difference such as an autosomal or sex-linked dominant or recessive mutation. Practically speaking the most interesting traits are those that cause disease, and examples of the genetic bases of some diseases will be given later. A modern application of pedigree analysis is called **genetic counseling** in which a human geneticist makes predictions about the probabilities of particular traits (deleterious or not) occurring among the progeny of a couple in whom one or both show evidence of that trait in their family. Also in this chapter we shall discuss some human traits that result from chromosomal aberrancies.

Pedigree analysis

Fig. 16.1 shows some of the symbols used in human pedigrees. There are other specialized symbols, but they do not concern us here. The symbols are put together in the pedigree as shown in Fig. 16.2.

Principles of pedigree analysis

There are 46 chromosomes in humans: 22 pairs of autosomes and a pair of sex chromosomes. For genes with a simple mechanism of inheritance, there are four different ways they can be inherited. **X-linked (sex-linked) recessive**, **X-linked (sex-linked) dominant, autosomal recessive**, or **autosomal dominant.** In the following we will consider some theoretical pedigrees and analyze them to see what type of inheritance is compatible with the data. As we shall see, this is not the exacting type of analysis we have applied for other eukaryotes. In particular, it is typically the case that the analysis of a single pedigree, or even several pedigrees, may be inconclusive.

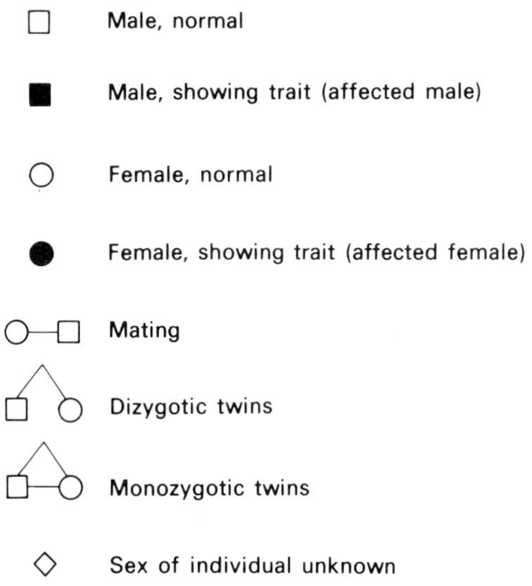

Male, normal
Male, showing trait (affected male)
Female, normal
Female, showing trait (affected female)
Mating
Dizygotic twins
Monozygotic twins
Sex of individual unknown

Fig. 16.1. Examples of symbols used in human pedigrees.

Generation
I
II
III
Marriage line
Offspring listed left to right in order of birth
1 2 3 4 5 6
1 2 3
etc.

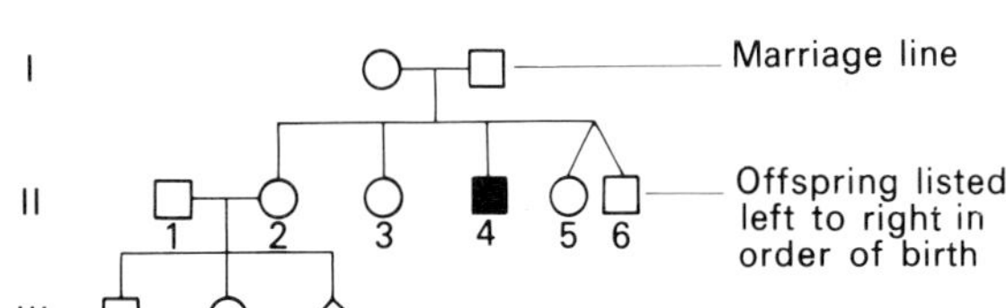

Fig 16.2. An hypothetical human pedigree showing the typical features of pedigrees.

One trait that is controlled by a single gene pair is the ability to taste phenylthiocarbamide (PTC). When tested, people will either find it to be bitter (they are "tasters"), or they will detect no taste at all (they are "nontasters"). Suppose a small family has been tested with the results shown in Fig. 16.3. In this case we will define the nontasters as having the trait and those are the people with the shaded symbols in the pedigree. Now we can determine which of the four mechanisms of inheritance could apply.

Nontaster
Taster
I
II
Taster son
Nontaster daughter

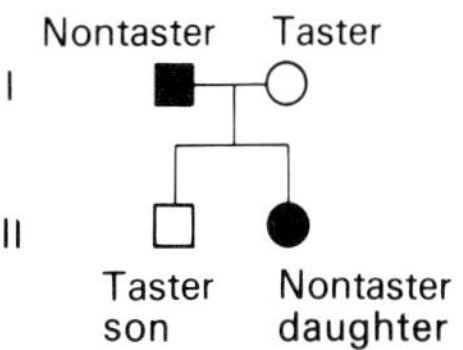

Fig. 16.3. An hypothetical pedigree for the trait PTC nontasting. Persons with the trait are shaded.

Sex-linked recessive. Nontasting could be the result of a sex-linked recessive gene if we make the assumption that the mother is heterozygous for the gene. That is reasonable in this case since nontasting is a fairly common trait affecting about 30% of the population. The nontaster father would be hemizygous for the mutant gene and the daughter homozygous for the mutant gene. Assigning t for the sex-linked recessive mutant gene, and t^+ for the wild type allele, the pedigree can be presented genotypically as in Fig. 16.4.

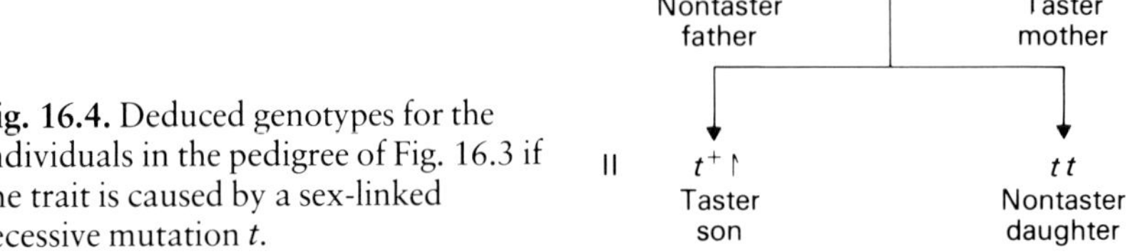

Fig. 16.4. Deduced genotypes for the individuals in the pedigree of Fig. 16.3 if the trait is caused by a sex-linked recessive mutation t.

Sex-linked dominant. Sex-linked dominance could also explain the inheritance of the nontasting trait in this particular pedigree. The father would be hemizygous for the dominant allele T and the mother would be homozygous for the wild type allele T^+ (Fig. 16.5).

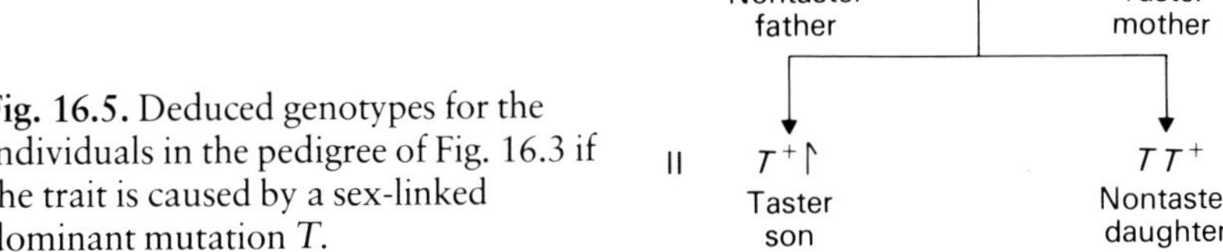

Fig. 16.5. Deduced genotypes for the individuals in the pedigree of Fig. 16.3 if the trait is caused by a sex-linked dominant mutation T.

Autosomal recessive. Autosomal recessiveness could explain the inheritance of the nontasting trait. The pedigree may then be explained if the father is homozygous tt and the mother is heterozygous t^+t (Fig. 16.6).

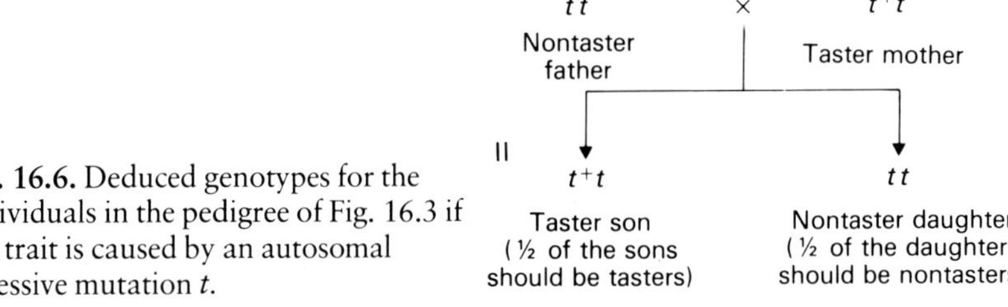

Fig. 16.6. Deduced genotypes for the individuals in the pedigree of Fig. 16.3 if the trait is caused by an autosomal recessive mutation t.

Autosomal dominant. Autosomal dominance could explain the inheritance of the nontasting trait. The pedigree may then be explained if the father was heterozygous TT^+ and mother homozygous wild type (T^+T^+) (Fig. 16.7). Then half of the sons and half of the daughters should be nontasters.

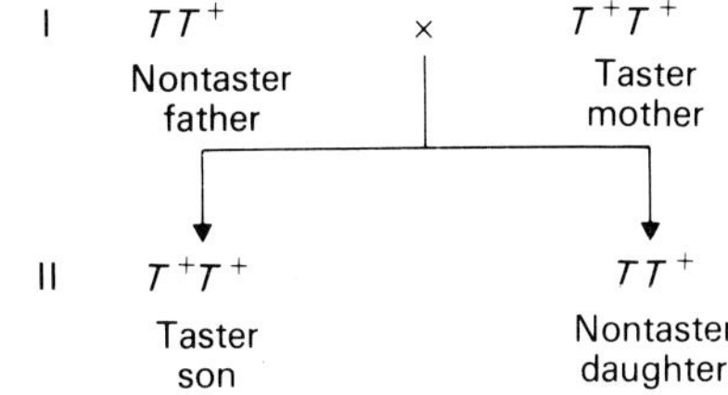

Fig. 16.7. Deduced gentoypes for the individuals in the pedigree of Fig. 16.3 if the trait is caused by an autosomal dominant mutation *T*.

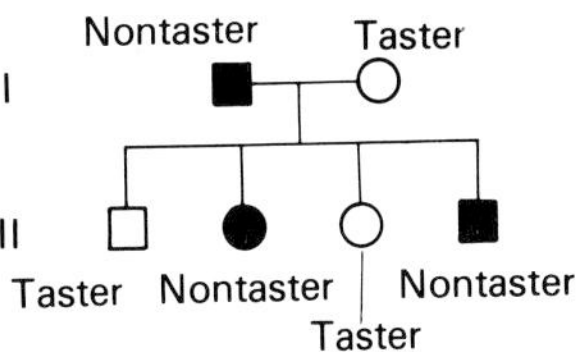

Fig. 16.8. A second hypothetical pedigree for the PTC nontasting trait.

Thus we cannot discern the exact mechanism of inheritance for PTC nontasting from the pedigree presented. In situations like this, it is sometimes helpful to analyze a larger pedigree. Thus two children later, the pedigree becomes as shown in Fig. 16.8.

Applying the same type of arguments as before, sex-linked recessive, autosomal recessive, and autosomal dominant are all possible mechanisms of inheritance for the trait (Fig. 16.9).

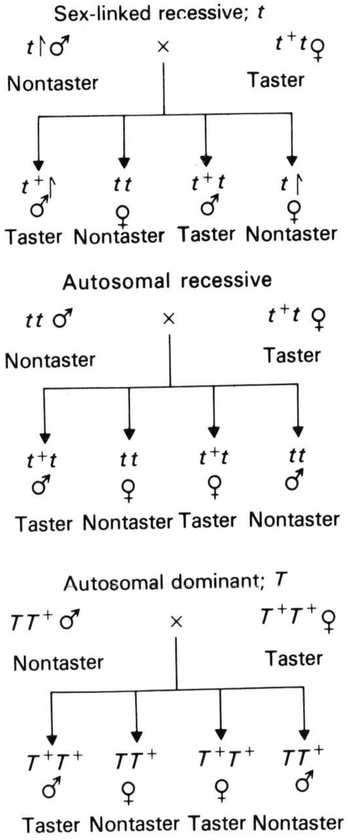

Sex-linked recessive; *t*

$t\upharpoonleft$ ♂ × t^+t ♀

Nontaster Taster

$t^+\upharpoonleft$ ♂ Taster; tt ♀ Nontaster; t^+t ♂ Taster; $t\upharpoonleft$ ♀ Nontaster

Autosomal recessive

tt ♂ × t^+t ♀

Nontaster Taster

t^+t ♂ Taster; tt ♀ Nontaster; t^+t ♀ Taster; tt ♂ Nontaster

Autosomal dominant; *T*

TT^+ ♂ × T^+T^+ ♀

Nontaster Taster

T^+T^+ ♂ Taster; TT^+ ♀ Nontaster; T^+T^+ ♀ Taster; TT^+ ♂ Nontaster

Fig. 16.9. Deduced genotypes for the individuals in the pedigree of Fig. 16.8 if the trait is caused by a sex-linked recessive, autosomal recessive, or autosomal dominant mutation.

However, the data do preclude the possibility that the trait results from a sex-linked dominant gene T. If this were the case the father would be T and the mother would be wild type (T^+T^+). From a pairing of this kind, all sons would be T^+Y, or tasters, and *all* daughters would be heterozygous and hence nontasters. The pedigree of Fig. 16.8 shows that these did not occur.

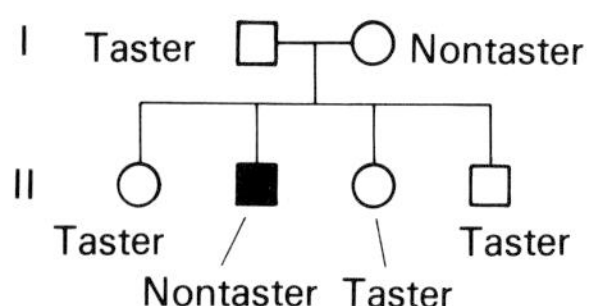

Fig. 16.10. A third hypothetical pedigree for the PTC nontasting trait. This pedigree rules out a dominant mutation as the basis for the trait.

To determine exactly the mechanism of inheritance here, more pedigrees would have to be analyzed. The pedigree shown in Fig. 16.10 rules out a dominant mutation since neither parent showed the trait. (It could be argued, though, that the mutation occurred in the germline of one or other of the parents—this is always something that must be considered.) This pedigree can be explained, however, if the trait is inherited as an autosomal recessive. In this case, both parents would be heterozygous t^+t, and we would expect, on the average, one-fourth of the children to be nontasters (tt) and that there would be no preference for either sex to exhibit the trait.

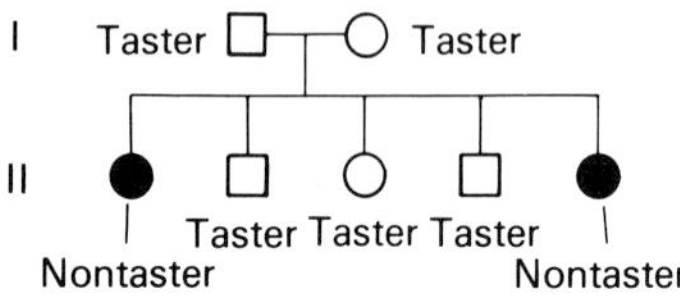

Fig. 16.11. A fourth hypothetical pedigree for the PTC nontasting trait. This pedigree rules out either a dominant mutation of a sex-linked recessive mutation as the basis for the trait.

In the pedigree shown in Fig. 16.11, a dominant mutation is again ruled out for the same reason as above. Also sex-linked recessive inheritance is not possible since for there to be a nontaster daughter (tt), the father must also be a nontaster (tY). Since both parents must donate an X chromosome bearing the t allele, this was not the case. The pedigree can be explained if the trait is inherited as an autosomal recessive using the same logic as for Fig 16.10. Thus, from all the pedigrees analyzed, we are left with the conclusion that the nontaster trait is the result of an autosomal recessive mutation. To be sure of this conclusion, it would be necessary to examine a larger number of pedigrees for unrelated families.

In conclusion, pedigree analysis is a rather tedious and sometimes tenuous procedure, especially when small families are involved. Moreover, the gene mutations causing diseases do not always act in a simple way, thereby leading to variations in the resulting phenotypes complicating the analysis. In a practical sense, it is the job of the genetic counselor to analyze pedigree data to provide estimates of the probability that the children of a particular couple will have a genetic defect. This is a difficult task in view of the complexities of gene expression.

Examples of human pedigrees

1. The pedigree shown in Fig. 16.12 illustrates the inheritance pattern that is characteristic of a **Y-linked** (*holandric*) trait, which presumably is the result of mutation on the Y chromosome. Y-linked traits should be readily recognizable since they show father-to-son transmission and no females should ever express them. The pedigree is the conventional pedigree given for the Lambert family of England that involves the so-called porcupine men. In 1716, Edward Lambert was born to two normal parents. Edward was one of

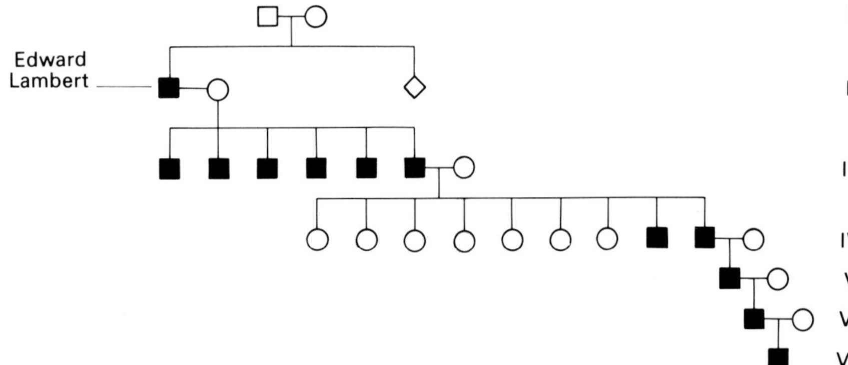

Fig. 16.12. The Lambert pedigree for the Y-linked trait, ichthyosis hystrix gravior, the "porcupine man" syndrome.

many children, and all but Edward remained normal throughout their lives. At 7–8 weeks of age, Edward's skin began to turn yellow, and it gradually became black and thickened until his whole body, with the exception of head, face, palms, and soles, was covered with rough, bristly scales and bristle like outgrowths—hence the name "porcupine man". The pedigree shows that all male descendants of Edward Lambert had this trait (called ichthyosis hystrix gravior), suggesting that it was Y-linked. Careful scrutiny of the records, however, indicates that the pedigree is not correct. In particular, it seems that only three rather than six generations had affected males, and only four or five rather than 12 persons exhibited the trait. Altogether, the reevaluation of the available data rules out this pedigree being a valid example of a Y-linked trait and suggests, instead, that the trait is the result of a rare autosomal dominant mutation.

One trait from present-day families that is considered to be Y-linked in humans is hypertrichosis of the ear. The phenotype is the presence of relatively long hairs on the pinnae of the ears.

One gene that has clearly been localized to the Y chromosome is the *H-Y locus,* the product of which is the H-Y antigen. (An antigen is any large molecule that stimulates the production of specific antibodies or that binds specifically to an antibody; an antibody is a protein molecule that recognizes and binds to a foreign substance—the antigen—introduced into the organism.) The evidence that this locus is on the Y chromosome came from work done in 1955 by E. J. Eichwald and C. R. Silmser to determine whether there were detectable differences between males and females that could be related to the sex chromosomes. Their experimental system was strains of inbred mice. The inbreeding (carried out by making brother–sister matings) was done for so many generations that the resulting individuals had essentially the same genotype. These strains are called *isogenic.* Eichwald and Silmser reasoned that the only genetic differences between isogenic brothers and sisters would be the result of the sex chromosome differences between the two sexes. They transplanted skin from females of an inbred mouse strain to male mice of the same strain and vice versa. In the former there was no rejection of the transplant, but in the latter experiment the skin transplanted to the female mice was eventually rejected. Since the only difference in males and females of the strain was the presence of the Y chromosome in the males, they concluded that the antigen responsible for the transplant rejection was coded for by Y-linked genes. Since the acceptance of tissue grafts is called *histocompatibility*, the Y-linked gene involved in this phenomenon is said to be a *histocompatibility locus*, called the H-Y locus. The H-Y antigen product of this locus is a protein found on cell surfaces in many mammals, including humans. It is not known how this protein functions in the sex-determining mechanism in mammals.

2. A example of a **sex-linked recessive** trait is hemophilia, a defect in blood coagulation, and the classical pedigree is that of Queen Victoria. Part of her pedigree is presented in Fig. 16.13. The assumed genotypes, where *h* is the recessive mutant gene, are given next to the people.

It seems that the *h* mutation arose in the germline of Queen Victoria. Note that only males show the trait since they are hemizygous *h*Y. Females would have to be *hh* to have hemophilia and, since it is a rare gene, this is unlikely as it requires the pairing of a hemophiliac male with a carrier (heterozygous) female. Such a pairing does occur from time to time, however.

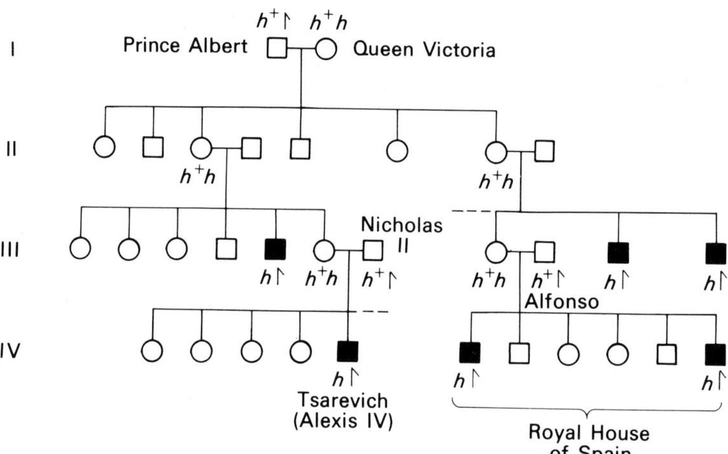

Fig. 16.13. A pedigree for a sex-linked recessive trait: part of Queen Victoria's pedigree showing the inheritance of hemophilia in the Royal families.

The characteristics of X-linked recessive inheritance are: (a) males are hemizygous and, therefore, if they express the trait, they transmit the mutant gene involved to all their daughters but to none of their sons; (b) for rare X-linked recessive traits, many more males than females exhibit the trait owing to the difference in the number of X chromosomes between the sexes; and (c) there should be approximately 1:1 ratio of normal individuals to individuals with the trait among the male progeny of heterozygous (carrier) mothers.

3. Not many **sex-linked dominant** traits have been identified. An example is brown enamel. Such traits show the same type of inheritance as sex-linked recessive traits except that heterozygous females express the trait. Since females have two X chromosomes and males have only one X, sex-linked dominant traits are more frequent in females than in males. For rare traits, most affected females in a pedigree would be heterozygous.

4. An example of an **autosomal dominant** trait is Huntington's chorea, a disease that results in involuntary movements, progressive central nervous system degeneration, and eventually death. The mean age of onset of this genetic trait is between 40 and 45 years, but the disease may appear early in life and sometimes after 60. The American folk singer Woody Guthrie was afflicted by this disease.

The characteristics of an autosomal dominant trait are: (a) every affected person in a pedigree must have at least one affected parent; (b) each generation in a pedigree should have individuals who express the trait; (c) since the X-chromosome is not involved, father-to-son and mother-to-daughter transmission should occur as frequently as father-to-daughter and mother-to-son transmission; and (d) approximately equal numbers of males and females in a pedigree should express the trait.

5. An example of an **autosomal recessive** trait is galactosemia. Galactosemia is a disease where the biochemical defect is known. The disease is manifested in *gg* babies when they are fed milk: they do not grow well and develop permanent brain damage as well as other problems. If they are removed from a milk diet, the children can develop normally. The defect in individuals with galactosemia is a nonfunctional enzyme, galactose 1-phosphate uridyl transferase (GUT). This enzyme is essential for the conversion of galactose to UDP-glucose (Fig. 16.14). Galactose is produced when the disaccharide lactose in milk is enzymatically cleaved to glucose and galactose. In the absence of the GUT reaction, galactose 1-phosphate accumulates and this damages cells.

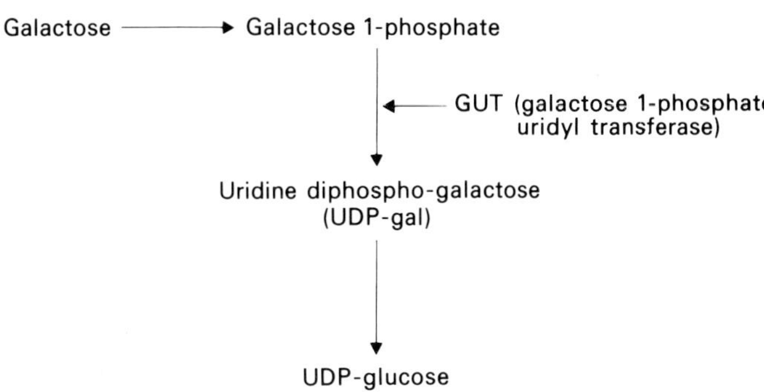

Fig. 16.14. Part of the pathway for the metabolism of galactose showing the reaction catalyzed by the enzyme GUT (galactose 1-phosphate uridyl transferase). GUT is nonfunctional in galactosemic individuals.

The characteristics of autosomal recessive inheritance for a relatively rare trait are: (a) the trait should be expressed in both sexes and transmitted by either sex to both male and female progeny individuals in approximately equal proportions; (b) most individuals expressing the trait should have two normal parents; (c) normal individuals and individuals exhibiting the trait should occur in a ratio of 3:1 in progeny of matings between carrier (heterozygous) parents; and (d) all progeny of two parents who express the trait will exhibit the trait.

For many autosomal recessive diseases the biochemical defect is known, and some of these defects can be tested for by following a procedure called *amniocentesis*. In amniocentesis a needle is inserted through a pregnant mother's abdomen and into the amniotic sac that surrounds the developing fetus. A sample of the amniotic fluid in the sac is collected. This fluid contains cells that have sloughed off the surface of the fetus and the cells may be cultured and then analyzed for possible evidence of a biochemical defect or of a chromosomal deficiency. One limitation of this procedure is that any defective gene must normally be expressed in fetal cells for it to be possible to detect a biochemical defect. Nonetheless, if familial pedigrees suggest that both parents are carriers, and biochemical analysis is feasible, then amniocentesis could be a worthwhile procedure to consider.

Chromosomal aberrancies and human diseases

In this section the types of chromosomal aberrancies that can occur in eukaryotic organisms and some human traits that have their bases in chromosomal aberrancies will be described.

Types of chromosomal aberrancies

There are two types of chromosomal aberrancies. The first involves changes in the entire set of chromosomes. A eukaryote, be it haploid or diploid in its normal state, is considered to be *euploid*, that is, to have an integer number of complete chromosome sets. Occasionally, by a breakdown in spindle fibre formation during meiosis, for example, gametes may be produced that have a diploid chromosome content while others may have no chromosomes. Fusion of these abnormal gametes with a normal gamete will produce triploid and monoploid progeny, respectively: they have three sets or one set of chromosomes instead of the normal two. Those individuals who have alterations in the number of sets of chromosomes are still considered to be euploids (Table 16.1). (Note that **monoploidy** is the term used for an *aberrant* state in a normally diploid cell or organism in which only one complete set of chromosomes is present. It should be distinguished from the term *haploidy*, which refers to the normal chromosome content of gametes of diploid organisms.)

Table 16.1. Terminology for variations in chromosome number.

Chromosome complement	Shorthand formula based on diploid cell	
Monosomic	2N−1	Aneuploidy
Trisomic	2N+1	
Tetrasomic	2N+2	
Monoploid	N	Euploidy
Diploid	2N	
Triploid	3N	
Tetraploid	4N	

In humans the development and functioning of the adult organism is dependent upon the correct gene dosage, that is, the diploid state. Thus, in general, the presence of fewer or extra sets of chromosomes results in a drastic perturbation of gene activities, and development it is aberrant. Thus monoploid, triploid, and **polyploid** (many sets of chromosomes) individuals are found only among aborted fetuses.

By contrast, the euploid chromosome anomalies of this kind can be, and are, tolerated by lower eukaryotes and higher plants. Since no human diseases are caused by monoploidy, triploidy, or polyploidy, this type of chromosomal aberrancy will not be considered further.

The second type of chromosomal aberrancy affects individual chromosomes rather than the complete set of chromosomes. These aberrancies may affect an entire chromosome or part of one. If the number of a particular chromosome present in an organism is abnormal, the individual is said to be **aneuploid** (Table 16.1). These result from defects in the segregation of chromosomes in meiosis. Thus a normally diploid individual such as man might have only one of a particular chromosome (**monosomy**) or he might have three copies (**trisomy**) instead of two. Again, since gene dosage is important in humans, monosomic and trisomic individuals, if they survive until birth, generally have serious defects.

Of less serious consequence to the human individual are the chromosomal aberrancies in which the basic diploid set of genes is maintained but their arrangement in the genome is altered. Examples of this are **inversions** and **translocations** (Fig. 16.15). In an inversion, a segment of chromosome is excised and then reintegrated in an orientation 180° from the original orientation. In a translocation a segment or segments of a chromosome changes position. They can be induced in organisms, for example, by treatment with ionizing radiation. These aberrations do not usually have drastic effects in the organism in which the chromosomal rearrangement has occurred unless the chromosome break point disrupts an important gene.

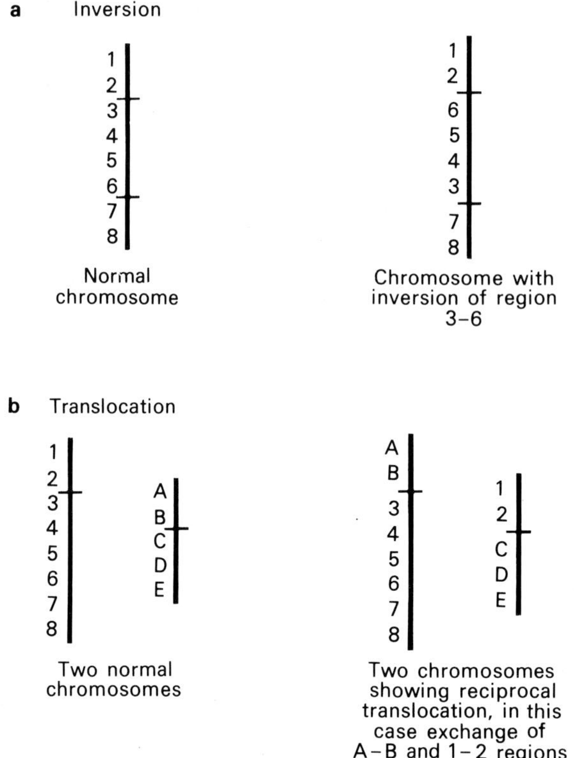

Fig. 16.15. Diagrammatic representation of (**a**) a chromosomal inversion and (**b**) a translocation, in this case a reciprocal translocation involving two chromosomes.

Serious consequences do result, however, when organisms carrying inversions or translocations produce gametes. Fig. 16.16 illustrates this for a heterozygous inversion in which the centromere is not included in the inverted chromosomal segment (a *paracentric inversion*), and Fig. 16.17 shows this for a heterozygous inversion in which the centromere is included in the inverted segment (a *pericentric inversion*). In both cases, gametes with normal sets of genes are produced by meiosis if no crossovers occur in the inverted segment. Half of the gametes in this instance will have the normal chromosome, and half will have the inverted chromosome. However, a single crossover in the inverted segment results in the formation of some gametes that have extra copies of genes (**duplications**) or which have some genes missing (**deletions** or **deficiencies**). Fusion with a normal gamete may well result in inviable zygotes in higher organisms owing to the gene dosage problem mentioned earlier. Thus only normal gametes are generally viable in these crosses. (Again, lower eukaryotes and plants tolerate duplications and deficiencies more readily than do higher eukaryotes.) A somewhat similar situation pertains to translocations, although this will not be illustrated here.

We now turn to a consideration of human traits that are the result of chromosomal aberrancies.

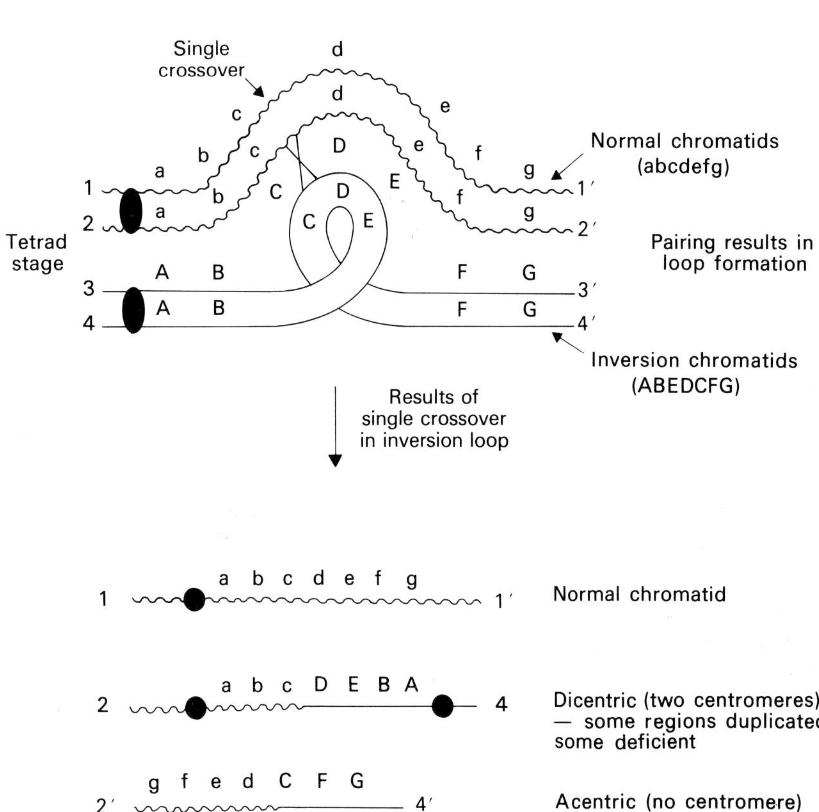

Fig. 16.16. Production of gametes when a single crossover takes place in the inversion loop in a heterozygote involving a normal and a paracentric inversion chromosome. Half of the gametes are defective.

Down syndrome

Individuals with **Down syndrome** (mongolism) are characterized by a number of abnormal attributes, including a very low IQ, epicanthal folds, a protruding furrowed tongue, short broad hands with incurving of the fifth finger, and below average stature. In the most common instances of Down syndrome (Fig. 16.18), these abnormalities are the result of an extra chromosome 21. In other words the individuals are aneuploid or more specifically they are trisomic for chromosome 21, hence the more formal name for the condition of the aberrant individuals, **trisomy-21**. That they survive at all is presumably related to the small size of that particular chromosome and the roles of the particular genes it contains.

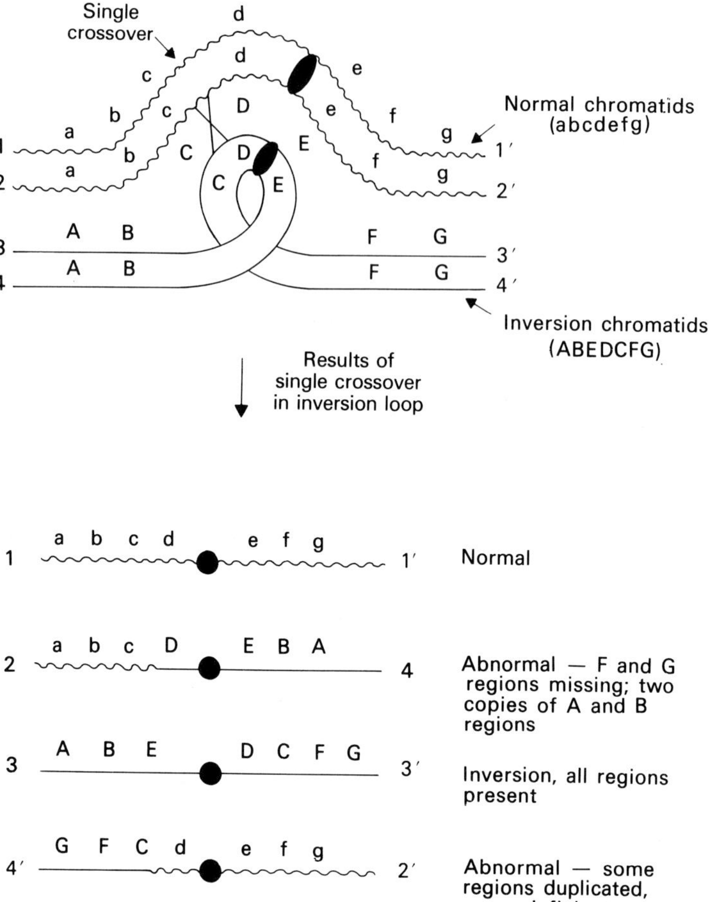

Fig. 16.17. Production of gametes when a single crossover takes place in the inversion loop in a heterozygote involving a normal and a pericentric inversion chromosome. Half of the gametes are defective.

Trisomic individuals can arise by fusion of a gamete with two copies of a particular chromosome and a gamete with one copy of the chromosome. The abnormal gamete with two copies of the chromosome occurs very rarely as a result of *nondisjunction* of chromosome pair during *either* of the two meiotic divisions; nondisjunction at the first division of meiosis is illustrated along with normal meiosis in Fig. 16.19. The molecular basis for the abnormal meiosis is not known but is presumably a consequence of spindle fiber failure. As the figure illustrates, there should be equal numbers of gametes produced that lack a particular chromosome as those which have the abnormal complement of two. Since no individuals have been born with monosomy-21 this condition is most probably lethal.

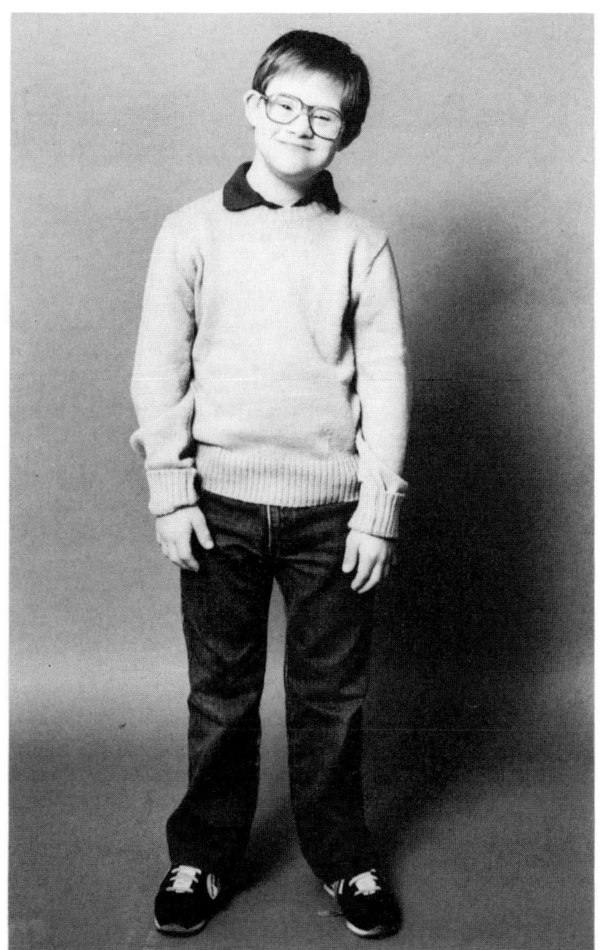

Fig. 16.18. An individual with Down syndrome (trisomy-21). (Courtesy of the Clinical Cytogenetics Laboratory, Oregon Health Sciences University.)

The incidence of trisomy-21 births is directly related to the mother's age (Table 16.2). The data presented in the table can be used by genetic counselors to advise older women of the potential risk of having trisomy-21

Table 16.2. The relationship between the age of the mother and the risk of a trisomy-21 child.

Age of mother	Risk of trisomy-21 in child
< 29	1/3000
30–34	1/600
35–39	1/280
40–44	1/70
45–49	1/40
All mothers combined	1/665

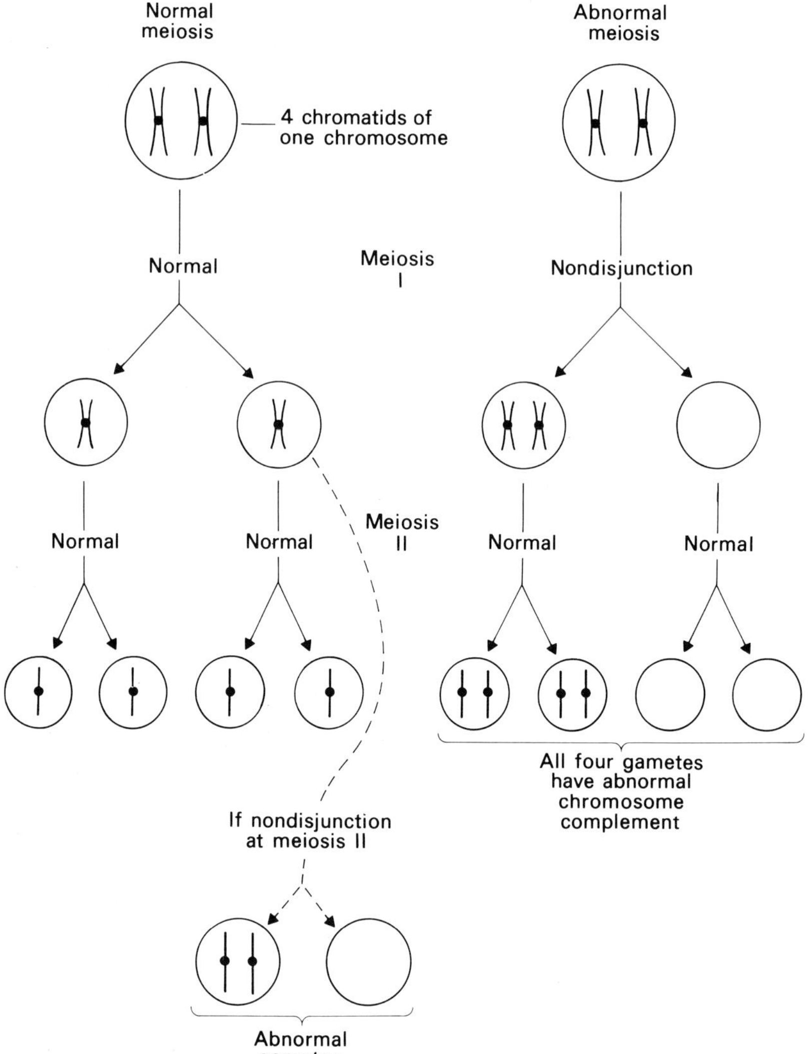

Fig. 16.19. Production of gametes with abnormal numbers of chromosomes when nondisjunction occurs at either the first division (right-hand diagram) or second division (left-hand diagram) of meiosis.

children. Once conceived, a child can be tested for trisomy-21 (or for other chromosomal aberrancies) by using aminocentesis or other prenatal tests, followed by karyotype analysis of the cultured cells.

The reason for the mothers' age dependency of trisomy-21 incidence is as follows. In mature females the ovaries contain the primary oocytes that are the progenitors of all the eggs that will be released by the ovary throughout the lifetime of the woman. The primary oocyte is a cell that is arrested in the first division of meiosis. Each month, starting at or near the onset of menstruation, at least usually one primary oocyte completes meiosis and the resulting egg is released into the fallopian tube. As would be expected, the probability of

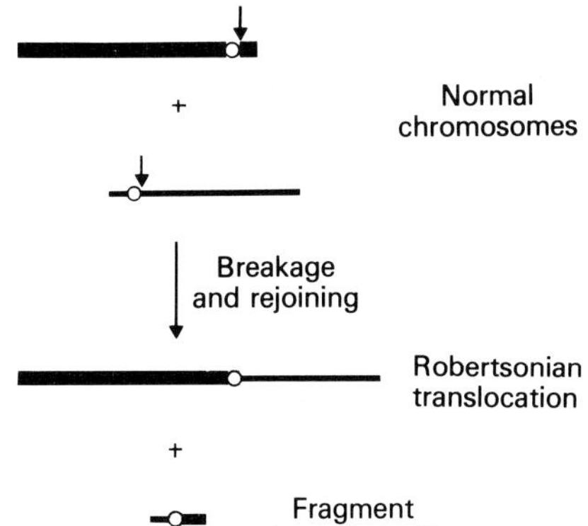

Fig. 16.20. Production of a Robertsonian translocation by chromosome breakage and rejoining involving two acrocentric chromosomes. The breakage points are indicated by arrows.

nondisjunction occurring during egg maturation increases with age since spindle fiber malfunctions are more likely to arise as the cell ages.

Down syndrome individuals can also result from a different sort of chromosomal aberrancy called a *Robertsonian translocation.* A Robertsonian translocation is a type of nonreciprocal translocation in which the long arms of two nonhomologous acrocentric chromosomes (chromosomes with centromeres near their ends) become attached to a single centromere (Fig. 16.20). The short arms of the two acrocentric chromosomes become attached to form the reciprocal product, which typically contains nonessential genes and is usually lost within a few cell divisions. In humans, when a Robertsonian translocation joins the long arm of chromosome 21 with the long arm of chromosome 14, the heterozygous carrier is phenotypically normal since there are two copies of all major chromosome arms and hence two copies of all essential genes involved. However, there is a high risk of Down syndrome among the offspring of these carriers. Fig. 16.21 shows the gametes produced by the heterozygous carrier parent and by the normal parent and the chromosome constitutions of the offspring generated. The normal parent produces gametes with one copy each of the relevant chromosomes 14 and 21. The heterozygous carrier parent produces three reciprocal pairs of gametes, each pair being produced as a result of different segregation of the three chromosomes involved. The zygotes produced by pairing of these gametes with gametes of normal chromosomal constitution are as follows (Fig. 16.21): one-sixth are normal with respect to chromosomes 14 and 21; one-sixth are heterozygous carriers like the parent and are phenotypically normal; three-sixths are inviable because of monosomy for chromosome 21 or chromosome 14, or trisomy for chromosome 14; and one-sixth are trisomy-21 and therefore give rise to a Down syndrome individual. (Note that these latter individuals actually have 46 chromosomes but because of the Robertsonian translocation, they have three copies of the long arm of chromosome 21, and this is sufficient to give the Down syndrome symptoms. Similarly, the trisomy-14 individuals shown in Fig. 16.21 have 46 chromosomes but they have three copies of the long arm of chromosome 14. Apparently the dosage of the genes involved on this larger chromosome is more critical, and consequently, these individuals are inviable.) In sum, one-half of the zygotes produced are inviable and, of the viable zygotes, one-third will give rise to a Down syndrome individual—a much higher risk than we discussed earlier with respect to the mother's age.

There are very few other examples of surviving trisomic individuals in humans, and these generally involve the small chromosomes. Many other human traits also have their basis in chromosome aberrancies such as partial deletions of chromosomes, rearrangements of chromosomes, etc. More drastic changes in chromosome complement such as polyploidy, monosomy,

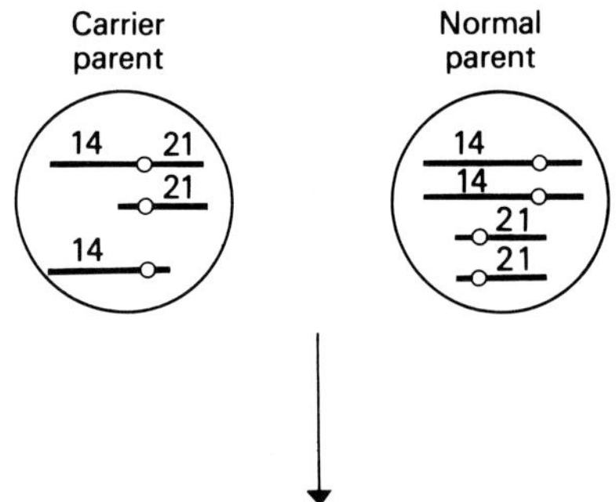

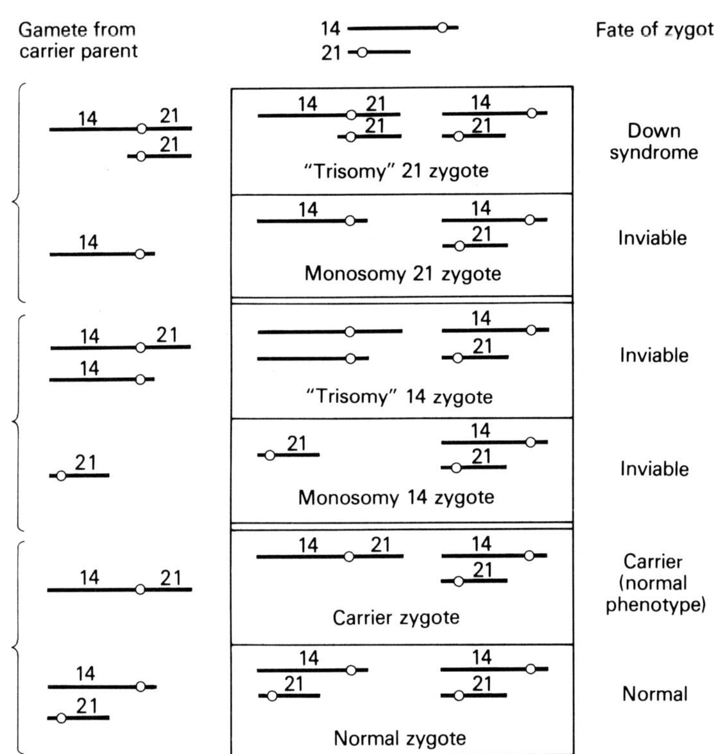

Fig. 16.21. The three segregation patterns of a heterozygous Robertsonian translocation involving the human chromosomes 14 and 21. Fusion of the resulting gametes with gametes from a normal parent produces zygotes with various combinations of normal and translocated chromosomes. A full discussion is given in the text.

and trisomy of the larger chromosomes have been detected in aborted fetuses, thus reinforcing the notion that correct gene dosage is very important to human development and function. Aberrancies of the X chromosome (which is a large chromosome in humans) are an exception to this and will be discussed next.

Anomalies of the sex chromosomes

In humans, sex determination is based on the presence (male) or absence (female) of the Y chromosome. Normally males are XY and females are XX, and since few if any genes on the Y are homologous to those on the X, there is a difference in gene dosage for X chromosome genes between males and females. However, microscopic examination of nuclei of human cells (and of the cells of many mammalian species) reveals a condensed mass of chromatin in females that is not apparent in males. This heterochromatinized material is called a **Barr body** after its discoverer Murray Barr. The Barr body is associated with the X chromosomes in that individuals who have extra X chromosomes also have extra Barr bodies (Table 16.3). A general formula is: number of Barr bodies = number of chromosomes − 1. What the Barr bodies represent is all but one of the X chromosomes in a genetically inactive or virtually inactive state. This means that adult human males and females both have only one active copy of most X chromosome genes. (Recent evidence suggests that not all the X chromosome in a Barr body is inactivated.) This point will be elaborated later in the discussion of the sex chromosome anomalies that follows.

Table 16.3. Relationship of number of Barr bodies to X chromosome complement.

Chromosomal constitution	Number of Barr bodies
XO	0
XY	0
XX	1
XXY	1
XXX	2
XXXX	3
XXXXY	3

Turner syndrome (XO). Individuals with **Turner syndrome** have a 45-chromosome karyotype with only one sex chromosome, an X (Fig. 16.22). They lack the Barr body and they arise by a nondisjunction mechanism with an occurrence of approximately 1 per 3000 live births. Since there is no Y chromosome, they are female. Turner's syndrome females have few major defects until puberty when they do not develop secondary sexual characteristics. Generally, these individuals are short and have a weblike neck, poorly developed breasts, and immature internal sexual organs. Sometimes they exhibit mental deficiencies and rarely they may be fertile.

Superfemale (XXX). These individuals have a 47-chromosome karyotype with three X chromosomes. Two Barr bodies can be seen in many cells.

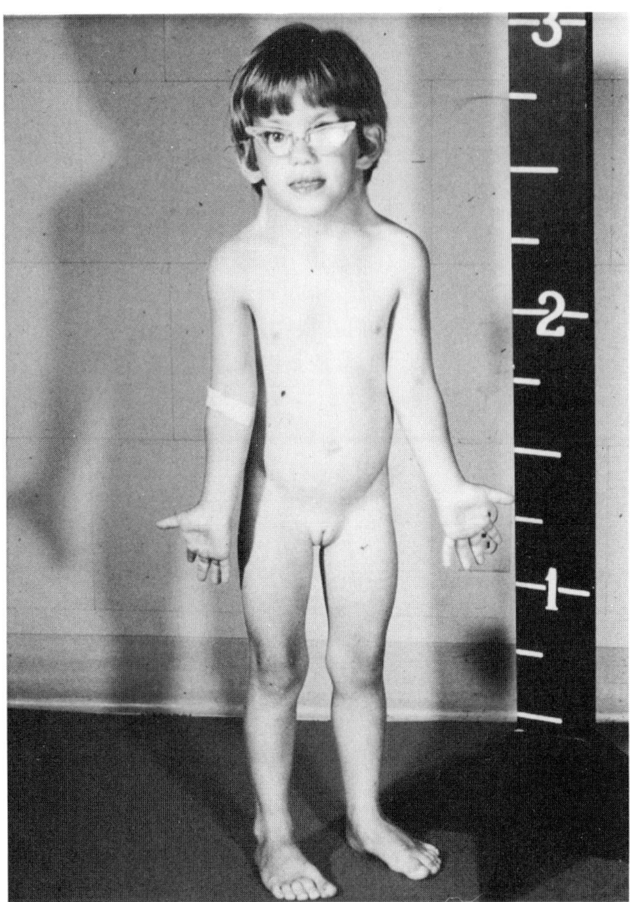

Fig. 16.22. Photograph of an individual with Turner syndrome (XO). (Photograph courtesy of Clinical Cytogenetics Laboratory, Oregon Health Sciences University.)

Superfemales are sometimes mentally deficient or infertile, hence their detection in hospitals for the retarded and in infertility clinics. Phenotypically most superfemales are reasonably normal with perhaps poorly developed secondary sexual characteristics.

The Klinefelter syndrome (XXY). **Klinefelter syndrome** individuals have an extra X chromosome and are male owing to the presence of the Y chromosome (Fig. 16.23). The incidence of Klinefelter syndrome is approximately 1 in 400 live births, and again nondisjunction is the culprit. Phenotypically many Klinefelter males are mentally deficient, have poorly developed testes, and are tall. These individuals have one Barr body in their cells. In more extreme cases of Klinefelter's syndrome, the chromosomal constitution is more irregular (e.g. XXXY, XXXYY, XXXXY, XXYY). In these cases the number of Barr bodies follows the general formula mentioned previously.

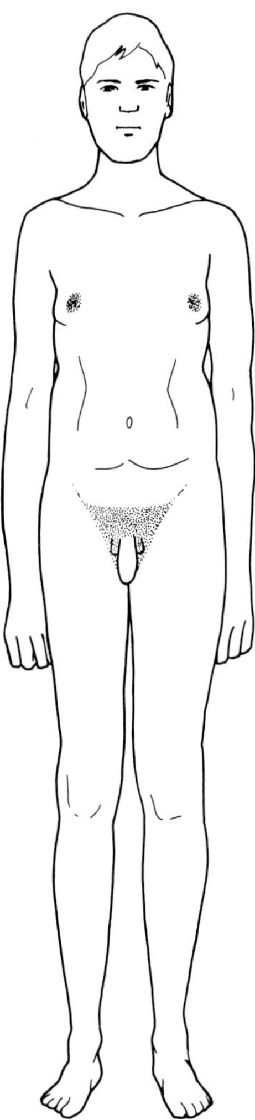

Fig. 16.23. Diagram of an individual with Klinefelter syndrome (XXY).

Other anomalies than those just described have been described in the literature and are beyond the scope of this text. In general, for individuals showing anomalies of the sex chromosomes, an individual with a Y chromosome has a male phenotype and an individual without a Y chromosome has a female phenotype. However, the degree of maleness or femaleness a particular individual has varies greatly. Basically the more abnormal the chromosome complement, the more aberrant the phenotype.

The Lyon hypothesis

Why can multiple X chromosomes be tolerated by humans with relatively minor consequences while extra copies of all but the smallest chromosome lead to lethality? The answer to this question has already been alluded to: the Barr body represents an inactive or mostly inactive X chromosome. In 1961 Mary Lyon formulated a hypothesis to explain, among other things, the survival of individuals with X chromosome anomalies. This has become known as the *Lyon hypothesis*, which states that:

1. The Barr body is a genetically inactivated X chromosome.
2. The inactivated X chromosome can be either of paternal or maternal origin in different cells of the same individual.
3. X chromosome inactivation (called **lyonization**) occurs early in embryonic life (the 16th day following fertilization in humans).

We now know that X inactivation is random and that, once inactivated in a cell, the descendants of that cell have the same inactive X, be it maternal or paternal.

Thus we can see that the Lyon hypothesis affords a simple explanation to why normal XY and XX have the same single set of active X chromosome genes. This is essentially a *gene dosage compensation mechanism.* Similarly the Lyon hypothesis can explain why, for example, XXY or XXX individuals are not so drastically different from normal individuals. That is to say, lyonization (inactivation) of the extra X chromosomes results in gene dosage compensation so that only one set of X chromosome genes is ultimately active no matter how many X chromosomes are present in the cell. The phenotypic differences from normality that the individuals with the anomalous sex chromosome have are presumably the result of the activity of the extra chromosomes during the first 16 days before chromosome inactivation is initiated.

In conclusion, we have presented an overview of two areas of human genetics—pedigree analysis and the occurrences of chromosome abnormalities. Present-day research is directed toward investigating whether particular human diseases have a genetic basis and, if so, what biochemical or chromosomal alterations are present. Once this is known, physicians and genetic counselors can advise the prospective parents as to the potential risks of the woman giving birth to a child with the genetic defect.

One important aspect of genetic counseling is the determination of whether a developing fetus is normal in the genetic or karyotypic sense. The fetus develops in an amniotic sac and is surrounded by amniotic fluid, a sample of which can be taken by inserting a syringe needle through the uterine wall and into the amniotic sac. The fluid contains fetal cells, which can be cultured and then examined for the possibility of an abnormal karyotype or for the presence of any enzyme defect presumed to be "running in the family". The latter is applicable only to those genetic enzyme deficiency diseases in which the enzyme involved is expressed in the developing fetus. It should be stressed, however, that amniocentesis is a fairly complicated procedure, and it is used, therefore, mostly for high risk cases.

Mapping genes to human chromosomes

Mapping using somatic cell hybridization

Genetic mapping experiments of the kind we have described for experimental organisms are clearly not possible with humans, and thus we must resort to other approaches for localizing genes on chromosomes. Pedigree analysis, for example, can show whether a gene is on an X chromosome or an autosome. Another method of map both X-linked and autosomal genes in humans involves fusion of human cells and rodent cultured cells; this method is called **somatic cell hybridization.**

Somatic cells can often be cultured in vitro to produce a *cell culture line*. If the cell culture line is derived from malignant tissues, the cells will grow and divide essentially indefinitely in culture, although the cells may have chromosomal constitutions that are different from that of normal cells. A cell culture line of this kind is called an *established cell line*.

A human cell can be fused with a cell from an established cell line from the mouse to produce a somatic cell hybrid. To facilitate the cell fusion, a suspension of the two cell types is incubated with inactivated Sendai virus, which binds to both types of cells and brings their membranes close enough so that membrane fusion occurs. The result is a single cell with two different nuclei; such a cell is called a binucleate heterokaryon. Once the two cells have fused, the nuclei often fuse to form a single nucleus containing the sets of chromosomes from the two organisms. A cell of this type is called a **synkaryon**, and by mitosis, the cell can give rise to a colony of cells.

In forming somatic cell hybrids, it is important to use conditions that enable the hybrid cells to grow while at the same time preventing the parental cells from growing. A convenient selection procedure for this is the *HAT* (*H*ypoxanthine-*A*minopterin-*T*hymidine) technique, in which conditions are established so that only hybrid cells can synthesize DNA and hence grow and divide.

The HAT technique is illustrated in Fig. 16.24 for the production of a human-mouse hybrid cell. Aminopterin is present in the medium to inhibit the de novo synthesis of purines and pyrimidines from precursors (see Chapter 3). The two cell lines have different genetic defects in the salvage pathways (see Chapter 3): in the example, the mouse cell line is defective in hypoxanthine phosphoribosyl transferase (HGPRT), a purine salvage pathway enzyme, and the human cell line is defective in thymidine kinase (TK), a pyrimidine salvage pathway enzyme. Conversely, the mouse cells have normal TK and the human cells have normal HGPRT. If aminopterin is present, neither cell type can grow alone, since the mouse cells cannot make purine monophosphates by the salvage pathway and the human cells cannot

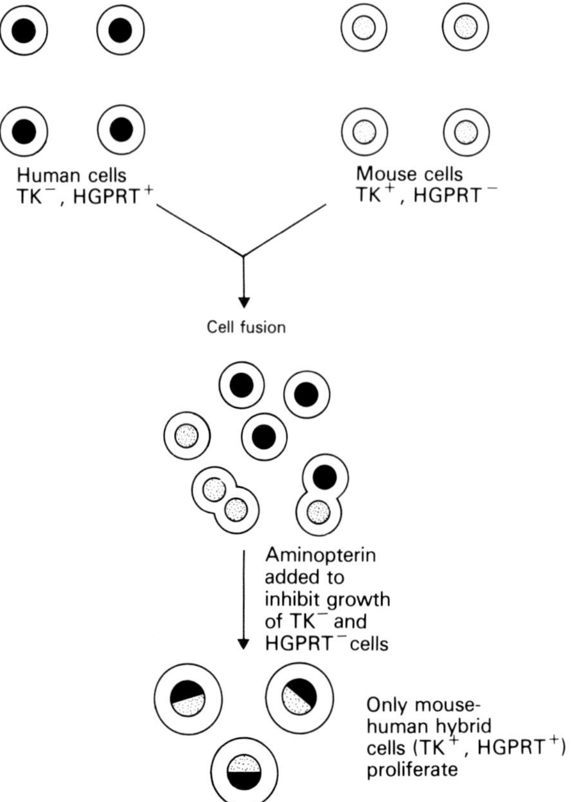

Fig. 16.24. HAT technique for the selection of human-mouse hybrid cells.

make the pyrimidine monophosphates by the salvage pathway. However, human-mouse hybrid cells are able to make these molecules, and hence to grow and divide, because the human chromosomes have a normal HGPRT gene and the mouse chromosomes have a normal TK gene. In other words, there is complementation of the two genetic defects.

Hybrid human-mouse somatic cells are useful for mapping human genes for two reasons: (1) human and mouse chromosomes are morphologically quite distinct; and (2) during reproduction of the hybrid cells, some chromosomes are lost, preferentially those of the human set. Eventually no more chromosomes are discarded, and a surviving stable hybrid cell line is produced that contains a complete set of mouse chromosomes plus some human chromosomes; the number and type of human chromosomes varies with the cell line. Of necessity, each surviving cell line contains the human chromosomes with the normal HGPRT gene.

To localize a human gene to a chromosome through the use of hybrid somatic cells, the initial human cell used to form the somatic hybrid cell must have one or more genetic markers, for example, for antibiotic resistance or nutritional requirements, that can readily be tested. The mouse line must be deficient for some genetic marker. Surviving stable cell lines (those no longer losing chromosomes) are analyzed for the presence or absence of the phenotypes controlled by the genetic markers. In addition, the human chromosome complement of each cell line is determined by karyotype analysis. By analyzing a number of surviving cell lines derived from the same somatic hybrid cell, a correlation can be made between the presence of a particular gene marker and the presence or absence of a particular chromosome. For example, if a drug-resistance phenotype is present in a number of surviving cell lines that have in common only chromosome 6, then the gene controlling the drug-resistance phenotype must be on chromosome 6. In this way, a number of human genes have been localized to particular chromosomes.

Other methods are used to map genes to human chromosomes, but they will not be described in detail here. The following two sections outline two such methods.

Deletion mapping

Deletion mapping is based on the principle that the absence of a chromosome region should be correlated with the loss of the expression of genes in that region. By comparison with experimental organisms such as *Drosophila*, the detection of deletions in human chromosomes is crude. Nonetheless, people do survive with visible deletions, and occasionally it is possible to recognize *hemizygosity* for one or more loci in such individuals. Heterozygosity for a locus in a person with a deletion, on the other hand, clearly localizes the gene elsewhere than in the deleted region. Practically speaking, deletions have in some cases been useful in ruling out a proposed position for a gene on a map, and even in the cases where a gene is localized to a particular region on the basis of deletion mapping, the size of the deletion usually precludes any fine detail localization of the gene. A complication prevents the use of deletion mapping for X-linked genes (which must be done in females) since, for reasons that are not understood, an X chromosome with a deletion is always the one that is inactivated. This special instance of nonrandomness of lyonization (discussed earlier in this chapter) means that genes on the X chromosome without the deletion are always expressed in deletion heterozygotes.

Mapping using nucleic acid hybridization

We learned in Chapter 2 that when DNA is heated, the hydrogen bonds holding the base pairs together are broken, causing the two strands to separate. Human cells grown in tissue culture can be arrested in their cell division cycle at metaphase by the use of the drug colchicine. When these cells are "squashed" and subjected to a stain such as acetocarmine, the resulting stained chromosomes may be visualized under the light microscope (see karyotypes in Chapter 2). If these chromosome spreads, as they are called, are heated to denature DNA under conditions where the chromosome shape is not disrupted (e.g. at 85°C for 10 min), we have the basis for localizing genes by *nucleic acid hybridization*, a process called *in situ hybridization*. That is, if we have cloned a gene, for example (see Chapter 12), the cloned DNA can be radioactively labeled, separated to single strands by heat denaturation, and then incubated with the heat-denatured chromosome preparation. The labeled single-strands of the cloned gene (called in general a *probe* in such experiments) will pair with the chromosomal gene by complementary base pairing (i.e. it hybridizes to the chromosomal DNA in situ), and then, using autoradiography (see Chapter 4), the chromosomal site of the gene can be defined. Following the hybridization step, the chromosomal spread on a microscope slide is placed against an X-ray film and the resulting "sandwich" placed in the dark. As a result of the decay of the radioisotope (usually ^{32}P), the film emulsion is changed so that when the film is developed the site of decay is seen as dark grains.

In situ hybridization, like deletion mapping, does not permit fine detail localization of genes on the chromosomes. Moreover, it can be difficult, although not impossible, to localize single-copy genes because the amount of radioactivity that can be bound to the chromosomes in this procedure is limited by the labeling procedures used. Multicopy genes, for example, ribosomal RNA genes, have been successfully localized using the in situ hybridization procedure (in this case ^{32}P-labeled rRNAs were used as probes). We can be optimistic, however, that with the increasing number of cloned single-copy (protein-coding) genes becoming available and with continuing improvements in labeling techniques, it should become more and more practical to map genes accurately by in situ hybridization.

Questions and problems

16.1 A man (A) suffering from defective tooth enamel, which results in brown-colored teeth, marries a normal woman. All their daughters have brown teeth, but the sons are normal. The sons of man A marry normal women, and all their children are normal. The daughters of man A marry normal men, and 50% of their children have brown teeth. Explain these facts.

16.2 Huntington's chorea is a human disease inherited as a Mendelian autosomal dominant. The disease results in choreic (uncontrolled) movements, progressive mental deterioration, and eventually death. The disease affects the carriers of the trait any time between 15 and 65 years of age. The American folk singer Woody Guthrie died of Huntington's chorea as did only one of his parents. Marjorie

Mazia, Woody's wife, had no history of this disease in her family. The Guthries had three children. What is the probability that a particular Guthrie child will die of Huntington's chorea?

16.3 In human genetics, the pedigree is used for analysis of inheritance patterns. The female is represented by a circle and the male by a square. The accompanying figure presents three family pedigrees for a trait in humans. Normal individuals are represented by unshaded symbols and people with the trait, by shaded symbols. For each pedigree (A, B, and C), state by answering "yes" or "no" in the appropriate blank space whether transmission of the trait can be accounted for on the basis of each of the listed simple modes of inheritance.

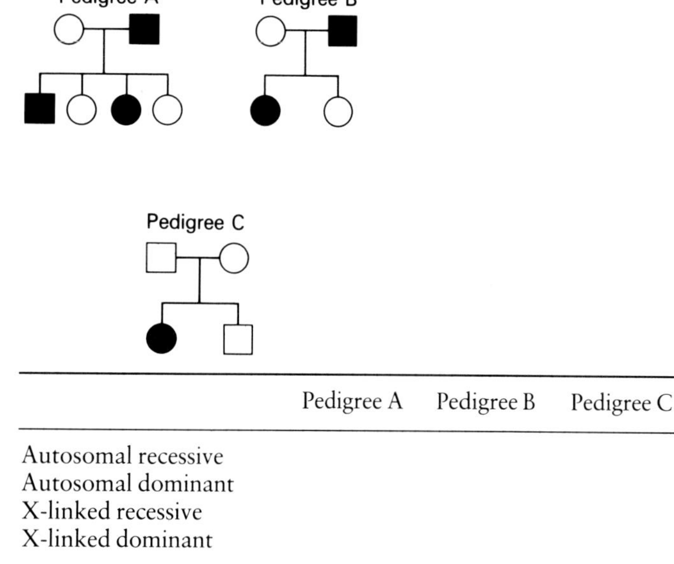

	Pedigree A	Pedigree B	Pedigree C
Autosomal recessive			
Autosomal dominant			
X-linked recessive			
X-linked dominant			

16.4 Phenylketonuria (PKU) is an inborn error in the metabolism of the amino acid phenylalanine. The characteristic feature of PKU is severe mental retardation many untreated patients with this trait have IQs below 20. The pedigree shown in the accompanying figure is of an affected family. The generations are labeled by roman numerals, and the individuals in each generation by arabic numerals.

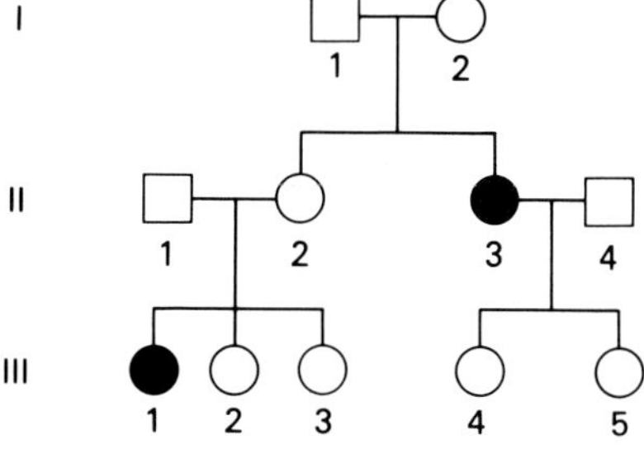

(a) What is the mode of inheritance of PKU?
(b) Which persons in the pedigree are known to be heterozygous for PKU?

(c) What is the probability that III-2 is a carrier (heterozygous)?
(d) If III-3 and III-4 marry, what is the probability that their first child will have PKU?

16.5 For the pedigrees shown in the accompanying figures, indicate the probable mode of inheritance: autosomal recessive, autosomal dominant, X-linked recessive, X-linked dominant, Y-linked.

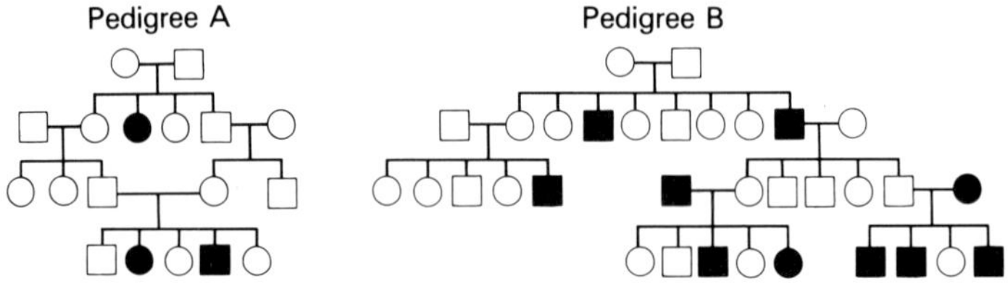

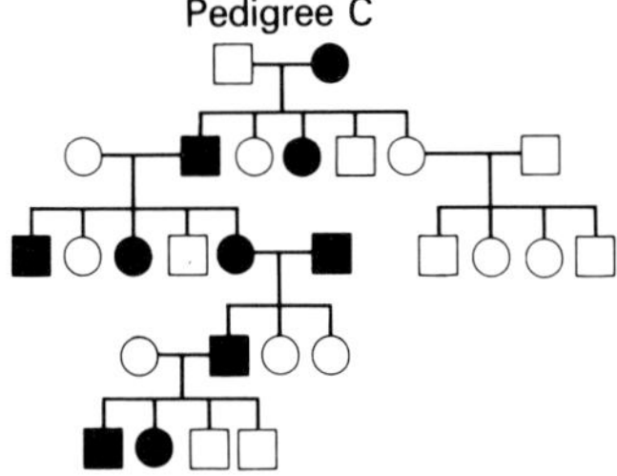

16.6 Define the terms aneuploidy, monoploidy and polyploidy.

16.7 If a normal diploid cell is 2N, what is the chromosome content of the following?
(a) a monosomic
(b) a double monosomic
(c) a tetrasomic
(d) a double trisomic
(e) a tetraploid
(f) a hexaploid

16.8 In man, how many chromosomes would be typical of nuclei of cells that are:
(a) monosomic
(b) trisomic
(c) monoploid
(d) triploid
(e) tetrasomic

16.9 A normal chromosome has the following gene sequences:

A B C D E F G H

Determine the chromosome mutation in each of the following chromosomes:

(a) *A B C F E D G H*

(b) *A D E F B C G H*

(c) *A B C D E F E F G H*

(d) *ABCD*○*EFFEGH*

(e) *ABD*○*EFGH*

16.10 Define pericentric and paracentric inversions.

16.11 Inversions are said to affect crossing over. Given the following homologs with the indicated gene order:

●*ABCDE*

○*ADCBE*

(a) Diagram how these chromosomes would align in meiosis.
(b) Diagram what a single crossover between genes in the inversion would result in.
(c) Considering the position of the centromere, what is this sort of inversion called?

16.12 Single crossovers within the inversion loop of inversion heterozygotes give rise to chromatids with duplications and deletions. What happens when, within the inversion loop, there is a two-strand double crossover in such an inversion heterozygote when the centromere is outside the inversion loop?

16.13 An inversion heterozygote possesses one chromosome with genes in the normal order:

—○*abcdefgh*

It also contains one chromosome with genes in the inverted order:

—○*abfedcgh*

A four-strand double crossover occurs in the areas *e-f* and *c-d*. Diagram and label the four strands at synapsis (showing the crossovers) and at the first meiotic anaphase.

References

Barr, M.L. and E.G. Bertram, 1949. A morphological distinction between neurones of the male and female, and the behavior of the nucleolar satellite during accelerated nucleoprotein synthesis. *Nature* **163**:676–677.

Bloom, A.D., 1972. Induced chromosome aberrations in man. *Adv. Human Genetics* **3**:99–153.

Bodmer, W.F. and L.L. Cavalli-Sforza, 1976. *Genetics, Evolution, and Man.* W.H. Freeman, San Francisco.

Boyer, S.H. (ed.), 1963. *Papers on Human Genetics.* Prentice Hall, Englewood Cliffs, New Jersey.

Dice, L.R., 1946. Symbols for human pedigree charts. *J. Hered.* **37**:11–15

Fraser, F.C., 1974. Current issues in medical genetics: Genetic counseling. *Amer. J. Human Genetics* **6**:636–659.

Garrod, A.E., 1909. *Inborn Errors of Metabolism.* Frowde, Hodder and Stoughton, London.

Harris, H., 1962. *Human Biochemical Genetics.* Cambridge University Press, Cambridge.

Levitan, M. and A. Montagu, 1971. *Textbook of Human Genetics*. Oxford University Press, London.

Lyon, M.F., 1961. Gene action in the X-chromosomes of the mouse (*Mus musculus L*). *Nature* **190**:372–373.

Lyon, M.F., 1962. Sex chromatin and gene action in the mammalian X-chromosome. *Amer. J. Human Genetics* **14**:135–148.

McKusick, V.A., 1965. The Royal hemophilia. *Sci. Amer.* **213**:88–95.

McKusick, V.A. and R. Claiborne (eds.), 1973. *Medical Genetics*. H.P. Publishing Co., New York.

Penrose, L.S., 1933. The relative effects of paternal and maternal age in mongolism. *J. Genetics* **27**:219–224.

Penrose, L.S. and G.F. Smith, 1966. *Down's Anomaly*. Little, Brown and Co., Boston.

Penrose, L.S. and C. Stern, 1958. Reconsideration of the Lambert pedigree (ichthyosis hystrix gravior). *Ann. Human Genetics* **22**:258–283.

Shaw, M.W., 1962. Familial mongolism. *Cytogenetics* **1**:141–179.

Stanbury, J.B., J.B. Wyngaarden and D.S. Fredrickson, 1972. *The Metabolic Basis of Inherited Disease*, 3rd ed. McGraw-Hill, New York.

Stern, C., 1973. *Principles of Human Genetics*, 3rd ed. W.H. Freeman, San Francisco.

Thompson, J.A. and M.W. Thompson, 1966. *Genetics in Medicine*. W.B. Saunders, Philadelphia.

Chapter 17 Extrachromosomal Genetics

Up to now we have considered the inheritance of traits that are under the control of the nuclear genome. In these cases the transmission pattern of the traits can be predicted from the known patterns of chromosome segregation and assortment. There are other traits whose inheritance does not follow these rules; these exhibit extrachromosomal inheritance. In the chapters on bacterial genetics and recombinant DNA, we discussed episomes and plasmids, and these are examples of traits showing extrachromosomal inheritance. Many of these have their basis in the genetic material found in the cell organelles: the mitochondria and chloroplasts. This chapter will concentrate on examples of extrachromosomal inheritance in eukaryotes to differentiate clearly this mode of inheritance from that shown by traits coded by the nuclear genome. In so doing we will present only a relatively small fraction of the information known about extrachromosomal elements that have genetic continuity.

Mitochondria

Mitochondria are essential constituents of all aerobic animal and plant cells. They are relatively complex organelles with a double membrane and contain several important respiratory enzymes of the electron transport chain (the *cytochromes*) that are involved with the generation of ATP by oxidative phosphorylation.

The cytochromes are proteins, each of which has an atom of iron held within a heme group. Their synthesis has been examined in studies involving the use of the previously mentioned antibiotics with the following interesting results:

1. *Cytochrome c*. The protein part of the molecule is synthesized entirely on cytoplasmic ribosomes, the heme group is attached, and then the whole molecule is transferred into the mitochondria. Thus apparently only nuclear genes are involved in its synthesis.
2. *Cytochromes* $a+a_3$. The protein part is synthesized on cytoplasmic ribosomes while the attachment of the heme group apparently occurs within the mitochondria.
3. *Cytochrome oxidase*. The assembly of this complex molecule requires some polypeptides synthesized on mitochondrial ribosomes and some synthesized on cytoplasmic ribosomes.

Mitochondrial genome

Mitochondria contain genetic material in the form of supercoiled, circular DNA. No histones are associated with this DNA. The chromosomes of higher animal mitochondria are much smaller than the chromosomes of plant and fungal mitochondria; for example, human mitochondrial (mt) DNA is about 14,000 base pairs, whereas *Neurospora crassa* mtDNA is 60,000 base pairs, and plant mtDNA is much larger. Despite the differences in the amount of mtDNA, mitochondria from all sources contain about the same amount of DNA that codes for functional mitochondrial products. The excess material in fungi and plants is spacer. Often the DNA has a different buoyant density from that of nuclear DNA, which facilitates its isolation by CsCl density gradient centrifugation.

Within mitochondria, there are nucleoid regions similar to those of bacterial cells. Several copies of mtDNA are found within each nucleoid region. In yeast, for example, there are 1 to 45 mitochondria per cell, 10 to 30 nucleoids per mitochondrion, and 4 to 5 mtDNA molecules per nucleoid.

The replication of mtDNA uses specific mitochondrial DNA polymerases. There is no synchrony with nuclear DNA replication. New mitochondria are produced by the growth and division of preexisting mitochondria.

Mitochondrial genes

Mitochondrial DNA codes for a number of mitochondrial components. Mitochondria contain ribosomes that carry out protein synthesis. The two rRNAs found in these ribosomes are encoded in the mtDNA, whereas most, if not all, the ribosomal proteins are specified by nuclear genes. The mtDNA also codes for 22 to 23 tRNA species and about 10 to 12 proteins that are components of the inner mitochondrial membrane.

By applying rapid DNA-sequencing techniques, much information has been obtained about mt genomes; some have been completely sequenced. In human DNA, for example, the genes for the 16 and 12S rRNAs of the large and small mitochondrial ribosomal subunits, respectively, are located adjacent to one another in the genome on the same strand. The tRNA genes are located at various positions on both strands: some are clustered, some are isolated, and one is between the two rRNA genes. The protein-coding genes are also found on both strands. The mRNAs transcribed from them do not become capped at the 5′ end, but a poly(A) tail is added to the 3′ end.

For comparisons of DNA sequences and the amino acid sequences of gene products, it has been shown that the mitochondrial genetic code is not the same universal code introduced in Chapter 9 (see Fig. 9.8). Fig. 17.1 shows

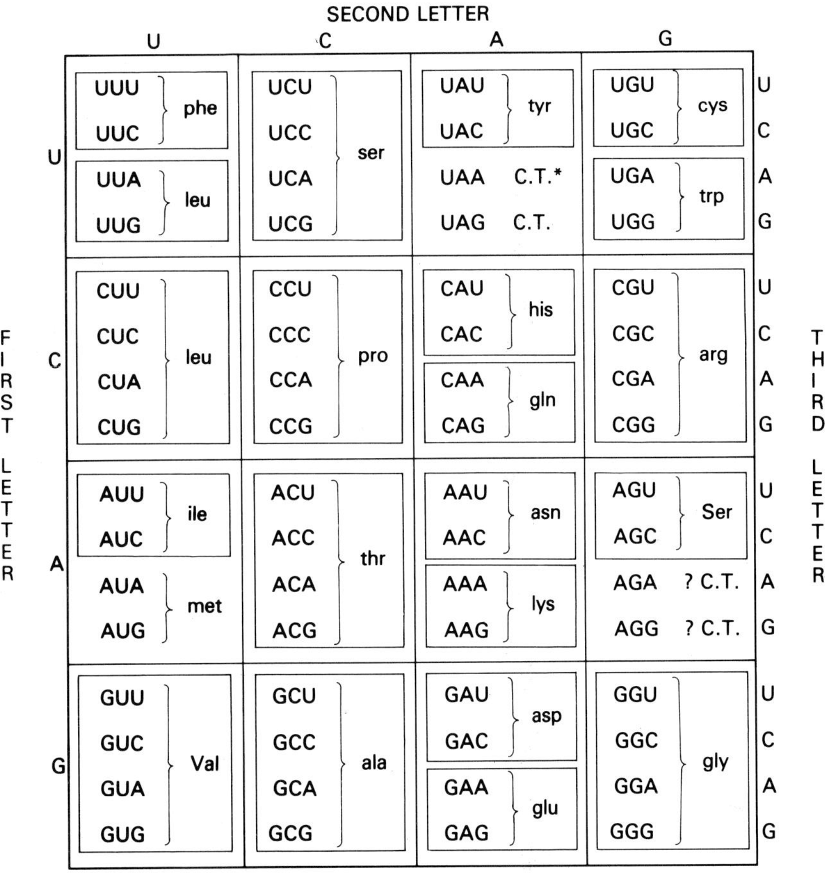

Fig. 17.1. Human mitochondrial genetic code. The coding properties of the known tRNAs are shown (boxed codons). One methionine-tRNA is known but, because there is uncertainty about the number present and their coding properties, the two met codons are not boxed.

the human mitochondrial genetic code. The codons UAA and UAG are chain-terminating codons as in the nuclear code. UGA, however, is not a chain terminating codon: it is read as tryptophan. Transfer RNAs have been identified for all of the other codons except AGA and AGG, so these may have a chain-terminating function. Unlike the nuclear code, methionine is coded for by AUA as well as by AUG. AUG is used as the initiation codon for protein synthesis and, at least in humans, it is found right at the 5′ end of the mRNA. Therefore the protein synthesis machinery must be quite different from that in the cytoplasm. Interestingly the yeast mitochondrial genetic code is slightly different from that of human mitochondria, so it is not possible to generalize about this organelle's code.

The human mitochondrial tRNAs also have some interesting features. While all the tRNAs can fold into a cloverleaf structure, a number of features considered invariable in cytoplasmic tRNAs, vary in mitochondrial tRNAs. For example, some tRNAs lack one of the loops. Considering the nuclear genetic code, with base-pairing wobble, a minimum of 32 tRNAs are needed to read all the 61 sense codons, and no tRNA can read more than three related codons. However, in mitochondria, a number of tRNAs can read all four members of a "box" (i.e. where the first two letters are the same and the third varies). As a result, only 22 different tRNA genes are needed in mammalian mitochondria.

Mitochondrial protein synthesis

Mitochondria have their own protein synthetic machinery. Mitochondrial ribosomes vary considerably among organisms. For example, in animals they are about the same size as *E. coli* ribosomes in terms of molecular weight and volume. Because of a high protein/RNA content, however, they have a relatively low sedimentation value of 55S. In fungi such as yeast and *Neurospora*, the mitochondrial ribosomes are larger with a sedimentation coefficient of 70 to 75S. Even larger ribosomes are found in human plant mitochondria.

In most cases the ribosomes lack the 5 and 5.8S rRNAs. Thus in most mt ribosomes there are two unequal-sized subunits with one rRNA in each subunit. For example, in animal mitochondria these rRNAs are 12 and 16S, and in *Neurospora* they are 19 and 25S. The number of ribosomal proteins is not as well defined, however. Generally, there are more than the number found in *E. coli* ribosomes. The mitochondrial ribosomal proteins are completely distinct from the ribosomal proteins of cytoplasmic ribosomes. Both classes are encoded by nuclear genes.

Mitochondrial protein synthesis is similar in some ways to bacterial protein synthesis. For example, an fmet-tRNA.fmet functions in initiation, and the initiation factors (IFs) are similar to those from bacteria. Further, many of the antibiotics that inhibit bacterial protein synthesis inhibit mitochondrial protein synthesis. In general, mt ribosomes are not sensitive to those antibiotics and inhibitors that inhibit cytoplasmic ribosome function. For example, mitochondrial ribosomes can be inhibited by chloramphenicol (to which cytoplasmic ribosomes are resistant) but not by cycloheximide (to which ribosomes are sensitive). Such differences in sensitivity have made it possible to determine whether proteins found in mitochondria are specified by nuclear or by mitochondrial genes.

Chloroplasts

Chloroplasts are found only in plant cells and they are the sites of photosynthesis, the transfer of light energy into chemical energy. Within each chloroplast is a series of flattened sacs called *thylakoids* whose membranes contain chlorophyll and the other pigments involved in the photosynthetic process. An average leaf cell of a higher plant contains 40–50 chloroplasts.

Chloroplast genome

The structure of the chloroplast (cp) genome is very similar to that of the mitochondrial genome. The cpDNA is circular, supercoiled, and no histones are associated with it. In many plants cpDNA has a distinct GC content from either nuclear or mtDNA.

The cpDNA is about eight to nine times larger than animal mtDNA, but whether there is a proportionally greater amount of genetic information than mtDNA is not known. Generally few loci have been identified in cpDNA, and only a very few have been correlated directly with any chloroplast phenotype. Similarly, very little is known about cpDNA replication other than it is semiconservative and that chloroplast-specific proteins and enzymes are used. Most of these proteins are encoded by nuclear DNA and transported into the chloroplast. In all cases there are multiple copies of cpDNA per chloroplast, with the molecules distributed among multiple nucleoids. For example, in some leaf cells of higher plants there can be 6000 cpDNA molecules per cell distributed among about 40 chloroplasts, with 4 to 18 nucleoids per chloroplast, and 4 to 8 cpDNA molecules per nucleoid.

Chloroplast genes

From a number of different types of experiments, we know that the chloroplast genome contains genes for the chloroplast rRNAs, for a number of tRNAs, and possibly for some ribosomal proteins. It also codes for three subunits of a multisubunit chloroplast enzyme and perhaps for some of the chloroplast membrane proteins that are needed for electron transport in photosynthesis.

Chloroplast protein synthesis

Protein synthesis in chloroplasts is similar to that of prokaryotes. However, the ribosomes are different from prokaryotic and mitochondrial ribosomes. The chloroplast ribosomes have a sedimentation coefficient of 70S and are composed of 50 and 30S subunits. The large subunit contains a 23S rRNA, as well as two small rRNAs with sedimentation values of 5 and 4.5S. The small subunit contains a 16S rRNA molecule. All of the rRNAs are specified by chloroplast genes. In general, there are 30 to 38 ribosomal proteins in the large subunit and 20 to 25 proteins in the small subunit. Most of these are specified by nuclear genes.

Protein synthesis in chloroplasts is similar to that directed by bacterial ribosomes. Formylmethionyl-tRNA is used to initiate all proteins, and the IFs (initiation factors) and EFs (elongation factors) are bacterial-like and distinct from their cytoplasmic counterparts. The sensitivity and resistance to antibiotics and inhibitors shown by chloroplast ribosomes parallel that of mitochondrial ribosomes.

Characteristics of extrachromosomal inheritance

Having described the genetic material found in mitochondria and chloroplasts, we now turn to a discussion of the inheritance patterns found for genes located outside of the nucleus in the cellular organelles. Such genes are called extranuclear genes and the inheritance of these genes is called extranuclear or extrachromosomal inheritance. Certain predictions can be made for a trait that shows extrachromosomal inheritance and they are:

1. Differences should be observed in the progeny of reciprocal crosses. A characteristic form of this is **maternal inheritance** where progeny show the phenotypes of the female parent. This is the case since in many organisms the female gamete provides vastly more cytoplasm to the zygote than does the male gamete and thus there is transmission through the cytoplasm. As we have shown before (with the exception of sex-linked genes), the results of reciprocal crosses are identical for nuclear gene mutations.

2. A second prediction is that the extrachromosomal mutation should be *nonmappable*. That is, in organisms that have well-mapped linkage groups, it should not be possible to find linkage of the mutation to any of the chromosomal genes.

3. A third prediction is that the presumed extrachromosomally coded characteristics should persist when the nuclei in the cells are substituted with nuclei of a different genetic constitution.

Therefore, for a trait to show extrachromosomal inheritance, the gene or genes involved must be nonnuclear for the above criteria to be met. Thus extrachromosomal inheritance should be distinguished from the phenomenon of *maternal effect* where the phenotype of the offspring is determined by the mother's nuclear genotype. The determination of coiling direction of the shell in the snail, *Limnaea peregra*, provides an illustration of maternal effect.

In the snail the direction of coiling is determined by a single pair of alleles. *D* for dextral coiling (to the right) and *d* for sinistral coiling (to the left). Genetic crosses have shown that *D* is dominant to *d* and that the direction of coiling is always determined by the genotype of the mother (Fig. 17.2). Specifically, reciprocal crosses can be performed between a homozygous *DD*, dextral-coiling snail and a homozygous *dd*, sinistral-coiling snail. When the dextral-coiling snail is the source of the egg (Fig. 17.2a) the F1 is *Dd* in

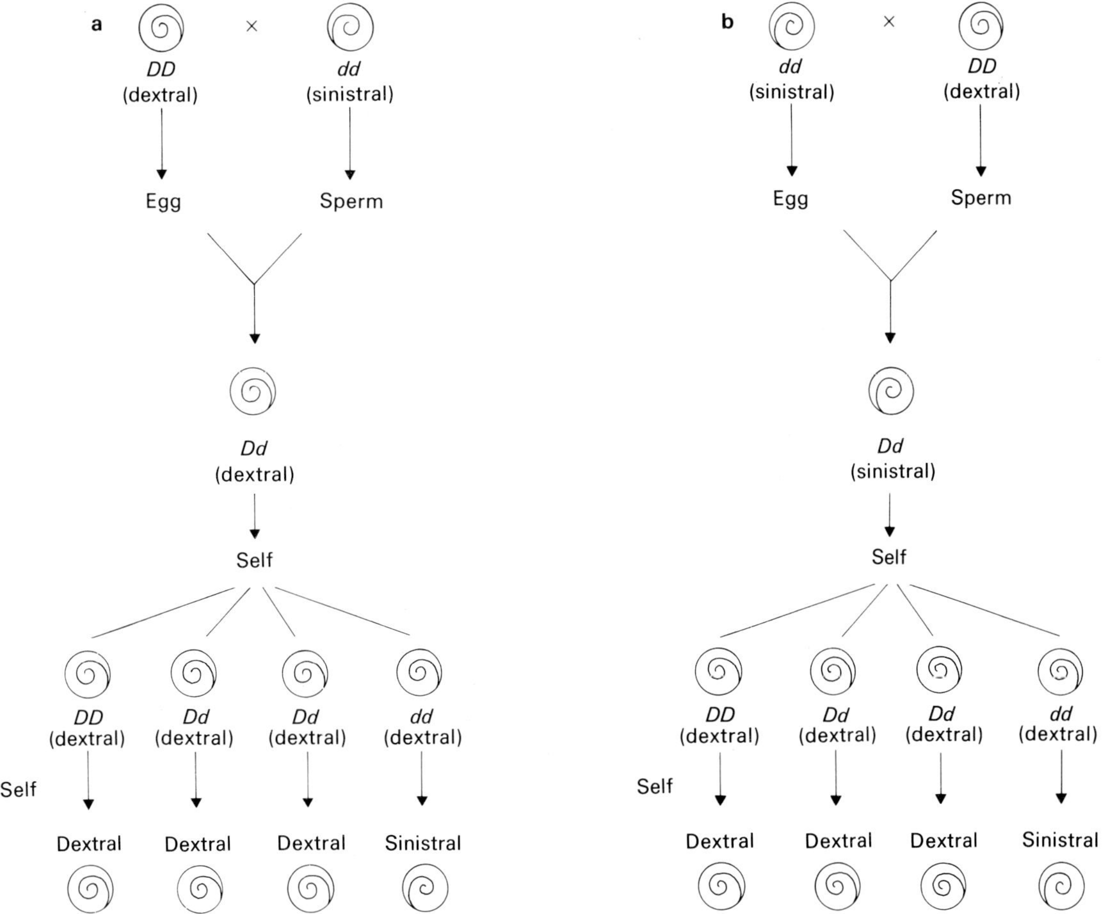

Fig. 17.2. An example of maternal effect: the direction of shell coiling in the snail *Limnaea peregra*. (**a**) Inheritance pattern when two homozygous individuals are crosses where a dextral shell snail is used as the female and successive generations are selfed. (**b**) Inheritance pattern for the reciprocal cross of **a**.

genotype and the snail has a dextral-coiling phenotype. Selfing of these snails produce a 1:2:1 ratio of *DD*:*Dd*:*dd* snail genotypes but *all* of the snails, even the *dd*, are dextral-coiling. When these snails are in turn selfed, the *DD* and *Dd* produce *all* dextral-coiling progeny, wheres the dextral-coiling *dd* snail gives rise to sinistral-coiling progeny. (Clearly this is the evidence that the parent was *dd* here and illustrates nicely the maternal effect concept.) Conversely if the sinistral-coiling *dd* snail is used as the maternal parent (Fig. 17.2b), the F1 is *Dd* as in the reciprocal cross. However, here the snails are sinistral-coiling since the mother was sinistral-coiling. Selfing of these snails results in the same distribution of genotypes as in the reciprocal case (Fig. 17.2a), and because the parent here was *genotypically Dd*, all of the progeny are dextral-coiling snails.

In general, then, maternal effects last only one generation and have as their basis extranuclear components that are encoded by nuclear genes. In the particular example given, the coiling phenotypes are dependent upon the spiraling cleavage patterns of the cells in the first few divisions after eggs are fertilized.

Examples of extrachromosomal inheritance

The following examples are taken from the numerous, well-studied instances of extrachromosomal inheritance in eukaryotes and illustrate the essential criteria of that mechanism of inheritance. As we shall see, unlike maternal effect, the differences in the results of reciprocal crosses do not disappear after one generation but rather occur as long as the extrachromosomal factor is maintained.

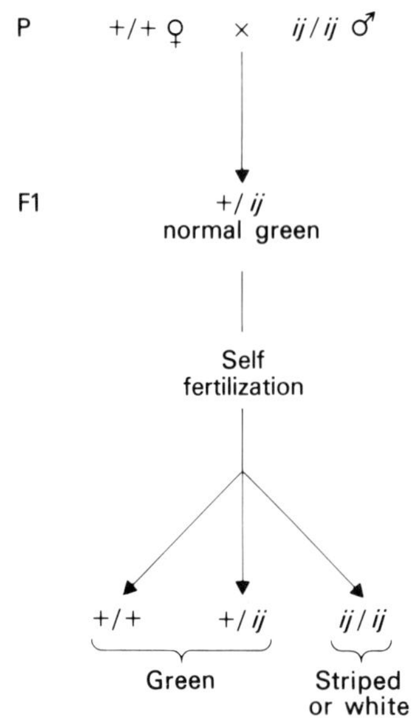

Fig. 17.3. Mendelian inheritance of the *iojap* trait in maize when *ij/ij* is used as the male parent.

The *iojap* trait in maize

In maize, *Zea mays,* there is a leaf-striping trait called "iojap" that is initiated by a nuclear chromosomal mutation but which then shows a non-Mendelian form of inheritance. The name iojap comes from the origin of the strain of maize used, Iowa, and the name of a similar striped variety of maize, 'japonica'. The gene mutation responsible for the iojap trait, *ij*, is recessive, and when it is homozygous the result is white-leaved plants. The genetic properties of the *ij* mutation will now be discussed.

In 1924, M. Jenkins pollinated a homozygous wild type plant with pollen from a homozygous *ij/ij* plant (Fig. 17.3). The heterozygous F1 plants all had normal green leaves. When these F1 plants were self-fertilized, the F2 plants consisted of 2498 green and 782 *iojap* (striped) plants. Twelve of the latter *iojap* plants had white levels. These results conform to the 3 : 1 ratio expected for Mendelian segregation. Further, the testcrosses of representative green F2 plants confirmed the expected 1:2 ratio of homozygotes to heterozygotes. Thus Jenkins concluded that the *iojap* trait shows Mendelian inheritance when it is used as the male parent. However, when he performed the reciprocal cross in which the *ij/ij* was the female parent, the F1 plants often had striped, typical *iojap* leaves in spite of their being +/*ij* in genotype where the + allele should be dominant. This phenomenon was examined in detail by M. Rhoades in 1943 with the results shown in Fig. 17.4. When he crossed female *ij/ij* plants with male +/+ plants, he found that the F1 plants showed a lot of variation depending on the experiment. They were either all green, all white, or some were green with the others striped (*iojap*) or white. To explain the results, Rhoades proposed that when the nuclear gene *ij/ij* is homozygous, the striping phenotype is initiated, but then the trait is inherited through the

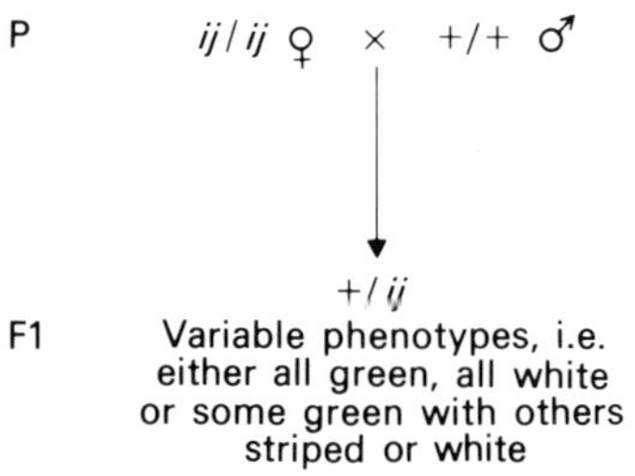

Fig. 17.4. Non-Mendelian inheritance of the maize *iojap* trait when *ij/ij* is used as the female parent. The F1s in this case show varying phenotypes.

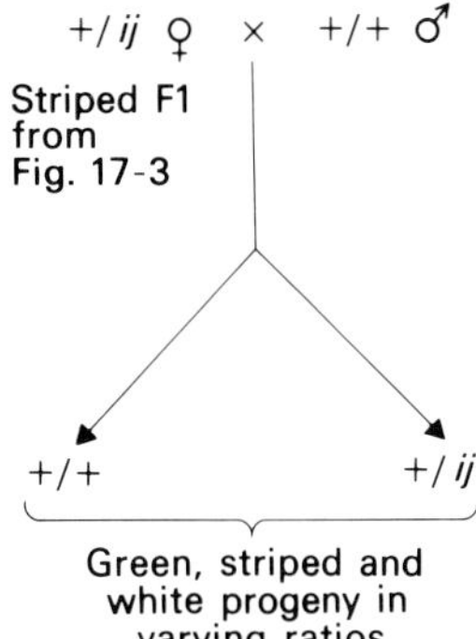

Fig. 17.5. Evidence for maternal inheritance of the *iojap* trait in maize. Here a striped F1 plant from the cross shown in Fig. 17.3 is pollinated with pollen from an unrelated +/+ plant. The resulting progeny show phenotypes like those of the female parent.

cytoplasm and thus only through the egg cell. In other words, *iojap* shows maternal inheritance. This hypothesis was confirmed by using an F1 striped plant as the female parent in a cross with an unrelated homozygous +/+ male (Fig. 17.5). In terms of gene segregation, half of the progeny were +/+ and half were +/*ij*. Phenotypically, however, the resulting plants consisted of green, striped, and white individuals in varying ratios. Some individuals were entirely green and others were entirely white. Clearly, the phenotypes observed did not follow classical Mendelian segregation patterns.

The simplest interpretation of *iojap* is that the phenotypes observed are related to the chloroplasts, which are under some nuclear control but are autonomously self-replicating structures. The striped and white plants, then, would be expected to have defective chloroplasts in the white or yellow areas, and indeed colorless, abnormal chloroplasts can be found there. However, the exact relationship between the chloroplast structure and the *iojap* mutation is not known.

Respiratory deficiencies of fungi

The poky mutant of Neurospora. In *Neurospora*, aerobic respiration is essential for the organism to survive. This aerobic respiration is a property of the mitochondria. Some slow-growing strains of this fungus that have defects in respiration have been characterized. For example, the slow-growing *poky* mutant lacks cytochromes $a+a_3$ and b and also has an excess of cytochrome c (Fig. 17.6). The cytochrome defect is manifested in all slow-growing progeny of a cross between *poky* and wild type, thus attesting to the genetic continuity of the mitochondria. In the following genetic evidence will be presented that shows that *poky* is the result of a mutation in the mitochondrial genome; it shows extrachromosomal inheritance.

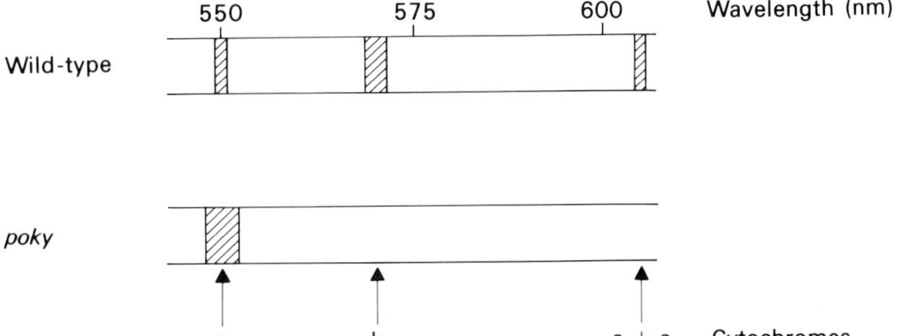

Fig. 17.6. Cytochrome spectrum of wild type and *poky* strains of *Neurospora* showing the cytochrome aberrancies characteristic of the latter.

When we discussed the life cycle of *Neurospora crassa* in an earlier chapter, we showed that the sexual cycle is initiated under nitrogen starvation conditions when there is fusion of nuclei from *A* and *a* strains. One added fact needs to be mentioned. If we inoculate the starvation medium with only one strain, it will produce a prefruiting structure called a *protoperithecium*. At this point, addition of conidia or mycelial fragments of a strain of the opposite mating type will lead to nuclear fusion and the formation of an *A/a* diploid zygote. The protoperithecial parent (female parent), then, is analogous to the female gamete in that it contributes the bulk of the cytoplasm to the cross. The conidial parent is equivalent to the male gamete. This makes it possible to make reciprocal crosses in *Neurospora* to examine the contribution of the cytoplasm to the trait under study.

Table 17.1. Pattern of extrachromosomal inheritance of the slow-growing *poky* mutant of *Neurospora* as shown by reciprocal crosses with the wild type.

Female (protoperithecial) parent		Male (conidial) parent	Segregation pattern of progeny tetrads
+	×	[poky]	8+:0[*poky*]
[poky]	×	+	0+:8[*poky*]

The results of reciprocal crosses between wild type and *poky* are shown in Table 17.1. In these crosses the mating-type genes and any other nuclear marker genes introduced into the cross, segregate 4:4 as is the characteristic of nuclear genes. By contrast the transmission of *poky* follows the cytoplasmic line in that the progeny *all* have growth rate phenotypes like that of the female parent in the cross. This is a clear example of extrachromosomal inheritance and, owing to the cytochrome deficiencies in poky, the simplest hypothesis is that the mutant strain carries a mutation in the mitochondrial genome. Support for this hypothesis has come from microinjection experiments in which a filament of a wild type strain was converted to a slow-growing state with mitochondria purified from the *poky* strain. More recently it has been shown that *poky* mitochondria have ribosomal subunits, which are present in disproportionate amounts compared with the wild type because of a virtual lack of the small ribosomal subunit. This leads to a slower rate of mitochondrial protein synthesis compared with the wild type. Since, as we have discussed, cytochromes or parts of cytochromes are made on the mitochondrial ribosomes, the abnormal ribosomal subunit complement in *poky* is presumably responsible for the cytochrome deficiencies and thus the slow-growth phenotype. There is some evidence that the abnormal ribosome phenotype of *poky* is the result of an altered or deficient ribosomal protein that is coded by the mitochondrial genome. (Most of the proteins of mitochondrial ribosomes are coded by the nuclear genome.)

Petite strains of yeast. When cells of the yeast, *Saccharomyces cerevisiae,* are plated onto a glucose-containing medium, a small percentage (up to 10%) of the colonies are *petite* (small) in that they are one-third to one-half the diameter of the other (normal) colonies. When cells from a normal-sized colony are then plated, up to 1% of those colonies will be *petite.* By contrast, *petite* cells breed true in that cells of *petite* colonies give rise to *petite* colonies only. The cells that make up the *petite* colonies are the same size as the normal cells. The reason for the small colony size is that the colonies developed from a *petite* mutant cell that lacked cytochromes b, c, $a+a_3$, and cytochrome oxidase—all of which are enzymes of the inner mitochondrial membrane. Unlike wild type, the *petite* mutants, then, cannot carry out oxidative phosphorylation for energy production (a mitochondrial function), and hence their growth rate and cell division rate are lower since their energy comes principally from glycolysis. Normal cells can be induced to grow at the same low rate if oxygen is removed from the environment.

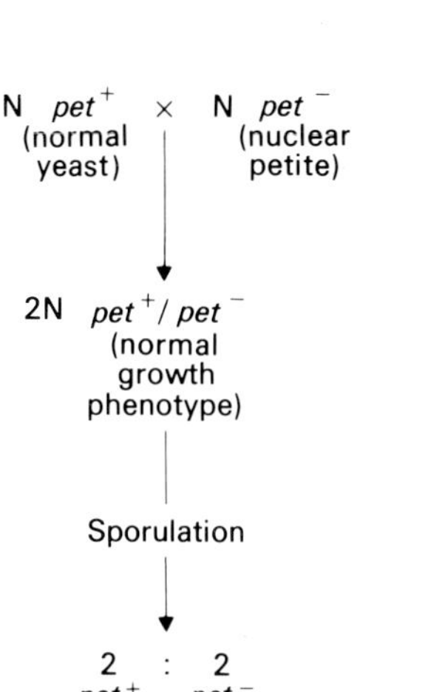

Fig. 17.7. Mendelian inheritance pattern for a nuclear *petite* mutant of yeast.

The *petite* strains can mate with normal strains, and analysis of the phenotypes of the progeny of such matings allows us to distinguish three types of *petite* strains, two of which show the characteristics of extrachromosomal inheritance.

1. Nuclear *petites.* The cytochrome deficiency (and hence slow-growth phenotypes) of these strains is the result of a nuclear gene mutation, *pet*$^-$. As illustrated in Fig. 17.7, when a *pet*$^+$/*pet*$^-$ diploid is formed and sporulated, the diploid is normal and the resulting four ascospores show a 2:2 segregation of normal (*pet*$^+$) to slow-growing (*pet*$^-$) colonies. This is indicative of typical Mendelian inheritance of a nuclear gene.

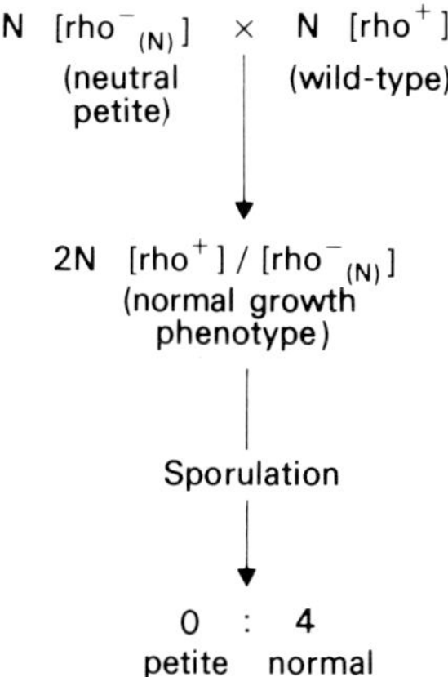

Fig. 17.8. Extrachromosomal inheritance pattern for a neutral *petite* mutant of yeast.

2. Neutral extrachromosomal *petites.* This class of *petites* shows extrachromosomal (non-Mendelian) inheritance for the slow-growing and cytochrome-deficiency phenotypes. The extrachromosomal factor in this case is called [$rho^-_{(N)}$] where the normal (wild type) factor is called [rho^+]. Fig. 17.8 shows the segregation properties of a neutral petite mutation. When a diploid is formed with a normal strain, the diploid grows normally as was the case with the nuclear petites. Sporulation of the [rho^+]/[$rho^-_{(N)}$] diploid results in four ascospores, all of which germinate to produce normally growing cells that have no cytochrome deficiencies. The petite phenotypes do not appear in future generations either. Nuclear gene markers segregating in the same cross all segregate 2:2. Thus the neutral petites show extrachromosomal inheritance. Since the two yeast cells that fuse to produce the diploid are the same size, it is not possible here to determine if maternal inheritance would occur with this type of *petites.*

At the molecular level, the mitochondria of the neutral petites have no DNA, and therefore they are unable to produce the cytochromes coded by the mitochondrial genome. As a consequence, aerobic respiration is nonexistent and the neutral *petite* strains grow slowly.

3. Suppressive extrachromosomal *petites*. Like the neutral *petites*, the suppressive *petites* show extrachromosomal inheritance for the slow-growth and respiration-deficient phenotypes. The extrachromosomal factor in this case is called [$rho^{-}_{(S)}$]. As their name indicates, however, they can suppress normal aerobic respiration when in the presence of normal cytoplasm (and thus normal mitochondria). The properties of suppressive *petites* are shown in Fig. 17.9. When a [rho^{+}]/[$rho^{-}_{(S)}$] diploid is formed, it has respiratory properties intermediate between those of normal and *petite* strains. In the presentation of the yeast life cycle in Chapter 15, it was pointed out that diploid formation and meiosis (sporulation) can be separated temporally. The [rho^{+}]/[$rho^{-}_{(S)}$] diploid can be sporulated immediately as it is formed, and the result of this is that the ascospores show 4:0 segregation of *petite* to normal phenotypes while nuclear marker genes segregate 2:2. In other words, the normal cytoplasm has been suppressed by the *petite* cytoplasm.

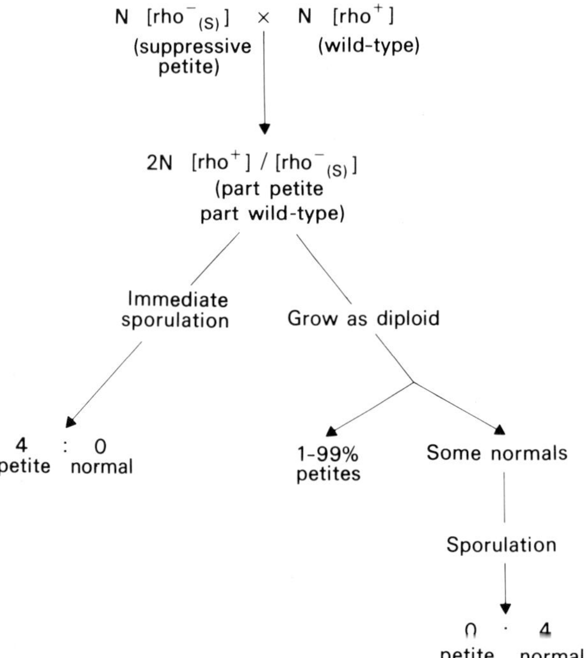

Fig. 17.9. Extrachromosomal inheritance pattern for a suppressive *petite* mutant of yeast.

If, on the other hand, the [rho^{+}]/[$rho^{-}_{(S)}$] diploid is grown vegetatively for a number of generations, they will produce diploid progeny in which the proportion of *petites* is between 1 and 99% depending upon the particular strain. Sporulation of the normal diploid cells produced by vegetative reproduction gives 4:0 normal to *petite* phenotypes.

At the molecular level, the suppressive *petites* have altered mitochondrial DNA. The more suppressive the *petite* is, the more mitochondrial DNA is altered in terms of its buoyant density. Thus the simplest explanation is that the suppressive *petites* show extrachromosomal inheritance owing to a defect in the mitochondrial genome.

Further evidence that the nonnuclear *petites* are the result of mutations residing in the mitochondria has come from work with two chemicals. As was pointed out earlier the spontaneous rate of formation of extrachromosomal petites is approximately 1%. If replicating wild type cells are treated with 10^{-6}M acriflavin, almost 100% of the cells are converted to the *petite* state. A similar result is obtained for nongrowing or growing cells if they are treated with ethidium bromide. The site of action of both of these chemicals has been shown to be the mitochondrial DNA, and the resulting *petites* have mitochondrial DNA that has altered buoyant density or no DNA depending on the type of petite induced.

In summary, there are distinctive properties of traits that show extrachromosomal inheritance and a few examples of mutants that show non-Mendelian inheritance have been discussed. Numerous examples of extrachromosomal inheritance have been found in many other organisms. Studies of these characters are leading to an understanding of the structure and function of the mitochondrial and chloroplast genomes.

Questions and problems

17.1 Compare and contrast the structure of the nuclear genome, the mitochondrial genome, and the chloroplast genome.

17.2 How do mitochondria reproduce?

17.3 Discuss the differences between the universal genetic code of the nuclear genes and the code found in mammalian and fungal mitochondria. Is there any advantage to the mitochondrial code?

17.4 When the DNA sequences for most of the mRNAs in human mitochondria are examined, no nonsense codons are found at their termini. Instead, either U or UA is found. Explain this.

17.5 Compare and contrast the cytoplasmic and mitochondrial protein-synthesizing systems.

17.6 What features of extranuclear inheritance distinguish it from the inheritance of nuclear genes?

17.7 Distinguish between maternal effect and extranuclear inheritance.

17.8 Distinguish between nuclear (segregational), neutral, and suppressive petite mutants of yeast.

17.9 The snail *Limnaea peregra* has been studied extensively with respect to the inheritance of the direction of shell coiling. A snail produced by a cross between two individuals has a shell with a right-hand twist (dextral coiling). This snail produces only left-hand (sinistral) progeny upon selfing. What are the genotypes of the F1 snail and its parents?

17.10 *Drosophila melanogaster* has a sex-linked, recessive, mutant gene called *maroon-like* (*ma-l*). Homozygous *ma-l* females or hemizygous *ma-l* males have light colored eyes because of the absence of an active enzyme, xanthine dehydrogenase, that is involved in the synthesis of eye pigments.

When heterozygous $ma\text{-}l^+/ma\text{-}l$ females are crossed to *ma-l* males, all of the offspring are phenotypically wild type. However, one-half of the female offspring from this cross when crossed back to *ma-l* males give all *ma-l* progeny. The other half of the females, when crossed to *ma-l* males, give all phenotypically wild type progeny. What is the explanation for these results?

17.11 When females of a particular mutant strain of *Drosophila melanogaster* are crossed to wild type males, all of the viable progeny flies are females. Hypothetically this result could be the consequence of either a sex-linked lethal mutation or a maternally inherited factor that is lethal to males. What crosses would you perform to distinguish between these alternatives?

17.12 Reciprocal crosses between two *Drosophila* species, *D. melanogaster* and *D. simulans*, produce the following results:

melanogaster female × *simulans* male → only females
simulans female × *melanogaster* male → males with few or no females.

Explain these results.

17.13 Some *Drosophila* flies are very sensitive to carbon dioxide; they become anesthetized when it is administered to them. The sensitive flies have a cytoplasmic particle called sigma, which has many properties of a virus. Resistant flies lack sigma. The sensitivity to carbon dioxide shows strictly maternal inheritance. What would be the outcome of the following two crosses?

(a) sensitive female × resistant male
(b) sensitive male × resistant female.

17.14 In yeast, a haploid nuclear (segregational) *petite* is crossed with neutral *petite*. Assuming that both strains have no other abnormal phenotypes, what proportion of the progeny ascospores are expected to be *petite* in phenotype if the diploid zygote undergoes meiosis?

17.15 When grown on a medium containing acriflavin, a yeast culture produces a large number of very small (*tiny*) cells that grow very slowly. How would you determine whether the slow growth phenotype was the result of a cytoplasmic factor or a nuclear gene?

17.16 In *Neurospora,* a chromosomal gene *F* suppresses the slow-growth characteristic of the *poky* phenotype and makes a *poky* culture into a *fast-poky* culture, which still has abnormal cytochromes. Gene *F* in combination with normal cytoplasm has no detectable effect. (Hint: since both nuclear and extranuclear genes have to be considered, it will be convenient to use symbols to distinguish the two. Thus, cytoplasmic genes will be designated in square brackets, e.g. [*N*] for normal cytoplasm, [*po*] for *poky*.)

(a) A cross in which *fast-poky* is used as the female (protoperithecial) parent and a normal wild type strain is used as male parent gives half *poky* and half *fast-poky* progeny ascospores. What is the genetic interpretation of these results?

(b) What would be the result of the reciprocal cross of that described in the previous question, i.e. normal female × *fast-poky* male?

17.17 Reciprocal crosses between two types of the evening primrose, *Oenothera hookeri* and *Oenothera muricata*, produce different effects on the plastids:

O. hookeri female × *O. muricata* male → yellow plastids
O. muricata female × *O. hookeri* male → green plastids

Explain the difference between these results, noting that the chromosome constitution is the same in both types.

17.18 A form of male sterility in maize is maternally inherited. Plants of a male-sterile line crossed with normal pollen give male-sterile plants. Some lines of maize carry a dominant so-called restorer (*Rf*) gene, which restores pollen fertility in male-sterile lines.

(a) If a male-sterile plant is crossed with pollen from a plant homozygous for gene *Rf*, what will be the genotype and phenotype of the F1?

(b) If the F1 plants of (a) are used as females in a testcross with pollen from a normal plant (*rf*/*rf*), what would be the result? Give genotypes and phenotypes, and designate the type of cytoplasm.

References

Aloni, Y. and G. Attardi, 1971. Expression of the mitochondrial genome in HeLa cells. III. Evidence for complete transcription of mitochondrial DNA. *J. Mol. Biol.* **55**:251–270.

Ashwell, M. and T.S. Work, 1970. The biogenesis of mitochondria. *Annu. Rev. Biochem.* **39**:251–290.

Attardi, B. and G. Attardi, 1971. Expression of the mitochondrial genome in HeLa cells. I. Properties of the discrete RNA components from the mitochondrial fraction. *J. Mol. Biol.* **55**:231–249.

Beale, G.H., A. Jurand and J.R. Preer, 1969. The classes of endosymbionts of *Paramecium aurelia*. *J. Cell Sci.* **5**:69–91.

Beisson, J., A. Sainsard, A. Adoutte, G.B. Beale, J. Knowles and A. Tait, 1974. Genetic control of mitochondria in *Paramecium*. *Genetics* **78**:403–413.

Boardman, N.K., A.W. Linnane and R.M. Smillie (eds.), 1971. *Autonomy and Biogenesis of Mitochondria and Chloroplasts.* North-Holland, Amsterdam.

Borst, P., 1972. Mitochondrial nucleic acids. *Annu. Rev. Biochem.* **41**:333–376.

Clayton, D.A., 1982. Replication of animal mitochondrial DNA. *Cell* **28**:693–705.

Ephrussi, B., 1953. *Nucleo-cytoplasmic Relations in Microorganisms.* Oxford University Press, New York.

Galper, J.B. and J.E. Darnell, 1971. Mitochondrial protein synthesis in HeLa cells. *J. Mol. Biol.* **57**:363–367.

Gillham, N.W., 1974. Genetic control of the chloroplast and mitochondrial genomes. *Annu. Rev. Genetics* **8**:347–392.

Gillham, N.W., 1978. *Organelle Heredity.* Raven Press, New York.

Gillham, N.W., J.E. Boynton and R.W. Lee, 1974. Segregation and recombination of non-Mendelian genes in *Chlamydomonas*. *Genetics* **78**:439–457.

Jinks, J.L., 1964. *Extrachromosomal Inheritance.* Prentice-Hall, Englewood Cliffs, New Jersey.

Kirk, J.T.O., 1971. Chloroplast structure and biogenesis. *Annu. Rev. Biochem.* **40**:161–196.

Kuroiwa, T., T. Suzuki, K. Ogawa and S. Kawano, 1981. The chloroplast nuclease: distribution, number, size, and shape, and a model for the multiplication of the chloroplast genome during chloroplast development. *Plant Cell Physiol.* **22**:381–396.

Lambowitz, A.M. and D.J.L. Luck, 1976. Studies on the *poky* mutant of *Neurospora crassa*. *J. Biol. Chem.* **251**:3081–3095.

Lambowitz, A.M., N.H. Chua and D.J.L. Luck, 1976. Mitochondrial ribosome assembly in *Neurospora*. Preparation of mitochondrial ribosomal precursor particles, site of synthesis of mitochondrial ribosomal proteins and studies on the *poky* mutant. *J. Mol. Biol.* **107**:223–253.

Levings, C.S., 1983. The plant mitochondrial genome and its mutants. *Cell* **32**:659–661.

Perlman, P.S. and C.W. Birky, 1974. Mitochondrial genetics in Baker's yeast: a molecular mechanism for recombinational polarity and suppressiveness. *Proc. Natl. Acad. Sci. USA* **71**:4612–4616.

Preer, J.R., 1971. Extrachromosomal inheritance: hereditary symbionts, mitochondria, chloroplasts. *Annu. Rev. Genetics* **5**:361–496.

Preer, J.R., L.B. Preer and A. Jurand, 1974. Kappa and other endosymbionts in *Paramecium. Bacteriol. Rev.* **38**:113–163.

Rifkin, M.R. and D.J.L. Luck, 1971. Defective production of mitochondrial ribosomes in the *poky* mutant of *Neurospora crassa*. *Proc. Natl. Acad. Sci. USA* **68**: 257–290.

Saccone, C. and A.M. Kroon (eds.), 1976. *The Genetic Functions of Mitochondrial DNA*. North-Holland, Amsterdam.

Sager, R., 1972. *Cytoplasmic Genes and Organelles*. Academic Press, New York.

Chapter 18 Biochemical Genetics (Gene Function)

Up to this point in the book, genes have been discussed as individual entities. We have described their structure in terms of nucleotide sequence, their transcription to an RNA copy (where the gene is DNA), and the translation of the RNA into the amino acid sequence of a polypeptide chain where the gene product is a protein. In the cell the polypeptide chains have either structural or enzymatic roles. However, the cell is very complex and its function depends upon the integration of the activities of many genes. The remainder of this book will consider how genes interact within the organism. In this chapter the historical accumulation of evidence for the relationship between genes and enzymes and the involvement of enzymes in biochemical pathways will be described. Subsequent chapters will discuss the regulation of expression of genes that have related functions in prokaryotes and in eukaryotes and the role of gene interactions as they pertain to population genetics.

Genetic control of metabolism in humans

Early evidence for the relationship between genes and enzymes came from the work of the physician Archibald Garrod in 1909, the results of which were published as a book called *Inborn Errors of Metabolism*. Garrod was interested in human diseases that had an apparent genetic basis and among those he studied was *alkaptonuria*. This is a rare disease characterized by a number of symptoms, including a hardening and blackening of the cartilage and a blackening of the urine when it is exposed to the air. The symptoms are the result of the accumulation of high amounts of homogentisic acid, which is not a usual component of the cartilage or urine. Garrod showed that the amount of homogentisic acid excreted was increased when patients with alkaptonuria were fed diets high in phenylalanine or tyrosine. He deduced that homogentisic acid is one of the intermediates in the degradation of those two amino acids, and he surmised that alkaptonuriacs (people with alkaptonuria) probably lacked an enzyme activity that in normal people breaks down homogentisic acid. In concomitant work, pedigree analysis by A. Garrod and W. Bateson showed that alkaptonuria is caused by a recessive mutation. Thus Garrod had obtained the first suggestion of a link between genes and enzymes in that a hereditary disorder lacked an enzyme activity. He argued that this sort of causal link probably explained other "inborn errors of metabolism".

Garrod's hypothesis was proved in 1958 when the biochemical pathways for phenylalanine and tyrosine metabolism were elucidated. Part of these pathways is shown in Fig. 18.1. Each reaction in the pathways is catalyzed by an enzyme, and each enzyme is encoded by a gene. Thus for normal metabolism of the two amino acids, the coordinated effort of a large number of gene products is required. The alkaptonuriacs of Garrod's studies cannot convert homogentisic acid to maleylacetoacetic acid since the enzyme homogentisic oxidase is nonfunctional in these individuals. This particular biochemical pathway also provides examples of other metabolic diseases such as phenylketonuria, albinism, and tyrosinosis. The positions of the enzyme deficiencies in these diseases are shown in Fig. 18.1.

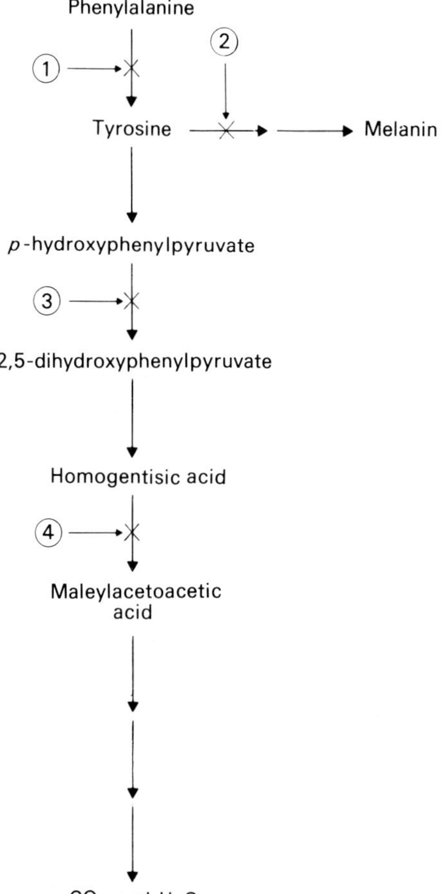

Fig. 18.1. Part of the biochemical pathway for the metabolism of phenylalanine and tyrosine. The sites of metabolic blocks associated with human diseases are indicated by crosses: (1) phenylketonuria; (2) albinism; (3) tyrosinosis; (4) alkaptonuria.

Genetic control of *Drosophila* eye pigments

Garrod's pioneering work provided the first clue to the link between Mendelian factors and proteins. In 1935 G. Beadle and B. Ephrussi obtained more evidence for gene-enzyme relationships in a biochemical pathway in their studies of the synthesis of the eye pigments in the fly, *Drosophila melanogaster*.

There are two pigment types in the *Drosophila* eye, the bright red pterins and the brown ommochromes. As we now know, these two pigments are made in two multistep biochemical pathways, each reaction of which is catalyzed by an enzyme. The bright red and brown pigments that are the end products of these pathways become attached to protein granules and are deposited in the eye cells. The combination of the two pigments results in the dull red eye color characteristic of the wild type. Further, in *Drosophila*, the larval stages contain groups of cells called *discs*, each of which develops into a particular adult structure during metamorphosis. Two discs can be identified as the progenitors of eyes. In their initial experiments, Beadle and Ephrussi showed that an eye disc from a larva can be transplanted into the abdomen of a second larva and the disc will develop into an identifiable eye structure that can be found in the abdomen of the adult following metamorphosis. This paved the way for a series of elegant experiments in which eye discs were transplanted between larvae of different genetic constitutions.

At the time of Beadle and Ephrussi's experiments, three gene loci were known that were implicated in the production of brown pigment: *scarlet, cinnabar,* and *vermilion*. Mutations in any one of these genes resulted in a bright orange eye (the shade of orange differs in each mutant type). Beadle and Ephrussi first addressed the question of whether embryonic eye tissue from the three mutant types *st* (scarlet), *cn* (cinnabar), or *v* (vermilion) transplanted into a wild type larval host would develop a wild type eye color. In the case of the *st* mutant the transplanted disc developed into a scarlet-colored eye (Fig. 18.2a) and is an example of *autonomous development* in which the wild type host was unable to provide substances to the disc to enable the brown pigment to be produced. By contrast, transplanted discs from either *v* or *cn* developed into normal-colored eyes in the wild type host and thus showed *nonautonomous development* (Figs. 18.2b and 18.2c, respectively). The conclusion in these cases was that the wild type provided a diffusible substance that the discs used to bypass the genetic block and to make brown pigment.

Of more importance historically were the results of reciprocal transplant experiments between *v* and *cn* larvae; these (along with the control results) are shown in Table 18.1. The experiments showed that *cn* discs transplanted into *v* hosts produced cinnabar-colored eyes, whereas *v* discs transplanted into *cn* hosts developed into wild type colored eyes. They concluded from these results that the production of the brown pigment involved a biochemical sequence with at least two precursors, the wild type having both, *cn* having one, and *v* having neither; the gene-reaction step relationships are shown in Fig. 18.3. Thus the production of a wild type colored eye when a *v*

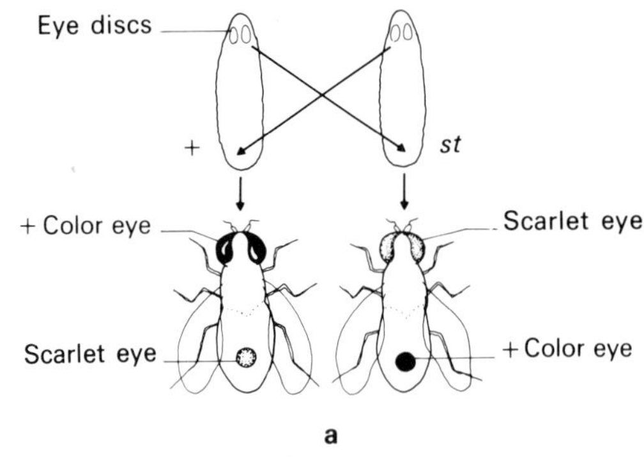

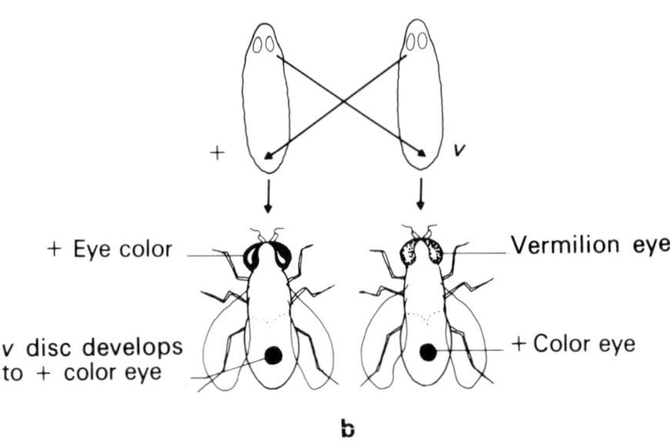

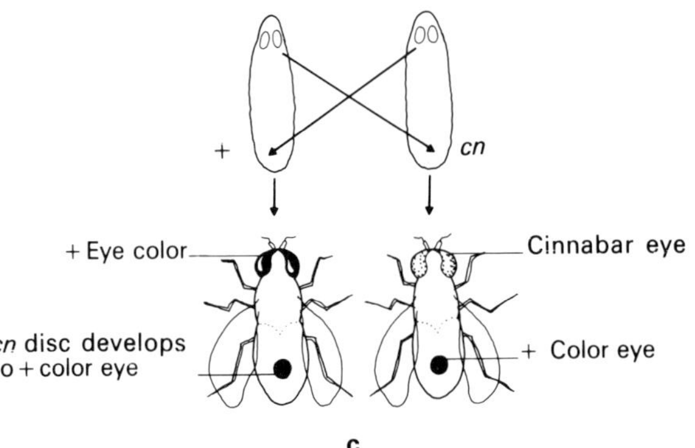

Fig. 18.2. Results of Beadle and Ephrussi's reciprocal transplant experiments with eye discs of *Drosophila*. (**a**) Reciprocal transplants between + and *st* (scarlet). The *st* disc develops into a scarlet-colored eye in the wild type host, indicating autonomous development for *st*. (**b**) Reciprocal transplants between + and *v* (vermilion). The *v* disc develops into a wild type–colored eye in the wild type host, indicating nonautonomous development of the *v* disc. (**c**) Reciprocal transplant between + and *cn* (cinnabar). Like *v*, the *cn* disc shows nonautonomous development in that it produces a wild type–colored eye in the wild type host.

disc is implanted into a *cn* host follows logically. That is, the *v* disc cannot make the v^+ substance (the product of the wild type allele of *v*) whereas the *cn* host can. This v^+ substance diffuses to the *v* disc where it is converted to the cn^+ substance (since genotypically the *v* disc carries the wild type allele of *cn*), and this substance is then converted to the brown pigment. In other words, the *cn* host makes up for the deficiency of the *v* disc by supplying it with a

Table 18.1. Results of eye disc transplantation experiments involving eye color mutants.

Source of eye disc	Host fly	Color of transplanted eye after metamorphosis
+	v	+
v	+	+
+	cn	+
cn	+	+
cn	v	cinnabar
v	cn	+

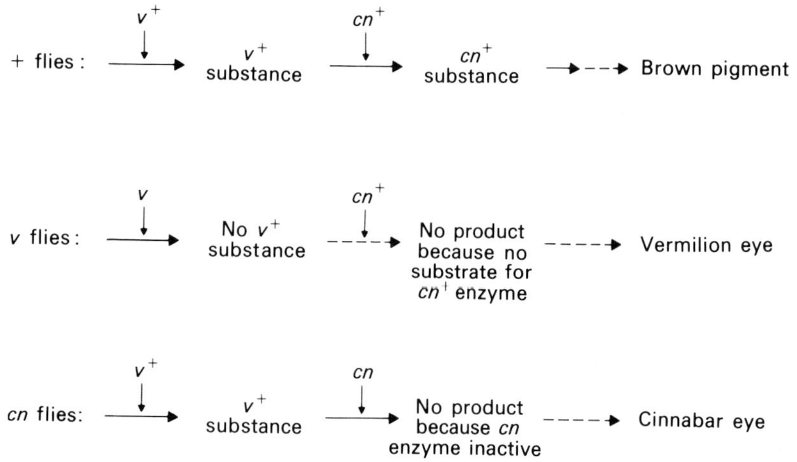

+ flies: —(v^+)→ v^+ substance —(cn^+)→ cn^+ substance →--→ Brown pigment

v flies: —(*v*)→ No v^+ substance --(cn^+)--→ No product because no substrate for cn^+ enzyme ----→ Vermilion eye

cn flies: —(v^+)→ v^+ substance —(*cn*)→ No product because *cn* enzyme inactive ----→ Cinnabar eye

Fig. 18.3. The biochemical sequence and gene-reaction step relationships for wild type, *v*, and *cn* flies as deduced from the results of Beadle and Ephrussi's experiments shown in Table 18.1.

diffusible substance so that it becomes wild type. In the reciprocal experiment, the *cn* disc cannot convert the v^+ substance to the cn^+ substance, and the *v* host is deficient in the production of v^+ substance so no brown pigment can be produced.

At the time of these experiments, the relationship between genes and enzymes was not known. Beadle and Ephrussi's work was important historically since it indicated a strong link between a phenotype (in this case eye color) and genotype (*v* and *cn*). E. Tatum further investigated this system by making extracts of *v* and *cn* flies by homogenization and by injecting the extracts into host larvae. He found that the same results were obtained as for the eye disc transplantation experiments: the injection of a *cn* extract into *v* larvae resulted in the production of wild type–colored eyes. This experiment was the first in a series designed to discover the nature of the v^+ and cn^+ substances. At that time, it was difficult to analyze the extracts he had prepared, and his limited analysis showed that the substances were water soluble and of low molecular weight. He proposed that the substances were

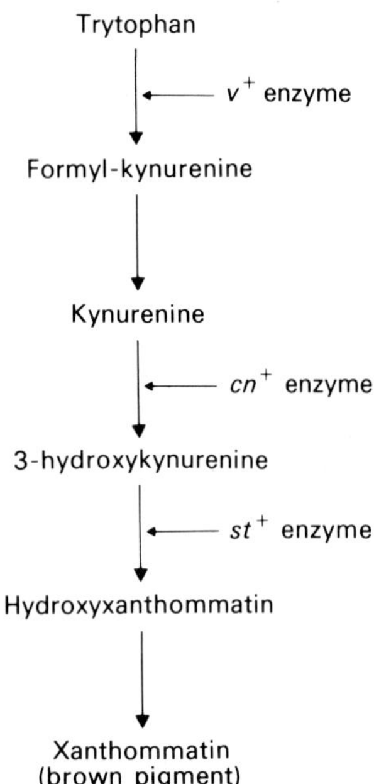

Fig. 18.4. The pathway for the production of brown eye pigment of *Drosophila* from the amino acid tryptophan. The steps catalyzed by the enzymes coded for by the v^+, cn^+, and st^+ genes are indicated on the diagram.

relatives of amino acids and set about trying to supplement the fly food in a defined way to try to bypass the genetic blocks in the brown pigment pathway and therefore to identify the biochemical precursors in the pathway in this way. Several complex media such as peptone resulted in brown pigment formation in adult eyes when fed to *v* and *cn* larvae. One exception was gelatin, which is deficient in tryptophan and tyrosine. Tatum then injected tryptophan into *v* and *cn* larvae and no brown pigment was produced in most experiments. However, in one experiment, brown pigment was produced, and it turned out that this occurred as a result of a bacterial contamination, which resulted in the metabolism of tryptophan into chemicals that are required by *v* and *cn* for the production of brown pigment. With the implication of tryptophan metabolism in the production of brown pigment, it was then a relatively simple matter to work out the biochemical pathway and to localize the steps catalyzed by the v^+ and cn^+ gene products (Fig. 18.4). As can be seen the *v* mutants are unable to convert tryptophan to formylkynurenine and the *cn* mutants cannot convert formylkynurenine to 3-hydroxykynurenine because the enzymes they have for these respective steps are nonfunctional.

The pathway shown in Fig. 18.4 can serve to illustrate a general point about the sequence of reactions and the consequences of genetic mutations. In the *v* mutations, for example, the flies accumulate high amounts of tryptophan since it is not converted to the brown pigment in this pathway. This is analogous to the accumulation of homogentisic acid in alkaptonuriacs. Here it provided further evidence that tryptophan metabolism is involved with eye pigment formation. Indeed, if one feeds wild-type larvae with radioactive tryptophan, the brown ommochrome pigment of the adult eye is found to be radioactive.

Biochemical mutants of *Neurospora*

The work of Beadle, Ephrussi, and Tatum led directly to studies of biochemical mutants of microorganisms. Of historical importance here is the work of G. Beadle and E. Tatum in the early 1940s on the relationships between genes and enzymes in the fungus *Neurospora crassa*. The switch from *Drosophila* to *Neurospora* for studies of gene function was made because the latter organism is relatively simple, it is haploid so dominance relationships do not have to be considered, and it can be handled like other microorganisms. Further, as indicated in the chapter on mutation, wild type *Neurospora* has simple nutritional requirements—a minimal medium consisting of inorganic salts, a carbon source, and the vitamin, biotin. When growing on this medium, *Neurospora* can synthesize all the other necessary components including amino acids, nucleotides, vitamins, etc., and, as described in the mutation chapter, it is relatively simple to isolate auxotrophic (nutritional) mutants that can no longer grow on the minimal medium but require a particular supplement to grow. This is illustrated in Fig. 18.5. The mutants obtained, then, include amino acid auxotrophs, purine auxotrophs, and so on.

Beadle and Tatum made the basic assumption that cells function by the interaction of the products of a large number of genes and that wild type *Neurospora* converted the constituents of minimal medium into the amino acids, nucleotides, etc. by a series of reactions organized into biochemical pathways. In this way, the synthesis of complex biochemical compounds occurs by a series of small steps, each catalyzed by its own enzyme; the product of each step is used as the substrate for the next enzyme (Fig. 18.6).

Beadle and Tatum then reasoned that if the enzymes of a biochemical pathway are specified by genes, it should be possible to obtain mutant strains that have a defect in one of the genes so the corresponding enzyme is no longer produced or is no longer active. In the hypothetical pathway shown in Fig. 18.6, a mutation affecting the production or activity of any one of four enzymes will lead to the common phenotype of a requirement for end product E. Genetic analysis would confirm that the mutations causing a requirement for E fall into four complementation groups and thus represent four distinct polypeptide products (enzymes in our example). How can we order the biochemical reactions by analyzing the gene mutations? We can do this by determining whether the mutant strains can grow if supplemented with any of the postulated intermediates in the pathway. For example, a mutation in gene 4 would result in the inability of the strain to carry out the conversion of D to E. Such a strain would only grow on minimal medium plus end product E. None of the intermediates A, B, C, or D would be a suitable supplement. By contrast a strain carrying a mutation in gene 2 will not be able to convert B or C, and this strain will be able to grow when supplemented with any of the compounds (C, D, or E) in the pathway after the lesion. It will not grow on minimal medium plus A or minimal medium plus B, since the lack of enzyme 2 will prevent the B to C conversion and no end product will result. Thus from the pattern of growth that mutant strains exhibit on presumed intermediates in a biochemical pathway, it is possible to order the sequence of reactions in the production of the end product.

The principles just described are illustrated by the analysis of arginine auxotrophs of *Neurospora* carried out by A. Srb and N. Horowitz in the 1940s. All of the strains grow on minimal medium plus arginine, some could grow on minimal medium plus citrulline, and others could grow on minimal medium plus ornithine. The patterns of growth are shown in Table 18.2. Based on the arguments already presented, a logical interpretation of the data is that the mutant genes represent sequential blocks in a biochemical pathway

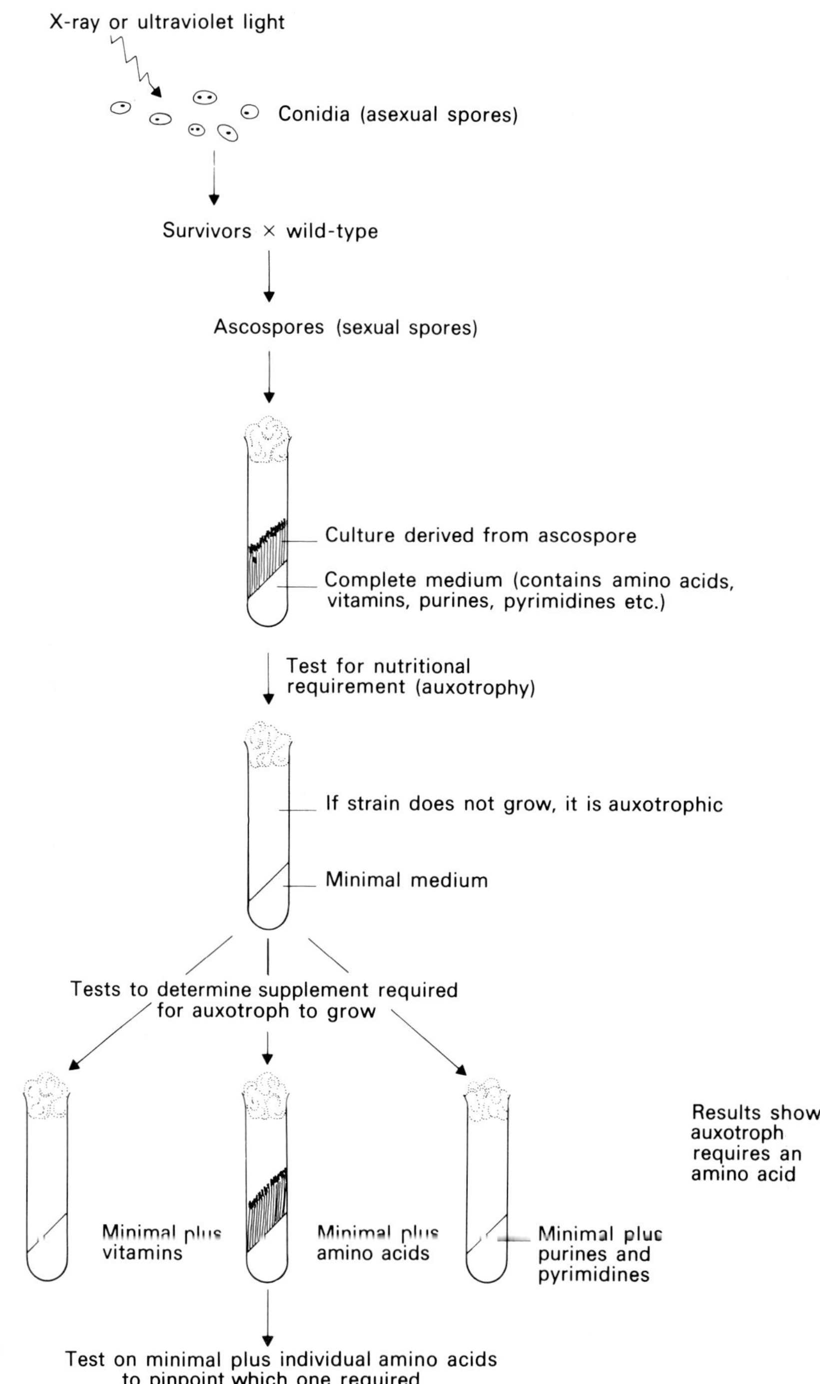

Fig. 18.5. Procedure used by Beadle and Tatum for the induction, isolation, and classification of nutritional mutants of *Neurospora*. The diagram illustrates the detection of an amino acid auxotroph. Similar steps would be used to detect strains auxotrophic for vitamins, purines, pyrimidines, and so on.

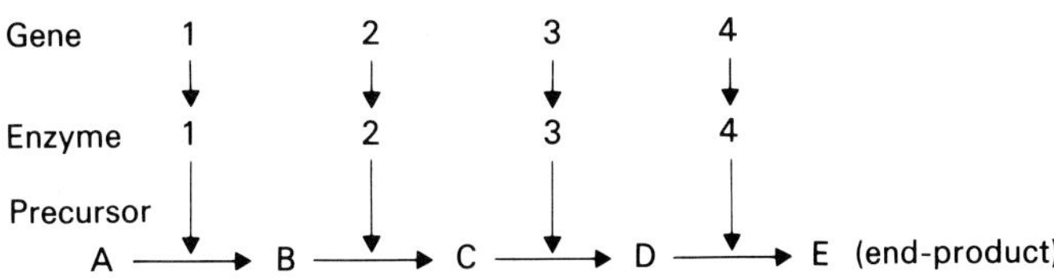

Fig. 18.6. An hypothetical biochemical pathway showing the gene–enzyme relationships.

Table 18.2. Growth responses of arginine auxotrophs of *Neurospora crassa* (After A. Srb and N. H. Horowitz, 1944. *J. Biol. Chem.* **154**:133).

	Growth response on minimal plus:			
Mutant strain	Nothing	Ornithine	Citrulline	Arginine
1	−	+	+	+
2	−	−	+	+
3	−	−	−	+

+ = growth on the medium; − = no growth on the medium.

with the end product arginine (Fig. 18.7). Thus strain (3) is blocked in the citrulline to arginine reaction and will grow if supplemented with arginine but not if supplemented with citrulline or ornithine. Strain (2) is blocked in the ornithine to citrulline reaction and thus can grow on citrulline or arginine. The logic for proposing a pathway, then, is straightforward; the earlier in the pathway a step is blocked, the greater the number of supplements that can be used to allow the strain to grow. The pathway discussed here is, in fact, part of a more complex series of reactions. Arginine, for example, can be broken down by the action of the enzyme arginase to ornithine and urea, hence completing the so-called ornithine cycle.

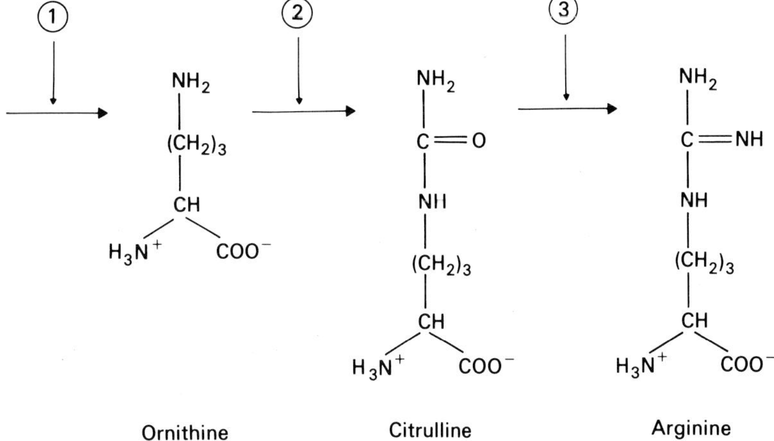

Fig. 18.7. Reaction steps in the arginine biosynthesis pathway as deduced from the growth patterns of arginine auxotrophs of *Neurospora* shown in Table 18.2. The steps blocked by the mutant strains 1, 2, and 3 are indicated by arrows.

From this kind of work, Beadle and Tatum concluded that there is a very definite, direct relationship between genes and enzymes. Thus they put forward the **one gene-one enzyme hypothesis**, which, simply stated, said that each biochemical reaction in the cell was catalyzed by an enzyme that is coded by one gene in the DNA. This concept is of very limited value now since the idea that each gene codes for one enzyme is too simple; genes also are known to code for single polypeptide chains that are parts of complex enzymes, for antibodies, for structural proteins, and for various types of nontranslated RNA. An updated version of Beadle and Tatum's hypothesis for DNA sequences that produce translatable RNA is the *one cistron-one polypeptide hypothesis* where a cistron (defined by the cis-trans test) codes for a single polypeptide chain. Enzymes with multiple, heterogeneous polypeptides, then, would be coded for by multiple cistrons. Nonetheless it is clear that Beadle and Tatum's work laid the foundations in the field of biochemical genetics.

Gene control of protein structure

So far we have learned that there is a direct relationship between genes and enzymes. Since an enzyme catalyzes one reaction in a biochemical pathway, a mutation in the gene for the enzyme will cause a block in the pathway. The consequences of this to the organism depend on the nature of the pathway, the effects of the compound accumulated prior to the block, and the effects of a deficiency of the end product.

However, while all enzymes are proteins, not all proteins are enzymes. Nevertheless, the functions of proteins are also affected by changes in amino acid sequence and tertiary structure. For example, the human disease sickle cell anemia is characterized by a sickling of the red blood cells. This leads to problems with the red blood cells passing through the capillaries, thereby causing tissue and organ damage, anemia, and possibly death. The life of the red cells is also greatly shortened. Sickle cell anemia is caused by homozygosity for a mutation in the gene for the β-polypeptide of hemoglobin. (Hemoglobin consists of two copies each of α- and β-polypeptides.) The mutation results in the substitution of the neutral amino acid valine for the normal acidic amino acid glutamic acid at the sixth position from the N-terminal end. This change is at an important part of the molecule, and the presence of a neutral amino acid instead of an acidic amino acid makes the region hydrophobic (water-hating) instead of hydrophilic (water-loving). As a result, the β-polypeptide folds to place the altered region away from the aqueous environment, and this causes an unusual stacking of hemoglobin molecules and sickling of the red blood cells.

Colinearity

A more precise understanding of the relationships between a gene (cistron) and the sequence of amino acids in the polypeptide for which it codes came from the work of C. Yanofsky and his group in 1967. Their studies centered on the enzyme tryptophan synthetase from *E. coli*. This enzyme consists of two copies each of two distinct polypeptides A and B, which are coded for by two adjacent genes. Here is an example of where two genes code for one enzyme—a clear exception to Beadle and Tatum's hypothesis. The reactions that the enzyme carries out are in the biosynthetic pathway for the amino acid tryptophan (Fig. 18.8).

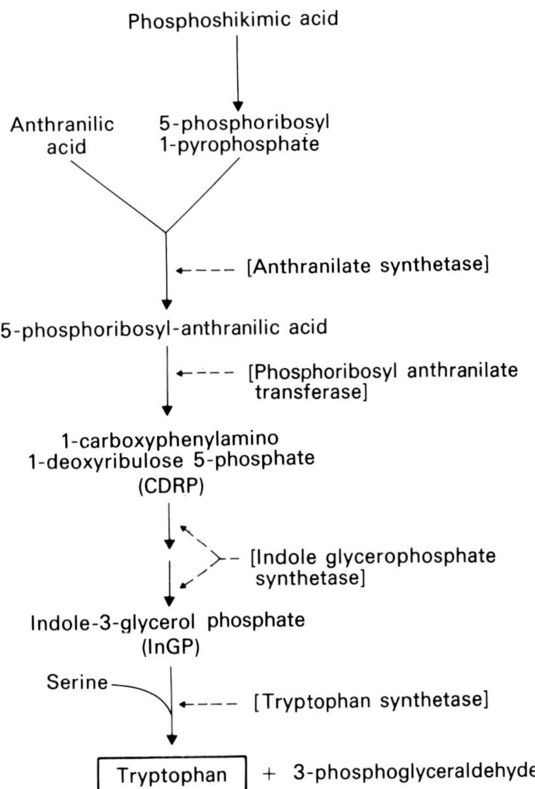

Fig. 18.8. The tryptophan biosynthetic pathway. The enzymes that catalyze the reaction steps are shown in square brackets. Tryptophan synthetase catalyzes the last step in the pathway.

Tryptophan synthetase is easily isolated and the two polypeptides A and B can be purified. Further, the amino acid sequence of the 267-amino acid-long A polypeptide was determined at the outset of their experiments. They then isolated a series of tryptophan auxotrophs and, with further tests, identified those that carried missense mutations in the *trpA* gene, which codes for the A polypeptide. The *trpA* mutants cannot carry out the production of tryptophan from indole glycerol phospate and serine. Fine-structure mapping was used to locate the various mutations within the gene and amino acid sequencing was used to pinpoint the amino acid substitutions in the A polypeptide resulting from the missense mutation. This enabled a comparison to be made of the relative locations of the corresponding amino acid substitutions in the polypeptide. The data showed that there is a complete correspondence between the sequence and relative positioning of the mutations and the amino acid substitutions, and this was termed *colinearity* (Fig. 18.9). In addition, the data indicated that no single mutation affected more than one amino acid and different amino acid substitutions at the same position in the A polypeptide. All in all, Yanofsky's work was highly significant since it confirmed the hypothesis that genes code for amino acid sequence in polypeptides.

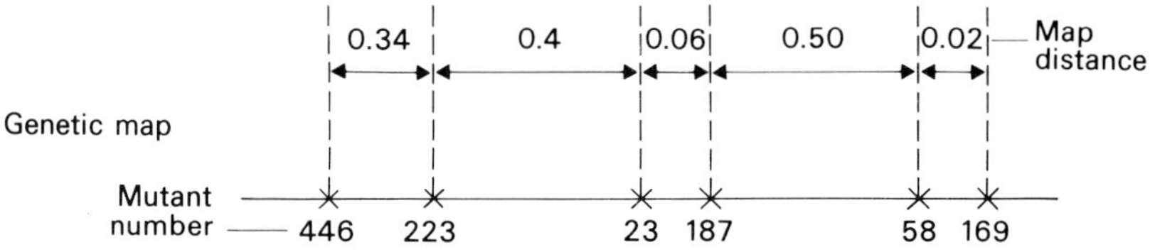

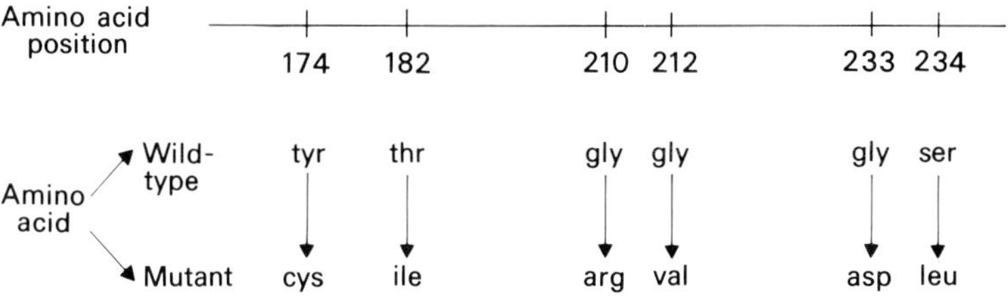

Fig. 18.9. Part of the *trpA* gene of *E. coli* and its product, the tryptophan synthetase A polypeptide, showing the colinearity of mutation positions and amino acid substitutions.

Summary

1. Many proteins consist of complexes of one or more polypeptide subunits.
2. Each polypeptide is specified by a single cistron.
3. A mutation in a gene coding for an enzyme or one of its subunits can result in a biochemical defect affecting one of the steps in a biochemical pathway. The strain carrying the mutations can grow if supplemented with any intermediate in the pathway that is past the block. The strain will also accumulate intermediates in the pathway that are prior to the block.
4. The sequence of mutations within a cistron is colinear with the sequence of amino acid substitutions in the polypeptide for which it codes.

Questions and problems

18.1 *A*, *B*, *C*, and *D* are independently assorting genes controlling the production of a black pigment. The alternate alleles that give abnormal functioning of these genes are *a*, *b*, *c*, and *d*. A black *AA BB CC DD* is crossed with a colorless *aa bb cc dd* to give a black F1. The F1 is then selfed. Assume that *A*, *B*, *C*, and *D* act in a pathway as follows:

$$\text{colorless} \xrightarrow{A} \xrightarrow{B} \xrightarrow{C} \text{brown} \xrightarrow{D} \text{black}$$

(a) What proportion of the F2 are colorless?
(b) What proportion of the F2 are brown?

18.2 Using the genetic information given in Question 18.1, now assume that *A*, *B*, and *C* act in a pathway as follows:

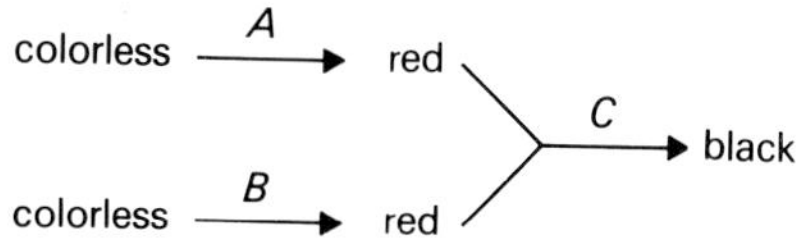

(a) What proportion of the F2 are colorless?
(b) What proportion of the F2 are red?
(c) What proportion of the F2 are black?

18.3 (a) Three genes on different chromosomes are responsible for three enzymes that catalyze the same reaction in corn:

$$\text{colorless compound} \xrightarrow{A,\, B,\, C} \text{red compound}$$

The normal functioning of any one of these genes is sufficient to convert the colorless compound to the red compound. The abnormal functioning of these genes is designated by *a*, *b*, and *c*, respectively. A red *AA BB CC* is crossed with a colorless *aa bb cc* to give a red F1, *Aa Bb Cc*. The F1 is then selfed. What is the proportion of the F2 that is colorless?
(b) It turns out that another step is involved in the pathway. It is controlled by gene *D*, which assorts independently of *A*, *B*, and *C*:

$$\text{colorless compound 1} \xrightarrow{A,\, B,\, C} \text{colorless compound 2} \xrightarrow{D} \text{red compound}$$

The inability to convert colorless compound 1 to colorless compound 2 is designated "d". A red *AA BB CC DD* is crossed with a colorless *aa bb cc dd*. The F1 are all red. The red F1's are now selfed. What proportion of the F2 is colorless?

18.4 In hypothetical diploid organisms called mings, the recessive *bw* causes a brown eye, and the (unlinked) recessive *st* causes a scarlet eye. Organisms homozygous for both recessives have white eyes. The genotypes and corresponding phenotypes, then, are as follows:

$bw^+\_\ \ st^+\_$	red eye
$bw\ bw\ st^+\_$	brown eye
$bw^+\_\ \ st\ st$	scarlet eye
$bw\ bw\ st\ st$	white eyes

Outline an hypothetical biochemical pathway that would give this type of gene interaction. Demonstrate why each genotype shows its specific phenotype.

18.5 The Black Riders of Mordor in the *Lord of the Rings* ride steeds with eyes of fire. As a geneticist you are very interested in the inheritance of the fire-red eye color. You discover that the eyes contain two types of pigments, brown and red, that are usually bound to core granules in the eye. In wild type steeds, precursors are converted by these granules to the above pigments, but in steeds homozygous for the recessive X-linked gene *w* (white eye), the granules remain unconverted and a white eye results. The metabolic pathways for the synthesis of the two pigments are shown below. Each step of the pathway is controlled by a gene. A mutation *v* gives vermilion eyes; *cn* gives cinnabar eyes; *st* gives scarlet eyes; *bw* gives brown eyes; and *se* gives black (sepia) eyes. All of the mutations are recessive to their wild type alleles and all are unlinked to one another.

For the following genotypes, show the phenotypes and proportions of steeds that would be obtained in the F1 of the following matings:

(a) $w\,w\,bw^+\,bw^+\,st\,st \times w^+Y\,bw\,bw\,st^+\,st^+$
(b) $w^+\,w^+\,se\,se\,bw\,bw \times wY\,se^+\,se^+\,bw^+\,bw^+$
(c) $w^+\,w^+\,v^+\,v^+\,bw\,bw \times wY\,v\,v\,bw\,bw$
(d) $w^+\,w^+\,bw^+\,bw\,st^+\,st \times wY\,bw\,bw\,st\,st$

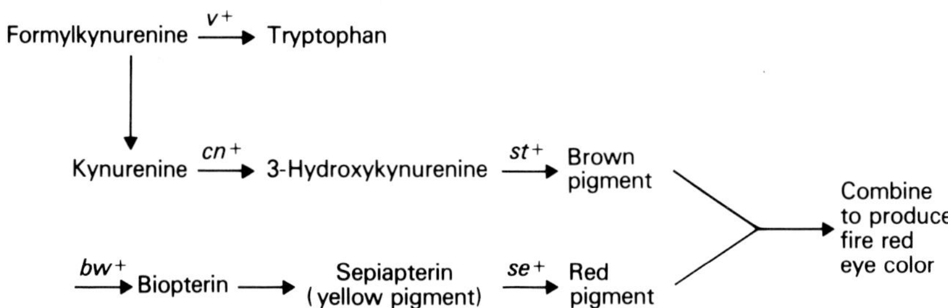

18.6 Upon infection of *E. coli* with bacteriophage T4, a series of biochemical pathways result in the formation of mature progeny phages. The phages are released following lysis of the bacterial host cells. Suppose that the following pathway exists:

$$A \xrightarrow[\text{enzyme}]{\downarrow} B \xrightarrow[\text{enzyme}]{\downarrow} \text{mature phages}$$

Also suppose that we have two temperature-sensitive mutants that involve the two enzymes catalyzing these sequential steps. One of the mutations is cold sensitive (cs) in that no mature phages are produced at 17°C. The other is heat sensitive (hs) in that no mature phages are produced at 42°C. Normal progeny phages are produced when phages carrying either of the mutations infect bacteria at 30°C. However, let us assume that we do not know the sequence of the two mutations in the biochemical pathway. Two models are therefore apparent:

(a) $A \xrightarrow{hs} B \xrightarrow{cs} \text{phages}$

(b) $A \xrightarrow{cs} B \xrightarrow{hs}$ phages

Outline how you would experimentally determine which is the correct model without artificially breaking phage-infected bacteria.

18.7 Two mutant strains of *Neurospora* lack the ability to make compound Z. When crossed, the strains usually yield asci of two types: (1) those with spores that are all mutant, and (2) those with 4 wild type and 4 mutant spores. The two types occur in a 1:1 ratio. (Refer to Chapter 15 for a discussion of genetic analysis in fungi.)

(a) Let *c* represent one mutant and *d* represent the other. What are the genotypes of the two mutant strains?

(b) Are *c* and *d* linked? Explain.

(c) Wild-type strains can make compound Z from the constituents of the minimal medium. Mutant *c* can make Z if supplied with X, but not if supplied with Y, while mutant *d* can make Z from either X or Y. Construct the simplest linear pathway for the synthesis of Z from the precursors X and Y, and show where the pathway is blocked by mutations *c* and *d*.

18.8 The following growth responses (where + = growth and 0 = no growth) of *Neurospora* mutants *1–4* were seen on the related biosynthetic intermediates A, B, C, D, and E. Assume all intermediates are able to enter the cell, that each mutant carries only one mutation, and that all mutants affect steps after B in the pathway.

Mutant	Growth on A	B	C	D	E
1	+	0	0	0	0
2	0	0	0	+	0
3	0	0	+	0	0
4	0	0	0	+	+

Which of the following schemes fits best with the above data, with regard to the biosynthetic pathway?

a $B \rightarrow A$; $A \rightarrow C$; $A \rightarrow D$; $A \rightarrow E$

b $B \rightarrow A \rightarrow D$; $D \rightarrow C$; $D \rightarrow E$

c $B \rightarrow A$; $A \rightarrow E$; $A \rightarrow C \rightarrow D$

d $B \rightarrow A \rightarrow E \rightarrow D$; $A \rightarrow C$

18.9 Four strains of *Neurospora*, all of which require arginine but are of unknown genetic constitution, have the following characteristics. The nutrition and accumulation characteristics are as follows (+ = growth; − = no growth):

	Growth on				
Strain	Minimal medium	Ornithine	Citrulline	Arginine	Accumulates
1	−	−	+	+	ornithine
2	−	−	−	+	citrulline
3	−	−	−	+	citrulline
4	−	−	−	+	ornithine

The pairwise complementation tests of the four strains gave the following results (+ = growth on minimal medium and 0 = no growth on minimal medium):

	4	*3*	*2*	*1*
1	0	+	+	0
2	0	0	0	
3	0	0		
4	0			

Crosses among mutants yielded prototrophs in the following percentages:

1×2: 25%
1×3: 25%
1×4: none detected among 1 million ascospores
2×3: 0.002%
2×4: 0.001%
3×4: none detected among 1 million ascospores

Analyze the data and answer the following questions:

(a) How many distinct mutational sites are represented among these four strains?

(b) In this collection of strains, how many types of polypeptide chains (normally found in wild type) are affected by mutations?

(c) Give the genotypes of the four strains, using a consistent and informative set of symbols.

(d) Give the map distances between all pairs of linked mutations.

(e) Give the percentage of prototrophs that would be expected among ascospores of the following types:

(i) strain 1×wild type (ii) strain 2×wild type
(iii) strain 3×wild type (iv) strain 4×wild type

References

Beadle, G.W. and B. Ephrussi, 1937. Development of eye colors in *Drosophila*: diffusible substances and their interrelationships. *Genetics* **22**:76–86.

Beadle, G.W. and E.L. Tatum, 1942. Genetic control of biochemical reactions in *Neurospora*. *Proc. Natl. Acad. Sci. USA* **27**:499–506.

Garrod, A.E., 1909. *Inborn Errors of Metabolism*. Oxford University Press, Oxford.

Srb, A.M. and N.H. Horowitz, 1944. The ornithine cycle in *Neurospora* and its genetic control. *J. Biol. Chem.* **154**:129–139.

Wagner, R.P. and H.K. Mitchell, 1964. *Genetics and Metabolism*, 2nd ed. Wiley, New York.

Yanofsky, C., 1967. Structural relationships between gene and protein. *Annu. Rev. Genetics* **1**:117–138.

Yanofsky, C., G.R. Drapeau, J.R. Guest and B.C. Carlton, 1967. The complete amino acid sequence of the tryptophan synthetase A protein (or subunit) and its colinear relationship with the genetic map of the A gene. *Proc. Natl. Acad. Sci. USA* **57**:296–298.

Chapter 19 Gene Regulation in Bacteria

Outline

In the previous chapter the notion was developed that the biosynthesis of a cellular component requires a number of steps in a biochemical pathway. Each step is catalyzed by a different enzyme and each enzyme is coded by one or more cistrons in the chromosome. In bacteria and phages the different genes controlling a particular pathway (and hence determining a particular trait) are often found clustered together in a group called an **operon**. This facilitates the regulation of expression of these genes as a single unit and contrasts sharply with the situation in eukaryotes where related genes are usually scattered throughout the genome. This chapter discusses some examples of bacterial operons and how their expression is regulated. As a prelude to this, we must remember that bacteria grow and divide rapidly in a nutrient medium, and to survive, they must be capable of making rapid adjustments when the environment changes, for example, when components of the medium are altered. To cope with this, bacteria have evolved effective ways of rapidly turning on and off the relevant sets of genes, which, when coupled with the short half-lives of the mRNAs, renders the organisms efficient in energy utilization. These regulatory systems operate at the transcription level.

The lactose operon of *E. coli*

Function in the wild type

The sugar lactose cannot be used as an energy source of *E. coli* unless it is first broken down into its glucose and galactose components. This reaction is catalyzed by the enzyme β-galactosidase, which has a tetrameric structure of identical 135,000-dalton polypeptides (Fig. 19.1).

In a wild type cell growing in a medium that does not contain lactose, there are only a few molecules of β-galactosidase. However, if the cell is growing in a medium containing lactose as the sole carbon and energy source, there are about 3000 copies of the enzyme. In other words, **induction** of enzyme synthesis occurs as a result of the presence of lactose. Lactose itself is not the actual inducer in this system, but rather induction is brought about by the action of allolactose, which is produced from lactose by the enzyme activity of the few molecules of β-galactosidase present in the uninduced cell (Fig. 19.2). As long as lactose is present, the system will remain induced so

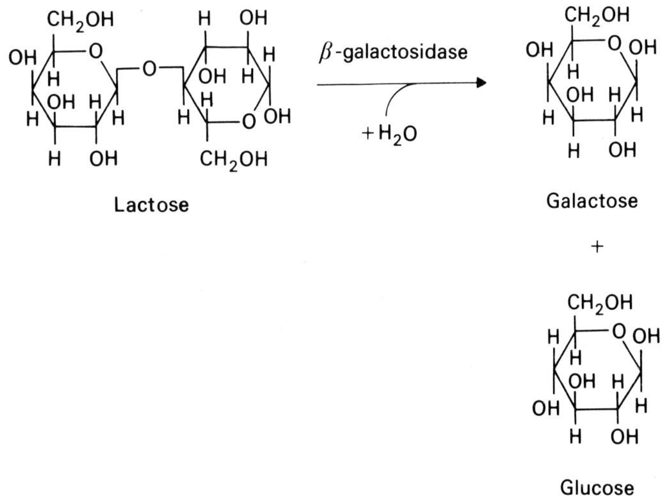

Fig. 19.1. The reaction catalyzed by the enzyme β-galactosidase: the cleavage of lactose to galactose and glucose.

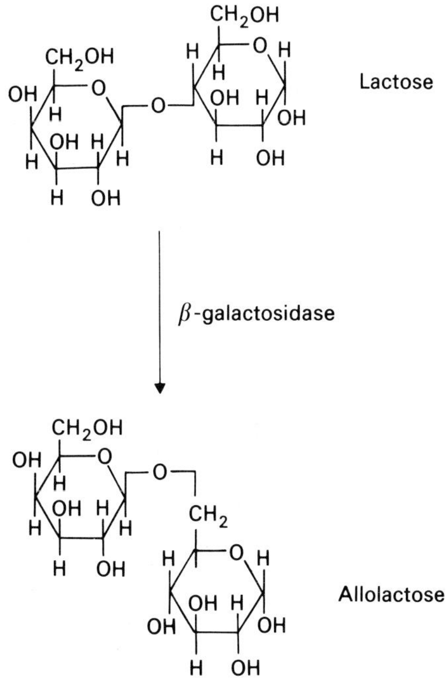

Fig. 19.2. The production of the actual inducer of the lactose operon, allolactose, from lactose by the action of β-galactosidase.

that the level of β-galactosidase will remain high, but once the lactose runs out the enzyme level will diminish rapidly. This is not an all-or-none phenomenon; the amount of enzyme produced is directly proportional to the amount of inducer present (up to the maximum amount found in the cell) (Fig 19.3).

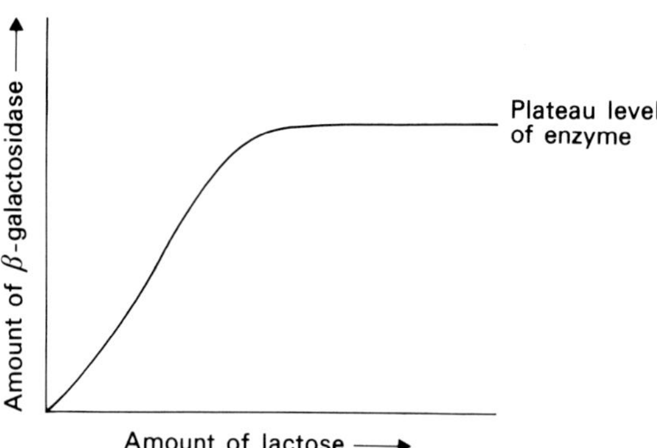

Fig. 19.3. Graph showing the direct proportionality between the amount of β-galactosidase produced and the amount of lactose added to the cells. The plateau level of enzyme amount reflects the maximum rate at which β-galactosidase can be synthesized.

The addition of lactose to the cell not only brings about the increase in β-galactosidase levels but also a rapid increase in the synthesis of β-galactoside permease and thiogalactoside transacetylase, two enzymes also needed for lactose breakdown. Genetic experiments have shown that the genes for all of the three proteins are linked on the chromosome and adjacent to them are two regulatory sites, the *operator* and the *promoter*. A short distance away is the *i* gene, which codes for a **repressor gene** involved in the regulation of the system. As a result of the phenotypic properties of regulatory mutations, F. Jacob and J. Monod proposed their classical operon model for the control of gene expression in bacteria, a more up-to-date description of which will be given here. By definition, an **operon** is a genetic unit consisting of contiguous genes that function coordinately under the joint control of an operator and a repressor. A diagram of the lactose operon and its regulatory sites and repressor gene in a wild type strain growing in the absence of lactose is given in Fig. 19.4.

The z^+, y^+, and a^+ structural genes code for β-galactosidase, permease, and transacetylase, respectively. Genetic experiments with mutant strains lacking enzyme activity showed that the three genes are adjacent on the chromosome. The i^+ gene (repressor gene) codes for a **repressor molecule**, which is a protein. The expression of this gene is *constitutive*; that is, the product is synthesized all the time. Control of the amount of gene product, then, is a function of the rate of which RNA polymerase can bind to the *i* gene promoter (p_i^+) and initiate the transcription of the repressor message. Translation of the message produces a polypeptide that aggregates to produce the functional repressor tetramer. If the cell is growing on a medium lacking lactose, the repressor will bind to the operator (o^+) region, which is located adjacent to the z^+ gene. When this complex is formed, RNA polymerase cannot bind to the structural gene promoter (p^+) region, and thus transcription of the gene is prevented.

If the cell is now placed in a medium containing lactose as the sole carbon

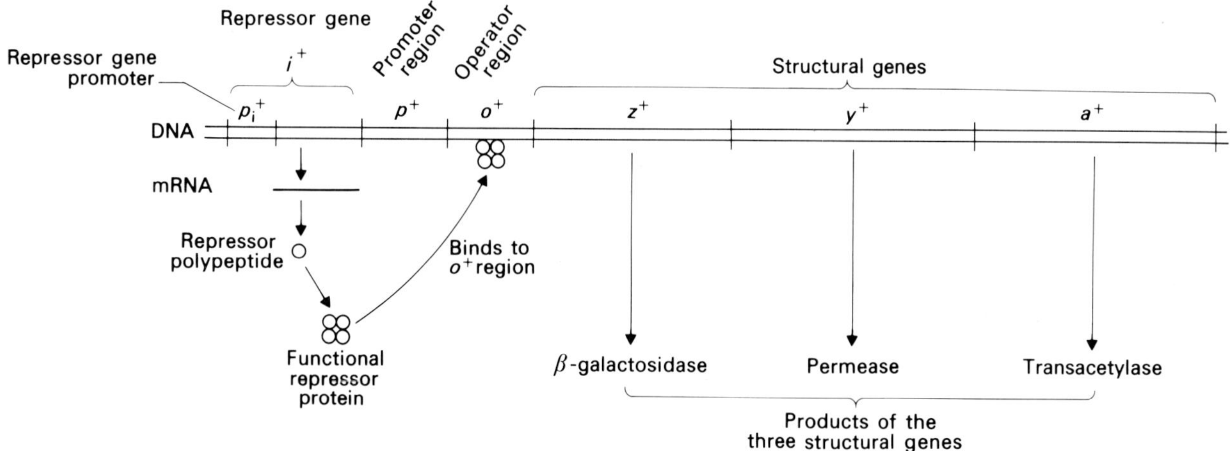

Fig. 19.4. Organizational features of the lactose operon and its regulatory sites in wild type *E. coli*. The diagram shows the condition when no lactose is present: the repressor protein produced binds to the operator region and translation of the structural genes is prevented.

source, lactose is transported into the cell by the activity of the few permease molecules present and then it is converted to the inducer, allolactose, by β-galactosidase. The inducer binds to the repressor, one molecule for each polypeptide, and causes a conformational change in the repressor such that it no longer has affinity for the DNA. As a consequence, the repressor falls off the operator and, once that has occurred, RNA polymerase can bind to the structural gene promoter and initiate transcription of the operon. The lactose operon is a single transcriptional unit so that the RNA polymerase transcribes the *z*, *y*, and *a* genes, in that order, onto a single **polygenic** (**polycistronic**) **mRNA**. This mRNA is translated by ribosomes attaching to the 5′ end (*z* gene end) and moving down the molecule. Thus the β-galactosidase is made first and, after the stop codon of that region, the ribosome continues moving toward the 3′ end of the message. Then after recognizing the initiation sequence for the permease gene, the ribosome begins to translate that part of the message to produce permease. The process is repeated at the boundary between the permease and transacetylase gene sequences. The general principle that pertains here is that ribosomes can *only* initiate translation at the 5′ end of a polycistronic mRNA. (This makes the coordinate production of proteins of related function easy to control.) It is not possible for ribosomes to bind and initiate translation at the start regions of the permease or transacetylase sequences, presumably because the correct initiation sequence is only found at the 5′ end of the message. All of this is summarized in Fig. 19.5. As long as lactose is present in sufficient amounts to bind with the repressor, the operon will be transcribed and the three enzymes will be produced.

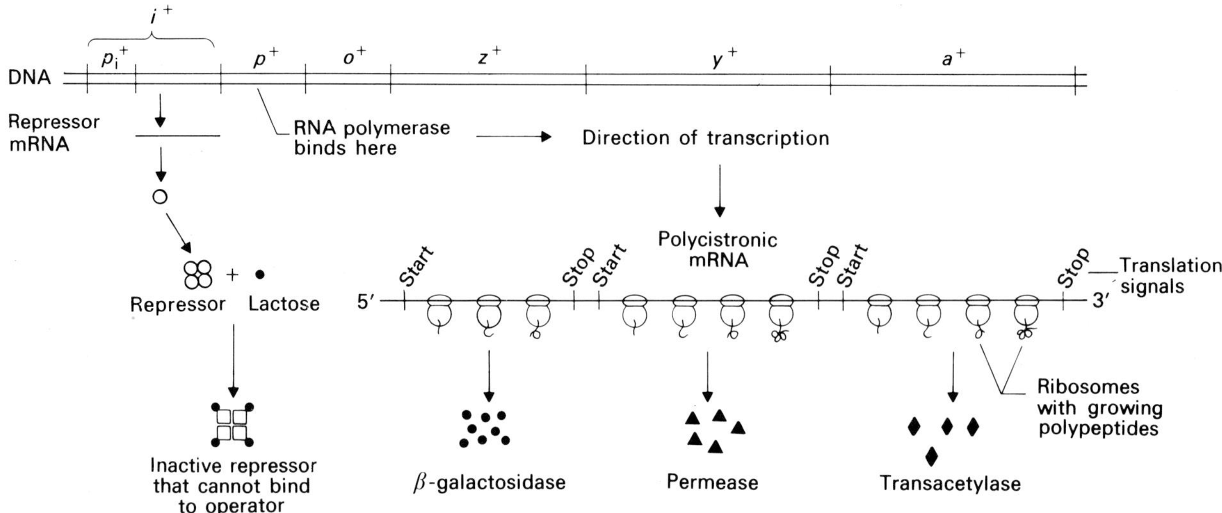

Fig. 19.5. Induction of the lactose operon in wild-type *E. coli* when lactose is present as the sole carbon source. Lactose inactivates the repressor thereby allowing transcription of the structural genes. Translation of the resulting polycistronic mRNA produces the three enzymes required for lactose utilization by the cell.

Elucidation of the regulation of the lactose operon by studies of mutants

The Jacob-Monod operon model of control of gene expression was based on studies of a number of regulatory mutants in which the control of the expression of the lactose operon was abnormal. An important part of Jacob and Monod's mutant studies involved partial diploid strains, and the construction of these strains will be described before the properties of the mutants themselves are discussed.

In the discussion of *Hfr* strains of *E. coli*, it was shown that *Hfr* strains could revert to the F^+ state if there was a reversal of the process for integration of the circular *F* factor. In most cases the *F* factor is excised correctly, but occasionally an error is made and part of the bacterial chromosome is looped out. The resulting episome is called an *F′* (F-prime) factor, and by conjugation, the genes picked up by the *F* can be transferred rapidly through an F^- population. Thus, by this process, episomes can be produced that carry the lactose region of the chromosome, and these are called *F′ lac* (Fig. 19.6).

Mutants of the structural genes. Both missense and nonsense mutants have been isolated for the three structural genes. Mapping these mutants provided evidence for the order *z-y-a* on the chromosome. A missense mutation in one of the genes results in the loss of activity of the enzyme for which the gene codes but does not affect the activities of the other two enzymes in the system.

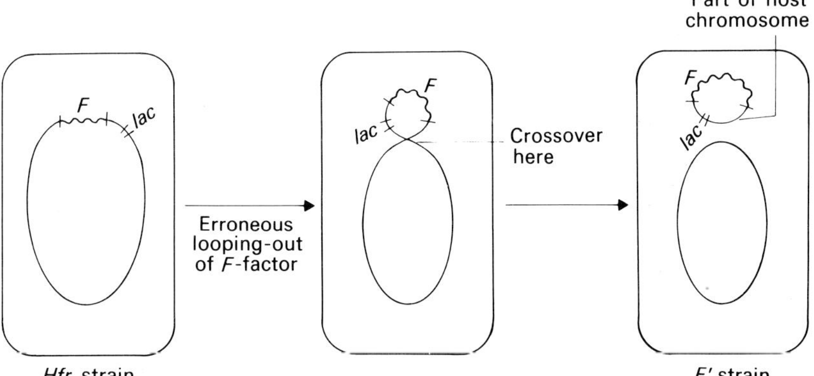

Fig. 19.6. Production of an episome (*F′ lac*) carrying the lac region of the *E. coli* chromosome by erroneous looping out of the *F* factor in an *Hfr* strain.

On the other hand, the nonsense mutations have some interesting properties. For example, a nonsense mutation in the *z* gene (z^-), providing it is not too near the end of the gene, will not only abolish *z* gene activity but also reduce or abolish the expression of the *y* and *a* genes. This is called **polarity,** or the *polar effect*, and thus the nonsense mutations are sometimes called *polar mutations*. Indeed there is a gradient of polarity of the effect of chain-terminating mutants in the *z* gene on the expression of the other two genes. The closer the mutation is to the operator, the more severe the polar effect; that is, the less likely permease and transacetylase will be produced. Polar mutations in the *y* gene affect the activity of the *y* gene and the *a* gene but not the activity of the *z* gene. The interpretation of these data was that the three genes are transcribed onto a single polycistronic mRNA, which is translated with 5′-to-3′ polarity in the order *z-y-a*, as discussed previously. The polar effect of chain-terminating mutations is the consequence of the distance ribosomes have to travel before recognizing a new translation initiation sequence—the farther it is, the more likely the ribosome will fall off the message.

Operator mutants. A series of mutants were isolated that were constitutive for enzyme production. In other words they had lost regulatory control such that the enzymes were produced whether or not the inducer was present. When the mutations were mapped, it was found that several were clustered next to the *z* gene in a region that is now called the operator. These so-called operator-constitutive (o^c) mutants have lost the ability to bind the repressor protein such that transcription cannot be prevented. This latter point is supported by genetic and biochemical evidence. We will discuss the former here.

The effect of o^c mutations on the contiguous structural genes and on *z*, *y*, and *a* genes located on a different piece of DNA was examined by haploid and partial diploid strains. In these studies, mutant strains were constructed with various combinations of operator and structural gene alleles, and the production of the enzymes was monitored in the absence and presence of the inducer, lactose. Representative data are shown in Table 19.1.

Table 19.1. Effect of an o^c mutation in the lactose operon on enzyme production in haploid and partial diploid strains grown in the presence and absence of lactose.

	No lactose		Lactose	
Genotype	β-galactosidase (z)	Permease (y)	β-galactosidase	Permease
1. $i^+p^+o^+z^+y^+$	−	−	+	+
2. $i^+p^+o^cz^+y^+$	+	+	+	+
3. $i^+p^+o^cz^+y^-/i^+p^+o^+z^-y^+$	+	−	+	+
4. $i^+p^+o^cz^-y^+/i^+p^+o^+z^+y^-$	−	+	+	+

In the table, class 1 is the wild type where neither enzyme is produced until lactose is added as the inducer. Class 2 is a haploid strain carrying an o^c mutation that leads to constitutive production of the enzymes. Class 3 is a partial diploid carrying o^c and y^- mutations on one chromosome and the o^+ and z^- alleles on the other chromosome. In this case β-galactosidase (z gene product) was produced constitutively, but permease (y gene product) was produced only in the presence of lactose. Class 4 is a similar case, with o^c and z^- on one piece of DNA and the o^+ and y^- alleles on the other DNA. Here the permease is produced constitutively and the β-galactosidase is inducible. The two latter cases show that the effect of the o^c mutation is limited to those genes adjacent to it on the same piece of DNA; that is, it is a cis-dominant mutation. Using class 3 as an example, z^+ is on the same DNA as o^c and hence is transcribed constitutively. The y^- mutation results in a nonfunctional permease. In the strain, the wild type y^+ allele is on the chromosome that also carries the o^+ allele. This DNA, then, is under normal regulatory control, and hence the y^+ is inducible. All of these data are consistent with the thesis that the operator region does not produce a diffusible product. Further, biochemical experiments showed that DNA from o^c strains will not bind the repressor, thus showing that the operator is the site of repressor action. The operator, then, is a controlling region that in the wild type can bind the repressor, which then prevents transcription. In the o^c mutants, the region has been altered (e.g. nucleotide pair change or deletion) so that the required protein–nucleic acid interaction does not occur and transcription cannot be stopped. Finally, since the o^c mutants do not affect the properties of β-galactosidase, it has been concluded that the operator is most probably a region distinct from the z cistron.

Repressor gene mutants. Mutations in the repressor (i) gene affect the control of expression of the lactose operon. The study of these types of mutants was also instrumental in the formulation of the operon model for gene regulation.

There are a number of classes of mutations mapping at the i locus that affect the regulation of the *lac* operon. One class, the i^- mutations, gives rise to a constitutive phenotype. Mapping experiments showed that these mutations are at a locus (the repressor gene locus) distinct from the operator region. Other classes of mutations in the repressor gene include the i^s (super-repressed) and the i^{-d} (trans-dominant), the properties of which are summarized in Table 19.2.

Table 19.2. Effect of mutations in the repressor gene (i) of the lactose operon on enzyme production in haploid and partial diploid strains grown in the presence and absence of lactose.

	No lactose		Lactose	
Genotype	β-galactosidase (z)	Permease (y)	β-galactosidase	Permease
1. $i^+p^+o^+z^+y^+$	–	–	+	+
2. $i^-p^+o^+z^+y^+$	+	+	+	+
3. $i^+p^+o^+z^+y^-/i^-p^+o^+z^-y^+$	–	–	+	+
4. $i^-p^+o^+z^+y^-/i^+p^+o^+z^-y^+$	–	–	+	+
5. $i^sp^+o^+z^+y^+$	–	–	–	–
6. $i^sp^+o^+z^+y^+/i^+p^+o^+z^+y^+$	–	–	–	–
7. $i^{-d}p^+o^+z^+y^+$	+	+	+	+
8. $i^{-d}p^+o^+z^+y^-/i^+p^+o^+z^-y^+$	+	+	+	+
9. $i^{-d}p^+o^+z^-y^+/i^+p^+o^+z^+y^-$	+	+	+	+

In the table, class 1 is the wild type as before and class 2 shows the constitutive nature of i^- mutations. Classes 3 and 4 involve partial diploids in which a z^- mutation is on one DNA and a y^- mutation is on the other DNA in both combinations with the i^- mutation. In both cases normal, inducible production of the enzymes is the case showing that the i^+ is dominant to i^- for genes either on the same or a different DNA. This is called *trans-dominance* and is interpreted to mean that the i^+ gene codes for a diffusible product that acts to prevent transcription in the absence of lactose. This we now know to be the repressor protein that binds to the operator. The i^- mutations produce an inactive repressor that cannot bind to the operator, and this results in constitutive enzyme production in haploid cells. In the i^+/i^- partial diploids, there are enough repressor molecules produced from the i^+ gene to bind to the two operators present and thus normal regulation is in effect.

An i^s mutation results in a completely negative phenotype with respect to *lac* enzyme production (class 5). In partial diploids, i^s is trans-dominant to the

i^+ allele (class 6). The interpretation is that i^s mutations result in the production of an altered repressor molecule (a super-repressor), which binds to the operator normally but is not capable of recognizing the inducer molecule, lactose. Hence transcription cannot be initiated since the defective repressors remain stuck on the operators.

The last class of *i* mutations are the i^{-d} mutations. Like the i^- mutations, these result in constitutive enzyme production in haploids (class 7), but unlike the i^- alleles, they are trans-dominant to the wild-type i^+ allele (classes 8 and 9). These i^{-d} mutants are very rare, and their phenotype is believed to be related to the tetrameric nature of the repressor. That is, the wild type produces a repressor protein of four identical polypeptides. This repressor molecule has only one binding site for the operator. In i^{-d} mutants the subunits appear not to combine normally so that no operator-specific binding site results. In the i^+/i^{-d} strains there is a mixture of normal and defective repressor subunits and the trans-dominant of i^{-d} is thought to be because repressors made from combinations of the two cannot bind to the operator. Only purely wild-type repressors could bind, and statistically, they would be quite rare. With this in mind, the i^- mutations discussed previously must either be nonsense mutations, which produce short polypeptides, or missense mutations, which result in polypeptides that do not participate in tetramer formation.

In summary, the *i* mutations provide evidence that the *i* gene produces a diffusible product—the repressor—which prevents transcription of the *lac* operon. The site of action of the repressor is the operator region. Also, the *i* mutations show that the lactose repressor has three recognition reactions coded into its structure:

1. With the inducer, lactose; this presumably alters its shape causing it to dissociate from the DNA.
2. With the operator region.
3. With itself in that it acts as a tetramer.

Promoter mutants. These mutants map to the left of the operator region and are characterized by the lack of mRNA production in the presence or absence of lactose. The properties of these mutations are shown in Table 19.3. The data show that p^- mutations are cis-dominant to p^+ in that their effect is limited only to the genes on the same piece of DNA. This is shown especially clearly by classes 5 and 6, where inducible enzyme production resulted for the wild-type gene on the p^+ DNA whereas no enzyme was produced from the wild-type structural gene on the p^- DNA. Classes 3 and 4 also indicate that the properties of p^- mutations are not affected by o^c or i^- constitutive mutants. These facts are interpreted if we assume the p^+ region is the binding site for RNA polymerase. If it is altered so that polymerase cannot bind, no transcription will occur.

Table 19.3. Effect of mutations in the promoter region (p) for the structural genes of the lactose operon on enzyme production in strains grown in the presence or absence of lactose.

	No lactose		Lactose	
Genotype	β-galactosidase (z)	Permease (y)	β-galactosidase	Permease
1. $i^+p^+o^+z^+y^+$	–	–	+	+
2. $i^+p^-o^+z^+y^+$	–	–	–	–
3. $i^+p^-o^cz^+y^+$	–	–	–	–
4. $i^-p^-o^+z^+y^+$	–	–	–	–
5. $i^+p^-o^+z^+y^-/i^+p^+o^+z^-y^+$	–	–	–	+
6. $i^+p^-o^+z^-y^+/i^+p^+o^+z^+y^-$	–	–	+	–

Positive control of the lactose operon (catabolite repression)

The lactose operon is under negative control in that a specific repressor molecule binds to the DNA to prevent transcription of the structural genes. There is also good evidence that a positive control signal must be present for the operon to function normally. In the discussions thus far it has been carefully stated that when lactose was present, it was the *sole* carbon and energy source. If both glucose and lactose are present in the growth medium, the cells will preferentially catabolize glucose and the lactose operon is not transcribed. A similar situation applies to a number of other operons involved in the catabolism of other sugars, for example arabinose and galactose. These are called *glucose-sensitive operons*, and the phenomenon is commonly called **catabolite repression** (the **glucose effect**).

The effect of glucose on transcription of the operons in question is the result of the action of a breakdown product of that sugar to lower the intracellular amount of cyclic AMP (cAMP: 3′, 5′-cyclic adenosine monophosphate). This molecule is made from ATP in a reaction catalyzed by adenyl cyclase and is broken down with the aid of the enzyme phosphodiesterase. Thus the (unknown) catabolite of glucose could bring about the decrease in cAMP levels either by inhibiting adenyl cyclase or by stimulating phosphodiesterase or perhaps by doing both (Fig. 19.7).

The importance of cAMP to the transcription of the lactose and other glucose-sensitive operons is that a complex of cAMP with a *catabolite gene activator* (CGA) protein (a dimer of molecular weight 44,000) must bind to the promoter before RNA polymerase can bind and initiate transcription. Schematically the events shown in Fig. 19.8 are needed for transcription of the operons. If insufficient cAMP is present to make the complex, transcription is blocked at the various operons.

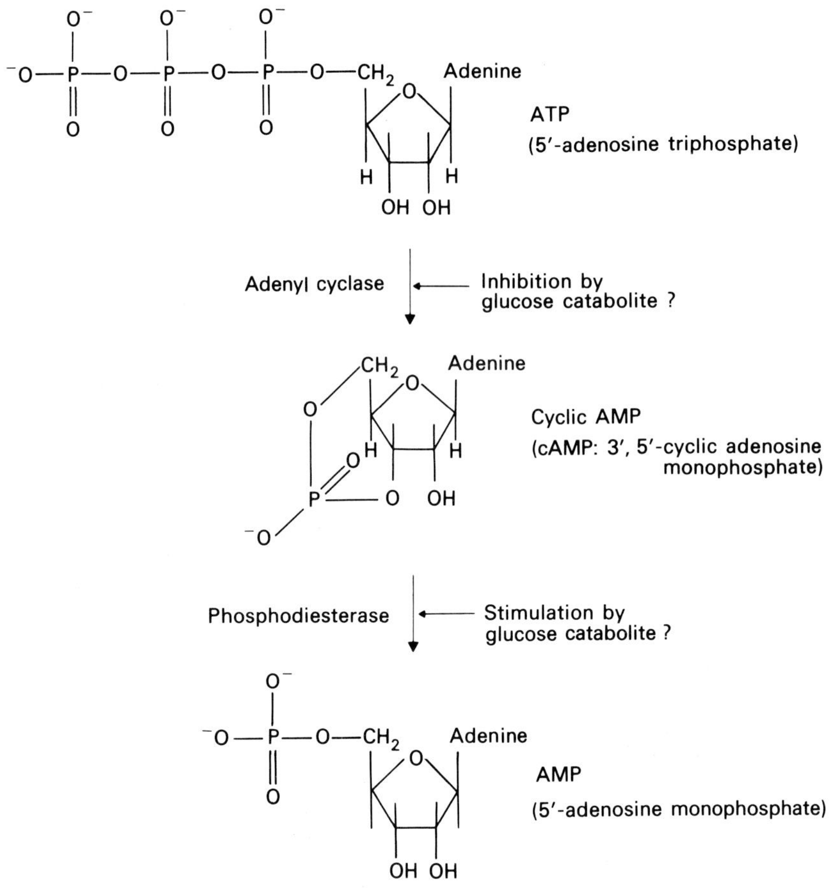

ATP
(5′-adenosine triphosphate)

Adenyl cyclase
Inhibition by glucose catabolite ?

Cyclic AMP
(cAMP: 3′, 5′-cyclic adenosine monophosphate)

Phosphodiesterase
Stimulation by glucose catabolite ?

AMP
(5′-adenosine monophosphate)

Fig. 19.7. The biosynthesis of cyclic AMP and ATP and its breakdown to AMP. In catabolite repression, a metabolic product of glucose inhibits adenyl cyclase activity or stimulates phosphodiesterase activity or it does both. The result is a decrease in cAMP level in the cell and this "turns off" glucose-sensitive operons in the cell.

Molecular details of lactose operon regulation

Information has been obtained about the sequence of nucleotide pairs involved in the promoter and operator regions of the lactose operon. Fig. 19.9 is a generalized diagram of the results.

The actual sequencing involved nucleic acid fingerprinting techniques. The repressor binding site was determined by sequencing the stretch of DNA protected from DNase digestion when the repressor protein is bound to the operator. The exact limits of the cAMP-CGA binding site and RNA polymerase interaction sites are not known, but genetic mutations that affect the level of expression of the lactose operon, and which presumably are in the promoter, show the promoter region to be about 80 nucleotide pairs long. It is noteworthy that the end of the *i* gene is immediately adjacent to the promoter sequence. Comparison of the sequence of the lactose operon

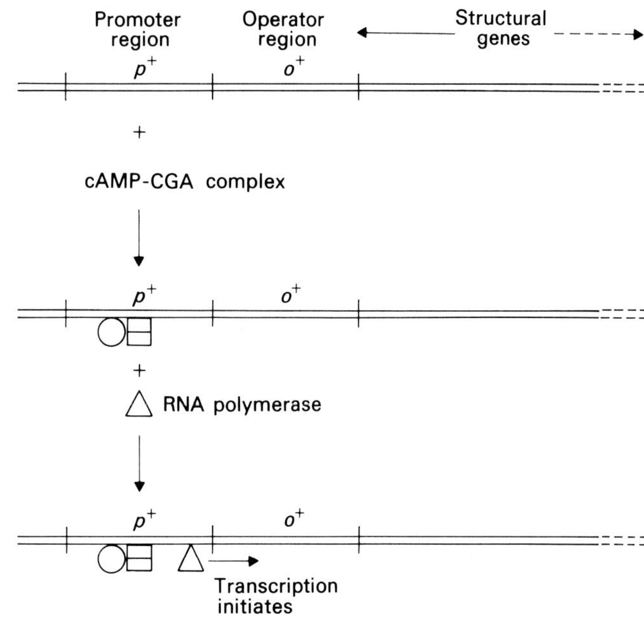

Fig. 19.8. The role of the cAMP in the functioning of glucose-sensitive operons. Cyclic AMP forms a complex with CGA (catabolite gene activator) protein, and this binds to the promoter, thereby facilitating binding of RNA polymerase and the initiation of transcription.

Fig. 19.9. Diagram of the promoter-operator region of the lactose operon showing the extents and positions of the binding sites for cAMP-CGA, RNA polymerase, and repressor protein. (After R. C. Dickson et al., 1975. *Sciences* **187**:27–35.)

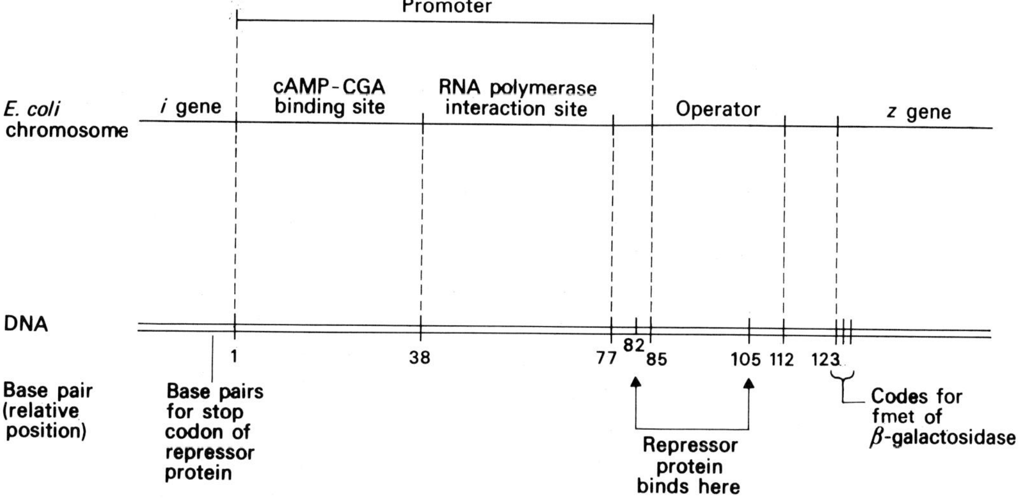

mRNA with the DNA sequence shows that transcription begins in the operator region within the region protected by the repressor. This is shown in more detail in Fig. 19.10, which also shows the region of the mRNA protected by the ribosome (the ribosome binding site) during the initiation step.

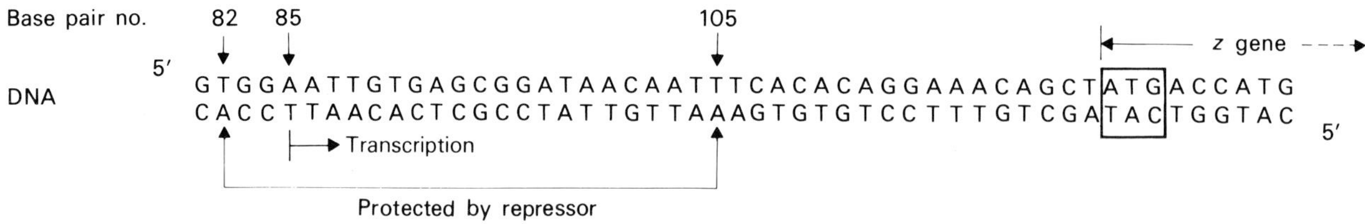

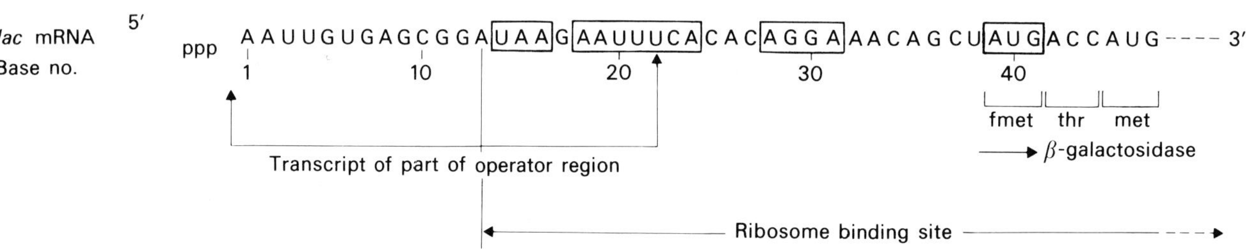

Fig. 19.10. The first 47 bases of the lac mRNA and the DNA sequence corresponding to it. The first 21 bases of the message are transcribed from the operator region. Other features of this region are described in the text. (After R. C. Dickson et al., 1975. *Science* **187**:27–35; and N. Maizels, 1974. *Nature* **249**:647–649.)

As can be seen, the first 38 bases of the mRNA are not translated, the codon for fmet starting at position 39. The first part of the mRNA is a copy of most of the operator region. The ribosome binding site was determined by sequencing the part of the mRNA remaining after RNase digestion while the ribosome is bound in its initiation configuration. This latter arrangement was achieved in vitro by having only fmet-tRNA in the reaction mixture so initiation but not elongation could occur. The ribosome binding site of the lactose operon covers 50 nucleotides of the mRNA and includes the codons for the first seven amino acids of the β-galactosidase (only three are given in the diagram). The boxed nucleotides in the diagram indicate sequences that have also been found in ribosome binding sites of other prokaryotic mRNAs. Specifically there is a nonsense codon (UAA here), a purine-purine-UUU-X-purine (where X is usually a purine), an AGGA sequence, and the start codon AUG.

In summary, the lactose operon has proved to be a model system for understanding gene regulation for a number of operons in prokaryotes.

Presumably, when a large number of promoter and operator sequences have been obtained, it will be possible to generalize about the molecular bases for the regulatory mechanisms at the DNA and RNA levels.

The arabinose operon of *E. coli*

The arabinose operon is another example of glucose-sensitive operon. As with lactose, when arabinose is absent, only a few molecules of the enzymes needed for arabinose catabolism are present in the cell. When arabinose is added (provided glucose is absent), there is a very rapid increase in the number of arabinose catabolic enzymes. This is controlled by a different mechanism from that described for the lactose operon.

The genes governing the metabolism of arabinose comprise what is called a *regulon*, which is composed of at least three operons. The *araBAD* operon contains the genes for the enzymes involved with the conversion of L-arabinose to D-xylulose 5-phosphate. The controlling sites for this operon are located adjacent to it. Two operons control the transport of arabinose into the cell: *araE*, which is the structural gene for the L-arabinose binding protein, and *araF*. The regulator gene for the system, *araC*, is located between the *araBAD* controlling site region and the *leu* operon. The *araC* gene controls the expression of *araBAD* by its positive and negative action in the controlling site region. Since *araBAD*, *araE*, and *araF* are inducible by L-arabinose and controlled coordinately by *araC*, it is assumed that the structures of the three controlling site regions are similar. The following discussion will focus on the *araBAD* operon, which is shown in Fig. 19.11.

The operon is thought to be controlled as follows. The *araC* gene codes for a P1 protein, which has repressor function and exerts its effect by binding to the adjacent *araO* (operator) controlling site and preventing the RNA polymerase from binding. Thus the operon is under negative control by P1. When L-arabinose is present, it stimulates the release of P1 from the DNA and the conversion of P1 to P2, which is an activator of the operon. The P2 binds to the *ara*I (initiator) site, facilitating RNA polymerase binding and the initiation of transcription. (All this, of course, requires the prior binding of cAMP-CGA complex, and this is thought to occur in the same vicinity.) The three structural genes are transcribed on a single polycistronic mRNA. The operon, then, is under positive control by P2.

Much of this is hypothesis, but there is good evidence for some parts of it. There is genetic evidence for the existence of the *ara*I site, and it is thought that part of this has promoter activity and another part is involved with cAMP-CGA binding. The function of the *araC* gene has been demonstrated by the study of genetic mutants. Mapping experiments with these mutants have shown that *araC* consists of only one cistron. Moreover, *araC* nonsense

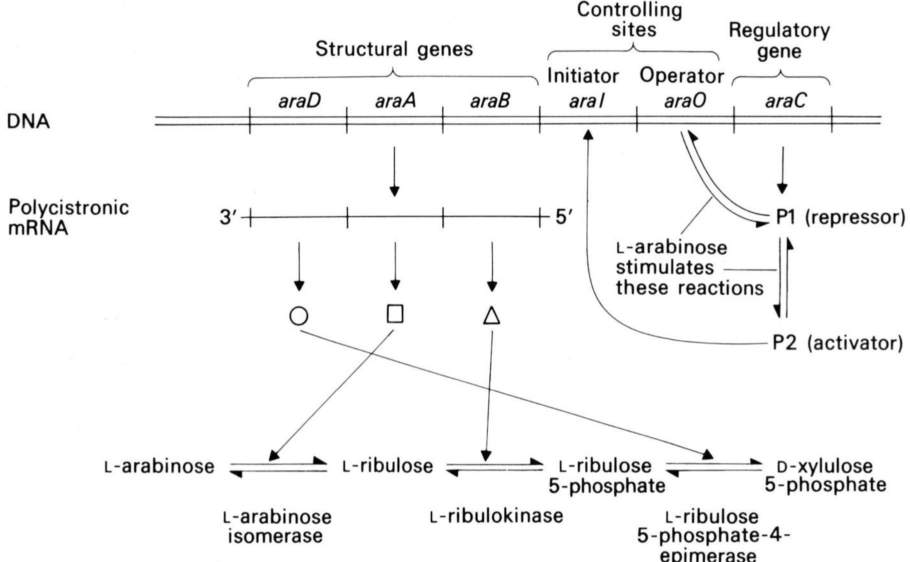

Fig. 19.11. The *araBAD* (arabinose) operon *E. coli* and the associated controlling sites and regulatory gene. The regulation of this operon is described in the text. (After the work of E. Englesberg.)

mutants have been shown to have no cis effect on the *araBAD* operon, thus indicating that *araC* is not in the *BAD* operon.

Three classes of *araC* alleles are known:

1. *araC*$^+$. The wild type allele renders the operon inducible; the three enzymes are induced by L-arabinose.
2. *araD*$^+$. These occur with high frequency and result in a pleiotropically L-arabinose-negative phenotype. In other words, the enzymes are not inducible by L-arabinose.
3. *araC*^{c}. These are quite rare and give a pleiotropically constitutive phenotype in that the enzymes are produced even in the absence of the inducer.

As with the regulatory mutants of the lactose operon, diploid studies have been used to obtain an understanding of the function of the *araC* gene. From these studies it was shown that *araC*$^-$ is recessive to C^+ in either the cis or trans arrangement. The C^- alleles are complemented by A^-, B^-, and D^- alleles, and thus the pleiotropic negative phenotype is not the result of a polarity effect on the *araBAD* operon. The conclusion from studies of C^- and C^+/C^- strains was that C^+ produces a protein that, in the presence of L-arabinose, is necessary for the expression of the region. This suggested some positive control in the system and contrasts with the lactose operon C^- mutants, which are constitutive because of the loss of negative control.

The C^c alleles are cis- and trans-dominant to C^-, suggesting that they produce the activator P2 in the absence of L-arabinose; this activator is able to turn on an operon that is either cis or trans to the C^c allele. On the other hand, C^+ is dominant to C^c. This suggests that there is some negative control of the operon. Based on the model for regulation of the operon that was presented (which was, of course, proposed on the basis of the data now being discussed), in the absence of arabinose, P1 acts as a repressor, preventing expression of the operon by *ara*C^c activator. That is, when P1 is on the operator, transcription ceases even if P2 is present.

There is some biochemical evidence to support the regulatory model. Studies of heat-sensitive *ara*C^- mutants have shown that *both* the repressor and activator functions are heat-labile, thereby indicating that *ara*C produces a protein that can serve both functions. The *ara*C protein has been purified and it has been shown to have both repressor and activator activity. Indeed there is some evidence that the P2 form of the protein is a dimer of the P1 form. There is also evidence that L-arabinose interacts directly with the *ara*C protein to bring about the necessary conversion. Also, there is direct evidence that the activator form of *ara*C protein, P2, is required for transcription of the operon. In an in vitro system, synthesis of *ara* mRNA shows an absolute requirement for *ara*C protein.

In conclusion, the L-arabinose regulon is under both positive and negative control, with the *ara*C protein playing a pivotal role in the regulatory process. The exact nature of the controlling sites remains to be worked out. In contrast to the lactose operon where inhibition must be removed for the genes to be expressed, the arabinose operon requires activation for transcription to begin.

The tryptophan operon of *E. coli*

The tryptophan operon of *E. coli* is an example of a *repressible operon* that is regulated by many of the basic features of the Jacob-Monod operon model. In this case, tryptophan is an amino acid that is needed for protein synthesis. When tryptophan is absent from the medium, the enzymes necessary for tryptophan genes are repressed. This makes sense from energetic consideration. Most operons controlling the synthesis of a compound are repressible, whereas operons involved with the breakdown of something (e.g. a carbon source) are usually inducible. Many but not all repressible operons have a regulatory gene that codes for an *apo-repressor* molecule that cannot bind to the operator. The end product of the biosynthetic pathway or a derivative of it (e.g. a charged tRNA molecule) acts as a *co-repressor* (and not as an inducer) in the system and, by binding with the apo-repressor, produces a functional repressor molecule capable of binding to the operator and blocking transcription. The general features of a repressible operon are shown in Fig. 19.12.

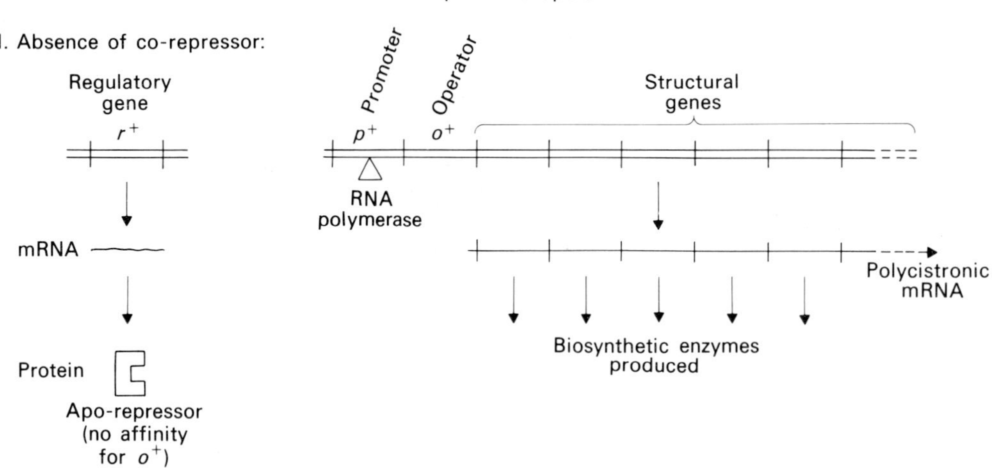

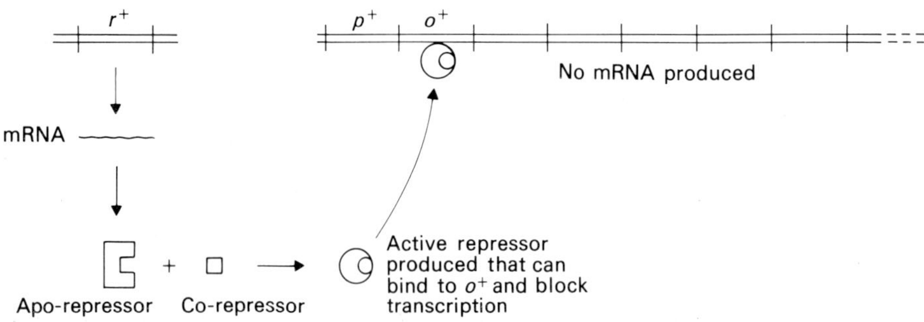

Fig. 19.12. The general functioning of a repressible operon in bacteria.

The tryptophan operon of *E. coli* is shown in Fig. 19.13 along with the estimated lengths of the structural genes and other regions. Most of the work that will be described is that of C. Yanofsky and colleagues. In this operon are two promoters, *p*1 and *p*2; the former is adjacent to the operator (*o*), and the latter is at the end of the *trpD* cistron and acts as a low-efficiency internal promoter. The function of the operon is controlled by the apo-repressor protein, which is the polypeptide product of the *trpR* gene. When tryptophan is present in the medium, a complex of the repressor with L-tryptophan binds to the operator region and represses transcription by preventing the attachment of RNA polymerase at the *p*1 promoter. In the absence of tryptophan, the operon is not repressed and mRNA transcription is initiated. For this

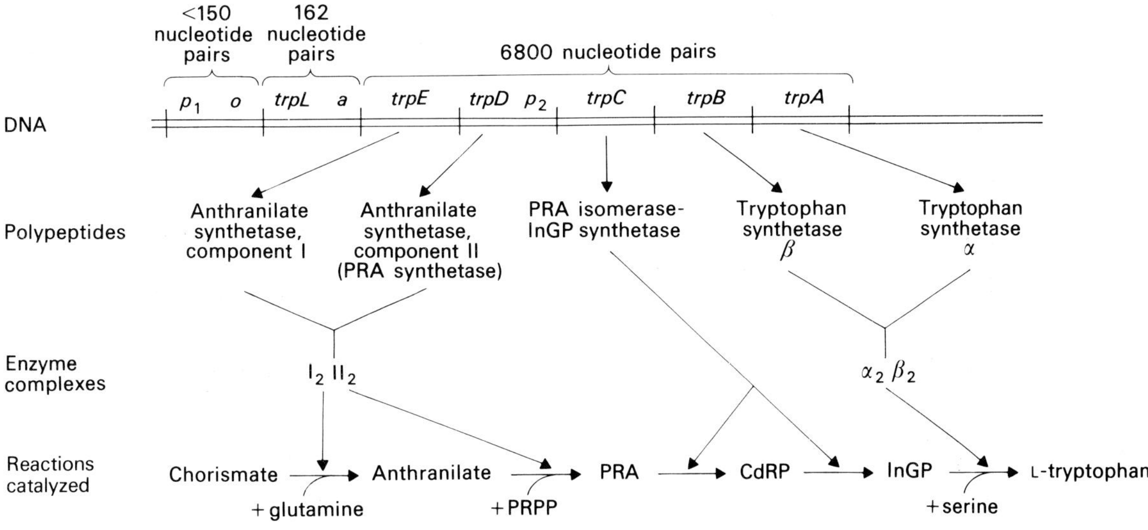

Key: PRPP = phosphoribosyl pyrophosphate
PRA = phosphoribosyl anthranilate
CdRP = 1-(o-carboxyphenylamino)-1-deoxyribulose 5-phosphate
InGP = indole-3-glycerol phosphate

Fig. 19.13. The tryptophan operon of *E. coli* showing the arrangement of structural genes, their polypeptide products, and the reactions they catalyze. *p*1 is the principal promoter; *o* is the operator; *p*2 is an internal promoter; *trpL* is the leader region, and *a* is the attenuator. (After the work of C. Yanofsky.)

operon, the increase of transcription brought about by the relief of repression is about 70-fold. Polycistronic mRNA transcribed from the operon contains a *leader sequence* of 162 nucleotides that is coded by the *trpL* region. This leader sequence precedes the translation initiation codon for the *trpE* polypeptide. A short transcript is also produced that contains only the first 142 nucleotides of *trp* mRNA. This latter RNA results from the termination of transcription at the *attenuator* (*a* in the figure), which is a regulatory site in the leader region of the operon. The attenuator plays an important role in the regulation of expression of the *trp* structural genes.

The existence of the leader sequence was shown by the sequencing of the 5′ segment of the *trp* operon mRNA synthesized either in vivo or in vitro. Internal deletions of the *trp* operon, which have one end between *p*1 and about nucleotide 130 of the leader sequence and the other end in one of the structural genes, generally result in an increase in operon expression. On the other hand, internal deletions that leave the leader region intact do not give this result. This indicated that there is a site called the attenuator, which normally limits operon expression in a process called **attenuation.** Quantitative hybridization experiments showed that the ratio of the number of copies of leader mRNA to structural gene *trp* mRNA is about 10:1, thus supporting the attenuator concept. Since short, 142-nucleotide-long transcripts are found, the attenuator must be at about that position in the leader region of the DNA. Indeed, in an in vitro transcription system, it has been shown that the RNA polymerase terminates transcription in the attenuator region.

How does the system function? Under repressing conditions (when cells are grown in the presence of excess tryptophan), the attenuator functions maximally to prevent distal transcription, thereby amplifying the effect of the repressor-operator interaction perhaps 10-fold. Transcription termination at the attenuator is regulated in response to changes in the number of charged tRNA.trp; the more charged tRNA.trp molecules there are, the fewer mRNAs are transcribed through the attenuator. Thus, when tryptophan is limited, repression is lifted, and attenuation is also relaxed. In extreme starvation conditions for tryptophan, the attenuation is completely relaxed. Thus it appears that the attenuator's function is to extend the range of expression of the operon beyond that possible by operator-repressor interaction alone.

Control at the attenuator is a function of the secondary structure of the RNA transcript of the leader region (Fig. 19.14). Some evidence for this has been obtained, for example, from RNA sequencing experiments, studies with genetic mutants, and investigations using an in vitro system. Specifically it has been found that there are two adjacent tryptophan codons (UGG) extremely close to the termination codon (UGA) on the leader mRNA. Between the UGA codon and the attenuator site (at about nucleotide 142 on the leader RNA), the nucleotide sequence has the potential to form various secondary structures as a result of hydrogen bonding between the bases. Thus, after RNA polymerase has synthesized this sequence, the RNA can fold up. Further, the leader RNA is being translated by ribosomes while transcription is proceeding, and the ribosomes are following the RNA polymerase closely. If high levels of tryptophan are present, the ribosome can read the tryptophan codons (since charged tRNA.trp molecules are present) and will terminate at the UGA codon, which is adjacent to the folded RNA. In this case, the ribosome's position will result in the formation of a particular secondary structure of the RNA, facilitating the base pairing of segments 3 and 4 (Fig. 19.14a). This in turn will affect the RNA polymerase's association with the DNA, and transcription will terminate at the attenuator. When the cells are starved for tryptophan, however, there are no charged tRNA.trp molecules present, and hence the ribosome will stall on the adjacent tryptophan codons

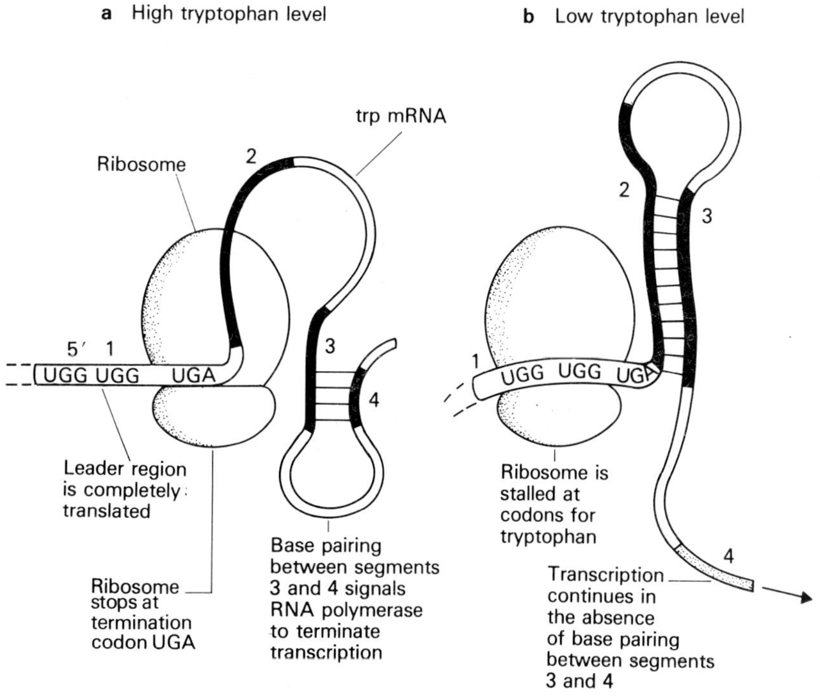

Fig. 19.14. Model for attenuation at the *trp* operon. (**a**) At high tryptophan levels, the ribosome stops at the chain-terminating codon of the leader transcript, resulting in base pairing of segments 3 and 4 of the transcript, which in turn stops further transcription. This phenomenon is attenuation. (**b**) At low tryptophan levels, attenuation is relieved since the ribosome stalls at adjacent tryptophan codons because of the lack of charged tRNA.trp molecules. As a result, segments 2 and 3 base pair and transcription is permitted to continue.

(Fig. 19.14b). In this event, a different secondary structure of the RNA occurs as a result of base pairing between segments 2 and 3. This will have no effect on RNA polymerase, and hence transcription will continue past the attenuator region and into the structural genes. Thus the continued transcription of a DNA segment can be controlled by the extent to which the RNA product is translated, and this with the repressor control at the operator results in a "fine-tuning" system for regulating the production of tryptophan biosynthetic enzymes. Apparently this mechanism is generally applicable to all bacterial operons that have an attenuator control region. (However, not all attenuator-controlled operons have repressor-operator systems. The *E. coli* histidine operon, for example, is controlled entirely by attenuation.)

Summary of operon regulation

In *E. coli* there are 100–200 operons that have been identified. Operons are ubiquitous throughout the prokaryotes (bacteria and phages), but with one possible exception (the galactose catabolism genes of yeast), there appear to be no operons in eukaryotes. As discussed earlier in this chapter, and as exemplified by the three operons described, the organization of genes and controlling regions into operons affords a simple and effective way of coordinating the activity of genes with related functions. As we have seen, with the basic operon organization of repressor gene and structural genes contiguous with promoter and operator regions, regulation of gene expression can be achieved in an inducible or repressible manner (negatively controlled operons) or by an activation mechanism (positively controlled operons). This regulation is at the transcriptional level, although for operons with attenuators the translational system may also play a role. In general, the control of gene expression by an operon system may be considered a fine control system, since it involves the regulation of one or a few transcriptional units by specific regulatory signals. This allows the organism to adapt rapidly to the changing environment. By contrast, in the next section we shall see how the overall growth and/or gene expression of a bacterium or phage is regulated by effects on (usually) several transcriptional units.

Major control of transcription and translation

The overall control of bacterial cell growth and division or of phage reproduction is related to macromolecular synthesis and particularly to RNA and protein synthesis. This section considers some examples of general regulation at the transcriptional and translational levels in prokaryotes.

Stringent control

Bacteria such as *E. coli* have evolved a general cellular regulatory system to survive under harsh conditions. In the laboratory such conditions can be imposed, for example, by depriving an amino acid autotrophic strain of the required amino acid or by moving the cells rapidly from a medium containing high amounts of amino acids to a medium containing no amino acids (a nutritional shift-down). Wild type *E. coli* cells respond in a stringent way to these conditions, and studies of this **stringent response** have shown that there is a regulatory mechanism that coordinates protein synthesis with the activities of a variety of other cellular activities. In other words, the starvation conditions described bring about a plethora of effects, including the inhibition of the rates of many disparate metabolic processes, including the accumulation of RNA, the replication of DNA, the biosynthesis of lipids, carbohydrates, nucleotides, peptidoglycans, and glycolytic intermediates; the stimulation of protein degradation processes; and the inhibition of many membrane transport processes. In the absence of these regulatory mechanisms, starved cells would grow in an uncontrolled manner, with all components (except proteins) accumulating in large amounts.

In a rapidly growing bacterium, a major proportion of the available energy is needed for the synthesis of ribosomes. Thus, blocking new ribosome synthesis in response to starvation or other environmental stresses is an effective means of energy conservation. When the nutritional deprivation is over, the stringent response mechanism facilitates rapid recovery and reinitiation of growth. The rapid change in the rate of ribosome biosynthesis in the stringent response results from a highly specific inhibition of stable RNAs (i.e. rRNAs and tRNAs) and of mRNAs for ribosomal proteins. The rate of total mRNA synthesis is only reduced a little after starvation, however. In other words, *E. coli* cells can selectively regulate the genes coding for the components of the translational apparatus. The principal control in this system appears to be exerted at the level of transcription initiation.

Bacterial cells that commence the stringent response upon amino acid starvation rapidly accumulate the unusual nucleotide, guanosine tetraphosphate (ppGpp; Fig. 19.15). That ppGpp plays a significant role in the stringent response is evidenced by the fact that relaxed (*relA*) mutants of *E. coli*, which do not exhibit the stringent response, do not accumulate ppGpp when cells are starved. The interpretation is that the *rel*$^+$ gene codes for a *stringent factor* (a protein of molecular weight 77,000 daltons) that is

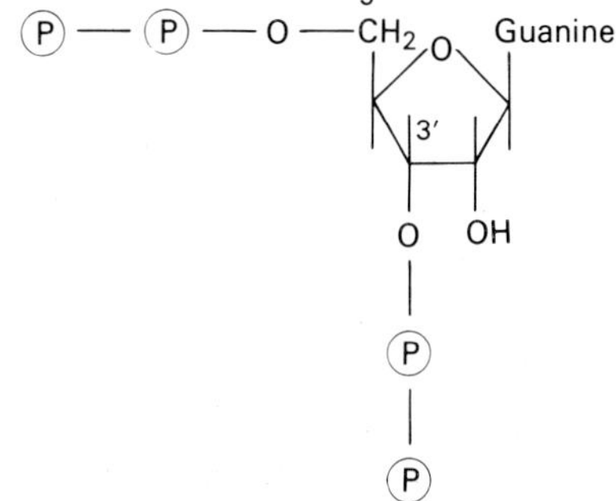

Fig. 19.15. Shorthand formula for the unusual guanine ribonucleotides implicated in the stringent response, guanosine tetraphosphate (5′-diphospho-guanosine-3′-diphosphate). (Ⓟ represents a phosphate group.)

responsible for the intracellular accumulation of ppGpp. The ppGpp molecules themselves are made on the ribosomes in response to the presence of nonaminoacylated (uncharged) tRNAs that accumulate during amino acid starvation. Thus, when uncharged tRNAs interact with ribosomes, the stringent factor can also bind, and this results in the formation of ppGpp.

There is evidence suggesting that RNA polymerase is a direct target of ppGpp and that the regulated process during the stringent response is transcription itself, most likely transcription initiation. In other words, interaction of RNA polymerase with the promoters of stringently controlled genes may be directly regulated by amino acid starvation. It seems that amino acid starvation selectively inhibits synthesis of certain RNA species by reducing, possibly through the involvement of ppGpp, the affinity of RNA polymerase for its binding sites so that those promoters for which RNA polymerase availability is limiting would no longer support efficient transcription initiation. Transcription initiation would not be impaired, however, at promoters with a higher affinity of RNA polymerase. Sequence analysis is beginning to reveal common features of stringently controlled promoters. For example, the structure of the −10 region of the RNA polymerase binding site does not appear to correlate with stringent regulation, whereas the RNA polymerase entry site at −35 deviates in sequence from nonstringently controlled promoters. Also, a GC-rich region—called a *discriminator region*—close to the transcription start point, may affect the ability of ppGpp-bound RNA polymerase to melt that region of DNA to single strands. Finally, with the exception of yeast, the stringent response is an exclusively prokaryotic phenomenon.

Control at the translational level

There are a number of ways by which gene expression can be modulated at the translational level. For example, one characteristic of the operon organization is that a polycistronic mRNA is produced when the structural genes are transcribed. For some operons, translation of the mRNA results in stoichiometrically equal amounts of the proteins coded for, whereas for other operons an unequal ratio of protein products is the case. For the lactose operon of *E. coli*, for example, there is a 10:5:2 ratio of *z*, *y*, and *a* gene products. The relative amounts of the products of a polycistronic mRNA in fact depend upon the primary structure of the mRNA. Specifically, at the end of each cistron on the message, there is at least one chain-terminating codon followed by an intercistron segment prior to the next translation initiation sequence. Thus the nucleotide sequences of these regions between stop and start codons can govern whether a ribosome "out of gear" will fall off the message before initiating the next polypeptide. Presumably in the lactose operon, approximately half of the ribosomes fall off between *y* and *a*. With the DNA cloning and sequencing techniques now available (see Chapter 12), it will be possible to obtain more detailed information about this regulatory control mechanism.

Some other examples of translational control have come from studies of the metabolic changes in a bacterial cell following phage infection. For instance, when T4 infects *E. coli*, there is rapid production of phage-specific mRNA molecules. These are translated selectively on the host-cell ribosomes because the latter become modified by the binding of a phage-specific protein. The exact nature of this functional modification is unknown, but it certainly serves as an effective means of channeling the host's metabolic activities toward phage reproduction.

A similar example of this type of translational channeling has been found in T2-infected *E. coli* cells. Messengers transcribed from the T2 genome contain no CUG (leucine) codons, and thus there is no need for the presence of leu-tRNAs with the anticodon for CUG in the host. The T2 phage codes for its own leu-tRNAs (for other leucine codons) and also codes for a ribonuclease that acts (with about 50% efficiency) to degrade the host's leu-tRNA with anticodons for CUG. This effectively favours the translation of phage-specific mRNAs over that of host-specific messengers.

In summary, gene expression can be regulated in prokaryotes in a number of ways. The regulatory mechanisms can operate at the transcriptional and/or the translational level. The examples presented illustrate the complexity of these control systems, and it is anticipated that more molecular details will be obtained when, for example, important controlling regions are cloned and sequenced.

Questions and problems

19.1 How does lactose bring about the induction of synthesis of β-galactosidase, M protein (permease), and transacetylase? Why does this not occur when glucose is also in the medium?

19.2 Operons produce polygenic (polycistronic) mRNA when they are active. What is a polygenic mRNA? What advantages, if any, do polygenic mRNAs confer upon a cell in terms of its function?

19.3 If an *E. coli* mutant strain synthesizes β-galactosidase whether or not the inducer is present, what genetic defect(s) might be responsible for this phenotype?

19.4 Distinguish the effects you would expect from (a) a missense mutation and (b) a nonsense mutation in the *z* (β-galactosidase) gene of the *lac* operon.

19.5 The elucidation of the regulatory mechanisms associated with the enzymes of lactose utilization in *E. coli* was a landmark in our understanding of regulatory processes in microorganisms. In formulating the operon hypothesis as applied to the lactose system, Jacob and Monod found that results from particular partial diploid strains were invaluable. Specifically in terms of the operon hypothesis, what information did the partial diploids provide that the haploids could not?

19.6 For the *E. coli lac* operon, write the partial diploid genotype for a strain that will produce β-galactosidase constitutively and permease by induction.

19.7 Mutants were instrumental in the elaboration of the model for the regulation of the lactose operon.

(a) Discuss why o^c mutants are cis-dominant but not trans-dominant.

(b) Explain why i^s mutants are trans-dominant to the wild-type i^+ allele, but i^- mutants are recessive.

(c) Discuss the consequences of mutations in the repressor gene promoter as compared with mutations in the structural gene promoter.

19.8 This question involves the lactose operon of *E. coli*. Complete the following table using + to indicate if the enzyme in question will be synthesized and – to indicate if the enzyme will not be synthesized.

	Inducer absent		Inducer present	
Genotype	β-galactosidase	Permease	β-galactosidase	Permease
1. $i^+p^+o^+z^+y^+$				
2. $i^+p^+o^+z^-y^+$				
3. $i^+p^+o^+z^+y^-$				
4. $i^-p^+o^+z^+y^+$				
5. $i^sp^+o^+z^+y^+$				
6. $i^+p^+o^cz^+y^+$				
7. $i^sp^+o^cz^+y^+$				
8. $i^+p^+o^cz^+y^-$				
9. $i^{-d}p^+o^+z^+y^+$				
10. $i^-p^+o^+z^+y^+/i^+p^+o^+z^-y^-$				
11. $i^sp^+o^+z^+y^-/i^+p^+o^+z^-y^+$				
12. $i^sp^+o^+z^+y^-/i^+p^+o^+z^-y^+$				
13. $i^+p^+o^cz^-y^+/i^+p^+o^+z^+y^-$				
14. $i^-p^+o^cz^+y^-/i^+p^+o^+z^-y^+$				
15. $i^sp^+o^+z^+y^+/i^+p^+o^cz^+y^+$				
16. $i^{-d}p^+o^+z^+y^-/i^+p^+o^+z^-y^+$				
17. $i^+p^-o^cz^+y^-/i^+p^+o^+z^-y^+$				
18. $i^+p^-o^+z^+y^-/i^+p^+o^cz^-y^+$				
19. $i^-p^-o^+z^+y^+/i^+p^+o^+z^-y^-$				
20. $i^-p^+o^+z^+y^-/i^+p^-o^+z^-y^+$				

19.9 What consequences would a mutation in the catabolite gene activator protein (CGA) gene of *E. coli* have for the expression of a wild-type *lac* operon?

19.10 The lactose operon is an inducible operon, whereas the tryptophan operon is a repressible operon. Discuss the differences between these two types of operons.

19.11 In the bacterium *Salmonella typhimurium*, seven of the genes coding for histidine biosynthetic enzymes are located adjacent to one another in the chromosome. If excess histidine is present in the medium, the synthesis in all seven enzymes is coordinately repressed, whereas in the absence of histidine all seven genes are coordinately expressed. Most mutations in this region of the chromosome result in the loss of activity of only one of the enzymes. However, mutations mapping to one end of the gene cluster result in the loss of all seven enzymes, even though none of the structural genes have been lost. What is the counterpart of these mutations in the *lac* operon sysetm?

19.12 What is the stringent control and how does this regulatory system work? Is stringent control found in any eukaryotic systems?

References

Lactose operon

Beckwith, J.R., 1967. Regulation of the lactose operon. *Science* **156**:597–604.

Beckwith, J.R. and D. Zipser, 1970. *The Lactose Operon.* Cold Spring Harbor Press, New York.

Beyreuther, K., 1978. Revised sequence for the *lac* repressor. *Nature* **274**:767.

Beyreuther, K., K. Adler, N. Geisler and A. Klemm, 1973. The amino acid sequence of *lac* repressor. *Proc. Natl. Acad. Sci. USA* **70**:3576–3580.

Calos, M.P., 1978. DNA sequence for a low-level promoter of the *lac* repressor gene and "up" promoter mutation. *Nature* **274**:762–765.

Dickson, R.C., J. Abelson, W.M. Barnes and W.S. Reznikoff, 1975. Genetic regulation: the *lac* control region. *Science* **187**:27–35.

Farabaugh, P.J., 1978. Sequence of the *lacI* gene. *Nature* **274**:765.

Gilbert, W., N. Maizels and A. Maxam, 1974. Sequences of controlling regions of the lactose operon. *Cold Spring Harbor Symp. Quant. Biol.* **38**:845–855.

Gilbert, W. and A. Maxam, 1973. The nucleotide sequence of the *lac* operator. *Proc. Natl. Acad. Sci. USA* **70**:3581–3584.

Gilbert, W. and B. Muller-Hill, 1966. Isolation of the *lac* repressor. *Proc. Natl. Acad. Sci. USA* **56**:1891–1898.

Jacob, F. and J. Monod, 1961. Genetic regulatory mechanisms in the synthesis of proteins. *J. Mol. Biol.* **3**:318–356.

Jacob, F. and J. Monod, 1965. Genetic mapping of the elements of the lactose region of *Escherichia coli. Biochem. Biophys. Res. Commun.* **18**:693–701.

Maizels, N., 1974. *E. coli* lactose operon ribosome binding site. *Nature New Biol.* **249**:647–649

Miller, J.H., C. Coulondre and P.J. Farabaugh, 1978. Correlation of nonsense sites in the *lacI* gene with specific codons in the nucleotide sequence. *Nature* **274**:770–775.

Ptashne, M. and W. Gilbert, 1970. Genetic repressors. *Sci. Amer.* **222**:36–44.

Arabinose operon

Englesberg, E., 1971. Regulation in the L-arabinose system. In *Metabolic Pathways,* H. Vogel (ed.), pp. 256–296, vol. 5. Academic Press, New York.

Englesberg, E., J. Irr, J. Power and N. Lee, 1965. Positive control of enzyme synthesis of gene *C* in the L-arabinose system. *J. Bacteriol.* **90**:946–957.

Englesberg, E., C. Squires and F. Meronk, 1969. The L-arabinose operon in *Escherichia coli B/r*: a genetic demonstration of two functional states of the product of a regulatory gene. *Proc. Natl. Acad. Sci. USA* **62**:1100–1107.

Englesberg, E. and G. Wilcox, 1974. Regulation: positive control. *Annu. Rev. Genetics* **8**:219–242.

Heffernan, L., R. Bass and E. Englesberg, 1976. Mutations affecting catabolite repression of the L-arabinose regulon in *Escherichia coli B/r*. *J. Bacteriol.* **126**:1119–1131.

Heffernan, L. and G. Wilcox, 1976. Effect of *araC* gene product on catabolite repression in the L-arabinose regulon. *J. Bacteriol.* **126**:1132–1135.

Irr, J. and E. Englesberg, 1970. Nonsense mutants in the regulator gene *araC* of the L-arabinose system of *Escherichia coli B/r*. *Genetics* **65**:27–39.

Lee, N., G. Wilcox, W. Gielow, J. Arnold, P. Cleary and E. Englesberg, 1974. In vitro activation of the transcription of *araBAD* operon by *araC* activator. *Proc. Natl. Acad. Sci. USA* **71**:634–638.

Sheppard, D.E. and E. Englesberg, 1967. Further evidence for positive control of the L-arabinose system by gene *araC*. *J. Mol. Biol.* **24**:443–454.

Wilcox, G., K.J. Clemetson, P. Cleary and E. Englesberg, 1974. Interaction of the regulatory gene product with the operator site in the L-arabinose operon of *Escherichia coli*. *J. Mol. Biol.* **85**:589–602.

Tryptophan operon

Bertrand, K., L. Korn, F. Lee, T. Platt, C.L. Squires, C, Squires and C. Yanofsky, 1975. New features of the structure and regulation of the tryptophan operon of *Escherichia coli*. *Science* **189**:22–26.

Bertrand, K., C. Squires and C. Yanofsky, 1976. Transcription termination in vivo in the leader region of the tryptophan operon of *Escherichia coli*. *J. Mol. Biol.* **103**:319–337.

Bertrand, K. and C. Yanofsky, 1976. Regulation of transcription termination in the leader region of the tryptophan operon of *Escherichia coli* involves tryptophan as its metabolic product. *J. Mol. Biol.* **103**:339–349.

Jackson, E.N. and C. Yanofsky, 1972. Internal promoter of the tryptophan operon of *Escherichia coli* is located in a structural gene. *J. Mol. Biol.* **69**:307–313.

Jackson, E.N. and C. Yanofsky, 1973. The region between the operator and first structural gene of the tryptophan operon of *Escherichia coli* may have a regulatory function. *J. Mol. Biol.* **76**:89–101.

Lee, F. and C. Yanofsky, 1977. Transcription termination at the *trp* operon attenuators of *Escherichia coli* and *Salmonella typhimurium*: RNA secondary structure and regulation of termination. *Proc. Natl. Acad. Sci. USA* **74**:4365–4369.

Miozzari, G.F. and C. Yanofsky, 1978. Translation of the leader region of the *Escherichia coli* tryptophan operon. *J. Bacteriol.* **133**:1457–1466.

Morse, D.E. and A.N.C. Morse, 1976. Dual-control of the tryptophan operon is mediated by both tryptophanyl-tRNA synthetase and the repressor. *J. Mol. Biol.* **103**:209–226.

Platt, T., C. Squires and C. Yanofsky, 1976. Ribosome protected regions in the leader-*trpE* sequence of *Escherichia coli* tryptophan messenger RNA. *J. Mol. Biol.* **103**:411–420.

Rose, J.K. and C. Yanofsky, 1974. Interaction of the operator of the tryptophan operon with repressor. *Proc. Natl. Acad. Sci. USA* **71**:3134–3138.

Squires, C., F. Lee, K. Bertrand, C.L. Squires, M.J. Bronson and C. Yanofsky, 1976. Nucleotide sequence of the 5′ end of tryptophan messenger RNA of *Escherichia coli*. *J. Mol. Biol.* **103**:351–381.

Squires, C.L., F. Lee and C. Yanofsky, 1975. Interaction of the *trp* repressor and RNA polymerase with the *trp* operon. *J. Mol. Biol.* **92**:93–111.

Yanofsky, C., 1976. Control sites in the tryptophan operon. In *Control of Ribosome Synthesis,* pp. 149–163. Alfred Benzon Symposium XI, Academic Press, New York.

Yanofsky, C., 1981. Attenuation in the control of expression of bacterial operons. *Nature* **289**:751–758.

Yanofsky, C., 1984. Comparison of regulatory and structural regions of genes of tryptophan metabolism. *Mol. Biol. Evol.* **1**:143–161.

Yanofsky, C. and R. Kolter, 1982. Attenuation in amino acid biosynthetic operons. *Annu. Rev. Genetics* **16**:113–134.

Yanofsky, C. and L. Soll, 1977. Mutations affecting tRNA.trp and its charging and their effect on regulation of transcription termination at the attenuator of the tryptophan operon. *J. Mol. Biol.* **113**:1457–1466.

Other examples of regulation

Calvo, J.M. and G.R. Fink, 1971. Regulation of biosynthetic pathways in bacteria and fungi. *Annu. Rev. Biochem.* **40**:943–968.

Cashel, M., 1975. Regulation of bacterial ppGpp and pppGpp. *Annu. Rev. Microbiol.* **29**:301–318.

Ihler, G. and D. Nakada, 1970. Selective binding of ribosomes to initiation sites on single stranded DNA from bacterial viruses. *Nature* **228**:239–242.

Lamond, A.I. and A.A. Travers, 1985. Stringent control of bacterial transcription. *Cell* **41**:6–8.

Miller, J.H. and W.S. Reznikoff, 1978. *The Operon.* Cold Spring Harbor Press, New York.

Summers, W.C., 1972. Regulation of RNA metabolism of T7 and related phages. *Annu. Rev. Genetics* **6**:191–202.

Chapter 20 Regulation of Gene Expression in Eukaryotes

Outline

General aspects of eukaryotic gene regulation

We are beginning to learn a fair amount about the logic of gene regulation systems in eukaryotic cells. For many protein-coding genes, transcriptional control is the most frequent level of control and regulation at this level is quite strict. Gene control can also occur at a number of post transcriptional levels. The mechanisms and molecules that are involved in the various types of control are unknown.

Potential sites for regulation of enzyme synthesis

A number of steps are involved in producing a functional mRNA from a protein-coding gene and then having it translated by the ribosomes. That is, the transcription units for mRNAs are often longer than the mature mRNAs. These long pre-mRNAs are processed in the nucleus to produce the final mRNA product. This includes:

1. The addition of a 5′ cap, usually before the rest of the RNA is completed. Essentially all RNA chains initiated by RNA polymerase II become 5′ capped, so no control operates at this level.
2. The addition of a poly(A) tail at the 3′ end.
3. The removal of specific spacer sequences (introns) and rejoining of the remaining RNA pieces (exons) to produce the mature mRNA.

With this in mind, the various levels of gene control in eukaryotes will now be discussed.

RNA transcription. The first level of control is the choice of RNA polymerases: RNA polymerase I transcribes rRNA genes, RNA polymerase II transcribes protein-coding genes, and RNA polymerase III transcribes 5S rRNA and tRNA genes. This specificity is clearly related to the nucleotide sequence of the promoter to which the appropriate RNA polymerase has affinity. Promoter sequences were discussed in Chapter 7. Apart from the strongly conserved regions necessary for the intiation of transcription, other sequences in the vicinity are presumably important in regulating access of RNA polymerases to DNA. These latter sequences undoubtedly control the rate at which transcription initiation occurs and, therefore, the amount of pre-RNA produced.

Thus, at least for RNA polymerase II, transcription initiation is believed to be the major controlling step in transcriptional control. However, relatively little is known about transcription termination and whether control is exerted at that level. For example, not all RNA polymerase II molecules that start a chain complete it.

RNA processing. While transcriptional control is the predominant level of control, it is not the only one. Some regulation occurs during RNA processing.

Processing of RNA polymerase II–catalyzed transcripts commences by the addition of the 5′ cap and 3′ poly(A) tail. No control appears to occur with respect to the addition of the 5′ cap. The situation regarding the poly(A) tail is less clear, however. Although the precision of transcription termination sites is not understood for RNA polymerase II transcription, the enzyme transcribes past the site for poly(A) addition in a number of well-studied genes. This means that poly(A) addition may need endonucleolytic cleavage and terminal addition of poly(A) to the cleaved 3′ end. Poly(A) addition usually precedes the splicing steps of mRNA processing.

Splicing of pre-mRNA to mature mRNA for most transcripts may be an automatic event for some transcription units; however, the primary transcript can give rise to two or more mRNAs that encode two or more different proteins. In these cases, differential processing occurs in response to particular regulatory signals.

Another regulatory choice at this level is whether a transcript will be processed or discarded in the nucleus. The latter is what is called *nuclear RNA turnover*. For many years it has been known that only a portion of the heterogeneous nuclear RNA is ever processed and found as cytoplasmic mRNA. While all transcripts contain a cap and over 90% of newly capped mRNAs found in polyribosomes (excluding the histone mRNAs) contain poly(A), poly(A) is added in the nucleus only to about 25% of all primary transcripts. Presumably the other 75% are discarded with time. What controls these events is unknown.

Cytoplasmic control. Once the mature mRNAs have been produced in the nucleus, at least a proportion of them are transported to the cytoplasm. The mechanism for this is unknown.

Once mRNAs have entered the cytoplasm, the fates of the molecules differ. Specifically there is a wide range of half lives for different specific mRNAs in the same cells and different half lives for the same mRNA in the same cell under different circumstances. These have different consequences to the cell in terms of the relative accumulation of the gene products involved, but the regulatory events operating here are essentially unknown.

Translational control. There is good evidence, for example, in embryonic development, for differential translation of completed mRNA molecules that are in the cytoplasm. It is also possible that tissue-specific preference for translation of specific mRNAs may occur.

Effector molecules. Regulation of the transcriptional and posttranscriptional events in the nucleus may be related to the types and concentrations of effector molecules (e.g. inducers, activators, repressors) that are transported into the nucleus from the cytoplasm. Controlling factors that alter the synthesis of these molecules and/or their transport into the nucleus can therefore affect the amount of enzyme ultimately synthesized.

In addition, other subtle control devices may be operating in the cell to regulate the amount of mRNA available for translation, and at present little is known about the regulatory signals for any of this. It is clear also that translational control is operant within eukaryotic cells. For example, the translation of a mRNA molecule may be affected by whether the ribosomes are membrane-bound or not, by specific factors that inhibit or stimulate ribosomes, by the availability of aminoacyl-tRNAs, or by the accessibility of initiation sequences of mRNAs to ribosomes.

Molecular aspects of transcription regulation

Two classes of proteins are associated with the DNA in chromatin: histones and nonhistones. The control of gene transcription must ultimately reside in the nucleotide sequences of the DNA so that the appropriate effector molecules can interact and control the amount and type of RNA produced. In addition, chromosomal proteins play a role in determining whether a region of DNA can be transcribed. As discussed in an earlier chapter, the histones are arranged in a regular fashion along the DNA, and thus it is unlikely that they play any specific role in the regulation of gene expression. Histones in vivo have been shown to be acetylated, phosphorylated, or methylated, but it is not known how these modifications relate to the transcriptional activity of chromatin.

On the other hand, there is much evidence implicating nonhistones in the regulation of gene expression in eukaryotes. For example, nonhistone proteins have tissue specificity and DNA binding specificity; they are present in higher amounts in transcriptionally active tissues as compared to inactive cell; they are much more diverse than histones; and certain specific classes of nonhistone proteins have, in fact, been linked with the induction of gene activity. In addition, if chromatin from transcriptionally active and inactive tissues is dissociated into DNA, histones, and nonhistones, it can be determined by reconstitution experiments which component confers the capacity for transcription. These have shown that the nonhistones are indeed the components that determine whether DNA can be transcribed. Thus it is currently believed that nonhistone proteins, presumably in response to specific signals, play a central role in the basic regulation of gene transcription in eukaryotes.

At this stage, however, we are still in the model-building stage for considering how nonhistones act specifically at the molecular level. For example, H. Weintraub and colleagues have obtained interesting information about the activation of the chick α-globin genes. They concluded that:

1. When the gene is activated, a whole region of the DNA in that area of the chromosome is activated, amounting to about 50 to 100 kbp. By comparison, the α-globin gene itself is about 1 kbp.

2. Within the activated region, the DNA is methylated to a lower level than in inactive DNA regions.

3. Within the activated region the DNA is significantly more sensitive to DNase I than is inactive DNA.

4. Within the activated region, specific nonhistone proteins are bound to the chromatin. These are called *high-mobility group* (HMG) *proteins* because they migrate rapidly in electrophoresis.

In summary, gene activation must involve an unwinding of the highly compact nucleosome secondary and tertiary structure so that RNA polymerase and other regulatory molecules can access the DNA.

Regulation of gene expression in lower eukaryotes

The demonstration that gene regulation in prokaryotes often involves operons that are controlled in ways analogous to the lactose operon of *E. coli* prompted researchers to investigate whether operons existed in eukaryotes. Indeed much of the early model building concerning the regulation of enzyme synthesis in eukaryotes was influenced by the regulatory models of enzyme synthesis in bacteria. However, as we shall see, enzyme synthesis is regulated in different ways in eukaryotes.

Eukaryotes have many basic similarities to prokaryotes. For example, the processes of DNA replication, transcription, and translation are more or less the same. However, eukaryotes are vastly more complex, with discrete cellular compartments (nuclei, mitochondria, chloroplasts, etc.) that determine the organization of the process. In this regard the lower eukaryotes, and particularly the fungi, have proved to be useful model systems for the study of gene regulation since they are typically eukaryotic in their cellular structure and genetic organization, yet they are microorganisms that can be handled in ways very similar to bacteria. Further, these organisms are simple and live in environments that are subject to rapid changes. Like bacteria, lower eukaryotes must be able to adapt rapidly at the gene expression level when such changes occur.

Yeast and *Neurospora* will be discussed in this section. These organisms have a genome complexity of approximately ten times that of *E. coli*. Early studies concentrated on determining whether operons exist in these fungi. Since they are highly amenable to genetic and biochemical investigations, it is relatively easy to isolate mutants affecting enzyme function and also the regulation of enzyme synthesis. This of course parallels the approach of Jacob and Monod in their studies of the regulation of the *E. coli* lactose operon. In general the fungal studies showed that, contrary to the findings in prokaryotes, genes with related function are *not* closely linked but rather tend to be scattered over the chromosomes in the genome. Nonetheless, it has been shown in both lower and higher eukaryotes that there is coordinate synthesis of all the enzymes in a particular biochemical pathway. This presumably involves a regulatory system different from that of prokaryotes. Characteristically, then, the gene products in eukaryotes consist of monocistronic mRNAs and not polycistronic mRNAs.

In some cases, however, evidence was obtained for apparent clustering of genes in fungi, and this raised the possibility of the existence of operons in eukaryotes. Three representative cases will now be discussed, and while the early information was highly suggestive of an operon organization, all the evidence to date runs counter to that possibility.

The galactose fermentation genes of yeast

The first three enzymes for galactose fermentation in yeast are galactokinase, α-D-galactose 1-phosphate uridyl transferase (c.f. galactosemia defect in humans; see Chapter 16), and uridine diphosphoglucose 4-epimerase. The genes for these enzymes, as defined by mutants, are *GAL1, GAL7,* and *GAL10*, respectively, and these are apparently closely linked on chromosome II in the order *GAL7-GAL10-GAL1*. The three genes are coordinately induced more than 5000-fold by the addition of galactose to the medium.

Regulatory mutants that affected the expression of the *GAL* genes were studied. One class of such mutants maps at a locus distinct from the structural genes, and these mutants are characterized by constitutive synthesis of the three *GAL* enzymes. These mutants are recessive to the wild-type allele. By analogy with the lactose operon of *E. coli*, the locus involved was called *i* and the mutants were designated i^-. A second class of regulatory mutants carry mutations that map to a locus, *GAL4*, which is unlinked to either the *i* locus or the *GAL* structural genes. These *GAL4* mutants are pleiotropically negative in that they are uninducible by galactose. A third class of regulatory mutants maps immediately adjacent to the *GAL4* locus at the *gal81* region, and these result in a constitutive production of galactose-fermenting enzymes. The *GAL81* mutants resemble o^c mutants of the lactose operon in that they behave as cis-dominants in diploids.

From the data H. Douglas and D. Hawthorne proposed a model for the regulation of expression of the *GAL* genes, and this is shown in Fig. 20.1.

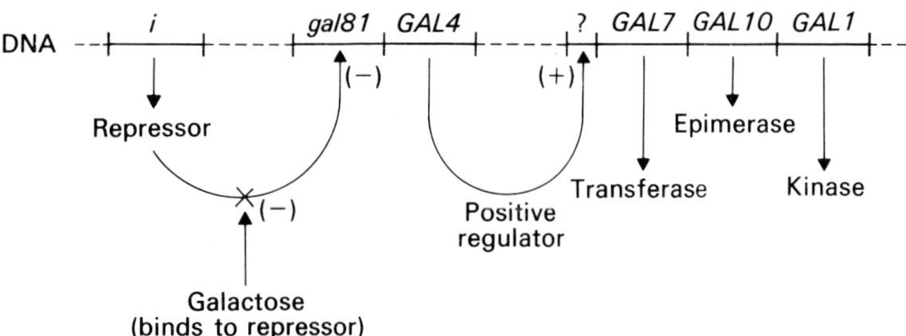

Fig. 20.1. The Douglas-Hawthorne model for the regulation of expression of the galactose fermentation genes in yeast.

Here the *i* gene produces a repressor that represses the expression of the *GAL4* gene by interacting with the adjacent *gal81* region if galactose is absent. If galactose is added, the repressor is inactivated and the *GAL4* gene can then be transcribed. Since *gal4* mutants are pleiotropically negative, the *GAL4* product is presumably a positive effector molecule required for the expression of the three *GAL* structural genes. How and where the *GAL4* effector interacts with the *GAL* gene cluster is not known. In other words, the i^--*gal81* relationship resembles the repressor-operator relationship of bacterial operons. Whether this is formally the case is a question for debate. The existence of i^s (super-repressible) mutations certainly supports the repressor concept in the model. Still unknown, however, is the exact nature of the *i* and *GAL4* gene products at the molecular level, how the *i* product interacts with the *gal81* region, and whether a polycistronic mRNA is produced from the *GAL* structural genes.

The genes for aromatic amino acid biosynthesis in *Neurospora*

Another potential candidate for an operon in eukaryotes are the genes for the early steps of the pathway for aromatic amino acid (phenylalanine, tyrosine, and tryptophan) biosynthesis in *Neurospora crassa*. These have been studied by N. Giles, M. Case and their colleagues for many years. The enzymes involved and the genes coding for them (as defined by mutants) are shown in Fig. 20.2.

Of particular significance was the discovery that a multienzyme aggregate (molecular weight 230,000 daltons) contained five different enzyme activities. These five activities are coded for by the so-called *arom* gene cluster of five adjacent genes: *aro-2*, *aro-4*, *aro-5*, *aro-9*, and *aro-1*. Genetic studies

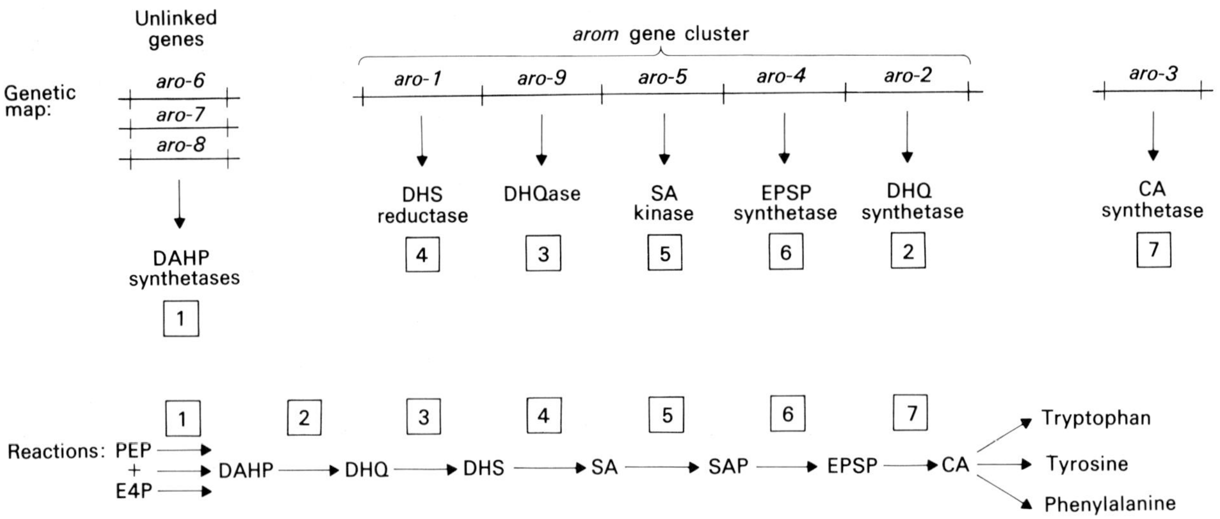

Key: PEP, phosphoenolpyruvate
E4P, erythrose 4-phosphate
DAHP, deoxy-arabinoheptulosonic acid phosphate
DHQ, dehydroquinic acid
DHQase, dehydroquinase
DHS, dehydroshikimic acid
SA, shikimic acid
SAP, shikimic acid phosphate
EPSP, enolpyruvyl shikimic acid phosphate
CA, chorismic acid

Fig. 20.2. The *arom* gene cluster of *Neurospora crassa* showing the enzymes encoded by the structural genes and the reactions they catalyze in the aromatic amino acid biosynthetic pathway (boxed numbers). (After the work of N. H. Giles, M. E. Case and colleagues.)

showed that mutations in a particular gene either affected the individual enzyme activity or caused the loss of two or more of the activities of the complex. These pleiotropic mutants were reminiscent of nonsense mutants in operons of bacteria where polar effects result during the translation of a single polycistronic mRNA. It was suggested, therefore, that the *arom* gene cluster coded for a polycistronic mRNA, and transcription commencing at the *aro*-2 gene. This supported the possibility that the *arom* gene cluster was an operon. However, F. Gaertner and Giles' group, working independently, have shown that the so-called *arom* gene cluster is actually a *single* structural gene that codes for a single polypeptide of molecular weight 115,000 daltons. This dimerizes to produce an enzyme that has the five enzyme activities just discussed. The separate polypeptides that were found in early investigations have been shown to be artifacts of the cellular fractionation techniques where the pentafunctional polypeptide is cleaved by endogenous protease activity.

Thus the *arom* system is not an operon but a fusion of five ancestral genes into one. This cluster gene has one promoter region. This is not the only example of a multifunctional polypeptide in eukaryotes; a number of other examples are known, particularly in the lower eukaryotes. The point it illustrates is that when one breaks open the cell and examines the contents, the results one obtains do not necessarily reflect the situation in vivo.

Regulation of quinic acid metabolism in *Neurospora*

N. Giles, M. Case, and their colleagues have also studied the regulation of expression of the *qa* genes of *Neurospora*, genes that are involved in the metabolism of quinic acid (QA) as a carbon source. From genetic and DNA-sequencing experiments, the *qa* genes have been shown to be clustered and to involve five structural genes and two regulatory genes (Fig. 20.3).

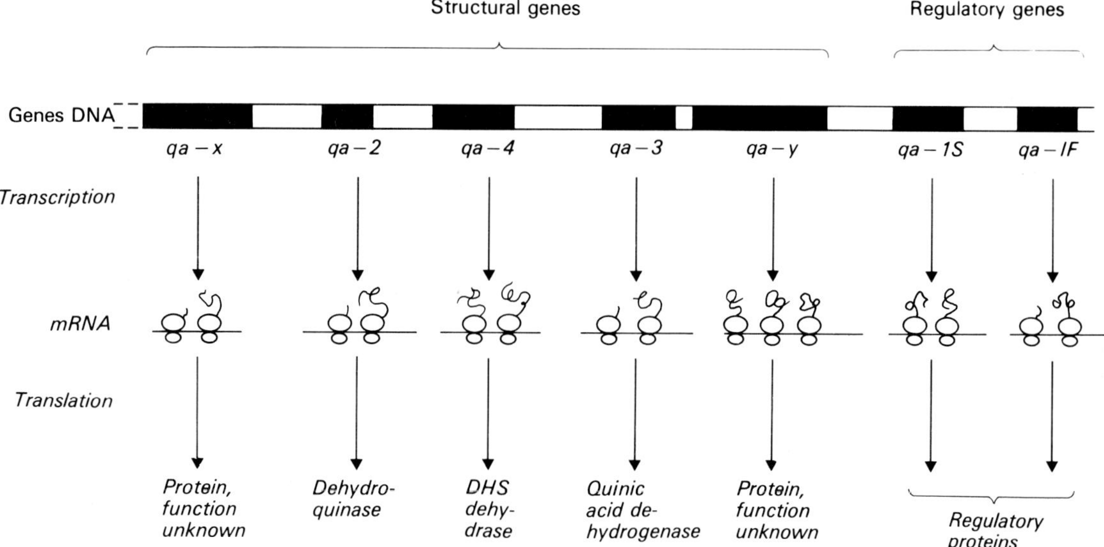

Fig. 20.3. Organization of the structural and regulatory genes in the *qa* gene cluster of *Neurospora*. (After the work of N. H. Giles, M. E. Case and colleagues.)

Transcription of the structural genes is induced 300- to 1000-fold by the addition of QA to the medium. The structural genes *qa-3*, *qa-2*, and *qa-4* code for the enzymes for the first three steps in the catabolism of QA (i.e. quinic acid dehydrogenase, catabolic dehydrogenase, and dehydroshikimic acid dehydrase, respectively) (Fig. 20.4). Note that both this catabolic pathway and the aromatic amino acid biosynthetic pathway just described have a dehydroquinase involved. There is no overlap between the pathways,

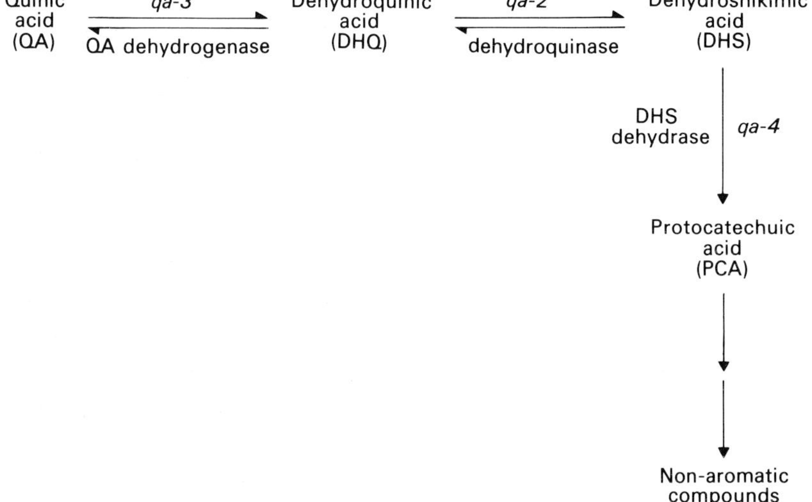

Fig. 20.4. The catabolic quinate-shikimate pathway showing the reaction steps, the enzymes involved, and the structural gene loci for the enzymes in *Neurospora crassa*. The pathway is induced when aromatic compounds reach a high level in the cell and the enzymes produced break them down to nonaromatic compounds. (After the work of N. H. Giles, M. E. Case and colleagues.)

however, since the multifunctional polypeptide of the biosynthetic pathway serves to channel the intermediates effectively.

The other two structural genes, *qa-x* and *qa-y*, were identified by analysis of the DNA sequence of the *qa* region. It is known that these two structural genes are transcribed and that they respond to QA induction, but the functions of their protein products are currently unknown.

At the *qa-y* end of the *qa* structural gene cluster are two regulatory genes *qa-1S* and *qa-1F* (Fig. 20.3). A regulatory gene adjacent to a cluster of structural genes is reminiscent of the organization of prokaryotic operons, but in the *qa* system there is no evidence either for an operator region or for a polycistronic mRNA. In fact, there is solid evidence that each structural gene codes for a distinct mRNA. Thus, even though the genes respond coordinately to QA induction, they do so by recognizing the induction signal independently.

Studies of mutations of the regulatory genes have led to some understanding of the regulation of expression of the *qa* structural genes. The regulatory gene mutations were studied both in the normal haploid cells and also in heterokaryons involving nuclei with different genetic constitutions in order to study dominance relationships. Mutants of the *qa-1S* gene are either constitutive and recessive (*qa-1S*^{c} alleles) or noninducible and semidominant (qa-*1S*$^{-}$ alleles). Mutants of the *qa-1F* gene (*qa-1F*$^{-}$ alleles) are noninducible and recessive even in the presence of a (constitutive) *qa-1S*^{c} allele. The following model has been proposed: *qa-1F* codes for an activator protein needed for transcription of itself and for all the structural genes except *qa-x*. *qa-1S* codes for a repressor protein that interacts with QA, in which state it has no effect on *qa-1F* transcription. In the absence of QA, however, the *qa-1S* protein blocks transcription of the *qa-1F* activator protein so that most of the structural genes are repressed. The exact role of *qa-x* in the *qa* system remains to be defined.

Thus the *qa* gene cluster is not like the *arom* gene cluster, since the *qa* genes clone for individual mRNAs and distinct protein products. The absence of a polycistronic mRNA or of an operator-like region in the *qa* gene system also distinguishes the *qa* gene cluster from prokaryotic operons.

In summary, in most instances genes with related functions in eukaryotes tend to be unlinked, although coordinately regulated. In lower eukaryotes in particular, where more extensive genetics has been possible, there are a number of examples of clustered genes but no definitive example of a bacterial-like operon. At least in some instances the gene cluster may actually be a single gene that codes for a multifunctional polypeptide.

Regulation of gene expression in higher eukaryotes

Higher eukaryotes are characterized by the differentiation of cells into tissues, organs, etc., that have specific functions. In this they differ markedly from the comparatively undifferentiated lower eukaryotes. The following section will concentrate on animal systems, in particular the vertebrates.

With the great complexity of cell specialization in higher eukaryotes come different problems in terms of the regulation of gene expression. For example, the specialized cells in these organisms are not subjected to drastic changes in the environment as is the case for lower eukaryotes and prokaryotes. This results in the fact that animals have homeostatic mechanisms that maintain relatively constant extracellular and intracellular environments. This is mediated by the blood, which has a fairly constant composition that is maintained by a variety of mechanisms. In vertebrates, for example, this is controlled by hormones. Thus animal cells are generally not exposed to large changes in the concentration of metabolites or substrates and therefore there is less need for rapid changes in the rates of enzyme synthesis. Characteristically, then, such changes are less frequent and of less magnitude than in lower eukaryotes or bacterial cells. For example, the enzyme ornithine decarboxylase, one of the most rapidly responding enzymes, exhibits a maximum increase of only 10- to 20 fold in four hours when induced. Contrast this with the 1000-fold induction of β-galactosidase within minutes in *E. coli*.

Before discussing the role of hormones in the regulation of enzyme synthesis, it must be stated that there are indeed enzyme induction and repression mechanisms operating in higher eukaryotes that are similar to those in prokaryotes. Because of the low number of regulatory mutants available in animal cells, there are relatively few systems that have been investigated in this regard. By contrast, the actions of hormones on gene expression have been studied extensively, and in the following section, some of the information that has been obtained will be considered.

Regulation of enzyme synthesis by hormones

A *hormone* may be defined as an effector molecule produced in low concentrations by one cell that evokes a physiological response in another cell. In vertebrates, a large number of classes of molecules have been shown to have hormonal activity, including polypeptides, amino acids, fatty acid derivatives, and steroids. Some of the hormones act directly on the cell's genome whereas others act at the cell surface, thereby activating membrane-bound adenyl cyclase to produce cyclic AMP (cAMP; 3′, 5′-cyclic adenosine monophosphate) from ATP (Fig. 20.5). The cAMP acts as a "second messenger" to evoke the intracellular effects observed following hormonal release.

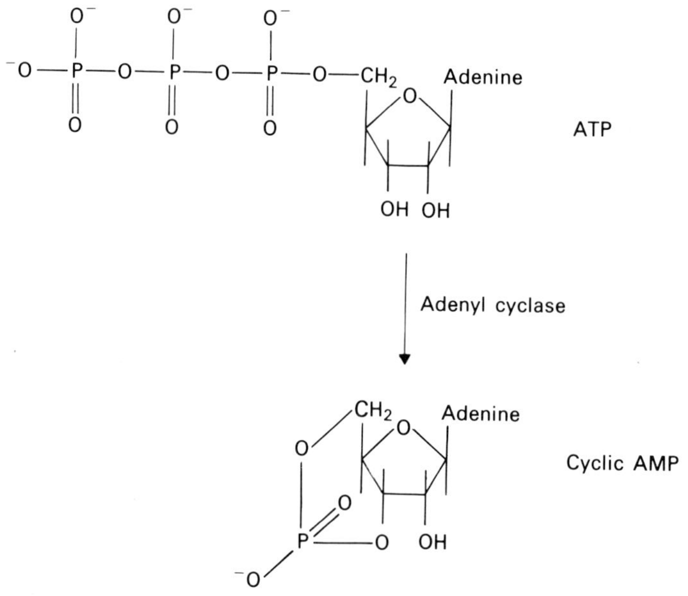

Fig. 20.5. The production of cyclic AMP from ATP in a reaction is catalyzed by adenyl cyclase.

Hormones act on specific target tissues that possess receptors capable of recognizing and binding to that particular hormone. For most of the polypeptide hormones, the receptors are on the cell surface, whereas the receptors for steroid hormones are in the cytoplasm.

Model for steroid hormone action

Steroid hormones are biosynthetically derived from sterols, which occur only in eukaryotic cells. Examples of steroid hormones are given in Fig. 20.6. In

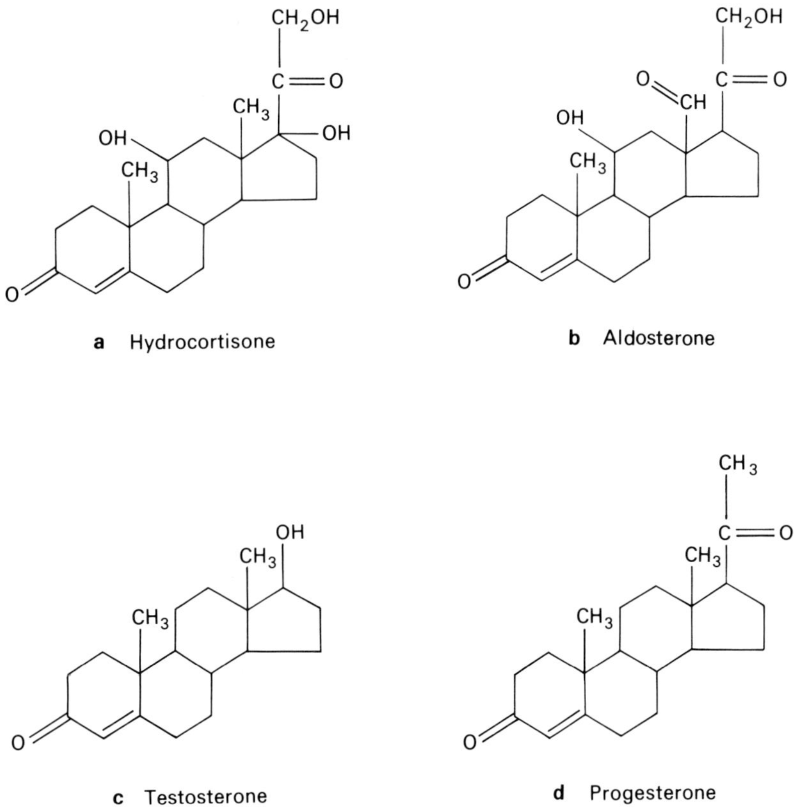

Fig. 20.6. Structures of some mammalian steroid hormones. All share a common four-ring structure. The small differences in the side groups are responsible for marked differences in the physiological effects in the animal. (**a**) Hydrocortisone—this is one of the glucocorticoid hormones produced by the cortex of the adrenal gland; it acts primarily to regulate carbohydrate and protein metabolism. (**b**) Aldosterone—this is a mineralocorticoid hormone secreted by the adrenal cortex, and it acts to regulate salt and water balance. (**c**) Testosterone—this is made by the testes and is responsible for the production and maintenance of male sexual characteristics and for the stimulation of sperm production. (**d**) Progesterone—this hormone is produced by the ovary and the placenta and, with estrogen, is needed to prepare and maintain the uterine lining for embryo implantation and for the ensuing pregnancy.

general, each class of steroid hormones mediates its biological response by binding to an intracellular receptor protein that is confined to target tissues. The interaction of the hormone with its receptor protein brings about a change in the structure of the protein such that there is increased affinity of the steroid-receptor complex for DNA. While steroid-receptor complexes bind to all DNAs, it is the high affinity of steroid-receptor complexes with *specific* DNA sequences that is important in bringing about the changes in gene expression. As a result of this specific binding, only a small number of genes within the target cell become transcriptionally activated as a result of interaction with the steroid hormone.

Relatively little is known about the structure of steroid receptors with regard to their action in affecting gene expression. Studies of unactivated receptors have indicated that there are great similarities among receptors for various classes of steroids. In all cases studied to date, the unactivated receptor exists as a multimer with a molecular weight of 200,000–300,000 daltons, and with sedimentation values of 8–10S. The activated form of steroid receptors always have lower S values than the unactivated forms. In the typical case, activation of an 8–10S receptor leads to the production of a

3–4S activated form, suggesting that the most stable activated form of steroid receptors is a monomeric structure. It should be pointed out, however, that there is evidence that a multimeric form is involved in gene activation.

There is no question that a great deal of progress has been made in recent years in establishing the structure of steroid receptors and the nature of their interaction with the steroid molecule itself. Further, in some cases information is coming to hand about the nature of the DNA sequences involved in high-affinity binding of the steroid-receptor complexes in the nucleus. However, much more information is needed before we will have an understanding of the precise mechanism(s) by which the steroid-receptor complex activates genes. Some possibilities of how this might occur are:

1. Steroid-receptor complexes might reverse a negative effect on gene transcription, perhaps by removing specific repressor proteins, thereby exposing promoters.

2. Direct binding of the steroid-receptor complex with RNA polymerase might stimulate polymerase activity.

3. Steroid-receptor complexes might alter the conformation of the chromatin (e.g. by removing nuclear proteins) in such a ways as to facilitate RNA polymerase binding to a promoter.

In summary, hormones act to integrate metabolism in higher eukaryotes. In some cases (e.g. in the liver) the coordination of metabolic activities involves the combined actions of several hormones. It is generally accepted that hormones act at the transcriptional level.

Long-term genetic regulation in higher eukaryotes

The examples of the previous section all show short-term regulation of gene expression in higher eukaryotes, that is, where adjustments are made in cellular activity in response to environmental changes (e.g. hormone release). There are, however, two properties of higher eukaryotes (and some lower eukaryotes) that reflect the long-term regulation of gene expression: development and differentiation. These processes are really outside the area of genetics and in the areas of developmental biology and embryology, and therefore only a very general discussion of them will be given in this text.

Definitions of development and differentiation

Development is the process of regulated growth that results from the interaction of the genome with the cytoplasm and the environment. Development involves a programmed sequence of phenotypic events that are usually irreversible. **Differentiation** is one aspect of development. It involves the formation of different cell, tissue, and organ types from a zygote through specific regulatory processes that control gene expression. In short, the genome carries the potential for the adult organism, but the final product results from complex gene-environment interactions.

In general, development is an irreversible or virtually irreversible process. We can consider development to involve at least three interacting processes:

1. The replication of the genetic material.
2. The growth of the organism as a result of cellular metabolic activity.
3. Cellular differentiation by which genetically identical cells diverge in their structure and function to give rise to organized tissues, which in turn associate to form organs.

Differentiation is the formation of different types of cells and tissues from a zygote by the specific regulation of gene activity in temporal and spatial ways.

General aspects of development and differentiation

A number of general attributes of development and differentiation can be related to the genetic concepts that have been presented in this text.

Nuclear DNA remains constant. Early models for gene involvement in development included one where there was a programmed loss of nuclear DNA as the organism developed, or rather that the development processes that occurred were the result of losses of particular genes in an ordered sequence. That is not true. Rather, cells of differentiated tissues contain the same genomic content of DNA as the fertilized egg (although some differentiated cells may be polyploid). On elegant experiment that showed this to be the case was performed by J. Gurdon. He transplanted a nucleus from the gut cell of a tadpole of *Xenopus laevis*, the South African clawed toad, into an unfertilized egg of that organism from which the nucleus had been removed. The result was that the egg, once stimulated, developed into a normal adult toad. Thus the differentiated cells of the tadpole exhibited **totipotency**; that is, they contained all of the genetic information required for the egg to develop and differentiate into an adult organism.

The DNA is transcribed in a programmed way. All of the available evidence shows that development and differentiation involve a detailed program of transcription of the DNA, which occurs in response to specific activator and repressor molecules. Two lines of evidence to support this will be considered here.

1. As mentioned previously, it is possible to quantify the extent to which RNA isolated from a cell hybridizes to the nuclear DNA. In general, the experiments involve RNA and DNA molecules that are radioactively labeled

with different isotopes. A refinement of this technique is competitive DNA : RNA hybridization where unlabeled RNA is first allowed to hybridize with the DNA before the radioactive RNA is added. If the RNAs are from the same tissue, the unlabeled RNA should effectively block all of the DNA sites to which the labeled RNA can bind, and this would be detected as 100% competition when the radioactivity is measured. If the RNAs are from two different tissues, however, the effect on the amount of radioactive RNA that will bind to the DNA will depend on how many of the RNA species were synthesized in common by both the tissues. One can do this experiment using mRNAs isolated from different tissues (e.g. lung, liver, kidney, muscle) of the same organism. Results of such an experiment show that there is limited competition between the mRNAs of the tissues in the hybridization and leads to the conclusion that differentiated cells reflect differences in the gene transcription activity. This correlates well with other studies showing different spectrums of enzymes and relative differences in enzyme amounts and activities in different tissues of the same organism. These differences must, of course, reflect differential gene activity.

2. In certain insects such as *Drosophila,* the chromosomes of the larval salivary gland cells undergo *polytenization*. That is, the chromosomes replicate up to a thousand times but without there being cell division. The replicated chromosomes remain together as the **polytene chromosomes**, which show characteristic banding patterns. A diagrammatic representation of a polytene chromosome is shown in Fig. 20.7. The bands are thought to represent the coding sequences of genes, while the function of the interband region is not known. The DNA is continuous along the length of these chromosomes. In *Drosophila*, there are three larval stages, each separated by a molting event. The last larval stage is followed by pupation. This is an interesting model system, therefore, for studying gene activity (since the genes are visible, if indeed the bands are genes) during development. In fact, specific bands "puff" in a particular pattern related to the time of larval development. The puffs are localized loosenings of the compact polytene chromosome that occurs so that RNA polymerase can initiate transcription. Indeed it can be shown that RNA is being actively synthesized in the puffs, and thus the puffs are visual evidence of gene activity. More importantly, the puffing patterns are reproducible, and, as has been stated, they are tissue and developmental stage specific.

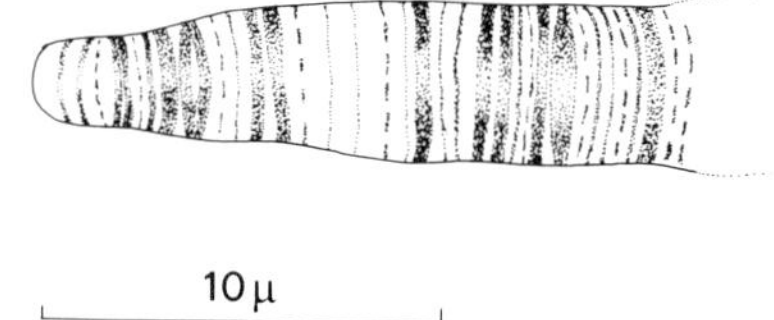

Fig. 20.7. Diagrammatic representation of part of a polytene chromosome of the fruit fly, *Drosophila melanogaster*. The bands (solid and dotted) may be seen under the light microscope and are thought to represent the genes. The diagram actually shows one end of the 414 μm long polytenized X chromosome. There are 1024 bands on this chromosome.

Gene-cytoplasm interactions. The gene activities of a cell are affected by the cytoplasm. Thus when certain genes are turned on during differentiation, particular proteins are synthesized, some of which have a regulatory role in maintaining the differentiated state of the cell. For example, in Gurdon's transplantation experiment discussed earlier, we made the point that the nucleus carried all the genetic information necessary for the egg to develop into an adult. However, the fact that the egg cell behaved as an egg cell and not as a tadpole gut cell is an example of how the activity of the nuclear genome is controlled by the cytoplasmic state.

In conclusion, development and differentiation involve long-term regulation of gene expression. Our discussion here has only scratched the surface of the information available for these processes. Nevertheless, we are still a long way from understanding them from a detailed molecular point-of-view and we have a lot to learn about nuclear-cytoplasmic, cell-cell, and cell-environment interactions as they relate to developmental processes.

Questions and problems

20.1 Are there operons in eukaryotes? Discuss the features of the *arom* and *qa* gene clusters in *Neurospora crassa* that suggested they might be eukaryotic operons. What characteristic feature or features of bacterial operons do these eukaryotic systems lack?

20.2 Discuss the regulation of gene expression of the *qa* gene cluster of *Neurospora crassa*. How does the functional arrangement of the gene product(s) differ from that of the *arom* gene cluster?

20.3 In what ways do the functions of eukaryotic multifunctional proteins (which have several enzyme activities for a biosynthetic pathway) resemble and differ from the functions of prokaryotic operons?

20.4 In eukaryotic organisms there are a large number of copies (usually more than a hundred) of the genes that code for ribosomal RNA, yet there is generally only one copy of each gene that codes for each ribosomal protein. Explain why.

20.5 What is a hormone?

20.6 How do hormones participate in the regulation of gene expression in eukaryotes?

20.7 Distinguish between the terms development and differentiation.

20.8 What is totipotency? Give an example of the evidence for the existence of this phenomenon.

20.9 What are polytene chromosomes? Discuss the molecular nature of the puffs that occur in polytene chromosomes during development.

20.10 In experiment A, ^{3}H-thymidine (a radioactive precursor of DNA) is injected into larvae of *Chironomus*, and, when the polytene chromosomes of the salivary glands are later examined by autoradiography, the radioactivity is seen to be distributed evenly throughout the polytene chromosomes. In experiment B, ^{3}H-uridine (a radioactive precursor of RNA) is injected into the larvae, and the polytene chromosomes are examined. The radioactivity is first found only around puffs; later, radioactivity is also found in the cytoplasm. In experiment C, actinomycin D (an inhibitor of transcription) is injected into larvae and then ^{3}H-uridine is injected. No radioactivity is found associated with the polytene chromosomes and few puffs are seen. Those puffs that are present are much smaller than the same puffs found in experiments A and B. Interpret these results.

References

Ashburner, M., C. Chihara, P. Meltzer and G. Richards, 1974. Temporal control of puffing activity in polytene chromosomes. *Cold Spring Harbor Symp. Quant. Biol.* **38**:655–662.

Baker, W., 1978. A genetic framework for *Drosophila* development. *Annu. Rev. Genetics* **12**:451–470.

Britten, R.J. and E.H. Davidson, 1969. Gene regulation of higher cells: a theory. *Science* **165**:349–357.

Brown, D.D., 1981. Gene expression in eukaryotes. *Science* **211**:667–674.

Brown, D.D. and I.B. Dawid, 1969. Developmental genetics. *Annu. Rev. Genetics* **3**:127–154.

Burgoyne, L., M.E. Case and N.H. Giles, 1969. Purification and properties of the aromatic (*arom*) synthetic enzyme aggregate of *Neurospora crassa. Biochim. Biophys. Acta* **19**:452–462.

Calvo, J.M. and G.R. Fink, 1971. Regulation of biosynthetic pathways in bacteria and fungi. *Annu. Rev. Biochem.* **40**:943–968.

Case, M.E. and N.H. Giles, 1971. Partial enzyme aggregates formed by pleiotropic mutants in the *arom* gene cluster of *Neurospora crassa. Proc. Natl. Acad. Sci. USA* **68**:58–62.

Case, M.E. and N.H. Giles, 1975. Genetic evidence on the organization and action of the *qa-1* gene product: a protein regulating the induction of three enzymes in quinate metabolism in *Neurospora crassa. Proc. Natl. Acad. Sci. USA* **72**:553–557.

Case, M.E. and N.H. Giles, 1976. Gene order in the *qa* gene cluster of *Neurospora crassa. Mol. Gen. Genetics* **147**:83–89.

Clever, U., 1968. Regulation of chromosome function. *Annu. Rev. Genetics* **2**:11–30.

Darnell, J.E., 1982. Variety in the level of gene control in eukaryotic cells. *Nature* **297**:365–371.

Davidson, E.H., 1976. *Gene Activity in Early Development,* 2nd ed. Academic Press, New York.

Davidson, E.H. and R.J. Britten, 1973. Organization, transcription and regulation in the animal genome. *Quart. Rev. Biol.* **48**:565–613.

Davidson, E.H. and R.J. Britten, 1979. Regulation of gene expression: possible roles of repetitive sequences. *Science* **204**:1052–1059.

Doerfler, W., 1983. DNA methylation and gene activity. *Annu. Rev. Genetics* **52**:93–124.

Douglas, H.C. and D.C. Hawthorne, 1966. Regulation of genes controlling synthesis of the galactose pathway enzymes in yeast. *Genetics* **54**:911–916.

Douglas, H.C. and D.C. Hawthorne, 1972. Uninducible mutants in the *gal i* locus of *Saccharomyces cerevisiae. J. Bacteriol.* **109**:1139–1143.

Gehring, W., 1979. Developmental genetics of *Drosophila. Annu. Rev. Genetics* **10**:209–252.

Giles, N.H., M.E. Case, C.W.H. Partridge and S.I. Ahmed, 1967. A gene cluster in *Neurospora crassa* coding for an aggregate of five aromatic synthetic enzymes. *Proc. Natl. Acad. Sci. USA* **58**:1453–1460.

Gurdon, J.B., 1968. Transplanted nuclei and cell differentiation. *Sci. Amer.* **219**:24–35.

Gurdon, J.B., 1974. *The Control of Gene Expression in Animal Development.* Harvard University Press, Cambridge, Massachusetts.

Hautala, J.A., J.W. Jacobson, M.E. Case and N.H. Giles, 1975. Purification and characterization of catabolic dehydroquinase, an enzyme in the inducible quinic acid catabolic pathway of *Neurospora crassa. J. Biol. Chem.* **250**:6008–6014.

Huiet, L., 1984. Molecular analysis of the *Neurospora qa-1* regulatory region indicates that two interacting genes control *qa* gene expression. *Proc. Natl. Acad. Sci. USA* **81**:1174–1178.

Lodish, H.F., 1976. Translational control of protein synthesis. *Annu. Rev. Biochem.* **45**:39–72.

Matsumoto, K., A. Toh-E and Y. Oshima, 1978. Genetic control of galactokinase synthesis in *Saccharomyces cerevisiae*: evidence for constitutive expression of the positive regulatory gene *gal 4*. *J. Bacteriol.* **134**:446–457.

Metzenberg, R.L., 1972. Genetic regulatory systems in *Neurospora. Annu. Rev. Genetics* **6**:111–132.

O'Malley, B.W., H.C. Towle and R.J. Schwartz, 1977. Regulation of gene expression in eukaryotes. *Annu. Rev. Genetics* **11**:239–275.

Reeves, R., 1984. Transcriptionally active chromatin. *Biochim. Biophys. Acta* **782**:343–393.

Revel, M. and Y. Groner, 1978. Post-transcriptional and translational controls of gene expression in eukaryotes. *Annu. Rev. Biochem.* **47**:1079–1126.

Ringold, G.M., 1985. Steroid hormone regulation of gene expression. *Ann. Rev. Pharmacol. Toxicol.* **25**:529–566.

Spelsberg, T.C., B.A. Littlefield, R. Seelke, G.M. Dani, H. Toyoda, P. Boyd-Leinen, C. Thrall and O.I. Kon, 1983. Role of specific chromosomal proteins and DNA sequences in the nuclear binding sites for steroid receptors. *Recent Prog. Horm. Res.* **39**:425–517.

Stein, G.S., T.C. Spelsberg and L.J. Kleinsmith, 1974. Nonhistone chromosomal proteins and gene regulation. *Science* **183**:817–824.

Tyler, B.M., R.F. Geever, M.E. Case and N.H. Giles, 1984. Cis-acting and trans-acting regulatory mutations define two types of promoters controlled by the *qa-1F* gene of *Neurospora. Cell* **36**:493–502.

Vardimon, L., D. Renz and W. Doerfler, 1983. Can DNA methylation regulate gene expression. *Recent Results Cancer Res.* **84**:90–102.

Yamamoto, K.R. and B.M. Alberts, 1976. Steroid receptors: elements for modulation of eukaryotic transcription. *Annu. Rev. Biochem.* **45**:721–746.

Chapter 21 Population Genetics

Outline

In previous chapters we have discussed the structure and function of genes as they have been studied in the laboratory. The organisms under investigation are often true-breeding strains so that differences seen in experiments are the result of the experimental treatments rather than the result of genotypic and hence phenotypic differences. Further, in laboratory experiments matings can be carried out with organisms whose genotypes are known. Indeed, from studies of this kind has come our basic conceptual understanding of the transmission of genetic material from one generation to the next.

By contrast the world outside the laboratory is very different. For instance, natural populations of an organism do not generally mate in the ordered way "preferred" by the geneticist. Also, in the "wild", the relative frequencies of alleles at a locus may vary over a wide range, whereas in the laboratory these frequencies are usually set at values convenient to the investigator. In this chapter we shall examine some of the basic principles of population genetics, the study of genes in natural (and sometimes laboratory) populations of organisms.

Definitions

Population. A Mendelian population is a group of interbreeding individuals. The largest possible population of a particular organism is the *species*. Of particular importance here is the gene exchange within the population from generation to generation. There are a number of factors that affect the genetic constitution of a population, including selection for particular alleles, mutation, migration into and out of the population of individuals, and genetic drift.

Gene pool. A gene pool is the total genetic information possessed by reproductive members of sexually reproducing individuals, that is, the sum of all alleles of all genes present in the population at a given time. Thus a Mendelian population, as distinct from any other group of individuals, is a group of individuals that share a common gene pool.

Allele frequencies. To analyze the genetics of populations, one must determine the frequencies of the alleles present so that changes can be detected over time and, hence, through evolution. An allele frequency (also called a *gene frequency*) is given by the number of copies of an allele divided by the sum of all alleles. In mathematical considerations of alleles in populations, the frequency of the dominant allele is commonly defined as p and that of the recessive allele as q. In a natural population in which only two alleles are being studied at a locus, the three possible genotypes may show any distribution, and hence p and q may have any values.

Hardy-Weinberg equilibrium

The **Hardy-Weinberg equilibrium** (or the Hardy-Weinberg law) is the situation where, in a large, randomly mating population with a closed gene pool, the allele frequencies remain constant from generation to generation. A closed gene pool means that there is no mutation, selection, drift, or migration. Let us consider an hypothetical case of two alleles *A* and *a* in a diploid organism. In a population of 300 individuals, suppose there are 148 *AA*, 125 *Aa*, and 28 *aa* individuals. From these values we can compute the frequencies of the *A* and *a* alleles. The frequency of the *A* allele is $[(2\times148)+125]/600 = 0.7$. (Here we are merely counting all the *A* alleles in the *AA*s, or 2×148, and the *A* alleles in the *Aa* individuals. The total number of alleles in 300 diploid individuals is 600.) Similarly, the frequency of the *a* allele is $[(2\times28)+125]/600 = 0.3$. If we now allow (or in this case require) this population to mate at random so that all possible pairings occur, we can predict the distribution of the three genotypes in the next generation from the calculated allele frequencies as shown in Fig. 21.1.

Allele of ♀	Allele of ♂	Frequency of pairing	Progeny Genotype	Progeny Frequency	Number in population of 300
A	*A*	0.7 × 0.7	*AA*	0.49	147
A	*a*	0.7 × 0.3	*Aa*	0.42	126
a	*A*	0.3 × 0.7			
a	*a*	0.3 × 0.3	*aa*	0.09	27

Frequency of *A* in progeny = 0.7

Frequency of *a* in progeny = 0.3

Fig. 21.1. Demonstration that random mating in a population will maintain genetic equilibrium if that population is in genetic equilibrium. The allele frequencies remain the same in the parental and progeny generations, as do the relative genotype frequencies.

As the figure shows, the relative numbers of the three genotypes closely match those of the parental generation. The frequency of the *A* allele is still 0.7 and that of the *a* allele is still 0.3. These relationships will hold at every successive generation if random mating is maintained.

It may be argued that the two sets of figures match closely since we calculated the allele frequencies from the original set of numbers. We can prove that it was not the case by considering a second hypothetical population with 300 individuals, 190 of which are AA, 40 *Aa,* and 70 *aa*. In this population also the allele frequency of *A* is 0.7 and of *a* 0.3. If this population were mated at random, the *progeny* genotypes would have the same relative frequencies that we calculated in Fig. 21.1, frequencies quite different from those of the parents in this second population.

To return to the first population, we showed that the allele and genotype frequencies remained the same as a consequence of random mating. In other words the population is in *genetic equilibrium*. This basic tenet was described independently in 1908 by G. H. Hardy and W. Weinberg, and it has become known as the *Hardy-Weinberg law* or the *Hardy-Weinberg equilibrium.*

For a population to remain at genetic equilibrium, there are a number of requirements:

1. Mating must be random through the population. This implies that all gametes are equally viable and able to participate in the fertilization event.

2. Mutation must not occur or, if it does, the forward (*A* to *a*) and back (*a* to *A*) mutation rates must be balanced, that is, in equilibrium.

3. The generations cannot overlap in the breeding sense.

4. There is no migration occurring to shift the relative frequency of alleles and genotypes.

5. The population is of infinite size so that sampling errors resulting from the finite size of the population do not exist.

6. Selection in the form of differential survival of fertility of the different genotypes does not occur.

All this would seem a tall order, but human and other populations have been studied that are at genetic equilibrium for particular alleles. It is worthwhile to note here, too, that if a population is not in Hardy-Weinberg equilibrium, it only requires one generation of random mating to establish an equilibrium that successive "rounds" of random mating will then maintain. This was demonstrated for the second hypothetical population previously discussed.

Formalization of the Hardy-Weinberg law

We start out with a gene pool of a randomly mating population at Hardy-Weinberg equilibrium of p *A* alleles and q *a* alleles. By definition $p+q = 1$. If these alleles pair at random, the frequencies of the three genotypes will be p^2 $AA+2pq\ Aa+q^2\ aa$ (a binomial distribution) and the allele frequencies are p $A+q\ a = 1$ (Fig. 21.2). We can now formerly prove that if we get this distribution in one generation, we shall get the same distribution in the next

	$p\ A$	$q\ a$
$p\ A$	$p^2\ AA$	$pq\ Aa$
$q\ a$	$pq\ Aa$	$q^2\ aa$

Genotype frequencies: $p^2\ AA + 2pq\ Aa + q^2\ aa = 1$

Allele frequencies: $p\ A + q\ a = 1$

Fig. 21.2. For a population in Hardy-Weinberg equilibrium with p *A* alleles and q *a* alleles, random pairing of alleles will give a genotype distribution of $p^2\,AA+2pq\,Aa+q^2\,aa = 1$.

generation if random mating has occurred. This is shown in Fig. 21.3 where all possible matings are considered in terms of the frequencies in which they will occur by random mating in the populations.

Mating ♀ × ♂	Frequency of mating		Progeny distribution *AA*	*Aa*	*aa*
AA × *AA*	$p^2 \times p^2$	$= p^4$	p^4	–	–
AA × *Aa* *Aa* × *AA*	$p^2 \times 2pq$ $2pq \times p^2$	$= 4p^3q$	$2p^3q$	$2p^3q$	–
Aa × *Aa*	$2pq \times 2pq$	$= 4p^2q^2$	p^2q^2	$2p^2q^2$	p^2q^2
AA × *aa* *aa* × *AA*	$p^2 \times q^2$ $q^2 \times p^2$	$= 2p^2q^2$	–	$2p^2q^2$	–
Aa × *aa* *aa* × *Aa*	$2pq \times q^2$ $q^2 \times 2pq$	$= 4pq^3$	–	$2pq^3$	$2pq^3$
aa × *aa*	$q^2 \times q^2$	$= q^4$	–	–	q^4

∴ Frequency of progeny is:

$$AA = p^4 + 2p^3q + p^2q^2 = p^2(p^2 + 2pq + q^2) = p2\,(1)$$

$$Aa = 2p^3q + 4p^2q^2 + 2pq^3 = 2pq(p^2 + 2pq + q^2) = 2pq\,(1)$$

$$aa = p^2q^2 + 2pq^3 + q^4 = q^2(p^2 + 2pq + q^2) = q^2\,(1)$$

Cancelling the term in brackets, the distribution is:

$$p^2\,AA + 2pq\,Aa + q^2\,aa$$

Fig. 21.3. Algebraic demonstration that, for a population in genetic equilibrium, random mating will produce a progeny population that has the same relative frequencies of *AA*, *Aa* and *aa* genotypes as the parental generation.

Applications of the Hardy-Weinberg law

For populations that are in genetic equilibrium , the Hardy-Weinberg Law is very useful in making predictions, for example, of genotype frequencies (and therefore phenotype frequencies) in ensuing generations. For instance, consider a human population where the inability to taste PTC (see Chapter 16) is caused by homozygosity for a recessive allele, *t*. Tasters are either *TT* or *Tt* in genotype since the *T* allele is dominant to the *t* allele. In the population 70% of the individuals are tasters and 30% are nontasters.

As a first step we calculate the frequencies of the *T* and *t* alleles. The frequency distribution of genotypes is:

$$p^2\, TT + 2pq\, Tt + q^2\, tt$$
$$q^2 \text{ (the frequency of nontasters)} = 0.3$$
$$\therefore q = \sqrt{0.3}$$
$$= 0.55, \text{ the frequency of allele } t$$

Since $p+q = 1$:

$\therefore p = 0.45$, the frequency of allele *T*.

Once we know the allele frequencies, we can calculate the genotype frequencies:

$$\begin{aligned} TT &= p^2 \\ &= (0.45)^2 \\ &= 0.2 \\ Tt &= 2pq \\ &= 2(0.45)(0.55) \\ &= 0.5 \\ \text{and } tt &= 0.3, \text{ from before} \end{aligned}$$

Using the calculated allele and genotype frequencies, we can calculate the frequency of nontaster children from marriages of parents neither of whom are nontasters. Random mating is assumed here. First, we will solve the problem using genotype frequencies we have calculated and then we shall derive a generalized formula for a problem of this kind.

Using genotype frequencies, we must consider all possible pairings and the frequencies of their occurrences (Fig. 21.4). Of the four types of marriages, only one, $Tt \times Tt$, can give *tt* (nontaster) children, and such children would be expected to constitute one fourth of the children of these marriages (according to basic Mendelian segregation). Thus the frequency of nontaster children among progeny of all marriages between taster parents is:

(Relative frequency of $Tt \times Tt$ marriages) $\times$ (Probability of nontaster child from $Tt \times Tt$ marriage)

Parental genotypes ♀	♂	Frequency of pairing
TT	TT	$0.2 \times 0.2 = 0.04$
TT	Tt	$0.2 \times 0.5 = 0.10$
Tt	TT	$0.5 \times 0.2 = 0.10$
Tt	Tt	$0.5 \times 0.5 = 0.25$
		Total $= 0.49$

$$\therefore \text{ Proportion of } Tt \times Tt \text{ pairings} = \frac{0.25}{0.49}$$

$$= 0.51$$

Fig. 21.4. Calculation of the proportion of marriages in which both parents are heterozygous tasters (Tt) among all marriages in which both parents are tasters. The population on which the calculation is based has a distribution of 0.2 TT, 0.5 Tt, and 0.3 tt individuals; that is the frequency of the T allele is 0.45 and that of the t allele is 0.55.

$$= \frac{0.25}{049} \times 0.25$$
$$= 0.128$$

In other words 128 of every 1000 children from such pairings should be nontasters.

Now we can derive a generalized formula for problems like this. Fig. 21.5 shows the frequencies of occurrence of marriages where both parents have at

Fig. 21.5. Algebraic derivation of a generalized formula for calculating the frequency of recessive phenotype children among progeny of all marriages in which each parent has at least one dominant allele.

Marriages	Frequency (if random)		Frequency of progeny phenotypes: Dominant	Recessive
$AA \times AA$	$p^2 \times p^2 = p^4$		p^4	–
$AA \times Aa$	$p^2 \times 2pq = 2p^3q$	$4p^3q$	$4p^3q$	
$Aa \times AA$	$2pq \times p^2 = 2p^3q$			
$Aa \times Aa$	$2pq \times 2pq = 4p^2q^2$		$3p^2q^2$	p^2q^2

$$\text{Proportion of recessive phenotype progeny} = \frac{p^2q^2}{p^4 + 4p^3q + 3p^2q^2}$$

least one dominant allele. From this we see that there are p^2q^2 progeny with the recessive phenotype. The frequency of recessive phenotype progeny from such pairings is:

$$\frac{p^2q^2}{p^4+4p^3q+4p^2q^2} = \frac{q^2}{p^2+4pq+4q^2}$$

$$= \left(\frac{q}{p+2q}^2\right)$$

Since $p+q = 1$, this becomes:

$$\left(\frac{q}{1-q+2q}\right)^2$$

$$= \left(\frac{q}{1+q}\right)^2$$

To apply this to the taster/nontaster example where $q = 0.55$, this would give:

$$\left(\frac{0.55}{1.55}\right)^2 = 0.126$$

which fits well with the frequency calculated before.

In conclusion, for populations in genetic equilibrium, the Hardy-Weinberg law is useful in making predictions about the upcoming generations or about subsets thereof.

Factors affecting genetic equilibrium

Preferential mating

The main feature of a population in Hardy-Weinberg equilibrium is that allele frequencies remain constant generation by generation. Thus there is no change in the relative genotype and phenotype frequencies. Populations such as this are static and are not evolving.

As has been mentioned repeatedly, the maintenance of Hardy-Weinberg equilibrium depends upon random mating in the population. In natural populations random mating is the exception, not the rule. Instead mating patterns such as like preferring like or like avoiding like are found commonly. In humans, for example, mating preference may occur with respect to height or to other phenotypes and thus the alleles involved are most likely not in Hardy-Weinberg equilibrium. Mating *is* essentially random, however, for many biochemical or physiological traits and the alleles involved are more likely to be in genetic equilibrium.

There are various types of preferential mating. When mating occurs more frequently among phenotypically similar organisms, this is **positive assortative mating.** In this case, parents are attracted by phenotypic similarities, which often involves genotypic similarities. Positive assortative mating is commonly used by animal and plant breeders to select for certain desirable traits. An extreme case of positive assortative mating is **inbreeding** in which matings occur only between related parents. Inbreeding occurs naturally in self-fertilizing plants, and it can be used to generate pure-breeding strains of animals or plants which are useful for certain genetic studies.

We can illustrate the consequences of extreme inbreeding by using the taster/nontaster example from before where 2/10 of the population were *TT*, 5/10 were *Tt*, and 3/10 were *tt*. The constraint here is then an individual of a particular genotype can only marry another of the same genotype: *TT* with *TT*, *Tt* with *Tt*, and *tt* with *tt*. In the first case only *TT* progeny will result and in the last case only *tt* progeny will be produced. However, from the $Tt \times Tt$ marriage one-fourth of the progeny will be *TT*, one-half will be *Tt*, and one-fourth will be *tt*. With this in mind, Fig. 21.6 shows the calculation of genotype frequencies for the progeny of these preferential matings. As can be seen, the frequency of *T* remains 0.45 and that of *t* remains 0.55. However, the genotype frequencies have changed since the $Tt \times Tt$ pairing result in a dispersion of alleles throughout the three possible genotypes, with the

Fig. 21.6. An example of how preferential mating (inbreeding) alters the distribution of the three genotypes in one generation. The starting population here is a taster/nontaster population with 0.2 *TT*, 0.5 *Tt*, and 0.3 *tt*. In this case only individuals of like genotype mate, with the consequence that the frequency of *Tt* individuals in the next generation is halved. The relative frequencies of *T* and *t* alleles is not changed by such a preferential mating.

Genotype frequencies:	*TT*	*Tt*	*tt*
	$2/10$	$5/10$	$3/10$

∴ Allele frequencies: $T = 0.45$ $t = 0.55$

Progeny frequencies if only preferential mating ($TT \times TT$; $Tt \times Tt$, and $tt \times tt$) occurs:

	TT	*Tt*	*tt*
	$2/10$ from $TT \times TT$	$(1/2 \times 5/10)$ from $Tt \times Tt$	$3/10$ from $tt \times tt$
	$(1/4 \times 5/10)$ from $Tt \times Tt$		$(1/4 \times 5/10)$ from $Tt \times Tt$
Total:	13/40	10/40	17/40

and allele frequencies are still $T = 0.45$ and $t = 0.55$

notable consequence that the frequency of the heterozygote, *Tt*, has been reduced by half from one-half to one-fourth and the frequencies of the two homozygotes have increased. This will occur at each successive generation as long as the same preferential mating constraints are in effect. Inbreeding, then, serves to result in homozygosity at all loci. The danger of inbreeding is that many loci will become homozygous for recessive deleterious genes, and this is a principal reason why cousin marriages are illegal in many countries as this is a form of inbreeding.

Another departure from random mating is **negative assortative mating** in which dissimilar individuals show preference for mating. Also in this case there is no effect on gene frequencies, only on the association of alleles in the formation of zygotes.

Selection against certain genotypes in reproduction

One of the major assumptions in the Hardy-Weinberg law is that all individuals in the population are equally able to reproduce, with no preference of one gamete over another and no differential mortality of the conceived offspring. In natural populations, the recessive alleles have often arisen by mutations and in many instances result in the production of a defective gene product. Depending on the genes involved, this could have effects on the overall ability of homozygous recessive individuals to have children. For many human traits, for example, the homozygotes for deleterious recessive alleles may be so severely affected that they die before reproductive age, and obviously this means that not all genotypes contribute to reproduction. Similar arguments can be made for dominant mutations that have lethal effects when homozygous. In addition, there are dominant and recessive mutations that have a range of effects on the ability of particular individuals carrying them to reproduce. This brings us to the concept of **fitness**, which is the relative reproductive success, or the relative numbers of offspring produced by given genotypes.

The relative fitness of an individual is dependent upon the environment since, if the environment changes, the relative fitnesses of different genotypes may change. This is related to how organisms adapt to their environment during evolution—a process called *adaptation*, or *adaptive evolution*. In algebraic calculations, fitness is symbolized by W, which can have a value between 0 and 1.

For the genotype that leaves no progeny (e.g. a sterile mutant), $W = 0$, and for the best adapted genotype, $W = 1$. Fitness is generally characterized by the *selection coefficient, s*, which expresses the fitness of a mutant relative to the wild type. The relationship here is $W = 1-s$ so that if fitness is 0, the selection coefficient is 1, meaning that the genotype being studied is sterile or dies before reproductive maturity.

As an example, suppose we have a population with genotype frequencies 0.25 *AA*, 0.5 *Aa*, 0.25 *aa*, where *A* is dominant to *a*. Phenotypically three-fourths of the population is *A* and one-fourth is *a* so that $q^2 = 0.25$ and thus $q = 0.5$. Since $p+q = 1$, $p = 0.5$. If *aa* individuals do not survive to reproductive age or are sterile, the effective breeding population (EBP) is composed of only *AA* and *Aa* individuals. In other words, the *aa* individuals have zero fitness; $W = 0$ and $s = 1$. As a result, in the EBP the relative frequency of *AA*s is $0.50/0.75 = 0.67$. These individuals then mate randomly, and thus we can calculate the frequency of matings and the distribution of progeny as shown in Fig. 21.7.

Original population: 0.25 *AA* + 0.5 *Aa* + 0.25 *aa*

∴ Allele frequencies: $p = 0.5$, $q = 0.5$

In this population the *aa* individuals do not reproduce.
The effective breeding population, then, consists of
$\frac{0.25}{0.75} = 0.33$ *AA* individuals and $\frac{0.50}{0.75} = 0.67$ *Aa* individuals

These pair at random:

Matings	Frequency	Progeny frequencies *AA*	*Aa*	*aa*
AA × *AA*	0.33 × 0.33 = 0.109	0.109	–	–
2 (*AA* × *Aa*)	2 (0.33 × 0.67) = 0.442	0.221	0.221	–
Aa × *Aa*	0.67 × 0.67 = 0.449	0.112	0.225	0.112
	Total = 1.000	0.442	0.446	0.112

Original population: 0.25 *AA* + 0.5 *Aa* + 0.25 *aa*

Progeny population: 0.442 *AA* + 0.446 *Aa* + 0.112 *aa*

Progeny allele frequencies: *A* = 0.665

a = 0.335

Fig. 21.7. Algebraic demonstration of the consequences of selection against recessive phenotype individuals in the population. Here we start with an hypothetical population of 0.25 *AA*, 0.50 *Aa*, and 0.25 *aa*, in which *aa* individuals are not part of the breeding population. Random mating of *AA* and *Aa* individuals result in a progeny population in which both the allele and genotype frequencies have changed from those of the parental population.

As can be seen, not only have the genotype frequencies shifted markedly in just one generation, but the allele frequencies have changed, that of *A* from 0.5 to 0.665 and that for *a* from 0.5 to 0.335.

However, selection does not always produce such large effects. In most cases there is not complete selection against one genotype, but there are different degrees of fitness for each genotype. This condition does lead to changes to allele frequencies, but there are smaller effects per generation than in the Fig. 21.7 example. This situation is further complicated by changes in the environment that may affect the relative fitness of the genotypes and hence the direction selection may go.

Selection is also dependent upon the actual allele frequencies. In a randomly mating equilibrium population the proportion of heterozygous individuals relative to the homozygous recessives increases as the frequency of the recessive allele decreases (Table 21.1). When *a* is frequent relative to *A* (e.g. when $q = 0.9$), there are several times more homozygous recessives than heterozygotes. At the other extreme, when *a* is vary rare (e.g. $q = 0.01$), there are 198 times as many heterozygotes than homozygous recessives. From these discussions one might propose that a way to remove a deleterious recessive allele from a population is to prevent the homozygous recessive individuals (if they survive to reproductive age) from participating in the procreation process. Unfortunately this does not work because with continued selection against *aa*s, selection becomes less effective as the frequency of *aa*s decreases (Table 21.2). Even at low *aa* frequencies there are many *a* alleles to be found in heterozygous *Aa* individuals. Therefore the eradication of a deleterious recessive allele from a population depends upon identifying the heterozygotes and restricting their contributions to the future gene pool. In human populations, this is where pedigree analysis and genetic counseling come into play.

Table 21.1. Relative distribution of a recessive allele between homozygotes and heterozygotes for different frequencies of that allele.

	Genotype frequencies			
q	*AA*	*Aa*	*aa*	*Aa*/*aa* value
0.9	0.01	0.18	0.81	0.22
0.8	0.04	0.32	0.64	0.50
0.7	0.09	0.42	0.49	0.86
0.6	0.16	0.48	0.36	1.5
0.5	0.25	0.50	0.25	2.0
0.4	0.36	0.48	0.16	3.0
0.3	0.49	0.42	0.09	4.67
0.2	0.81	0.18	0.01	18.0
0.01	0.9801	0.0198	0.0001	198.0

Even so, the mutation process will continue to pump deleterious alleles into the population, albeit at a low frequency.

On the other hand, deleterious dominant mutations are not hidden by the

Table 21.2. Effects of constant selection against *aa* individuals on the frequency of *aa*s in the population.* Six different starting frequencies of *aa* individuals are shown.

Generation	Frequency of *aa* individuals					
0	0.990	0.750	0.500	0.250	0.100	0.010
1	0.249	0.215	0.172	0.112	0.058	0.008
2	0.112	0.100	0.086	0.062	0.038	0.007
3	0.062	0.058	0.051	0.040	0.026	0.006
4	0.040	0.038	0.034	0.028	0.019	0.005
5	0.028	0.026	0.024	0.020	0.015	0.004

*This is calculated for each generation as follows: effective breeding population = p^2 *AA*$+2pq$ *Aa*. Matings that give *aa* progeny are *Aa*×*Aa*. In the breeding population the frequency of *Aa*s is

$$\frac{2pq}{p^2+2pq}$$

The frequency of *Aa*×*Aa* pairings $= \left(\frac{2pq}{p^2+2pq}\right)^2$

Dividing through by p we have $\left(\frac{2q}{p+2q}\right)^2$

Substituting $1-q$ and p, this gives $\left(\frac{2q}{1-q+2q}\right)^2 = \left(\frac{2q}{1+q}\right)^2 = \frac{4q^2}{(1+q)^2}$

From *Aa*×*Aa*, 1/4 of the progeny will be *aa*.

∴ Frequency of *aa* here $= 1/4\times \frac{4q^2}{(1+q)^2} = \frac{q^2}{(1+q)^2} = \left(\frac{q}{1+q}\right)^2$

heterozygotes' condition as are the deleterious recessive alleles. Therefore selection is effective in removing deleterious dominant mutations from a population.

Mutation

Mutation is a primary source of variation in a population, and it will disrupt the genetic equilibrium of a population. The rate of spontaneous mutations at a locus may only be 10^6 or less, but even so mutations are essential if evolution (that is, adaptation to, for example, new environments) is to take place. If mutation generates an allele that is deleterious to the organism in the particular environment in which it lives, there will be selection against the allele. In general, a balance will occur between the rate of occurrence of the mutation and the rate of loss of the mutation from the population selection. Thus the mutation can then be distributed throughout the population by the processes of mating and recombination.

Let us consider mutation in a more formal sense. There are two kinds of mutation that will be dealt with here: **forward mutation**, the change from the (now) wild-type form of the gene to the mutated form (e.g. *A* to *a*, for the purposes of discussion); and **reverse** (back) **mutation**, the change from the mutated to the wild-type form (*a* to *A*). For any gene locus the rate of forward mutation is likely to be different from the rate of back mutation, and usually the former is higher than the latter. These relationships can be represented as:

$$A \underset{v}{\overset{u}{\rightleftarrows}} a$$

where the mutation rate for *A* to *a* is u and that for *a* to *A* is v. Only where $u = v$ will there be no change in allele frequencies in the population.

Assume that if in one generation the frequency of *A* is p and that of *a* is q. The change in the *A* frequency, Δp, for the next generation as a result of mutation is computed by the addition of *A* alleles produced by the mutation of *a* alleles (vq), and the subtraction of *A* alleles that mutate to *a* alleles (up). Thus:

$$\Delta p = vq - up$$

The recurrent production of an allele by mutation tending to increase its frequency in the gene pool is called *mutation pressure*. The gene frequencies will reach equilibrium under mutation pressure when the additions equal the subtractions:

$$\Delta p = vq - up = 0$$

$$\therefore vq = up$$

and $vq = u(1-q)$, since $p = 1-q$.

This is solved to give the equilibrium value of q, $\hat{q}$:

$$vq = u - uq$$

$$vq + uq = u$$

$$q(v+u) = u$$

$$\hat{q} = u/u+v$$

The equilibrium value for p, $\hat{p}$, is $1-\hat{q}$, or if it was calculated directly:

$$\hat{p} = v/u+v$$

Consider an example in which the initial values of $p_i = 0.8$ and $q_i = 0.2$, and u and v 4×10^{-5} and 1×10^{-5}, respectively. In other words, the forward mutation rate is four times that of the back mutation rate, or $u = 4v$.

$$\text{Since } \hat{q} = \frac{u}{u+v}$$

$$\text{here } \hat{q} = \frac{4v}{5v}$$

$$\text{Therefore, } \hat{q} = \frac{4}{5} = 0.80$$

$$\text{and } \hat{p} = \frac{1}{5} = 0.20$$

This means that the population will reach an equilibrium with respect to the frequencies of the mutating alleles when the *A* frequency is 0.20 and the *a* frequency is 0.80. (Note that this equilibrium is distinct from the concept of the Hardy-Weinberg equilibrium discussed earlier.) These value are markedly different from the starting values of 0.80 and 0.20, respectively. The new equilibrium values are reached only after many generations; in one generation only a small change occurs:

If Δp is the net change in *A* frequency in one generation,
then $\Delta p = vq - up$

$$= (1\times10^{-5})(0.2)-(4\times10^{-5})(0.8)$$
$$= (0.2\times10^{-5})-(3.2\times10^{-5})$$
$$= -(3.0\times10^{-5})$$

The new frequency of A, p_n, is the sum of the initial frequency, p_i, and the change in the *A* frequency, Δp:

$$p_n = p_i + \Delta p$$
$$\therefore p_n = 0.8 + (-3.0\times10^{-5})$$
$$= \underline{0.79997}$$
$$\text{and } q_n = 1.0 - 0.79997$$
$$= \underline{0.20003}$$

In other words, even though the mutation rate from *A* to *a* is four times the rate from *a* to *A*, there is only a very small effect on the gene pool in one generation.

Before leaving this subject, it is worth noting that mutations can be advantageous, disadvantageous or neutral to the organism at a given time and in a given environment. In the first two cases, the mutations are the source of variation in the population, and selection acts on them, but selection does not act on neutral mutations. Charles Darwin proposed the theory that evolution occurs through natural selection; that is, the present-day diversity of life has evolved from common ancestors rather than having a divine origin. Evolution by selection on advantageous and disadvantageous mutations is therefore usually called *Darwinian evolution*. It has been proposed by others that evolution might also occur by the accumulation of neutral mutations, or without selection. This form of evolution is called *non-Darwinian evolution*.

Migration

Another assumption of the Hardy-Weinberg law is that the population is a closed one in that there is no loss of individuals from it or additions of new individuals from the outside. In natural populations (except those that are geographically isolated), migration of individuals commonly occurs. From all that we have said concerning genetic equilibrium, it should be obvious that the introduction of new alleles into the gene pool by individuals entering the population and interbreeding with it will shift the equilibrium. This process is presumably essential for evolution. In fact, together with natural selection, the fastest way by which allele frequencies can be changed is by the introduction of genetically different individuals into the population by migration. Migration may enhance the effect of natural selection, or it may reduce the effect of selection by replacing genes removed by that process.

Genetic drift

The theoretical population considered at the outset of this chapter to be in Hardy-Weinberg equilibrium was of infinite size. Even though natural populations are smaller than that, many of them are large enough for random mating to occur in effect, and genetic equilibrium is maintained through the generations. On the other hand, if the sample of alleles contributing to the zygotes for the next generation is not representative of the overall allelic composition of the population's gene pool, deviations from genetic equilibrium can occur. This will be observed either as chance variations of allele frequencies in the population or possibly as chance fixation of an allele (i.e. p or $q = 1$) in the population. Such a random change in allele frequencies is called **genetic drift**. The effect of genetic drift is very small in large population but can be large in a small population. Fixation of one allele or the other becomes more likely the smaller the population is.

In natural populations, the risk of loss of an allele from a population as a result of genetic drift is very high for alleles that are present at very low frequencies. This results in fixation of the frequent allele. Clearly, if the rare allele is lost in the sampling process in one generation, it cannot be replaced by a compensating variation of sampling in another generation. Instead, it can only be replaced by mutation. When that occurs, the new allele is again very rare in the population and is again subject to loss through genetic drift. Those alleles that remain in the population may spread through a population as a result of genetic drift, regardless of their selective advantage or disadvantage. Further, in small populations, genetic drift will act on all loci represented by two or more alleles, but the direction and magnitude of the effect is likely to be different at each locus.

There can be other effects of population size on allele frequencies. For example, in human populations, rare alleles are often found in particular areas at relatively high frequencies. The reason for this is that small isolates (self-contained breeding units) within a larger population are particularly susceptible to genetic drift. A good illustration of this is found in the Dunkers' population of eastern Pennsylvania, USA. The Dunkers are a small religious sect who are descended from 28 West German immigrants who arrived in the United States over 250 years ago. The current population consists of about 300 individuals, of whom about 90 are parents in each generation. This is undoubtedly a very small breeding population and one which is isolated from other populations because of their religious beliefs. From studies of a variety of genetic traits, it has been concluded that the frequencies of some genes are very different in the Dunker population compared with either the United States or West German population. This illustrates what is called the **isolate effect** or **founder effect**; that is, the present allele frequencies reflect a chance sampling of alleles in the original immigrants and pairings occurring only within the population.

Conclusions

We have seen in this chapter a little of how genes in populations are distributed and the effects of various factors, such as mutation, selection, and migration, on the gene pool of a population. All of the genes of an organism, in association with the environment, are responsible for the phenotype of that organism. Therefore, as the environment changes, different combinations of gene frequencies in the population will result over many generations by the forces described. This is the process of evolution, at least in simple terms. However, evolution is an extremely complex process in which numerous factors are intertwined, and a lot remains to be learned about it. This chapter has presented a simplified view of the genetics of populations and it is hoped that the reader will extrapolate the basic concepts that have been discussed to natural populations and to the evolutionary process.

Questions and problems

21.1 In a large interbreeding population, 81% of individuals are homozygous for a recessive character. In the absence of mutation or selection, what percentage of the next generation would be homozygous recessives? Homozygous dominants? Heterozygotes?

21.2 Let A and a represent dominant and recessive alleles whose respective frequencies are p and q in a given interbreeding population at equilibrium (with $p+q = 1$).

(a) If 16% of individuals in the population are of recessive phenotype, what percentage of the total number of recessive genes exist in heterozygous condition?

(b) If 1% of individuals were homozygous recessive, what percentage of the recessive genes would occur in heterozygotes?

21.3 A population has eight times as many heterozygotes as homozygous recessives. What is the frequency of the recessive gene?

21.4 In a large population of range cattle, the following ratios are observed: 49% red (*RR*), 42% roan (*Rr*), 9% white (*rr*).

(a) What percentage of the gametes that give rise to the next generation of cattle in this population will contain allele *R*?

(b) In another cattle population, only 1% of animals are white, and 99% are either red or roan. What is the percentage of *r* alleles in this case?

21.5 In a population gene pool the alleles A and a have initial frequencies of p and q, respectively. Prove that the gene frequencies and zygotic frequencies do not change from generation to generation as long as there is no selection, mutation, or migration, the population is large, and the individuals mate at random.

21.6 The *S*-*s* antigen system in humans is controlled by two codominant alleles, *S* and *s*. In a group of 3146 individuals the following genotypic frequencies were found: 188 *SS*, 717 *Ss*, and 2241 *ss*.

(a) Calculate the frequency of the *S* and *s* alleles.

(b) Test whether the genotypic frequencies conform to the Hardy-Weinberg equilibrium using the chi-square test. (The chi-square test is in Appendix I).

21.7 Refer to the previous Question 21.6. A third allele is sometimes found at the *S* locus. The allele S^u is recessive to both the *S* and the *s* alleles and can only be detected in the homozygous state. If the frequencies of the alleles, *S*, *s*, and S^u are p, q, and r, respectively, what would be the expected frequencies of the phenotypes: *S*__, *Ss*, *s*__, and $S^u S^u$?

21.8 A selectively neutral recessive character appears in 0.40 of males and in 0.16 of females in a randomly interbreeding population. What is the gene's frequency? How many females are heterozygous for it? How many males are heterozygous for it?

21.9 Suppose you found two distinguishable types of individuals in wild populations of some organism in the following frequencies:

	Type 1	Type 2
Females	99%	1%
Males	90%	10%

The difference is known to be inherited. What is its genic basis?

21.10 Red-green color blindness is due to a sex-linked recessive gene. About 64 women out of 10,000 are color-blind. What proportion of men would be expected to show the trait, if mating is at random?

21.11 About 8% of men in a population of red-green color-blind (due to a sex-linked recessive gene). Assuming random mating in the population, with respect to color blindness:

(a) What percentage of women would be expected to be color-blind?

(b) What percentage of women would be expected to be heterozygous?

(c) What percentage of men would be expected to have normal vision two generations later?

21.12 If two alleles of locus, *A* and *a*, can be interconverted by mutation:

$$A \underset{v}{\overset{u}{\rightleftarrows}} a$$

and u is a mutation rate of 6×10^{-7}, and v is a mutation rate of 6×10^{-8}, what will be the frequencies of *A* and *a* at mutation equilibrium, assuming no selective difference, no migration, and no random fluctuation caused by genetic drift?

21.13 Upon sampling three populations and determining genotypes, you find three different genotype distributions. What would each of these distributions imply in regard to selective advantages or population structure?

Population	*AA*	*Aa*	*aa*
1	0.04	0.32	0.64
2	0.12	0.87	0.01
3	0.45	0.10	0.45

21.14 The frequency of two adaptively neutral alleles in a large population is 70% *A*: 30% *a*. The population is wiped out by an epidemic, leaving only four individuals, who produce offspring. What is the probability that the population several years later will be 100% *AA*? (Assume no mutations.)

21.15 Because of changed environmental circumstances, a completely recessive gene becomes lethal in a certain population. It was previously neutral, and its frequency was 0.5.

(a) What was the genotype distribution when the recessive genotype was not selected against?

(b) What will the gene frequency be after one generation in the altered environment?

(c) After two generations?

21.16 Humans homozygous for a certain recessive autosomal gene die before reaching reproductive age. In spite of this removal of all affected individuals, there is no indication that homozygotes occur less frequently in succeeding generations. To what might you attribute the constant rate of appearance of recessives?

References

Ayala, F.J. (ed), 1976. *Molecular Evolution.* Sinauer, Sunderland, Massachusetts.

Bodmer, W.F. and L.L. Cavalli-Sforza, 1976. *Genetics, Evolution, and Man.* W.H. Freeman, San Francisco.

Crow, J.F. and M. Kimura, 1970. *An Introduction to Population Genetics Theory.* Harper and Row, New York.
Darwin, C., 1859. *The Origin of Species.* John Murray, London.
Dobzhansky, T., 1947. Adaptive changes induced by natural selection in wild populations of *Drosophila. Evolution* **1**:1–16
Dobshansky, T., 1955. A review of some fundamental concepts and problems of population genetics. *Cold Spring Harbor Symp. Quant. Biol.* **20**:1–15.
Falconer, D.S., 1960. *Introduction to Quantitative Genetics.* Oliver and Boyd, Edinburgh.
Fisher, R.A., 1930. *The Genetical Theory of Natural Selection.* Clarendon Press, Oxford.
Ford, E.B., 1971. *Ecological Genetics*, 3rd ed. Chapman and Hall, London.
Haldane, J.B.S., 1931. *The Causes of Evolution.* Harper and Row, New York.
Harland, S.C., 1936. The genetic conception of the species. *Biol. Rev.* **11**:83–112.
Kettlewell, H.B.D., 1961. The phenomenon of industrial melanism in Lepidoptera. *Annu. Rev. Entomol.* **6**:245–262.
Lewontin, R.C., 1974. *The Genetic Basis of Evolutionary Change.* Columbia University Press, New York.
Li, C.C., 1955. The stability of an equilibrium and the average fitness of a population. *Amer. Nat.* **89**:281–295.
Mather, K., 1953. The genetical structure of populations. *Symp. Soc. Exp. Biol.* **7**:66–95.
Mayr, E., 1963. *Animal Species and Evolution.* Harvard University Press, Cambridge, Massachusetts.
Merrell, D.J., 1953. Selective mating as a cause of gene frequency changes in laboratory populations of *Drosophila melanogaster. Evolution* **7**:287–298.
Ohta, T., 1974. Mutational pressure as the main cause of molecular evolution and polymorphism. *Nature* **252**:351–354.
Ohta, T. and M. Kimura, 1971. Functional organization of genetic material as a product of molecular evolution. *Nature* **233**:118–119.
Powell, J.R. and R.C. Richmond, 1974. Founder effects and linkage disequilibrium in experimental populations. *Proc. Natl. Acad. Sci. USA* **71**:1663–1665.
Simpson, G.G., 1953. *The Major Features of Evolution.* Columbia University Press, New York.
Speiss, E., 1977. *Genes in Populations.* Wiley, New York.
Wallace, B., 1968. *Topics in Population Genetics.* Norton, New York.
Wright, S., 1951. The genetic structure of populations. *Ann. Eugenics* **15**:323–354.
Wright, S., 1969. *Evolution and the Genetics of Populations.* University of Chicago Press, Chicago, Illinois.

Appendix: Chi-Square Test

The phenotypic ratios observed among progeny of a cross typically do not match exactly the expected or predicted phenotypic ratios even though the hypothesis on which the expected ratios are based is correct. The discrepancy between the observed and expected phenotypic ratios may result from, for example, small sample size, sampling error, decreased viability of one or more of the genotypes involved, or other chance events.

A common way to decide in a genetic analysis when the deviation between the observed and the expected results is large enough to question the hypothesis under scrutiny is to use a statistical test called the **chi-square** (χ^2) **test**, which is a goodness of fit test. The following example illustrates the use of the chi-square test for the results of a dihybrid testcross.

In rabbits, the "English" type of coat (white-spotted) is dominant over non-English (unspotted), and short hair is dominant over long hair (Angora). When homozygous English, short-haired rabbits are crossed with non-English Angoras, the F1 rabbits are all English short-haired. When these doubly heterozygous F1 rabbits were testcrossed with non-English Angoras, the following offspring were obtained: 114 English short-haired, 111 English Angora, 106 non-English short-haired, and 121 non-English Angora.

If the two genes are unlinked, we hypothesize a 1:1:1:1 ratio of the above phenotypes. To test this hypothesis, we employ the chi-square test, as shown in the following table:

(1) Phenotypes	(2) Observed number (o)	(3) Expected number (e)	(4) d	(5) d^2	(6) d^2/e
English short-haired	114	113.00	1.00	1.00	0.00885
English Angora	111	113.00	−2.00	4.00	0.03540
non-English short-haired	106	113.00	−7.00	49.00	0.43363
non-English Angora	121	113.00	8.00	64.00	0.56637
Total	452	452.00	0.00	118.00	1.04425

(7) $\chi^2 = 1.044$ (8) df = 3

Column 1 lists the four phenotypes expected in the F2 of the testcross and column 2 lists the observed numbers (o) (not percentages or proportions) for each phenotypic class. The expected number (e) in each phenotypic class is calculated from the total number of progeny counted in the F2 and the hypothesis under test, in this case 1:1:1:1, giving the figures shown in column 3. Each expected number is subtracted from the respective observed number to give the deviation value, d, for each class (column 4). The sum of the d values for all the phenotypic classes is always zero. Each d value is squared to give d^2 (column 5) and each d^2 value is divided by the respective expected number, e, to give d^2/e (column 6). The chi-square (χ^2) value ([7] in the table) is the total of the d^2/e values in column 6. In the particular example being discussed, χ^2 is 1.044, and the general formula is:

$$\chi^2 = \Sigma(d^2/e).$$

The last value in the table, 8, is the degrees of freedom (df) for the set of data being analyzed. In general, the degrees of freedom in a test involving n classes is usually equal to $n = 1$. The rationale is that, if the total number of progeny (452 in the example) is divided among n classes (four in the example), once the expected numbers have been calculated for $n = 1$ classes, the remainder must fall into the last class. In our example, then, there are only 3 degrees of freedom in the analysis.

The χ^2 value and the degrees of freedom are used to determine the probability, P, that the deviation from the observed values from the expected values is due to chance. If P is greater than 5 in 100 ($P > 0.05$), the deviation of the observed data from the expected data is considered not to be statistically significant and could have occurred by chance alone. In other words, the hypothesis being tested could apply to the data being analyzed. If $P = 0.05$, however, the deviation from the

Table A.1. Chi-square probabilities.

	Probabilities									
df	0.95	0.90	0.70	0.50	0.30	0.20	0.10	0.05	0.01	0.001
1	0.004	0.016	0.15	0.46	1.07	1.64	2.71	3.84	6.64	10.83
2	0.10	0.21	0.71	1.39	2.41	3.22	4.61	5.99	9.21	13.82
3	0.35	0.58	1.42	2.37	3.67	4.64	6.25	7.82	11.35	16.27
4	0.71	1.06	2.20	3.36	4.88	5.99	7.78	9.49	13.28	18.47
5	1.15	1.61	3.00	4.35	6.06	7.29	9.24	11.07	15.09	20.52
6	1.64	2.20	3.83	5.35	7.23	8.56	10.65	12.59	16.81	22.46
7	2.17	2.83	4.67	6.35	8.38	9.80	12.02	14.07	18.48	24.32
8	2.73	3.49	5.53	7.34	9.52	11.03	13.36	15.51	20.09	26.13
9	3.33	4.17	6.39	8.34	10.66	12.24	14.68	16.92	21.67	27.83
10	3.94	4.87	7.27	9.34	11.78	13.44	15.99	18.31	23.21	29.59
11	4.58	5.58	8.15	10.34	12.90	14.63	17.28	19.68	24.73	31.26
12	5.23	6.30	9.03	11.34	14.01	15.81	18.55	21.03	26.22	32.91
13	5.89	7.04	9.93	12.34	15.12	16.99	19.81	22.36	27.69	34.53
14	6.57	7.79	10.82	13.34	16.22	18.15	21.05	23.69	29.14	36.12
15	7.26	8.55	11.72	14.34	17.32	19.31	22.31	25.00	30.58	37.70
20	10.85	12.44	16.27	19.34	22.78	25.04	28.41	31.41	37.57	45.32
25	14.61	16.47	20.87	24.34	28.17	30.68	34.38	37.65	44.31	52.62
30	18.49	20.60	25.51	29.34	33.53	36.25	40.26	43.77	50.89	59.70
50	34.76	37.69	44.31	49.34	54.72	58.16	63.17	67.51	76.15	86.65

← accept | reject →
at 0.05 level

From Table IV of Fisher and Yates, *Statistical Tables for Biological, Agricultural and Medical Research*. Oliver and Boyd, Edinburgh, Reprinted by permission of the authors and publishers.

expected values is considered to be statistically significant and not due to chance alone and, in this case, the hypothesis may be rejected. If $P = 0.01$ or less, the deviation from the expected values is highly significant and some nonchance factor must have been involved; in this instance, the hypothesis will certainly be rejected. The *P* value for a set of data is read from tables of χ^2 values for various degrees of freedom. Table 1 shows part of a table of chi-square probabilities. For $\chi^2 = 1.044$, with 3 df, the *P* value is greater than 0.70 and less than 0.90. This means that 70 to 90 times out of 100, chance deviations from the expected values of this magnitude would be expected. Since this *P* value is much greater than 0.05, the borderline between accepting and rejecting a hypothesis, the hypothesis that the two genes are unlinked is not invalid.

Answers

Chapter 1

1.1 (d)

1.2 (a) Lived
(b) Died
(c) Lived
(d) Died; in this case, DNA from the S bacteria transformed the R bacteria to a virulent form.

1.3 (a) They showed that transformation could occur in vitro using extracts of S cells. The transforming principle copurified with DNA, indicating the genetic material was DNA.
(b) This result showed that the transforming principle was DNA.
(c) Neither protease nor ribonuclease destroyed the transforming principle, but deoxyribonuclease did abolish transforming activity. This substantiated the fact that DNA was the genetic material.

1.4 See text, pp. 3–4.

1.5 (b), (c), and (d)

1.6 More information is necessary since such a ratio says nothing about complementarity of bases. The bases in the numerator are not complementary to the bases in the denominator, and their ratio therefore may be equivalent to 1 or not equivalent to 1 in both single- and double-stranded DNA.

1.7 In double-stranded DNA the (A+C)/(G+T) ratio is expected to be equal to 1 because of the base pairing between A and T and between G and C. That is, A = T, G = C; A+C = G+T; therefore (A+C)/(G+T) = 1. However, the (A+T)/(G+C) ratio may or may not be equal to 1 since there is no pairing between the bases in the numerator and the bases in the denominator. For example, a double-stranded molecule of DNA may have many A-T base pairs and few G-C base pairs.

1.8 We can make no predictions regarding the base content of single-stranded DNA. Any base-pair equality would depend on the overall sequence of the chromosome: A might be equal to T, but that result would be unlikely given all the other possible sequences that might be the case.

1.9 (a) 200,000 (b) 10,000 (c) 3.4×10^4 nm

Chapter 2

2.1 

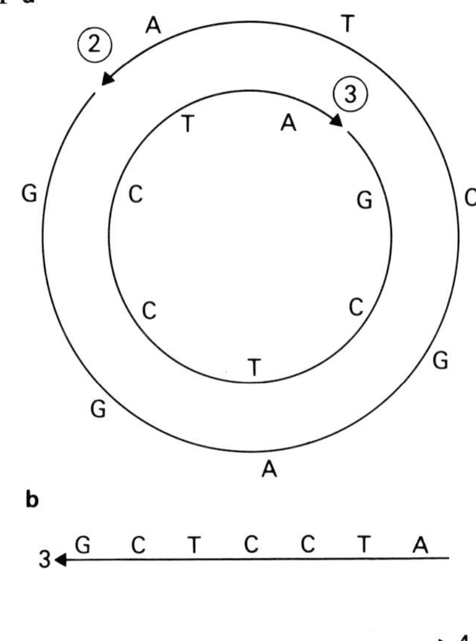

b

3 ← G C T C C T A

C G A G G A T → 4

2.2 (c)

2.3 See text, pp. 13–23.

2.4 See text, pp. 22–25.

Chapter 3

3.1 In the semiconservative model, the two strands of the double helix separate and each serves as a template for new DNA synthesis. Each daughter DNA double helix has one old and one new DNA strand. In the conservative model, the two strands remain together to serve as the template for new DNA synthesis. One daughter helix consists of both parental strands, and the other consists of two new strands.

3.2 See text, pp. 44–46. The key result in the experiment was the presence in the first generation of only DNA with intermediate density.

3.3 Key: ^{15}N-^{15}N DNA = HH; ^{15}N-^{14}N DNA = HL; ^{14}N-^{14}N DNA = LL.
(a) Generation 1: all HL; 2: 1/2 HL, 1/2 LL; 3: 1/4 HL, 3/4 LL; 4: 1/8 HL, 7/8 LL; 6: 1/32 HL, 31/32 LL; 8: 1/128 HL, 127/128 LL.
(b) Generation 1: 1/2 HH, 1/2 LL; 2: 1/4 HH, 3/4 LL; 3: 1/8 HH, 7/8 LL; 4: 1/16 HH, 15/16 LL; 6: 1/64 HH, 63/64 LL; 8: 1/256 HH, 255/256 LL.

3.4 See pp. 46–47.

3.5 A primer strand is a nucleic acid sequence that is extended by DNA synthesis activities. A template strand directs the base sequence of the DNA strand being made; for example, an A on the template causes a T to be inserted on the new chain, and so on.

3.6 (a) One base pair is 0.34 nm and the chromosome is 1100 μm, so the number of base pairs is $(1100/0.34)\times1000 = 3.24\times10^6$.
(b) There are 10 base pairs per turn in a normal DNA double helix and therefore a total of 3.24×10^5 turns.
(c) 3.24×10^5 turns and 60 min for unidirectional synthesis; therefore, $(3.24\times10^5/60)$ turns per minute = 5392 revolutions per minute.
(d)

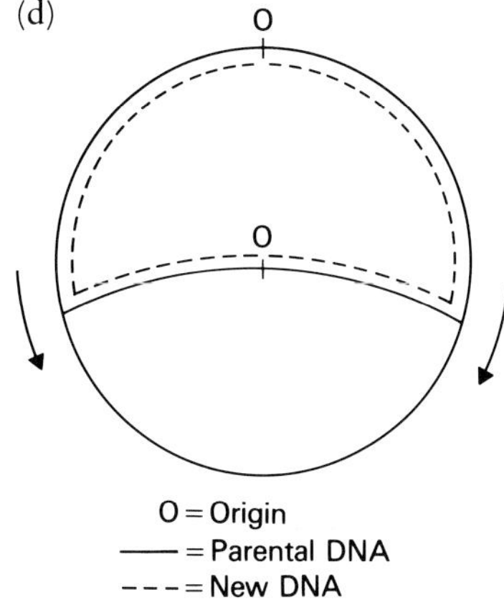

3.7 Develop a way to assay for gene products of genes on either side of origin and for various other points around the chromosomes. If replication is bidirectional, there should be a doubling of the gene products both clockwise and counter-clockwise from the origin. The "wave" of replication in both directions can be followed in time by such assays.

3.8 See text pp. 48–49 and Table 3.2; they differ in associated exonuclease activities.

3.9 See text, pp. 49–52.

3.10 DNA ligase seals single-stranded gaps in a DNA double helix in which there is a 3′-OH and a 5′-monophosphate. The enzyme is used to join Okazaki fragments produced by the discontinuous mechanism of DNA replication, so a temperature-sensitive DNA ligase mutant would not be able to join these fragments at high temperature. The accumulation of such fragments in a temperature-sensitive ligase mutant strain grown at the nonpermissive temperature was used as evidence for the discontinuous model of replication.

3.11

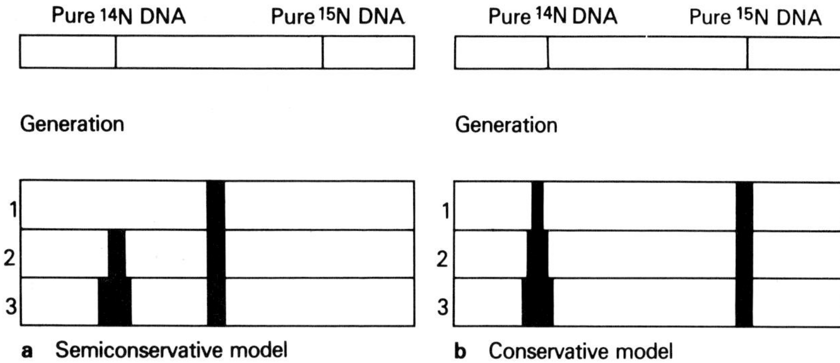

3.12

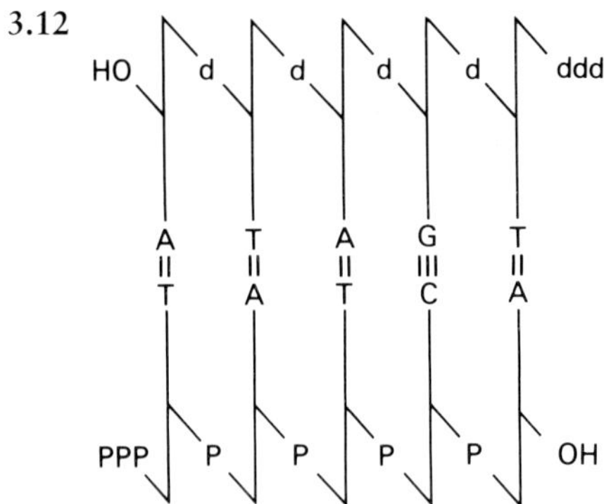

3.13 See text pp. 40–41. DNA polymerase; intact, high-molecular-weight DNA; dATP, dGTP, dTTP, and dCTP; and magnesium ions are needed.

3.14 There are three principal lines of evidence that the Kornberg enzyme is not the enzyme involved in *E. coli* chromosome replication in vivo. First, a mutant that is deficient in the Kornberg enzyme nonetheless grows, replicates the DNA, and divides. Second, other mutants that are temperature sensitive for DNA synthesis and that do not replicate the DNA or divide have Kornberg enzyme activity at the nonpermissive temperature. Finally, its rate of polymerization is too slow.

Chapter 4

4.1 See text, pp. 48–49 and 58–59.

4.2 See text, pp. 61–62.

Chapter 5

5.1 (c)

5.2 (b)

5.3 (c)

5.4 (a)

5.5 (c)

5.6 (e)

5.7 (a) metaphase (b) anaphase

5.8 See text, pp. 69–71.

Chapter 6

6.1 Photoreactivation needs photolyase enzyme and a photon of light in 320–370 nm wavelength. It causes the direct cleavage of the thymine dimer. Dark repair requires a number of enzymes and does not depend on light. First, an endonuclease makes a single-stranded nick on the 5′ side of the dimer; then an exonuclease trims away part of one strand, including the dimer; next, DNA polymerase fills in the single-stranded region in the 5′-to-3′ direction. Finally, the gap is sealed by ligase.

6.2 (a) 5BU in its normal form is a T analog; in its rare form, it resembles C. Thus the mutation is an AT-to-GC transition.

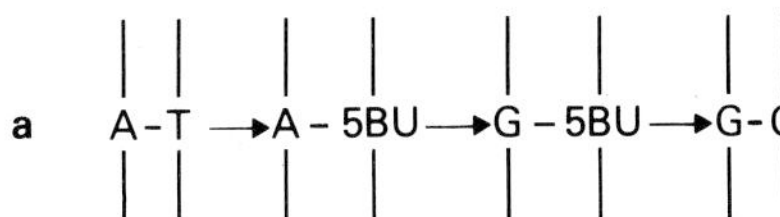

(b) Nitrous acid can deaminate C to U, so the mutation is a GC-to-AT transition. Thus nitrous acid can revert a 5BU-induced mutation.

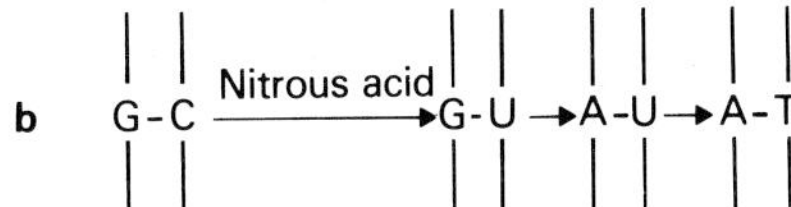

6.3 (a) and (b)

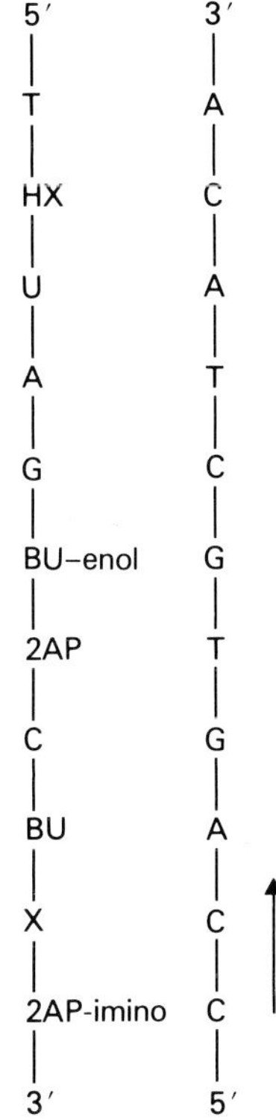

(c) One strand is affected by the mutagen. At the first mitotic division, one daughter helix consists of two normal strands, and the other daughter helix consists of two mutated strands. Mosaics may be formed when populations of cells arise from these two types by repeated mitoses.

6.4

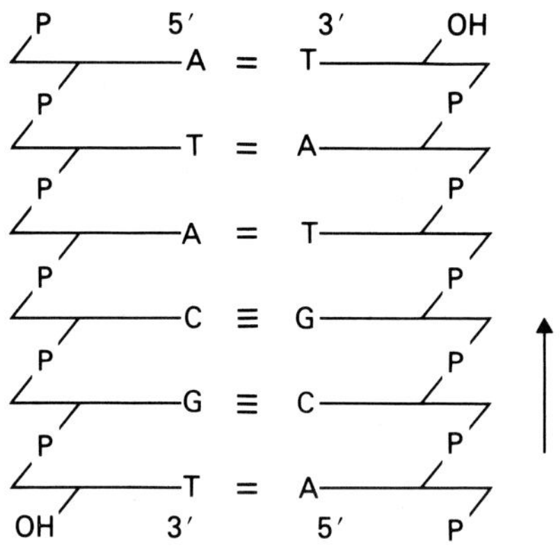

```
P      5′      3′      OH
       A   =   T
P                      P
       T   =   A
P                      P
       A   =   T
P                      P
       C   ≡   G
P                      P
       G   ≡   C
P                      P
       T   =   A
OH     3′      5′      P
```

6.5 The reaction stops when it needs dGTP; only dGMP is present.

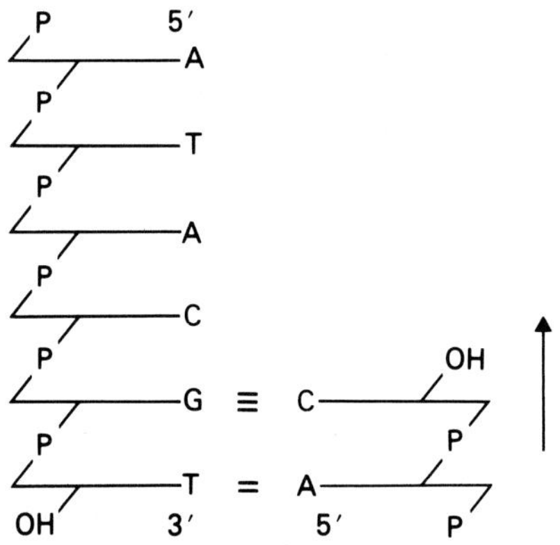

```
P      5′
       A
P
       T
P
       A
P
       C
P                      OH
       G   ≡   C
P                      P
       T   =   A
OH     3′      5′      P
```

6.6 dHTP substitutes for the lack of dGTP, but there is no dCTP.

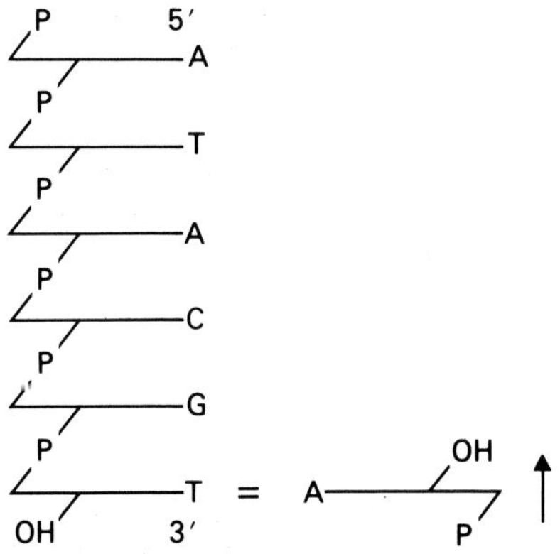

```
P      5′
       A
P
       T
P
       A
P
       C
P
       G
P                      OH
       T   =   A
OH     3′              P
```

6.7 HNO_2 converts A to H, C to U, and G to X.

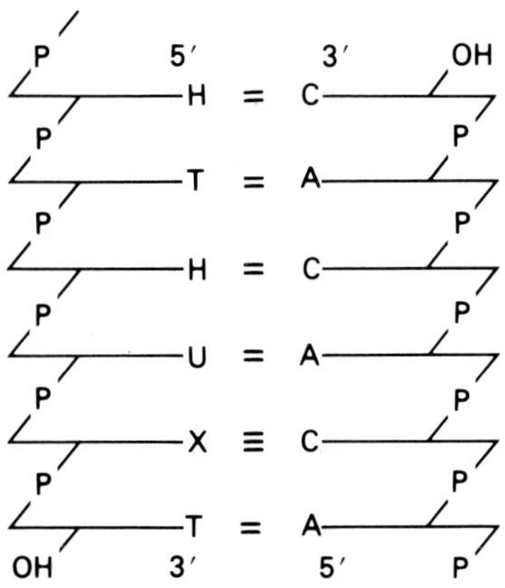

P 5′ 3′ OH
H ≡ C
T = A
H ≡ C
U = A
X ≡ C
T = A
OH 3′ 5′ P

6.8 dHTP substitutes for the absence of dGTP.

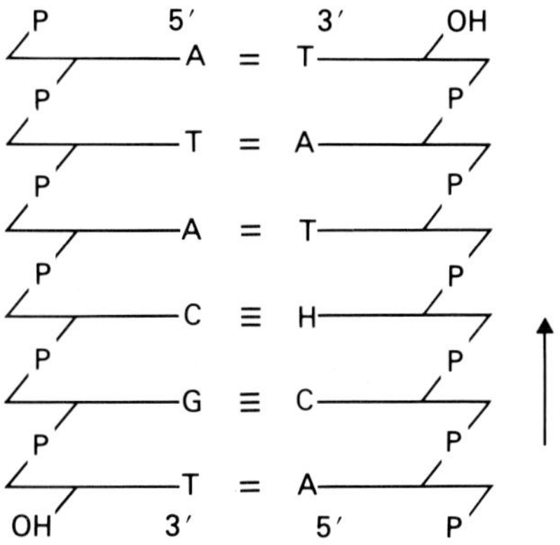

P 5′ 3′ OH
A = T
T = A
A = T
C ≡ H
G ≡ C
T = A
OH 3′ 5′ P

6.9

Mutation induced by	Proportion of mutations reverted by BU	AP	NA	HA	Base-pair substitution inferred
BU	–	+	+	–	GC → AT
AP	+	–	+	+	AT → GC
NA	+	+	+	+	GC ⇄ AT
HA	–	+	+	–	GC → AT

6.10 *ara*$^+$ to *ara-1*: This mutation is CG to AT since it is reverted by base analogs but not by HA or the frameshift mutagen.
ara$^+$ to *ara-2*: This mutation is AT to GC since it is reverted by base analogs and HA but not by frameshift mutagen.

ara$^+$ to *ara-3*: This mutation is AT to GC for the same reasons as given for the second mutant.
Mutagen X causes transition mutations in both directions because mutants are revertible by base analogs, some are revertible by HA, and none (if this is a representative sample) by frameshift mutagens.

Chapter 7

7.1 See text, pp. 1–3. The DNA contains deoxyribose and thymine, whereas RNA contains ribose and uracil, respectively. Also, DNA is usually double stranded while RNA is usually single stranded.

7.2 See text for this chapter, pp. 99–104, and Chapters 3 (pp. 48–49) and 4 (pp. 58–59). Both DNA polymerases and RNA polymerases catalyze the synthesis of the nucleic acids in the 5′-to-3′ direction. Both must recognize DNA and ensure the correct base pairing during the synthesis process. The RNA polymerases usually recognize specific base-pair sequences as signals for where to start transcription, which is generally not the case for DNA polymerases. The DNA polymerases cannot initiate a DNA chain whereas RNA polymerases can.

7.3 See text, pp. 99–102.

7.4 See text, pp. 99–104.

7.5 (a) See text, pp. 101, 104. They differ in size, number of polypeptide subunits, and susceptibility to inhibitors.
(b) The rate of initiation of transcription at different genes by the same enzyme varies considerably. This rate is related to the affinity of the polymerase for the promoter that is determined by the specific base-pair sequence in the promoter region. Along this same line, mutations that cause known single base-pair changes in a promoter have been identified, and these changes alter the rate of initiation compared with the wild type situation.

7.6 See text, pp. 99–102.

7.7 Bind the eukaryotic RNA polymerase to DNA under conditions where initiation of transcription cannot occur. Digest away the DNA not protected by the enzyme with DNase. Isolate the remaining DNA and sequence it by a rapid DNA-sequencing method.

7.8 See text, pp. 102–104. Specific base-pair sequences are involved. In prokaryotes, transcription termination involves a complex between the RNA polymerase and the *nusA* protein, which interacts with the termination sequence. The protein factor *rho* is also involved in transcription termination; it plays a role in RNA transcript release from the DNA. In eukaryotes the RNA polymerase reads the termination signal(s).

7.9 See text, pp. 104–113. Eukaryotic mRNAs have a 5′ cap and 3′ poly(A) tail; these features are absent in prokaryotic mRNAs.

7.10 See text, pp. 97–98. The three classes of RNA are mRNA, tRNA, and rRNA.

7.11 See text, pp. 110–113. Intervening sequences (ivs) are detected, for example, by comparing sizes and sequences of the genes and the mature mRNAs. The ivs are removed from precursor-mRNA by specific nuclease action. This action presumably involves a looping action, bringing together the two junction sequences, and then specific cleavage and relegation of the two parts of the mRNA at each loop.

7.12 See text, pp. 105, 107–114, 117, 121–124. For mRNA: 5′ capping and 3′ polyadenylation in eukaryotes. For tRNA: modification of a number of the bases, removal of 5′-leader and 3′-trailer sequences (if present) and addition of a three-nucleotide sequence (5′-CCA-3′) at the 3′ end of the tRNA. For rRNA: in prokaryotes, a pre-rRNA molecule is processed to the mature 16, 23, and 5S rRNAs. In eukaryotes, a precursor molecule is processed to the 18, 5.8, and 28S rRNA molecules; 5S rRNA is made elsewhere. The large rRNAs in both types of organisms are methylated. Additionally, the same rRNAs in eukaryotes are pseudoridylated.

7.13 See text, pp. 106, 106, 106, 110, 125, 125 respectively.

7.14 See text, pp. 124–125.

Chapter 8

8.1 Initiation and the formation of the first peptide bond occur, but translocation of the ribosome to the next codon is inhibited. The evidence for this is that a dipeptide is produced. This rules out blocks in initiation, in the first step of elongation (i.e. the binding of a charged tRNA molecule in the A site), and in the formation of the first peptide bond catalyzed by peptidyl transferase.

8.2 See text, pp. 137–138.

8.3 See text, pp. 139–151.

8.4 See text, pp. 139–140, 149–150. One Met-tRNA. Met is used for initiation, and the other is used in all elongation steps. In prokaryotes but not in eukaryotes, the methionine on the initiator-tRNA is formulated on the amino group. The anticodon loop also differs between the two tRNAs so that the initiator tRNA shows 5′ wobble in contrast to 3′ wobble.

8.5 A probable reason is that actinomycin D might block the transcription of a gene that codes for an inhibitor of an enzyme activity.

Chapter 9

9.1 Three-letter code; universal; nonoverlapping; degenerate; comma-less; specific start and stop signals.

9.2 31

9.3 (a) 4 A : 6 C

AAA = (4/10) (4/10) (4/10) = 0.064, or 6.4% Lys
AAC = (4/10) (4/10) (6/10) = 0.096, or 9.6% Asn
ACA = (4/10) (6/10) (4/10) = 0.096, or 9.6% Thr
CAA = (6/10) (4/10) (4/10) = 0.096, or 9.6% Gln
CCC = (6/10) (6/10) (6/10) = 0.216, or 21.6% Pro
CCA = (6/10) (6/10) (4/10) = 0.144, or 14.4% Pro
CAC = (6/10) (4/10) (6/10) = 0.144, or 14.4% His
ACC = (4/10) (6/10) (6/10) = 0.144, or 14.4% Thr

In summary, 6.4% Lys, 9.6% Asn, 9.6% Gln, 36.0% Pro, 24.0% Thr, and 14.4% His.

(b) 4 G : 1 C

GGG = (4/5) (4/5) (4/5) = 0.512, or 51.2% Gly
GGC = (4/5) (4/5) (1/5) = 0.128, or 12.8% Gly
CCG = (4/5) (1/5) (4/5) = 0.128, or 12.8% Ala
CGG = (1/5) (4/5) (4/5) = 0.128, or 12.8% Arg
CCC = (1/5) (1/5) (1/5) = 0.008, or 0.8% Pro
CCG = (1/5) (1/5) (4/5) = 0.032, or 3.2% Pro
CGC = (1/5) (4/5) (1/5) = 0.032, or 3.2% Arg
CCC = (4/5) (1/5) (1/5) = 0.032, or 3.2% Ala

In summary, 64.0% Gly, 16.0% Ala, 16.0% Arg, 4.0% Pro.

(c) 1 A : 3 U : 1 C

The same logic is followed here, using 1/5 as the fraction for A, 3/5 for U, and 1/5 for C:
AAA = 0.008, or 0.8% Lys
AAU = 0.024, or 2.4% Asn
AUA = 0.024, or 2.4% Ile
UAA = 0.024, or 2.4% chain terminating
AUU = 0.072, or 7.2% Ile
UAU = 0.072, or 7.2% Tyr
UUA = 0.072, or 7.2% Leu
UUU = 0.216, or 21.6% Phe
AAC = 0.008, or 0.8% Asn
ACA = 0.008, or 0.8% Thr
CAA = 0.008, or 0.8% Gln
ACC = 0.008, or 0.8% Thr
CAC = 0.008, or 0.8% His
CCA = 0.008, or 0.8% Pro
CCC = 0.008, or 0.8% Pro
UUC = 0.072, or 7.2% Phe
UCU = 0.072, or 7.2% Ser
CUU = 0.072, or 7.2% Leu
UCC = 0.024, or 2.4% Ser
CUC = 0.024, or 2.4% Leu
CCU = 0.024, or 2.4% Pro
UCA = 0.024, or 2.4% Ser
UAC = 0.024, or 2.4% Tyr
CUA = 0.024, or 2.4% Leu
CAU = 0.024, or 2.4% His
AUC = 0.024, or 2.4% Ile
ACU = 0.024, or 2.4% Thr

In summary, 0.8% Lys, 3.2% Asn, 12.0% Ile, 2.4% chain terminating, 9.6% Tyr, 19.2% Leu, 28.8% Phe, 4.0% Thr, 0.8% Gln, 3.2% His, 4.0% Pro, and 12.0% Ser. The likelihood is that the chain would not be long because of the chance of the chain-terminating codon.

(d) 1 A : 1 U : 1 G : 1 C; all 64 codons will be generated. The probability of each codon is 1/64, so there is a 3/64 chance of the codon being a chain-terminating codon. With those exceptions, the relative proportion of amino acid incorporation is directly dependent upon the codon degeneracy for each amino acid and that can be determined by inspecting the code word dictionary.

9.4 (a) UAG: CAG Gln; AAG Lys; GAG Glu; UUG Leu; UCG Ser; UGG Trp; UAU Tyr; UAC Tyr; UAA chain terminating
(b) UAA: CAA Gln; AAA Lys; GAA Glu; UUA Leu; UCA Ser; UGA chain terminating; UAU Tyr; UAC Tyr; UAG chain terminating
(c) UGA: CGA Arg; AGA Arg; GGA Gly; UUA Leu; UCA Ser; UAA chain terminating; UGU Cys; UGC Cys; UGG Trp

9.5 (1) GCG; (2) GAG; (3) GUG; (4) AUG; (5) GGG; (6) AAG; (7) ACG; (8) UCG; (9) AAU or AUC; (10) GAU or GAC; (11) CAU or CAC; (12) UAU or UAC; (13) CAG; (14) CCG

9.6

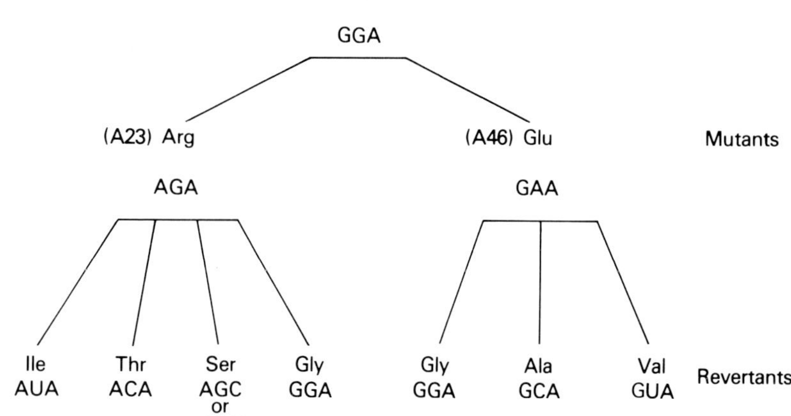

9.7 Wild type mRNA:

(39) AUG-UUU-GCU-AAC-CAU-AAG-AGU-GUA-GGX (47)

The second codon (for amino acid 40) could be UUC. The last codon could have A, G, C, or U at the third position.
For mutant mRNA:

(39) AUG-UUG-CUA-ACC-AUA-AGA-GUG-UAG

This is generated by the loss of one of the Us in the second codon (or, if it was UUC, by the loss of the C). This is a frameshift mutation and leads to the production of a premature chain-terminating codon at what would have been amino acid 46.
Partial revertant mRNA:

(39) AUG-UUG-CUA-ACC-AUA-AGA-GGU-GUA-GGX (47)

The reading frame is restored by the addition of a G before or after the first G in the codon before the chain-terminating codon. This removes the premature chain-terminating codon and leaves a stretch of six codons that are different from that found in the wild type mRNA.

Chapter 10

10.1

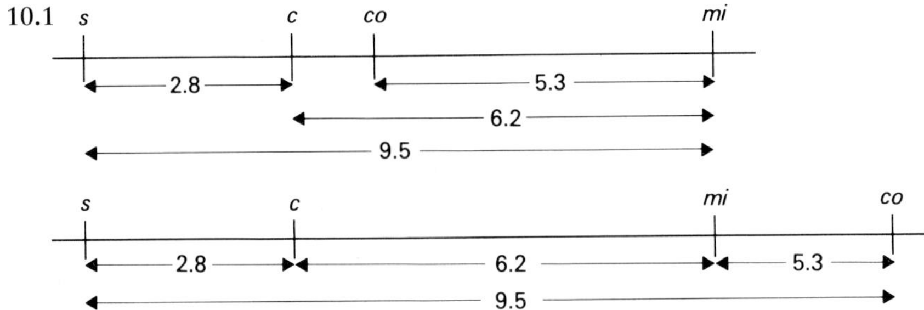

10.2 The phage progeny values are handled just as any other three-point mapping cross. The order of the genes as established by identifying the reciprocal pair of classes that represent the double crossover progeny is *m-r-tu*. The *m-r* distance = (162+520+474+172)/10,342×100% = 12.84%. The *r-tu* distance = (853+162+172+965)/10,342×100% = 20.81%. This gives the following map:

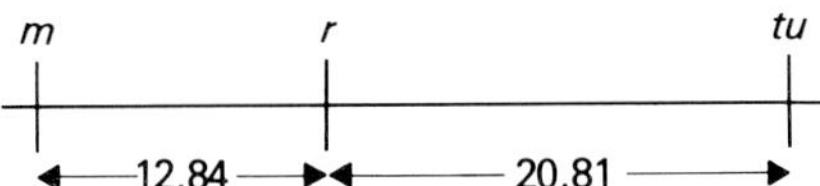

The observed double crossover frequency is (162+172)/10,342×100% = 3.23%. The expected double crossover frequency derived from the map distances just calculated is (0.1284×0.2081)×100% = 2.67%. The coefficient of coincidence is 3.23/2.67 = 1.21. This value indicates the absence of interference in this cross. In fact, since the value is higher than one, we may hypothesize in this case that the presence of one crossover in a region actually enhances the occurrence of a second crossover event nearby. This phenomenon is called negative interference.

10.3 0.5% recombination (the numbers of plaques counted are so small that this is a rather rough approximation).

10.4 0.07 mu. The plaques produced on K12(λ) are wild type, r^+, phage, and in the undiluted lysate there are 470×5/ml (since 0.2 ml was plated) = 2350/ml. The r^+ phage was generated by recombination between the two *rII* mutations. The other product of the recombination event is the double mutant and it does not grow on K12(λ). Therefore the true number of recombinants in the population is actually equivalent to twice the number of r^+ phage, since for every wild type phage produced there ought to be a double mutant recombinant produced. Thus there are 4700 recombinants per milliliter. The total number of phage in the lysate is 672×(dilution factor)×(1 ml divided by the sample size plated) per milliliter, or 672×1000×10 = 6,720,000/ml. The map distance between the mutations is (4700/6,720,000)×100% = 0.07%.

10.5

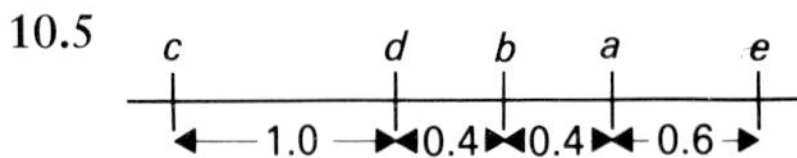

Note that the reciprocal recombinant class is the double *rII* mutant, which is indistinguishable from the single *rII* mutants in phenotype. Therefore the frequencies of wild type recombinants between sites are doubled to give map distances.

10.6

1
2
7
5
6
4
3

b e a c d

Chapter 11

11.1 See F^- (p. 186); F^+ (p. 186); *Hfr* (p. 187).

11.2 The whole chromosome would have to be transferred in order for the recipient to become a donor in a $Hfr \times F^-$ cross; that is, the *F* factor in the *Hfr* strain is transferred to the F^- cell last. This takes approximately 100 minutes and usually the conjugal unions break apart by then.

11.3 See text, pp. 192–193, for descriptions of lysogenic and lytic cycles.

11.4 See text, pp. 193–199, for descriptions of generalized and specialized transduction.

11.5

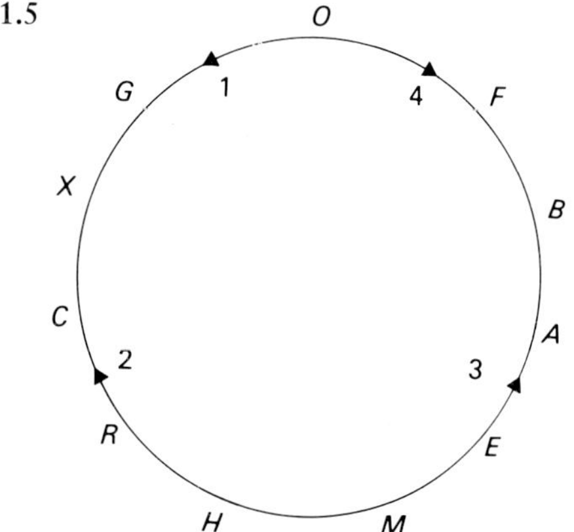

11.6 (a) When the prophage enters the recipient (F^-), there are no repressor molecules present so the prophage is induced, it goes through the lytic cycle and progeny phages are released. It is called zygotic induction since the initiation of the lytic cycle is dependent upon the formation of a zygote (conjugal union) between the donor and recipient. As long as the prophage (the λ gene set) is still within the donor, there are enough repressor molecules present to keep the genome in the prophage state.

(b) The λ genome can be mapped just like any other gene by determining the time at which it enters the recipient. As we have seen, as soon as it enters the recipient, the lytic cycle is induced and progeny phages are released from the lysed cell. The way to map the prophage site, then, is to do an interrupted mating experiment and, at various sample times, plate exconjugants on suitable plates that will select for those F^- that have received donor markers, that will select against both parentals, and that contain a lawn of sensitive bacteria on which plaques of λ can be detected. Up to a certain time, no plaques will be seen and then, once the λ is transferred, plaques will be seen.

11.7 The total number of transductants is 446, and the transductants that were generated where one crossover was between *leu* and *trp* (the second two classes) total 77. The percentage recombination is 77/446×100% = 17.3%. This is the map distance between the two genes.

11.8 Strain A is *thr* leu^+ and B is thr^+ *leu*. Transformed B should be thr^+ leu^+ and its presence should be detected on a medium containing neither threonine nor leucine.

11.9 Percentage recombinants in the transformation experiment is (228+141)/1000×100 = 36.9%, which is the map distance between *p* and *m*. The rationale is that classes I and II represent the transformants that have arisen as a result of a crossing-over event between *p* and *m*.

Chapter 12

12.1 At any one position in the DNA, there are four possibilities for the base pair: A-T, T-A, G-C, and C-G. Therefore the length of the base-pair sequence that the enzyme recognizes is given by the power to which four must be raised to equal (or approximately equal) the average size of the DNA fragment produced by enzyme digestion. The answer in this case is 6; that is, 4 to the power of 6 = 4096. If the enzyme instead recognized a four base-pair sequence, the average size of the DNA fragment would be 4 to the power of 4 = 256.

12.2 The complete digestion and incomplete digestion experiments with enzyme I indicate that the fragments are connected in the following order: a-d-b-c, with c also adjacent to fragment a since it is a circular molecule. The best evidence for this is the incomplete digestion experiments since various combinations of fragments are produced. Inspection of the fragments produced following digestion both with I and II suggest that the 1400 and 900 base-pair fragments are the same as those produced after digestion with I alone. Further, the data suggest that enzyme II digests fragment a into 1200 and 800 base-pair fragments, and that it digests fragment d into 400 and 300 base-pair fragments. Thus enzyme II cuts the fragments asymmetrically, but without data from cutting with enzyme II alone, it cannot be determined exactly where II cuts relative to the enzyme I sites. The following diagram shows the restriction map as we have deduced it:

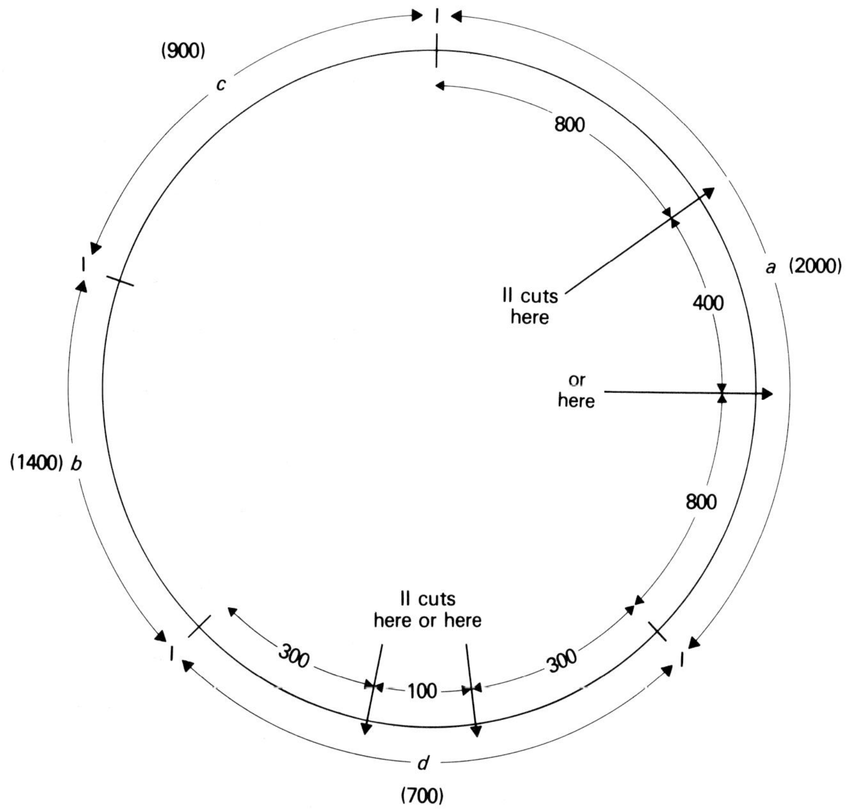

12.3 The sequencing gel banding pattern is as follows:

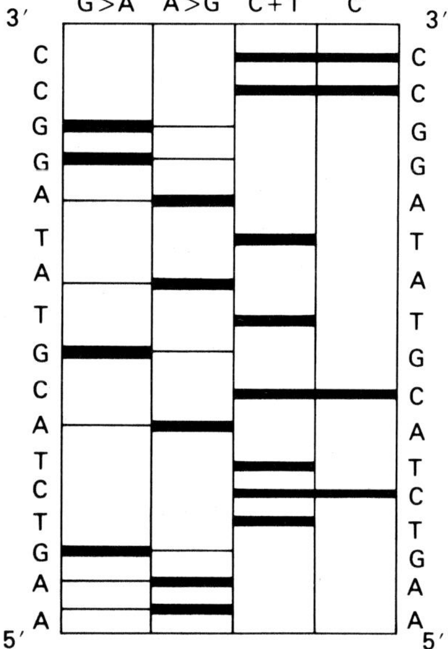

Chapter 13

13.1 (a) red (b) 3 red : 1 yellow (c) all red (d) 1/2 red, 1/2 yellow

13.2 If *R* is the factor for red fruit and *r* for yellow fruit, the red parent is *Rr* and the yellow parent is *rr*.

13.3 (a) Parents are *Rr* (rough) and *rr* (smooth); F1 are *Rr* (rough) and *rr* (smooth). (b) $Rr \times Rr \rightarrow 3/4$ rough, 1/4 smooth.

13.4 Bull is *PP* or *Pp*. With A a horned calf *pp* is produced, so bull must be *Pp*. Cow A has to be *pp*, so there should be 1/2 polled and 1/2 horned in the progeny. Cow B is polled and gives a horned calf, so the cow is *Pp*. Offspring from this cross is 3/4 polled and 1/4 horned. Cow C is horned and must be *pp*, and the polled calf produced is *Pp*. Offspring are 1/2 polled and 1/2 horned.

13.5 Progeny ratio approximates a 3 : 1, so the parent is heterozygous. Of the dominant progeny there is a 1 : 2 ratio of homozygous to heterozygous, so 1/3 will breed true.

13.6 Black is dominant to brown. If *B* is the allele for black and *b* for brown, female X is *Bb* and female Y is *BB*. The male is *bb*.

13.7 (a) *Pp Ss*, so purple, spiny
(b) 9/16 purple, spiny : 3/16 purple, smooth : 3/16 white, spiny : 1/16 white, smooth
(c) 1/2 purple, spiny : 1/2 white, spiny
(d) 1/2 purple, spiny : 1/2 purple, smooth

13.8 (a) all *Pp Ss*, purple, spiny
(b) 1/2 purple, spiny (*Pp Ss*) : 1/2 white, spiny (*pp Ss*)
(c) 3/4 purple, spiny : 1/4 white, spiny
(d) 9/16 purple, spiny : 3/16 purple, smooth : 3/16 white, spiny : 1/16 white, smooth
(e) 3/8 purple, spiny : 3/8 purple, smooth : 1/8 white, spiny : 1/8 white, smooth
(f) 1/4 purple, spiny : 1/4 purple, smooth : 1/4 white, spiny : 1/4 white, smooth

13.9 (a) *WW Dd* × *ww dd*
(b) *Ww dd* × *Ww dd*
(c) *ww DD* × *WW dd*
(d) *Ww Dd* × *ww dd*
(e) *Ww Dd* × *Ww dd*

13.10 The cross is *tt GG ss* × *TT gg SS*.
(a) The F1 is triply heterozygous *Tt Gg Ss*, which will show the dominant phenotype of each gene pair and hence will have a tall stem, green pods, and smooth seeds.
(b) This is a trihybrid cross. The number of expected phenotypic classes is $2^3 = 8$. For each gene pair, we expect a 3 : 1 ratio of F2 progeny with dominant and recessive characteristics, respectively. Each gene pair segregates independently. Therefore the answer is 27/64 tall, green, smooth; 9/64 tall, green, wrinkled; 9/64 tall, yellow, smooth; 9/64 short, green, smooth; 3/64 tall, yellow, wrinkled; 3/64 short, green, wrinkled; 3/64 short, yellow, smooth; 1/64 short, yellow, wrinkled.
(c) Cross is *Tt Gg Ss* × *tt GG ss*. Half of the progeny will be tall, and half will be short. All the progeny will be green. Half the progeny will be smooth, and half will be wrinkled. The compiled answer is: 1/4 tall, green, smooth; 1/4 tall, green, wrinkled; 1/4 short, green, smooth; 1/4 short, green, wrinkled.

(d) Cross is *Tt Gg Ss*×*TT gg SS*. All of the progeny will be tall and smooth. Half the progeny will be green, and half will be yellow. The compiled answer is 1/2 tall, green smooth; 1/2 tall, yellow, smooth.

13.11 (a) The F1 has the genotype Aa Bb Cc^h and is all agouti, black. The F2 is 27/64 agouti, black; 9/64 agouti, black, Himalayan; 9/64 agouti, brown; 9/64 black; 3/64 agouti, brown, Himalayan; 3/64 black, Himalayan; 3/64 brown; 1/64 brown, Himalayan.
(b) 4/27 (c) 1/4 (d) 3/4

13.12 (a) All agouti
(b) 3/4 agouti, 1/4 chinchilla
(c) 3/4 agouti, 1/4 albino
(d) 1/2 agouti, 1/2 Himalayan
(e) 1/2 agouti, 1/2 Himalayan
(f) 1/2 chinchilla, 1/2 Himalayan
(g) 1/2 Himalayan, 1/2 albino
(h) 3/4 agouti, 1/4 Himalayan
(i) 3/4 agouti, 1/4 chinchilla

13.13 (a) Yes. Both As could be I^Ai heterozygotes so that an *ii* child could result.
(b) If A is I^Ai and B is I^Bi, an O child can result.
(c) One parent must donate an I^A allele, and the other an I^B allele for a child to be AB. The O parent is *ii*, so it is not possible to produce an O child.
(d) The parents are $I^AI^B \times I^A\_$. An O child cannot result from this marriage.
(e) Yes, if the B parent is I^Bi, the A child would be I^Ai.

13.14 (a) pink
(b) 1/4 red, 1/2 pink, 1/4 white
(c) 1/2 red, 1/2 pink
(d) 1/2 pink, 1/2 white

13.15 Half will be *Rr* and hence will resemble the parents.

13.16 (a) 1/2 red, 1/2 roan
(b) 1/4 red, 1/2 roan, 1/4 white
(c) 1/2 roan, 1/2 white

13.17 (a) The cross is *FF oo* × *ff OO*. The F1 is *Ff Oo*, which is fuzzy with round leaf glands.
(b) Interbreeding the F1 gives: 3/16 fuzzy, oval-glanded; 6/16 fuzzy, round-glanded; 3/16 fuzzy, no-glanded; 1/16 smooth, oval-glanded; 2/16 smooth, round-glanded; 1/16 smooth, no-glanded.
(c) Cross is *Ff Oo*×*ff OO*. The progeny are: 1/4 fuzzy, oval-glanded; 1/4 fuzzy, round-glanded; 1/4 smooth, oval-glanded; and 1/4 smooth, round-glanded.

13.18 The parents were *Ll Ww* and *ll ww*. The baby was *ll Ww*. From a cross of *ll Ww*×*Ll Ww*, the phenotypic classes expected are: 1/8 short, yellow; 2/8 short, cream; 1/8 short, white; 1/8 long, yellow; 2/8 long, cream; 1/8 long, white.

13.19 The cross is *LL RR* × *ll rr*. The F1 is *Ll Rr* (i.e. oval, purple). Selfing the doubly heterozygous F1 gives the following phenotypic classes in the F2: 1/16 long, red; 2/16 long, purple; 1/16 long, white; 2/16 oval, red; 4/16 oval, purple; 2/16 oval, white; 1/16 round, red; 2/16 round, purple; 1/16 round, white.

13.20 (a) 3/4 walnut (*R_P_*) and 1/4 single (*R_pp*)
(b) 1/2 walnut and 1/2 pea (*rr P_*)
(c) 1/4 walnut, 1/4 pea, 1/4 rose, 1/4 single
(d) 9/16 walnut, 3/16 rose, 3/16 pea, 1/16 single
(e) 3/4 rose, 1/4 single

13.20 (a) 3/4 walnut ($R\_P\_$) and 1/4 single (R_pp)
(b) 1/2 walnut and 1/2 pea ($rr\ P\_$)
(c) 1/4 walnut, 1/4 pea, 1/4 rose, 1/4 single
(d) 9/16 walnut, 3/16 rose, 3/16 pea, 1/16 single
(e) 3/4 rose, 1/4 single

13.21 (a) $Rr\ Pp$ (walnut)$\times rr\ pp$ (single)
(b) $Rr\ pp$ (rose)$\times Rr\ Pp$ (walnut)
(c) $RR\ pp$ (rose)$\times rr\ Pp$ (pea)
(d) $Rr\ Pp$ (walnut)$\times Rr\ Pp$ (walnut)

13.22 (a) If *Y* governs yellow and *y* governs agouti, *YY* are lethal, *Yy* are yellow, and *yy* are agouti. Let *C* determine colored coat, and *c* determine albino. The parental genotypes, then, are *Yy Cc* (yellow) and *Yy cc* (white).
(b) The proportion is 2 yellow: 1 agouti: 1 albino. None of the yellows breed true since they are all heterozygous, homozygous *YY* individuals being lethal.

13.23 (a) All wild type: the flies are $+/e\ +/bl$, and the mutant alleles are recessive.
(b) The ratio is $9\ +\ +\ :\ 3\ e\ +\ :\ 3\ +\ bl\ :\ 1\ e\ bl$; that is, 9 wild type : 7 black body, since both mutants produce a black body color.
(c) (i) 1 wild type : 1 ebony; (ii) 1 wild type : 1 black.

13.24 (a) *YY RR* (crimson)$\times$*yy rr* (white) gives *Yy Rr* F1 plants, which have magenta-rose flowers. Selfing the F1 gives an F2 as follows: 1/16 crimson (*YY RR*), 2/16 orange-red (*YY Rr*), 1/16 yellow (*YY rr*), 2/16 magenta (*Yy RR*), 4/16 magenta-rose (*Yy Rr*), 2/16 pale yellow (*Yy rr*), and 4/16 white (*yy RR, yy Rr,* and *yy rr*). Crossing the F1 (*Yy Rr*) with the crimson parent (*YY RR*) gives 1/4 crimson (*YY RR*), 1/4 orange-red (*YY Rr*), 1/4 magenta (*Yy RR*), and 1/4 magenta-rose (*Yy Rr*).
(b) *YY Rr*$\times$ *Yy rr* gives 1/4 orange-red, 1/4 yellow, 1/4 magenta-rose, 1/4 pale yellow.
(c) *YY rr* (yellow)$\times$*yy Rr* (white) gives 1/2 magenta-rose (*Yy Rr*) and 1/2 pale-yellow (*Yy rr*).

13.25 The results of the cross are 2/16 *YY RR,* 2/16 *YY Rr,* 4/16 *Yy RR,* 4/16 *Yy Rr,* and 4/16 *yy*__. Since individuals homozygous for *y* are found, both parents must have been heterozygous *Yy*. From the ratios of the phenotypes, one parent must have been *RR* and the other *Rr*. The parents, therefore, were *Yy RR* (magenta) and *Yy Rr* (magenta-rose).

13.26 The plum-eyed allele is a dominant allele that gives plum eyes when heterozygous with the wild type allele; it is lethal when homozygous. That is, *Pm Pm* is lethal, $Pm\ Pm^+$ gives plum eyes, and $Pm^+\ Pm^+$ gives red eyes. A similar situation prevails for the stubble phenotype: *Sb Sb* is lethal, $Sb\ Sb^+$ gives stubble bristles, and $Sb^+\ Sb^+$ gives normal bristles. The cross is $Pm\ Pm^+\ Sb\ Sb^+ \times Pm\ Pm^+\ Sb\ Sb^+$, and the offspring are as follows:

1/16 $Pm\ Pm\ Sb\ Sb$	lethal
2/16 $Pm\ Pm\ Sb\ Sb^+$	lethal
1/16 $Pm\ Pm\ Sb^+\ Sb^+$	lethal
2/16 $Pm\ Pm^+\ Sb\ Sb$	lethal
4/16 $Pm\ Pm^+\ Sb\ Sb^+$	plum, stubble
2/16 $Pm\ Pm^+\ Sb^+\ Sb^+$	plum, normal bristles
1/16 $Pm^+\ Pm^+\ Sb\ Sb$	lethal
2/16 $Pm^+\ Pm^+\ Sb\ Sb^+$	red, stubble
1/16 $Pm^+\ Pm^+\ Sb^+\ Sb^+$	red, normal bristles

13.27 (a) 40 cm
(b) *AA bb, aa BB, Aa Bb*
(c) 6/16

13.28 (a) 8 cm
(b) 1 (2 cm) : 6 (4 cm) : 15 (6 cm) : 20 (8 cm) : 15 (10 cm) : 6 (12 cm) : 1 (14 cm)
(c) Proportion of F2 like *AA BB CC* = $(1/4)^3$; proportion of F2 like *aa bb cc* = $(1/4)^3$; proportion of F2 with heights equal to one another = $(1/4)^3+(1/4)^3$ = 2/64.
(d) The F1 height of 8 cm can be produced only by maintaining heterozygosity at no less than one gene pair (e.g. *AA Bb cc*). Thus a pure-breeding 8 cm strain would not be expected to occur.

13.29 The cross is *aa bb cc* (4 gm)×*AA BB CC* (10 gm), giving an F1 that is *Aa Bb Cc* (7 gm). The F2 will show a range of phenotypes from 4 to 10 gm, with relative proportions of the seven types being the coefficients in the binomial expansion of $(a+b)^6$: 1 (4 gm) : 6 (5 gm) : 15 (6 gm) : 20 (7 gm) : 15 (8 gm) : 6 (9 gm) : 1 (10 gm).

Chapter 14

14.1 (a) The original cross is *ww* × w^+Y. The F1 female is w^+w and is crossed with the w^+Y parental male. All female progeny of this cross will have red eyes; half the male progeny will have white eyes, and one-half will have red eyes. The F1 male is *w*Y and is crossed with the *ww* parental female. All progeny of this cross will have white eyes.
(b) The F1 cross is *ww*×w^+Y. The F1 is w^+w female and *w*Y male. Interbreeding the F1 gives females 1/2 of which have red eyes and 1/2 of which have white eyes. One-half of the males in the F2 have red eyes and 1/2 have white eyes. Interbreeding the F2 involves two types of females with two types of males. In the F3, 5/16 of the progeny are red-eyed females, 3/16 are white-eyed females, 2/16 are red-eyed males, and 6/16 are white-eyed males.

14.2 The woman is heterozygous for the recessive color-blindness allele, let us say c^+c. The man is not color blind and thus has the genotype c^+Y. (It does not matter what phenotype his mother and father have since males are hemizygous for sex-linked genes; thus the genotype can be assigned directly from the phenotype.) All of the female offspring will have normal color vision, and 1/2 of the male offspring will be color blind and 1/2 will have normal color vision.

14.3 The parentals are *ww* vg^+vg^+ ♀ and w^+Y *vg vg*♂, respectively.
(a) The F1 males are all *w*Y vg^+vg; white eyes, long wings. The F1 females are all w^+w vg^+vg; red eyes, long wings.
(b) 3/8 red, long, females; 3/8 white, long, females; 1/8 red, vestigial, females; 1/8 white, vestigial, females. Same ratio of the respective phenotypes for the males.

(c) Cross of F1 male with parental female is $wY\ vg^+vg \times ww\ vg^+vg^+$. All of the progeny, male and female, will have white eyes and long wings. Cross of F1 female with parental male is $w^+w\ vg^+vg \times w^+Y\ vg\ vg$. Female progeny: all have red eyes; 1/2 have long wings and 1/2 have vestigial wings. Male progeny: 1/4 red, long; 1/4 red, vestigial; 1/4 white, long; 1/4 white, vestigial.

14.4 Female was $w^+w\ vg^+vg$ and male was $w^+Y\ vg^+vg$.

14.5 (a) Cross is $AA \times aY$, so all progeny are $A\_$; thus, probability of $A\_$ is 1.
(b) 0.
(c) All females will have the phenotype required.
(d) There are three genotypes for each of the autosomal genes, and four genotypes for the sex-linked gene (AA, Aa, AY, aY), giving $4(3)^3 = 108$.
(e) Only females can be heterozygous for the sex-linked gene, and the probability of a female offspring is 1/2. The probability of heterozygosity for any particular gene pair is 1/2, and therefore for four gene pairs the answer is $(1/2)^4 \times 1/2$ (for the probability of femaleness) $= 1/32$.
(f) (i) 1/64 of females (1/128 of total progeny); (ii) 1/32 of males (1/64 of total progeny); (iii) 1/128 of males (1/256 of total progeny); (iv) 0.

14.6 They are linked; if they were independent, the expected numbers would be 90 : 30 : 30 : 10.

14.7 From the χ^2 test, $\chi^2 = 15.10$; P is less than 0.01 at 3 degrees of freedom. This test reveals that the two genes do not fit a 1 : 1 : 1 : 1 ratio. It does not say why. Linkage might seem reasonable until it is realized that the minority classes are not reciprocal classes (both carry the *aa* phenotype). If the segregation at each locus is considered, however, the $B\_ : bb$ ratio is about 1 : 1 (203 : 197) while the $A\_ : aa$ ratio is not (240 : 160). The departure, then, is specifically the result of a deficiency of *aa* individuals. This departure should be confirmed in other crosses that test the segregation at locus *A*. In corn, further evidence would be that the *aa* deficiency might show up as a class of ungerminated seeds or seedlings that die early.

14.8 $400/1000 \times 100 = 40\%$

14.9 There are 158 progeny rabbits. The recombinant classes are the English plus Angora, and the non-English plus short-haired so the map distance between the genes is $(11+6)/158 \times 100\% = 10.8\% = 10.8$ map units (mu).

14.10 The cross is $b\ vg^+/b\ vg^+ \times b^+\ vg/b^+\ vg$, which gives a $b^+vg/b\ vg^+$ F1. An F1 female is crossed with a homozygous recessive $b\ vg/b\ vbg$. The female is used as the double heterozygous parent since no crossing over occurs in male *Drosophila*. Since the genes are in repulsion, the recombinant phenotypes from the cross are grey plus long and black plus vestigial, and there are 283 and 241 flies in these classes, respectively. The total number of progeny is 3236, so the map distance is $(283+241)/3236 \times 100\% = 16.2\% = 16.2$ map units (mu).

14.11

14.12 The following map distances are given by the data:

black, curved	22.7	curved, purple	20.0
black, purple	6.18	curved, speck	30.2
black, speck	47.6	curved, vestigial	8.20
black, vestigial	17.8	purple, speck	45.7
purple, vestigial	11.8	speck, vestigial	35.9

From the lower map distance values, the following unambiguous map can be drawn:

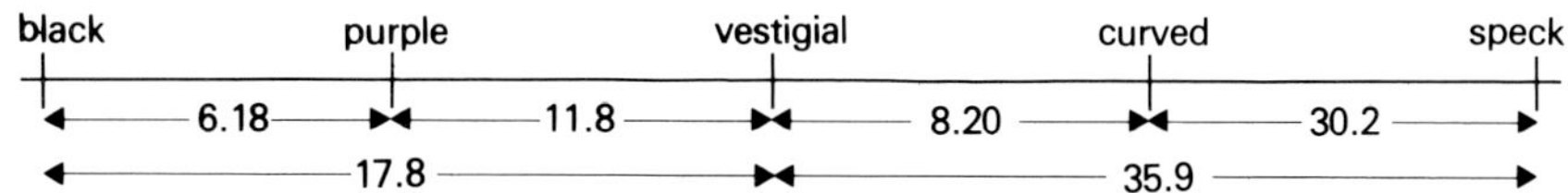

14.13 45% $a\ b^+$, 45% $a^+\ b$, 5% $a^+\ b^+$, 5% $a\ b$

14.14 (a) $0.035+0.465 = 0.50$
(b) Zero—all the daughters have *A B* phenotype

14.15 (a) 40% colorless, starchy; 40% colored, waxy; 10% colorless, waxy; 10% colored, starchy
(b) 45% colorless, starchy; 20% colored, waxy; 30% colorless, waxy; 5% colored, starchy

14.16 (a) the genotype of the F1 is *A B/a b*. The gametes produced by the F1 are: 40% *A B*; 40% *a b*; 10% *a b*; 10% *a B*. From a testcross of the F1 with *a b/a b*, we get: 40% *A B/a b*; 40% *a b/a b*; 10% *A b/a b*; 10% *a B/a b*.
(b) the F1 genotype is *A b/a B*. The gametes produced by the F1 are: 40% *A b*; 40% *a B*; 10% *A B*; 10% *a b*. From a testcross of the F1 with *a b/a b*, we get: 40% *A b/a b*; 40% *a B/a b*; 10% *A B/a b*; 10% *a b/a b*. (This question illustrates that map distance is computed between sites in the chromosome and it does not matter whether the genes are in coupling or in repulsion; the percentage of recombinants will be the same, even though the particular phenotypic classes constituting the recombinants differ.)

14.17 In the first cross, four classes of progeny were produced, with about 20% tall plus pear and dwarf plus spherical progeny. Thus the tall, spherical parent was doubly heterozygous and the genes were in coupling; that is, *D P/d p*, for *D* tall, *d* dwarf, *P* spherical, and *p* pear. In the other cross, there are also four classes of progeny, with about 20% tall plus spherical and dwarf plus pear progeny. Thus the tall, spherical plant was also doubly heterozygous with the genes this time in repulsion (i.e. *D p/d P*). The cross between the two tall, spherical parents is *D P/d p*×*D p/d P*. The offspring produced and their phenotypes are given in the following figure:

D P/*d p* × *D p*/*d P*
tall, spherical   tall, spherical

	D p 0.4	*d P* 0.4	*D P* 0.1	*d p* 0.1
D P 0.4	*D P*/*D p* tall, spherical 0.16	*D P*/*d P* tall, spherical 0.16	*D P*/*D P* tall, spherical 0.04	*D P*/*d p* tall, spherical 0.04
d p 0.4	*D p*/*d p* tall, pear 0.16	*d p*/*d P* dwarf, spherical 0.16	*d p*/*D P* tall, spherical 0.04	*d p*/*d p* dwarf, pear 0.04
D p 0.1	*D p*/*D p* tall, pear 0.04	*D p*/*d P* tall, spherical 0.04	*D p*/*D P* tall, spherical 0.01	*D p*/*d p* tall, pear 0.01
d P 0.1	*d P*/*D p* tall, spherical 0.04	*d P*/*d P* dwarf, spherical 0.04	*d P*/*D P* tall, spherical 0.01	*d P*/*d p* dwarf, spherical 0.01

Phenotypes:

tall, spherical → 0.16 + 0.16 + 0.04 + 0.04 + 0.04 + 0.04 + 0.01 + 0.04 + 0.01 = 0.54
tall, pear → 0.16 + 0.04 + 0.01 + 0.01 = 0.22
dwarf, spherical → 0.16 + 0.04 = 0.20
dwarf, pear → 0.04

14.18 Each chromosome pair segregates independently. We can compute the relative proportions of gametes produced for each homologous pair of chromosomes separately from the known map distances (P = parental; R = recombinant):

P		R		P		R		P		R	
A B	0.4	*A b*	0.1	*C D*	0.45	*C d*	0.05	*E f*	0.35	*E f*	0.15
a b	0.4	*a B*	0.1	*c d*	0.45	*c D*	0.05	*e f*	0.35	*e F*	0.15

To answer the questions, simply multiply the probabilities of getting the particular gamete from the F1 multiple heterozygote.

(a) *ABCDEF* = 0.4×0.45×0.35 = 0.063 (6.3%)
(b) *ABCdef* = 0.4×0.05×0.35 = 0.007 (0.7%)
(c) *AbcDEf* = 0.1×0.05×0.15 = 0.00075 (0.075%)
(d) *aBCdef* = 0.1×0.05×0.35 = 0.00175 (0.175%)
(e) *abcDeF* = 0.4×0.05×0.15 = 0.003 (0.3%)

14.19 (a) A is linked to B, but neither is linked to C. The cross involves A and B in repulsion:

$$\frac{A\ b\ \ C}{A\ b\ \ C} \times \frac{a\ B\ \ c}{a\ B\ \ c}$$

This gives the following F1:

$$\frac{A\ b\ \ C}{a\ B\ \ c}$$

The F1 is crossed with a triply homozygous recessive. There are 20 map units (mu) between A and B, so the total of all recombinants involving A and B add up to 20% of all progeny. The progeny and their proportions are as follows:

Parentals:	*A b C* amiable, active, crazy	800 (20%)
	A b c amiable, active, sane	800 (20%)
	a B C nasty, benign, crazy	800 (20%)
	a B c nasty, benign, sane	800 (20%)
Recombinants:	*A B C* amiable, benign, crazy	200 (5%)
	A B c amiable, benign, sane	200 (5%)
	a b C nasty, active, crazy	200 (5%)
	a b c nasty, active, sane	200 (5%)

(b) If A and B were completely linked, none of the recombinant classes would be produced.
(c) If A and B were unlinked, none of the classes would be missing, although there would be equal numbers of each of the eight phenotypic classes, or 500 each.
(d) If all three genes are unlinked, the results of selfing a triple heterozygote, the F1, are as for a trihybrid cross. That is, in the F2 there are eight phenotypic classes in a ratio 27:9:9:9:3:3:3:1. The nasty bippies must be all *aa* in genotype, so then the distribution is essentially a subset of the previous classes, that is, 9 nasty, benign, crazy:3 nasty, benign, sane:3 nasty, active, crazy:1 nasty, active, sane.

14.20 (a) a is in the middle. The genotype of the triple heterozygote is *B A C*/*b a c*.
(b) d is in the middle. The genotype is *C d E*/*c D e*.
(c) g is in the middle. The genotype is *F G h*/*f g H*.

14.21 (a) *F m W* and *f M w* will be the least frequent (because of double crossovers).
(b) *M f W* and *m F w* will be the least frequent.
(c) *M w F* and *m W f* will be the least frequent.

14.22

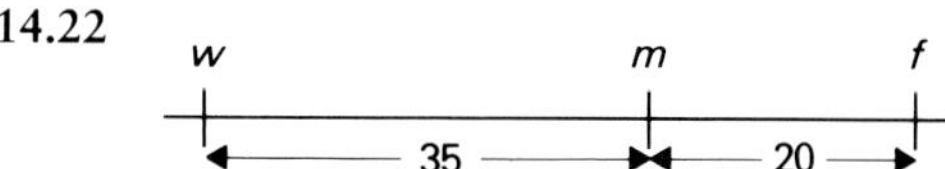

14.23 (a) By doing a three-point mapping analysis as described in the chapter, we find that the order of genes in the chromosome is *dp-b-hk*, with 35.5 mu between *dp* and *b*, and 5.4 mu between *b* and *hk*.

(b) (i) The frequency of observed double crossovers is 1.4%, and the frequency of expected double crossovers is 1.9%. The coefficient of coincidence, therefore, is 1.4/1.9 = 0.73. (ii) The interference value is given by 1−coefficient of coincidence = 0.27.

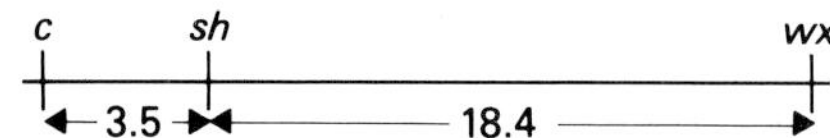

14.24 The original cross was:

$$\frac{L\ R\ Rs}{L\ R\ Rs} \times \frac{l\ r\ rs}{l\ r\ rs}$$

The testcross of the F1 is a typical three-point mapping cross. The total number of progeny is 3684. The double crossovers are the classes with 39 and 54 representatives. Along with the genotypes of the F1, this result tells us that the order of genes is *l r rs*. From the methods developed in the chapter, the map is as shown:

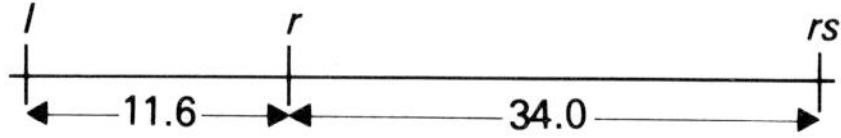

14.25 (a) *a* + *c* and + *b* + (least frequent)
(b) + *b c* and *a* + + (least frequent)
(c) locus *c*

14.26 (a) $a^+\ c\ b/a\ c^+\ b^+$
(b) $a^+\ c^+\ b^+/Y$
(c) a c b; 5, 10

14.27 All three genes are X-linked. The parents are *b* + *c*/+ *a* + and +++/Y. Map distances are *b-a* = 9 mu; *a-c* = 8 mu (progeny totals for linkage calculations is 2000 males). The coefficient of coincidence is 0.0020/0.0072 = 0.278.

14.28 The recombination percentages of the four crosses are calculated by taking into account the absence of certain classes owing to the recessive lethality. The percentages for the four crosses are: (1) 1%; (2) 4%; (3) 12%; (4) 8%. The two possible maps to be drawn from these figures (it cannot be assumed automatically that the terminal markers are involved in a single cross, although, as it turns out, they are) are as shown in the following:

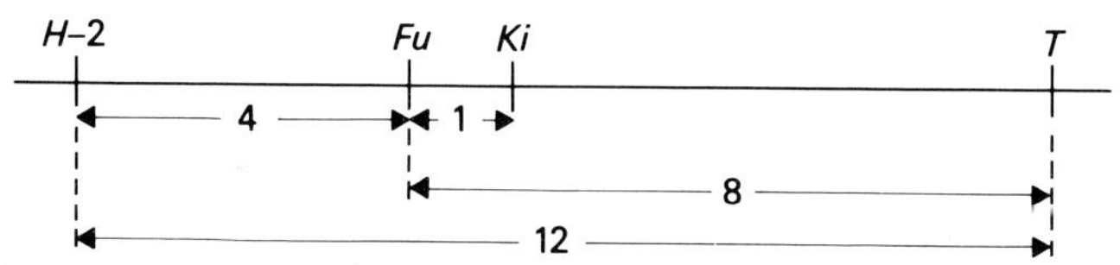

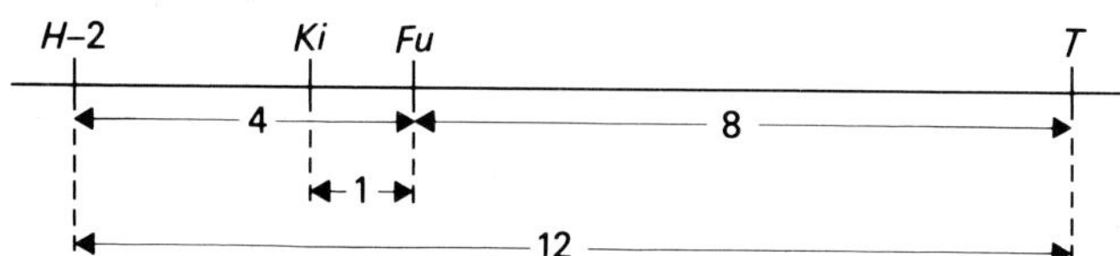

14.29 (a) a, b, c, and d are linked on the same chromosome, and e is on a separate chromosome. The map is as shown:

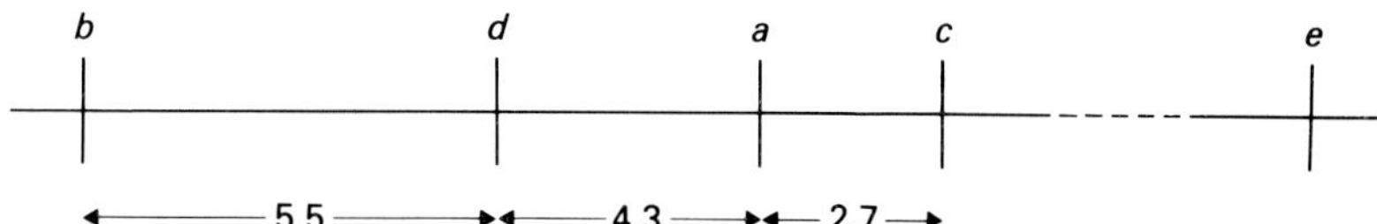

(b) 1.6×10^{-5}, that is, $0.055 \times 0.043 \times 0.027 \times 0.5$ (for the half of the progeny produced by the triple crossover that have all the wild type alleles)$\times 0.5$ (to give the proportion that are e^{+}).

Chapter 15

15.1 PD: noncrossovers and two-strand double crossovers; NPD: four-strand double crossovers; T: single crossovers and three-strand double crossovers

15.2 The 30 colonies represent half of the recombinants since the double mutants cannot grow on minimal medium. Map distance between *pan* and *lys* is given by $(60/750) \times 100\% = 8\%$ (8 map units).

15.3

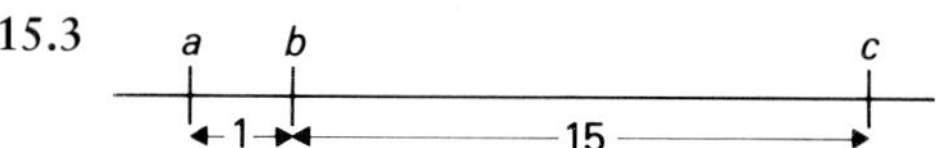

15.4 Genes 1 and 2 are linked, since they yield much less than 25% wild type. 25% wild type is expected on the basis of independent assortment, the other recombinants—the double mutants—being among the albino progeny. The 25 individuals who are wild type in the 1×2 cross are presumably accompanied by the albino, reciprocal class (double mutants) in approximately equal frequency. The true recombination frequency is thus about 50/1000, or 5%. The expectation of independent assortment is realized in the 1×4 cross. The cross 1×3 indicates, as far as one can tell, that the two mutations are allelic or at least very closely linked; they fail to produce any recombinants among a sizable number of progeny.

15.5 *arg* ——— 31 ——— *met* ——— 19 ——— *thi*

15.6

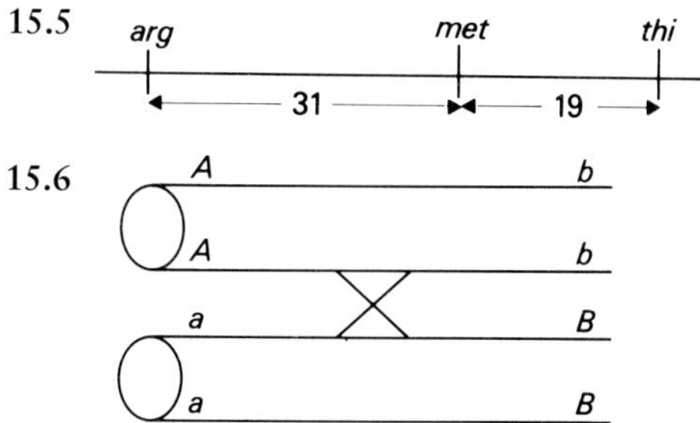

15.7 (a) The second exchange should involve the top and bottom strands to give four recombination products.

(b) The second exchange should involve the same two middle strands as the first to produce four nonrecombination products.

(c) 50%. At this value, the number of parentals equals the number of recombinants, representing the production of equal proportions of all four possible gamete types by the parent and the random combination of these gametes.

15.8 The two genes can be considered to assort independently since the number of parental ditype (20) is approximately equal to the number of nonparental ditype (18).

15.9 Two types of tetrads are expected:
pe col$^+$, *pe col*$^+$, *pe col*$^+$, *pe col*$^+$, *pe*$^+$ *col*, *pe*$^+$ *col*, *pe*$^+$ *col*, *pe*$^+$ *col*
and
pe col, *pe col*, *pe col*, *pe col*, *pe*$^+$ *col*$^+$, *pe*$^+$ *col*$^+$, *pe*$^+$ *col*$^+$, *pe*$^+$ *col*$^+$

15.10

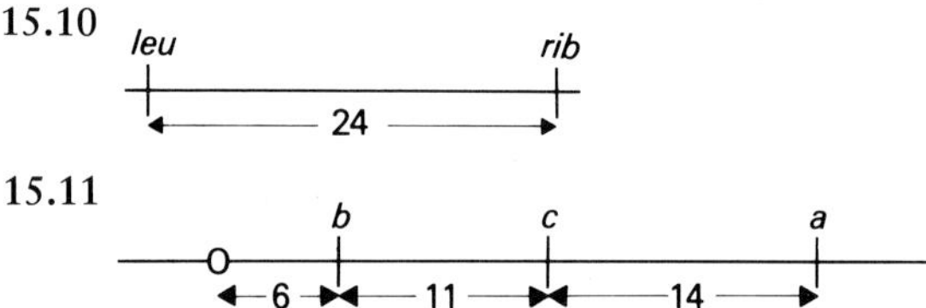

15.11

15.12 Haploidization gives an indication of linkage groups as discussed in the text. The data indicate that there are three separate linkage groups involved, since they appear in haploids as independent "blocks" of genes: *y*, *w-pu-ad*, and *sm-phe*.

15.13 There are three linkage groups as determined by which "blocks" of genes stay together during the haploidization process: *sm-phe*, *pu-w-ad*, and *y-bi*.

15.14 The last class in each segregant type (*ribo*, *an*, *pro*, *paba*, *y*, *bi* in each case) represents haploid segregants and indicates that all of the genes under investigation are in the same chromosome. The rest of the classes are diploid segregants representing various mitotic recombinants.
Consider the adenine-independent segregants: Following the logic developed in the chapter for ordering genes in a chromosome arm—that is, all genes distal to a crossover become homozygous—we can deduce that the genes *an* and *ribo* are on the same chromosome arm as *su-ad* and that the order of genes is centromere-*an-ribo-su-ad*.
Similarly analysis of the yellow diploid segregants shows that the order of genes in that chromosome arm is centromere-*pro-paba-y*-(*ad-bi*). Since all yellow segregants are *ad* and *bi*, the latter two loci must be distal to *y*, but their relative order cannot be determined from the data given.

15.15 I is in 10; II is in 2; III is in 4; IV is in 19

Chapter 16

16.1 The simplest hypothesis is that brown-colored teeth is determined by a sex-linked dominant mutant allele. Man A was *B*Y and his wife was *bb*. All sons will be *b*Y normals and cannot pass on the trait. All of the daughters receive the X chromosome from their father so that they are *Bb* with brown enamel. Half of their sons will have brown teeth since half of their sons receive the X chromosome with the *B* mutant allele.

16.2 Since only one of Woody's parents died of this disease, Woody must have been heterozygous *Cc* for the dominant mutant allele. His wife is *cc*. From *Cc*×*cc*, the probability of having a child that will die of Huntington's chorea is 1/2.

16.3

	Pedigree A	Pedigree B	Pedigree C
Autosomal recessive	Yes	Yes	Yes
Autosomal dominant	Yes	Yes	No
X-linked recessive	Yes	Yes	No
X-linked dominant	No	Yes	No

16.4 (a) Autosomal recessive
(b) I.1 and I.2; II.1 and II.2; III.4 and III.5
(c) 2/3
(d) III.3: 2/3 probability of being heterozygous; III.4 is known to be heterozygous; 1/4 of children of two heterozygotes will express an autosomal recessive trait; answer is $2/3 \times 1/4 = 1/6$.

16.5 Pedigree A: autosomal recessive; Pedigree B: X-linked recessive; Pedigree C: autosomal dominant

16.6 See text, pp. 308–309.

16.7 (a) 2N−1 (b) 2N−1−1 (c) 2N+2 (d) 2N+1+1 (e) 4N (f) 6N

16.8 (a) 45 (b) 47 (c) 23 (d) 69 (e) 48

16.9 (a) Pericentric inversion (*D* o *E F* inverted)
(b) Nonreciprocal translocation (*B C* moved to other arm)
(c) Tandem duplication (*E F* duplicated and inserted in same orientation as original)
(d) Reverse tandem duplication (*E F* duplicated and duplicate inserted in reverse orientation to original)
(e) Deletion (*C* deleted)

16.10 See text, p. 310.

16.11 (a) **a**

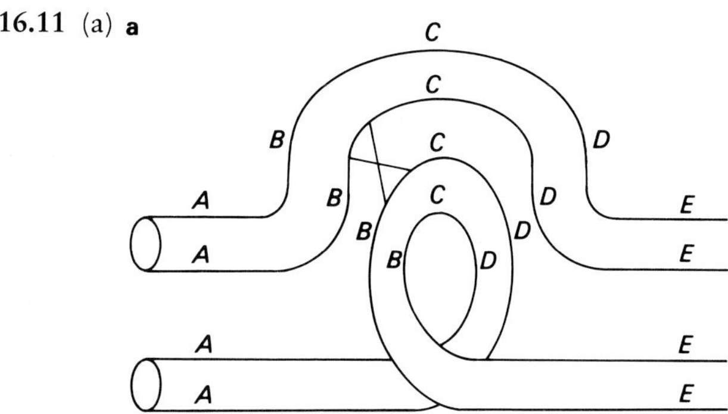

(b) From part (a) the crossover between *B* and *C* is as follows:

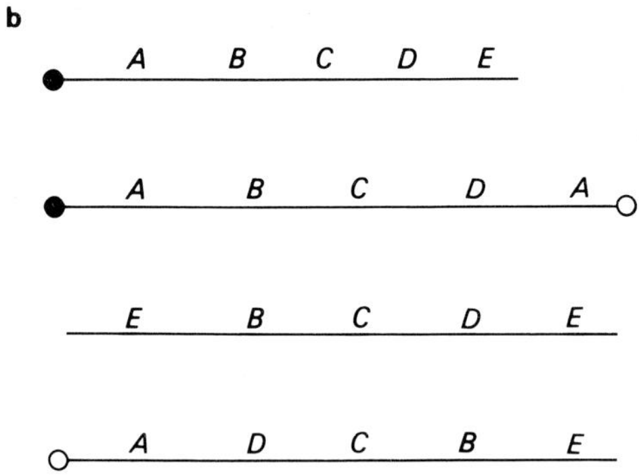

(c) Paracentric inversion because the centromere is not included in the inverted DNA segment

16.12 An example of a two-strand double crossover and the resulting meiotic products is shown in the following:

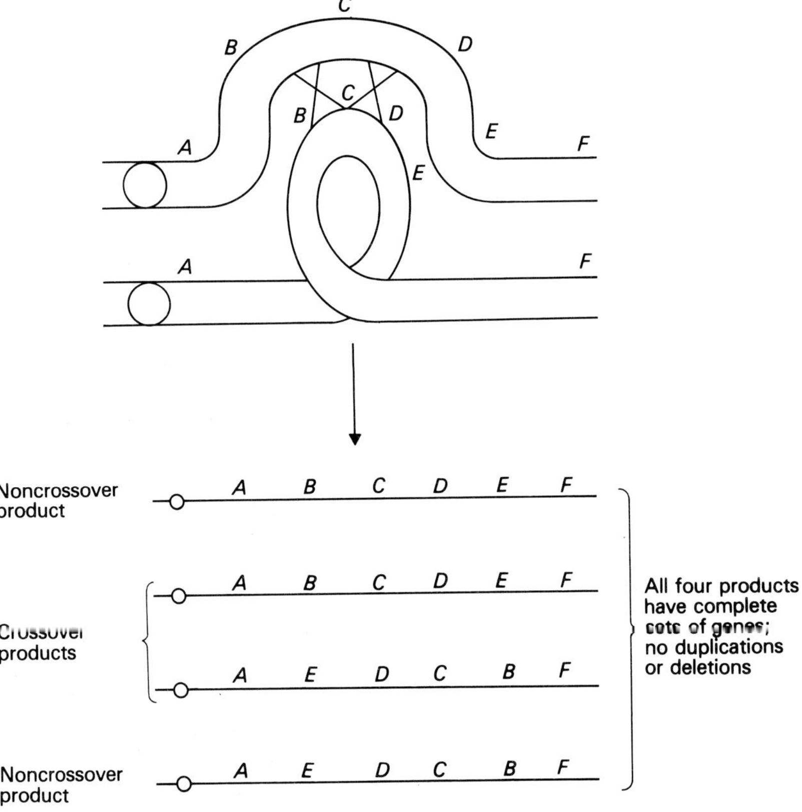

In this case, no dicentric bridge structure is formed as is the case with single crossovers, and all four meiotic products are viable.

16.13 In the following, the crossover between *c* and *d* involves strands 2 and 4, and the crossover between *e* and *f* involves strands 1 and 3.

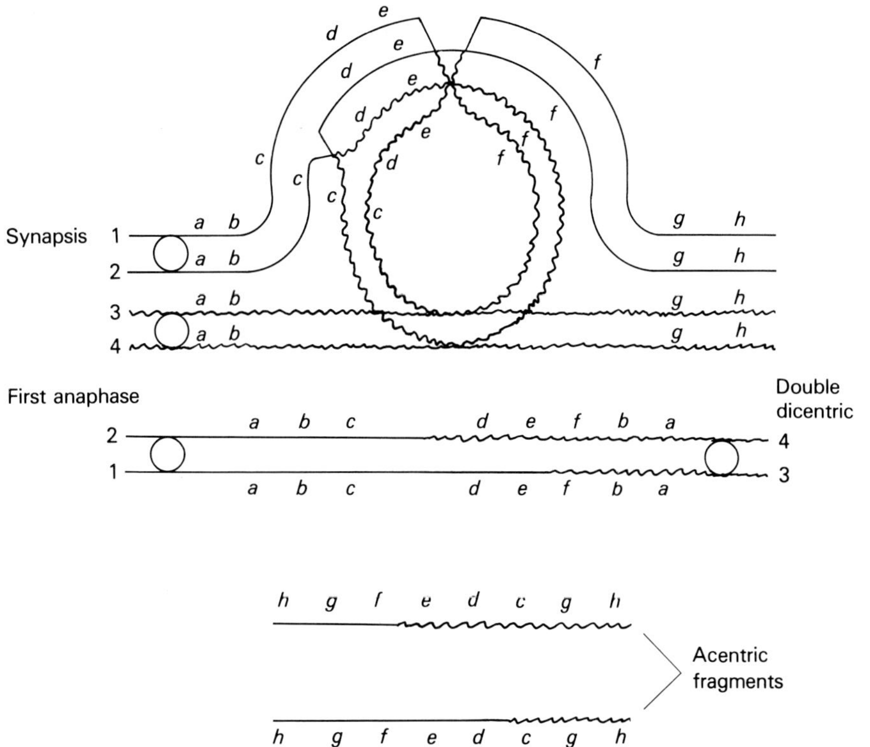

Chapter 17

17.1 The nuclear genome is organized into linear structures called chromosomes, which are composed of double-stranded DNA complexed with histones and nonhistone proteins. The nuclear chromosomes replicate at a specific time in the cell cycle and are segregated to progeny cells in an orderly fashion during the processes of mitosis and meiosis. The mitochondrial and chloroplast genomes consist of circular, naked, double-stranded DNA that is replicated independently of the nuclear genome and in many phases of the cell cycle. Segregation of progeny double helices to "daughter" organelles does not involve either mitotic or meiotic processes.

17.2 New mitochondria are generated by the growth and division of "old" mitochondria.

17.3 The two codes are different in codon designations. The mitochondrial code has more extensive wobble so that fewer tRNAs are needed to read all possible sense codons. As a consequence, fewer mitochondrial genes are needed. The advantage is that fewer tRNAs are needed and hence fewer tRNA genes need to be present than is the case for cytoplasmic tRNAs.

17.4 All mitochondrial mRNAs have a poly(A) tail added at their 3′ ends after transcription. The string of As added to the ending U or UA of the transcript completes the chain-terminating codon UAA.

17.5 See text, pp. 330–332, and Chapters 7 and 8.

17.6 The features of extranuclear inheritance are: differences in reciprocal cross results (not related to sex), nonmappability, Mendelian segregation not followed, and indifference to nuclear substitution.

17.7 Maternal effect is the determination of gene-controlled characters by the maternal genotype prior to the fertilization of the egg cell. The genes involved are nuclear genes, whereas the genes involved in extranuclear inheritance are extranuclear, being found in the mitochondria and chloroplasts.

17.8 Nuclear *petites* owe their phenotype to nuclear gene mutations, whereas neutral and suppressive *petites* owe their phenotypes to mutational changes in the mitochondrial genome.

17.9 The parental snails were *Dd* female and *d*_ male. The F1 snail is *dd*. (Given the F1 genotype, the male can either be homozygous *d* or heterozygous, but the determination cannot be determined from the data given.)

17.10 There is a maternal effect. In the $ma\text{-}l^{+}/ma\text{-}l$ female heterozygote, the xanthine dehydrogenase mRNA made by the wild-type allele is pumped into all of the eggs so all of the offspring are phenotypically wild type. (The mRNA must be stable.) Genotypically the progeny from the cross are half homozygous *ma-l* and half heterozygous. The former give all maroon-like progeny in the backcross with *ma-l* males, and the latter give all wild type since the cross is the same as the original cross.

17.11 The first possibility is that the results are the consequence of a sex-linked lethal gene. The females would be homozygous for a dominant gene, *L*, that is lethal in males but not in females. In this case, mating the F1 females of a $L/L \times +/\mathrm{Y}$ cross to +/Y males should give a sex ratio of 2 females : 1 male in the progeny flies. The second possibility is that the trait is cytoplasmically transmitted via the egg and is lethal to males. If this is so, the same F1 females should continue to have only female progeny when mated with +/Y males.

17.12 The simplest explanation is that a *D. simulans* X chromosome is needed for hybrid survival in *D. melanogaster* cytoplasm. However, in *D. simulans* cytoplasm, the presence of a *D. melanogaster* X chromosome is generally lethal.

17.13 (a) All progeny flies will be sensitive.
(b) All progeny flies will be resistant.

17.14 1/2 *petite*, 1/2 wild type

17.15 This problem is formally very similar to determining the inheritance mode for yeast *petite* mutants. If the *tiny* phenotypes are due to a nuclear gene, then from a cross of *tiny*×normal, meiotic segregation should generate a 1:1 ratio of *tiny* : normal cells in the progeny (c.f. properties of segregational *petites*). If, in contrast, an extranuclear gene is involved, segregation will not be evident and all progeny will be normal.

17.16 (a) Parents: $[po]F$ ♀×$[N]^{+}$ ♂; progeny: $[po]F$ and $[po]^{+}$ in equal numbers. Since standard *poky* $[po]^{+}$ is found among the offspring, gene *F* must not effect a permanent alteration of *poky* cytoplasm. The 1:1 ratio of *poky* to *fast-poky* indicates that all progeny have the *poky* mitochondrial genotypes (by maternal inheritance) and that the *F* gene must be a nuclear gene segregating according to Mendelian principles. Thus the *poky* progeny are $[po]^{+}$ and the *fast-poky* are $[po]F$.

(b) Parents: $[N]^+$♀ × $[po]F$♂; progeny: $[N]^+$ and $[N]F$ in equal numbers. These two types are phenotypically indistinguishable.

17.17 The plastids are primarily derived from the egg cytoplasm. There is a difference in the effect of the hybrid *hooker/muricata* nuclear gene combination on different plastids. When *hookeri* is the maternal parent, the *hookeri* plastids that are maternally inherited become yellow in the *hookeri/muricata* nuclear genetic background. When *muricata* is the maternal parent, the maternally inherited *muricata* plastids remain green in the *hookeri/muricata* nuclear genetic background.

17.18 (a) If normal cytoplasm is [*N*] and male-sterile cytoplasm is [*Ms*], the F1 genotype is [*Ms*]*Rf/rf*, and the phenotype is male-fertile.
(b) The cross is [*Ms*]*Rf/rf* female × [*N*]*rf/rf* male, giving 50% [*N*]*Rf/rf*, and 50% [*N*]*rf/rf*, all of which are phenotypically male-fertile.

Chapter 18

18.1 (a) The simplest approach is to calculate the proportion of the F2 that are colored and then to subtract that value from 1. In this case, the noncolorless are the brown or black progeny. The proportion of progeny that make at least the brown pigment is given by the probability of having the following genotype: *A_B_C_*(*DD*, *Dd*, or *dd*). The answer is 3/4×3/4×3/4×1 = 27/64. Therefore, the proportion of colorless is 1−27/64 = 37/64.
(b) The brown progeny have the following genotype: *A_B_C_dd*. The probability of getting individuals with this genotype is 3/4×3/4×3/4×1/4 = 27/256.

18.2 (a) To be colorless, an individual must be *aa bb* (*C_* or *cc*). The probability of this genotype is 1/4×1/4×1 = 1/16.
(b) To be red, an individual must be either *A_* or *B_* or both of these, and it must be *cc*. With regard to the *a* and *b* loci, it is easier to calculate the probability of colorless (*aa bb*) and then to subtract that value from 1 to get the probability of getting red color with just those two loci. The probability of colorless is 1/4×1/4 = 1/16; therefore the probability of red is 1-1/16 = 15/16. Now considering the 1/4 change of being *cc* and hence not converting red to black, the overall proportion of the F2 that is red is 15/16×1/4 = 15/64.
(c) To be black an individual must be *A_B_C_*, and the proportion of individuals that have this genotypic constitution is 3/4×3/4×3/4 = 27/64.

18.3 (a) 1/64 (b) 67/256.

18.4 The simplest scheme is that there are two distinct biochemical pathways, one producing a brown pigment, and the other producing a scarlet pigment, as shown in the accompanying figure. The two pigments then combine or are combined to produce a red pigment that gives the eye its color. Individuals homozygous for *bw* have a defective enzyme in the scarlet pigment pathway so that only brown pigment is produced; that is, they are scarletless. Similarly, individuals homozygous for *st* have a defective enzyme in the brown pathway so that only scarlet pigment is produced; that is, they are brownless. Double mutants are scarletless and brownless, giving a white eye.

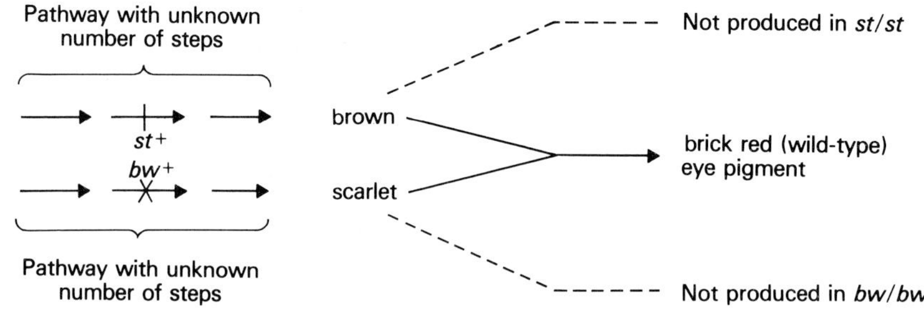

18.5 (a) Half have white eyes (the sons) and half have fire red eyes (the daughters).
(b) All have fire red eyes.
(c) All have brown eyes.
(d) All are w^+; a fourth are $bw^+\_st^+\_$, red; a fourth are $bw^+\_$ *st st*, scarlet; a fourth are *bw bw* $st^+\_$, brown; a fourth are *bw bw st st*, the color of 3-hydroxy-kynurenine plus the color of the precursor to biopterin, or colorless.

18.6 Wild type T4 will produce progeny phages at all three temperatures. Suppose that model (a) is correct. If cells infected with the double mutant are first incubated at 17°C and then shifted to 42°C, progeny phages will be produced and the cells will lyse. The explanation for this is as follows: The first step, *A* to *B*, is controlled by a gene whose product is heat-sensitive. At 17°C, the enzyme works and *A* is converted to *B*, but *B* cannot then be converted to mature phages since that step is cold sensitive. When the temperature is raised to 42°C, the *A*-to-*B* step is now blocked, but the accumulated *B* can be converted to mature phage since the enzyme involved with that step is cold sensitive and thus the enzyme is functional at the high temperature. If (b) is the correct pathway, progeny phages should be produced using a 42 to 17°C temperature shift, but not vice versa. In general, two gene-product functions can be ordered by this method whenever one temperature shift allows phage production and the reciprocal shift does not, according to the following rules. (i) If low-to-high temperature results in phages but high-to-low does not, then the hs step precedes the cs step (c.f. model (a)), (ii) if high-to-low temperature results in phages but low-to-high does not, the cs step precedes the hs step (c.f. model (b)).

18.7 (a) $c\ d^+$ and $c^+\ d$
(b) The genes are not linked since parental ditype (PD) and nonparental ditype (NPD) tetrads are in equal frequency.
(c) The pathway is Y to X to Z, with *d* blocking the synthesis of Y and *c*, the synthesis of X from Y.

18.8 The scheme is part (d); see the accompanying diagram.

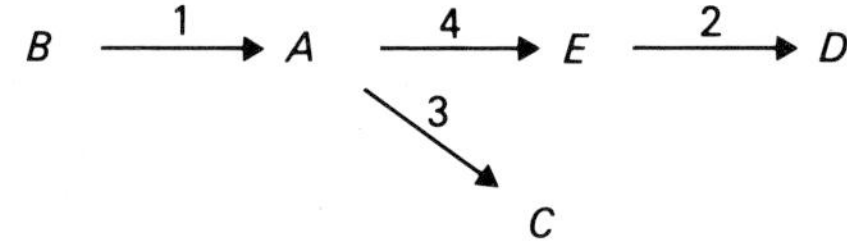

18.9 *Analysis of data*: The nutrition and accumulation data suggest that strain *1* is blocked between ORN and CIT because it accumulates the former and grows on the latter. Strains *2* and *3* are blocked, by the same reasoning, between CIT and ARG. Strain *4* is rather strange since it has the growth properties of strains *2* and *3* and the accumulation phenotype of strain *1*. On these grounds, it is doubtful that it is a single mutant.

Complementation indicates that the mutants *2* and *3* carry mutations in the same gene. That gene is not the same as in strain *1*. Strain *4* looks like a double mutant or a deletion mutant lacking two adjacent genes.

The question about strain *4* is resolved by the genetic tests. While mutations in strains *1* and *3* assort independently (25% prototrophs), the matings *1*×*4* and *3*×*4* yield no prototrophs. All data are consistent with the notion that strain *4* carries the mutation found in strain *1 and* the distinct mutation found in strain *3*. (A mating of strain *4*×wild type should therefore give only 25% wild type, instead of the 1:1 segregation expected of single mutants.)

Consistent with the allelism of mutations in strains *2* and *3* (note complementation) is their tight linkage in crosses. They are at different mutational sites, 0.004 map units apart, but in the same gene. The remaining question about the difference in prototroph yield in the *2*×*3* cross, and the *2*×*4* cross is deferred until answers to the specific questions are given, as follows:

(a) Three

(b) Two

(c)

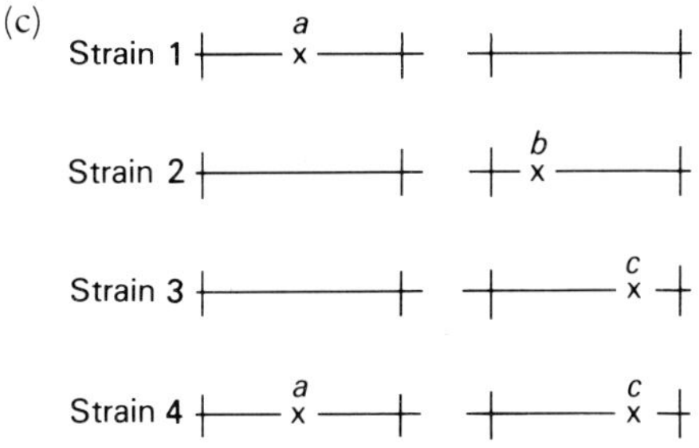

Two chromosomes are shown, with the ends of the relevant genes indicated. Mutations *b* and *c* are at distinct sites within one gene.

(d) *b* and *c* are 0.004 map units apart.

(e) (i), (ii), (iii) Strains *1*, *2*, or *3*, when mated with wild type, will give 50% prototrophs and 50% auxotrophs. (iv) Strain *4*, as noted above, will give 25% prototrophs, since only one of the four equally frequent genotypes expected (i.e. a^+ c^+ from the cross a^+ c^+ $\times a$ c) will grow without arginine.

Chapter 19

19.1 Lactose induces synthesis of the three enzymes as described on pp. 362–365, 371. The lactose causes a change in repressor shape, resulting in the repressor being released from the operator, thereby allowing initiation of transcription from the adjacent promoter.

In the presence of glucose, catabolite repression occurs, which prevents the transcription of catabolite-sensitive operons. A catabolite of glucose results in a decrease in the amount of cAMP in the cell, thereby reducing the amount of cAMP-CGA (catabolite gene activator protein) complex that must bind to promoters of catabolite sensitive operons before RNA polymerase can bind.

19.2 Polycistronic mRNA contains information for more than one protein, and since these mRNAs are transcribed from operons, the proteins are for related functions, such as catalyzing steps in a biochemical pathway. An advantage of polycistronic mRNA is that it provides a convenient package for the coordinate production of proteins with related functions.

19.3 A constitutive phenotype can be the result of either an i^- or o^c mutation.

19.4 (a) A missense mutation results in partial or complete loss of β-galactosidase activity, but there would be no loss of permease and transacetylase activities.
(b) A nonsense mutation is likely to have polar effects unless the mutation is very close to the normal chain-terminating codon for β-galactosidase. Thus permease and transacetylase activities would be lost in addition to the loss of β-galactosidase activity.

19.5 Partial diploids were able to show (a) cis- and trans-dominant effects, (b) that repressor action was the result of a diffusible substance, (c) that operator function did not require a diffusible substance, and (d) that promoter function did not require a diffusible substance.

19.6 $i^+ \; o^c \; z^+ \; y^- / i^+ \; o^+ \; z^- \; y^+$. (It cannot be ruled out that one of the repressor genes is i^-.)

19.7 (a) The o^c mutants have an altered base pair in the operator that prevents the repressor from recognizing the operator. As a result, the structural genes cis (adjacent) to the o^c are transcribed constitutively; o^c has no effect on the other lactose operons in the same cell because o^c codes for no product that could diffuse through the cell and affect other DNA sequences.
(b) The i^s mutants produce super-repressor molecules that have lost their ability to recognize the inducer. As a result, super-repressor molecules bind to normal lactose operators in the cell, no matter where they are, and prevent the transcription of the structural genes. In this case, trans-dominance is seen because the i^s mutants produce a diffusible product. Even with normal repressor molecules (from an i^+ gene) in the cell, the super-repressor molecules are "stuck" on the operators. i^- mutants are recessive for the following reasons: i^- mutations cause a change in the repressor molecule such that it cannot bind to the operator. However, if an i^+ gene is present in a partial diploid with i^-, a diffusible, functional repressor will be made that can bind to the operator and block transcription.
(c) Mutations in the repressor gene promoter result either in an increase or a decrease in the level of expression of the repressor gene. Such mutations are likely to have little effect on the control of expression of the *lac* operon structural genes since the repressor molecule itself is unaltered. (The only possible effect would be if the number of repressor molecules made was low enough to cause increased structural gene expression in the absence of inducer.) Promoter mutations for the structural genes can also increase or decrease the level of expression of the (induced) operon. Most known promoter mutations cause an almost complete loss of expression of the three genes.

19.8

Genotype	Inducer absent		Inducer present	
	β-galactosidase	Permease	β-galactosidase	Permease
1. $i^+ p^+ o^+ z^+ y^+$	−	−	+	+
2. $i^+ p^+ o^+ z^- y^+$	−	−	−	+
3. $i^+ p^+ o^+ z^+ y^-$	−	−	+	−
4. $i^- p^+ o^+ z^+ y^+$	+	+	+	+
5. $i^s p^+ o^+ z^+ y^+$	−	−	−	−
6. $i^+ p^+ o^c z^+ y^+$	+	+	+	+
7. $i^s p^+ o^c z^+ y^+$	+	+	+	+
8. $i^+ p^+ o^c z^+ y^-$	+	−	+	−
9. $i^{-d} p^+ o^+ z^+ y^+$	+	+	+	+
10. $i^- p^+ o^+ z^+ y^+ / i^+ p^+ o^+ z^- y^-$	−	−	+	+
11. $i^s p^+ o^+ z^+ y^- / i^+ p^+ o^+ z^- y^+$	−	−	−	−
12. $i^s p^+ o^+ z^+ y^- / i^+ p^+ o^+ z^- y^+$	−	−	−	−
13. $i^+ p^+ o^c z^- y^+ / i^+ p^+ o^+ z^+ y^-$	−	−	+	+
14. $i^- p^+ o^c z^+ y^- / i^+ p^+ o^+ z^- y^+$	+	−	+	+
15. $i^s p^+ o^+ z^+ y^+ / i^+ p^+ o^c z^+ y^+$	+	+	+	+
16. $i^{-d} p^+ o^+ z^+ y^- / i^+ p^+ o^+ z^- y^+$	+	+	+	+
17. $i^+ p^- o^c z^+ y^- / i^+ p^+ o^+ z^- y^+$	−	−	−	+
18. $i^+ p^- o^+ z^+ y^- / i^+ p^+ o^c z^- y^+$	−	+	−	+
19. $i^- p^- o^+ z^+ y^+ / i^+ p^+ o^+ z^- y^-$	−	−	−	−
20. $i^- p^+ o^+ z^+ y^- / i^+ p^- o^+ z^- y^+$	−	−	+	−

19.9 CGA, in a complex with cAMP, is required to facilitate RNA polymerase binding to the *lac* promoter. RNA polymerase binding occurs only in the absence of glucose and only if the operator is not occupied by repressor (i.e. if lactose is absent). A mutation in the CGA gene, then, would render the *lac* operon incapable of expression since RNA polymerase would not be able to recognize the promoter.

19.10 Inducible and repressible operons are basically similar; they differ in the details of the control of transcription. For example, in the lactose operon, the system is off when lactose is absent and on when lactose is present. Mechanistically this is brought about by the fact that the lactose operon repressor can bind to the operator in the absence of lactose, but when lactose is added, lactose binds to the repressor, preventing it from binding to the operator. The operon is then transcribed. In a repressible operon, such as one that codes for the enzymes that catalyze the steps in a biochemical pathway (e.g. for tryptophan biosynthesis), the strategy is the opposite. When the amino acid is present in the medium, the operon should be turned off. To do this, the amino acid binds to a co-repressor protein and changes its shape so that it can bind to the operator and block transcription. In the absence of the amino acid the co-repressor has no affinity for the operator and the operon is active.

19.11 p^- (promoter) mutations

19.12 See text, pp. 382–384. Stringent control is the regulatory mechanism that operates in bacteria to stop essential cellular biochemical activities when the cell is in a harsh nutritional condition, such as when a carbon source runs out or when an auxotroph is starved for the nutrient involved. Characteristically the unusual nucleotide ppGpp accumulates rapidly when the stringent-control system is in effect. A form of stringent control has been described in yeast, but in general no similar systems involving the accumulation of ppGpp are found in eukaryotes.

Chapter 20

20.1 There are no operons in eukaryotes. The *arom* and *qa* gene clusters in *Neurospora* have some properties that are similar to those of bacterial operons. However, it is now known that the *arom* gene cluster encodes a multifunctional protein rather than a number of distinct protein products that participate in a related function. The *arom* cluster, then, is more "supergene" rather than a tandem gene array encoding a polycistronic mRNA. The *qa* cluster, however, does encode distinct protein products, although a polycistronic mRNA is not involved. The characteristic feature of bacterial operons that has so far not been found in eukaryotic systems is a controlling site formally analogous to an operator.

20.2 See text, pp. 397–399. The *arom* gene cluster encodes a pentafunctional protein product, whereas the *qa* gene cluster encodes distinct protein products that function together.

20.3 An operon regulates transcription using signals from the environment and provides for the coordinate production of proteins (e.g. enzymes) with related functions. The first property may or may not be found in a eukaryotic gene that encodes a multifunctional protein; the gene may not be inducible or repressible. However, the existence of a single multifunctional protein encoded by a single gene ensures the coordinate production of the related enzyme activities. Unlike the protein products of a prokaryotic operon (which are separate after synthesis), the multifunctional eukaryotic protein product has all of the necessary functions at one location in the cell and thus is not susceptible to changes in the cellular microenvironment.

20.4 The final product of the rRNA genes is an rRNA molecule. Hence a large number of genes are required to produce the large number of rRNA molecules required for ribosome biosynthesis. Ribosomal proteins, on the other hand, are the end-products of the translation of mRNAs, which can be "read" repeatedly to produce the large numbers of ribosomal protein molecules required for ribosome biosynthesis.

20.5 A hormone is a chemical messenger transmitted in body fluids from one part of the organism to another, which produces a specific effect on target cells that may be remote from its point of origin and which functions to regulate gene activity, physiology, growth, differentiation, or behaviour.

20.6 See text, pp. 400–402. Some hormones exert their effects by binding directly to the cell's genome and causing changes in gene expression. Other hormones act at the cell surface to activate a system that produces cAMP. The cAMP then acts as a secondary messenger molecule to activate the cellular events normally associated with the hormone involved.

20.7 In brief, development is a process of regulated growth and differentiation that results from the interaction of the genome with cytoplasm, internal cellular environment, and external environment. In multicellular organisms, differentiation is the aspect of development that involves the formation of distinctly different types of cells and tissues from a zygote through processes that are specifically regulated by genes.

20.8 Totipotency refers to the ability of a differentiated cell to go through all of the stages of development. In other words, a cell taken from a differentiated tissue is totipotent if it can be isolated and if a complete functional organism can develop from it. The implication is that the cell contains all of the genetic information present in a zygote so that the developmental program for the complete organism can be executed.

20.9 See text, p. 404. The puffs in polytene chromosomes are a visual representation of differential gene activity. The puffs are regions where the multiple copies of the chromosomes have loosened their tightly packed arrangement to expose sites for transcription. Puffs are active regions of RNA synthesis.

20.10 Experiment (a) results in all the DNA becoming radioactively labeled. Since DNA is a fundamental and major component of polytene chromosomes, radioactivity is evident throughout the chromosomes. Experiment (b) results in radioactive labeling of RNA molecules. Since radioactivity is first found only around puffs and later in the cytoplasm, we can hypothesize that the puffs are sites of transcriptional activity. Initially, radioactivity is found in RNA that is in the process of being synthesized, and the later appearance of radioactivity in the cytoplasm reflects the completed RNA molecules that have left the puffs and are being translated in the cytoplasm. Experiment (c) provides additional support for the hypothesis that transcriptional activity is associated with puffs. That is, the inhibition of RNA transcription by actinomycin D blocks the appearance of ^{3}H-uridine (which would be in RNA) at puffs; in fact, puffs are much smaller, indicating that the puffing process is intimately associated with the onset of transcriptional activity for the gene(s) in that region of the chromosome.

Chapter 21

21.1 Let p = probability of A, q = probability of a. Then $q^2 = 0.81$, $q = 0.9$, $p = 0.1$. Therefore we expect in the next generation, 0.81 *aa*, 0.01 *AA*, and (2)(0.9)(0.1) = 0.18 *Aa*.

21.2 (a) $\sqrt{0.16} = 0.4 = 40\%$ = frequency of recessive alleles; $1-0.4 = 0.6 = 60\%$ = frequency of dominant alleles; $2pq = (2)(0.4)(0.6) = 0.48$ = probability of heterozygous diploids. Then, $(0.48)/[(2\times0.16)+0.48] = 0.48/0.80 = 60\%$ of recessive alleles are in heterozygotes.
(b) If $q^2 = 1\% = 0.01$, then $q = 0.1$, and $p = 0.9$, and $2pq = 0.18$ heterozygous diploids. Therefore, $(0.18)/[(2\times0.01)+0.18] = 0.18/0.20 = 90\%$ of recessive alleles in heterozygotes.

21.3 $2pq/q^2 = 8$, so $2p = 8q$; then $2(1-q) = 8q$, and $2 = 10q$, or $q = 0.2$

21.4 (a) $49\% + 1/2(42\%) = 70\%$
(b) $\sqrt{0.01} = 0.1 = 10\%$

21.5 The zygotic frequencies generated by random mating are p^2 $AA+2pq$ $Aa+q^2$ aa = 1. All of the gametes of the AA individuals and half of the gametes of heterozygotes will bear the A allele. Remembering that $p+q = 1$, the frequency of A in the gene pool of the next generation is $p^2+pq = p(p+q) = p$. Thus each generation of random mating under Hardy-Weinberg conditions fails to change either the allelic or zygotic frequencies. The more lengthy proof of this, considering all possible matings, their frequencies, and the relative frequencies of different genotypes in their offspring, is in the text.

21.6 (a) Let p = the frequency of S, and q = the frequency of s. Then

$$p = \frac{2(188)\,SS+717\,Ss}{2(3146)} = \frac{1093}{6292} = 0.1737$$

$$q = \frac{717\,Ss+2(2241)\,ss}{2(3146)} = \frac{5199}{6292} = 0.8263$$

(b)

Class	Observed	Expected	d	d^2/e
SS	188	94.9	+93.1	91.235
Ss	717	903.1	−186.1	38.361
ss	2241	2147.9	+93.1	4.032
	3146	3145.9	0	133.628

There is only 1 degree of freedom because the three genotypic classes are completely specified by two gene frequencies, p and q. Thus the number of degrees of freedom = number of genes −1. The χ^2 value for this example is 133.628 which, for 1 degree of freedom, gives a p value of < 0.0001. Therefore the distribution of genotypes differs significantly from the Hardy-Weinberg equilibrium.

21.7 Frequency of S^- = p^2+2pr; frequency of Ss = $2pq$; frequency of s^- = q^2+2qr; frequency of S^U S^U = r^2

21.8 Since the frequency of the trait is different in males and females, it suggests that the character might be caused by a sex-linked recessive gene. If the frequency of this gene is q, females would occur with the character at a frequency of q^2, and males with the character would occur at a frequency of q. Frequency of males q = 0.4, and thus we may predict that the frequency of females would be $(0.4)^2 = 0.16$, if this is a sex-linked gene. This fits the observed data. Therefore frequency of heterozygous females $2pq = 2(0.6)(0.4) = 0.48$. For sex-linked genes, no heterozygous males exist.

21.9 A sex-linked pair of alleles, occurring with frequencies of 0.1 recessive and 0.9 dominant

21.10 0.064 = q^2, so q = 0.08 = probability of color-blind male

21.11 Let p = frequency of recessive (c) = 0.08 and let q = frequency of dominant (C) = $(1-p)$ = 0.92.
(a) Frequency of color-blind women is given by $p^2 = (0.08)^2 = 0.064$; that is, 0.64% of women are color-blind.
(b) Frequency of heterozygous women is given by $2pq = (2)(0.08)(0.92) =$ 0.1472; that is, 14.72% of women are heterozygotes.

(c) Frequencies of alleles will not change in two generations given random mating in a population in Hardy-Weinberg equilibrium. Thus frequency of men with normal vision two generations later is given by $q = 0.92$; that is 92% of men will have normal vision in two generations.

21.12 Since $q = u/(u+v)$, then $q = 0.9$. Thus the frequencies are 0.01 *AA*, 0.18 *Aa*, and 0.81 *aa*.

21.13 Population 1 shows a binomial distribution of genotypes, indicating selectively neutral alleles. Population 2 has excess heterozygotes. If this condition is stable, it is an example of what is called adaptive polymorphism, or hybrid vigor. Population 3 has excess homozygotes, suggesting that interbreeding is occurring or that many heterozygotes are inviable.

21.14 100% *AA* would be expected if all four individuals were homozygous dominant. The probability of this is $(0.7)^2$ to the fourth power = $(0.49)^4$ [= about $(1/2)^4$ or 1/16]. The result of 100% *AA* might be achieved even if not all the four survivors were homozygous dominant, and the precise answer would be a probability somewhat greater than $(0.49)^4$.

21.15 (a) When selectively neutral, the genes distribute themselves binomially, so that 0.25 are *AA*, 0.50 are *Aa*, and 0.25 are *aa*.
(b) $q = 0.33$.
(c) $q = 0.25$.

21.16 Mutation of *A* to *a*.

Glossary

adenine (A) A purine base found in RNA and DNA; in double-stranded DNA, adenine pairs with the pyrimidine thymine.

allele One of two or more alternative forms of a single gene locus. Different alleles of a gene each have a unique nucleotide sequence, and their activities are all concerned with the same biochemical and developmental process, although their individual phenotypes may differ.

amino acids The building blocks of polypeptides; 20 different amino acids are normally found in polypeptides.

aminoacyl-tRNA A tRNA molecule covalently bound to an amino acid. This complex brings the amino acid to the ribosome so that it can be used in polypeptide synthesis.

aminoacyl-tRNA synthetase An enzyme that catalyzes the addition of a specific amino acid to a tRNA molecule. Since there are 20 amino acids, there are 20 aminoacyl-tRNA synthetases.

anaphase The stage in mitosis or meiosis during which the sister chromatids (mitosis) or homologous chromosomes (meiosis) separate and migrate toward the opposite poles of the cells.

aneuploidy The abnormal condition in which one or more whole chromosomes of a normal set of chromosomes either are missing or are present in more than the usual number of copies.

anticodon A three-nucleotide sequence in a tRNA that pairs with a codon in mRNA by complementary base pairing.

asexual reproduction Reproduction in which a new individual develops from either a single cell or from a group of cells in the absence of any sexual process.

attenuation In certain bacterial biosynthetic operons, a regulatory mechanism that controls gene expression by causing RNA polymerase to terminate transcription.

autosomal dominant inheritance A mechanism of inheritance of a trait that is due to a dominant mutant gene carried on an autosome.

autosomal recessive inheritance A mechanism of inheritance of a trait that is due to a recessive mutant gene carried on an autosome.

autosome A chromosome other than a sex chromosome.

auxotroph A mutant strain of a given organism that is unable to synthesize a molecule required for growth and therefore must have the molecule supplied in the growth medium for it to grow.

auxotrophic mutation (nutritional, biochemical) A mutation that affects an organism's ability to make a particular molecule essential for growth.

bacteriophage See **phage.**

Barr body A highly condensed mass of chromatin found in the nuclei of normal females, but not found in nuclei of normal male cells, which represents a cytologically condensed and inactivated X chromosome.

base analog A chemical whose molecular structure is extremely similar to the bases normally found in DNA.

biochemical mutation See **auxotrophic mutation.**

biotechnology/genetic engineering Recombinant DNA technology with significant applications in many areas, including plant breeding, animal breeding, and medicine.

capping The addition of a methylated guanine nucleotide (a "cap") to the 5′ end of a premessenger RNA molecule in eukaryotes; the cap is retained on the mature mRNA molecule.

catabolite repression (**glucose effect**) Inactivation of an inducible bacterial operon in the presence of glucose even though the operon's inducer is present.

cell cycle Life cycle of an individual cell. In proliferating cells the cell cycle consists of four phases: the mitotic phase (M) and a three-stage interphase (G1, S, and G2).

centi-Morgan (**cM**) The map unit. It is sometimes called a centi-Morgan in honor of T. H. Morgan.

centromere A specialized region of a chromosome seen as a constriction under the microscope. This region is important in the activities of the chromosomes during cellular division.

chiasma (**pl. chiasmata**) Place on a homologous pair of chromosomes at which a physical exchange occurs; it is the site of crossing-over.

chiasma interference (**chromosomal interference**) The physical interference caused by the breaking and rejoining chromatids that reduces the probability of more than one crossing-over event occurring near another one in one part of the meiotic tetrad.

chi-square test A statistical procedure to determine whether there is a significant difference between observed results and results expected on the basis of a particular hypothesis.

chloroplast Organelle found in the cytoplasm of green plant cells that is the site for photosynthesis.

chromatid One of the two visibly distinct longitudinal subunits of all replicated chromosomes that becomes visible between early prophase and metaphase of meiosis.

chromatin Fragments of eukaryotic chromosomes, consisting of DNA plus histones plus nonhistones.

chromosomal interference See **chiasma interference**.

chromosome A structure, either linear or circular, consisting of a linear collection of genes. In prokaryotes the chromosomes consist of DNA or RNA with no proteins associated. In eukaryotes the chromosomes consist of DNA complexed with protein.

cis-trans (**complementation**) **test** A test used to determine whether two mutations are within the same cistron (gene).

cloning The generation of many copies of a DNA molecule (e.g., a recombinant DNA molecule) by replication in a suitable host.

cloning vehicle, cloning vector Double-stranded DNA molecule that is able to replicate autonomously in a host cell and with which a DNA fragment or fragments can be bonded to form a recombinant DNA molecule for cloning.

coding sequence The part of an mRNA molecule that specifies the amino acid sequence of a polypeptide during translation.

codominance The situation in which the heterozygote exhibits the phenotypes of both homozygotes.

codon A group of three adjacent nucleotides in an mRNA molecule that specifies either one amino acid in a polypeptide chain or the termination of polypeptide synthesis.

coefficient of coincidence A number that expresses the extent of **chiasma interference** throughout a genetic map; the ratio of observed double-crossover frequency to expected double-crossover frequency. Interference is equal to 1 minus the coefficient of coincidence.

complementary base pairing The hydrogen bonding between a particular purine and a particular pyrimidine in double-stranded nucleic acid molecules (DNA-DNA, DNA-RNA, or RNA-RNA). The major specific pairings are guanine with cytosine and adenine with thymine or uracil.

conditional mutant A mutant organism that is normal under one set of conditions but that becomes seriously impaired or dies under other conditions.

conjugation A process having a unidirectional transfer of genetic information through direct cellular contact between a donor ("male") and a recipient ("female") bacterial cell.

conservative model A DNA replication scheme in which the two parental strands of DNA remain together and serve as a template for the synthesis of a new daughter double helix.

coupling An arrangement in which the two wild type alleles are on one homologous chromosome and the two recessive mutant alleles are on the other.

cross See **cross-fertilization.**

cross-fertilization (**cross**) A term used for the fusion of male and female gametes from different individuals; the bringing together of genetic material from different individuals for the purpose of genetic reproduction.

crossing-over Process of chromosomal interchange by which recombinants (new combinations of linked genes) arise.

cytokinesis The division of the cytoplasm. The two new nuclei compartmentalize into separate daughter cells, and the mitotic cell division process is completed.

cytological marker Cytologically distinguishable feature of a chromosome.

cytosine (**C**) A pyrimidine base found in RNA and DNA. In double-stranded DNA, cytosine pairs with the purine guanine.

deletion (**deficiency**) A chromosome mutation resulting in the loss of a segment of the genetic material and the genetic information contained therein from the chromosome.

deoxyribonuclease (**DNase**) An enzyme that specifically catalyzes the degradation of DNA.

deoxyribonucleic acid (**DNA**) A polymeric molecule consisting of deoxyribonucleotide building blocks that in a double-stranded double-helical form is the genetic material of most organisms.

deoxyribonucleotide The basic building block of DNA consisting of a sugar (deoxyribose), a base, and a phosphate.

development The process of regulated growth that results from the interaction of the genome with cytoplasm and the environment. It involves a programmed sequence of phenotypic events that are typically irreversible.

differentiation An aspect of development that involves the formation of different types of cells, tissues, and organs from a zygote through the processes of specific regulation of gene expression.

diploid A eukaryotic cell with two sets of chromosomes.

diploidization Fusion of two haploid nuclei in a heterokaryon to produce a diploid nucleus.

discontinuous DNA replication A DNA replication scheme involving the synthesis of short DNA segments that are subsequently linked to form a long polynucleotide chain.

disjunction The process in anaphase during which sister chromatid pairs undergo separation.

DNA Shorthand for deoxyribonucleic acid, the genetic material of most organisms.

DNA ligase (**polynucleotide ligase**) An enzyme that catalyzes the formation of a covalent bond between free single-stranded ends of DNA molecules during DNA replication and repair.

DNA polymerase Enzyme that catalyzes DNA synthesis.

DNA polymerase I An *E. coli* enzyme that catalyzes DNA synthesis, originally called the Kornberg enzyme.
Down syndrome (trisomy 21) Human clinical condition characterized by various abnormalities that is caused by the presence of an extra copy of chromosome 21.
duplication A chromosome mutation that results in the doubling of a segment of a chromosome.

elongation factor Protein involved in bringing aminoacyl-tRNA molecules to the ribosome so that the amino acid can be added to the growing polypeptide chain.
epistasis A form of gene interaction in which one gene interferes with the phenotypic expression of another nonallelic gene.
eukaryote Organism with cells that possess a "true" nucleus, that is, one with a nuclear membrane.
excision repair (dark repair) An enzyme-catalyzed, light-independent process of repair of UV light-induced pyrimidine dimers in DNA that involves the removal of the dimers and the synthesis of a new piece of DNA.
expressivity The degree to which a particular genotype is expressed in the phenotype.
extrachromosomal inheritance Inheritance of characters determined by genes not located on the nuclear chromosomes but on mitochondrial or chloroplast chromosomes. Such genes show inheritance patterns distinctly different from those of nuclear genes.

forward mutation Mutation that occurs in the wild type to mutant direction.
founder effect A phenomenon that occurs when the isolate effect is exhibited by a small breeding unit that has formed by migration of a small number of individuals away from a large population.
frameshift mutation A mutational addition or deletion of a base pair in a gene that disrupts the normal reading frame of an mRNA.

gene (Mendelian factor) The determinant of a characteristic of an organism. Genetic information is coded in the DNA, which is responsible for species and individual variation. A gene's nucleotide sequence specifies a polypeptide or RNA and is subject to mutational alteration.
gene frequency (allele frequency) Proportion of one particular type of allele to the total of all alleles at this genetic locus in a Mendelian breeding population.
gene pool Total genetic information encoded in the sum total of the genes in a breeding population existing at a given time.
genetic counseling The procedures whereby the risks of prospective parents having a child who expresses a genetic disease are evaluated and explained to them; the genetic counselor typically makes predictions about the probabilities of particular traits (deleterious or not) occurring among children of a couple.
genetic map Representation of the genetic distances separating gene loci in a linkage structure (the genes of one linkage group or chromosome).
genetic recombination A process whereby parents with different genetic characters give rise to progeny so that genes in which the parents differed are associated in new combinations. For example, from *A B* and *a b*, the recombinants *A b* and *a B* are produced.
genome The total amount of genetic material in a cell; in eukaryotes, the haploid set of chromosomes of an organism.
glucose effect See **catabolite repression**.
Goldberg-Hogness box Found approximately at position −30 from the transcription initiation site. The Goldberg-Hogness sequence is considered to be the likely eukaryotic

promoter sequence. The consensus sequence for the box is TATAAAAA.

haploid A cell or an individual with one copy of each nuclear chromosome.
haploidization Breakdown of a diploid nucleus to produce haploid nuclei.
Hardy-Weinberg equilibrium (**Hardy-Weinberg law, Hardy-Weinberg law of genetic equilibrium**) An extension of Mendel's laws of inheritance that describes the expected relationship between gene frequencies in natural population and the frequencies of individuals of various genotypes in the same population.
helicase An enzyme that catalyzes the unwinding of the DNA double helix during replication in *E. coli*; product of the *rep* gene.
hemizygous The condition of X-linked genes in males. Males that have an X chromosome with an allele for a particular gene but do not have another allele of that gene in the gene complement are hemizygous.
heterogeneous nuclear RNA (**hnRNA**) The RNA molecules of various sizes that exist in a large population in the nucleus in eukaryotes. Some of the RNA molecules are precursors to mature mRNAs.
heterokaryon Result of fusion of two cell types in which the nuclei do not fuse with each other.
heterozygous A term describing a diploid organism having different alleles of one or more genes and therefore producing gametes of different genotypes.
Hfr (**high-frequency recombination**) **cell** A male cell in *E. coli*, with the *F* factor integrated into the bacterial chromosome. When the *F* factor promotes conjugation with a female (F^-) cell, bacterial genes are transferred to the female cell in high frequency.
highly repetitive sequence A DNA sequence that is repeated between 10^5 and 10^7 times in the genome.
histone One of a class of basic proteins that are complexed with DNA in eukaryotic chromosomes and that play a major role in determining the structure of eukaryotic nuclear chromosomes.
homologous chromosomes The members of a chromosome pair that are identical in the arrangement of genes they contain and in their visible structure.
homozygous A term describing a diploid organism having the same alleles at one or more genes and therefore producing gametes of identical genotypes.

incomplete dominance The condition resulting when one allele is not completely dominant to another allele so that the heterozygote has a phenotype between that shown in individuals homozygous for either allele involved.
independent assortment, principle of See **Mendel's second law.**
induced mutation A mutation that results from treatment with mutagens.
inducer For bacterial operons, a chemical or environmental agent that brings about the transcription of an operon.
induction Synthesis of a gene product or products in response to the action of an inducer (a chemical or environmental agent).
initiation factor Protein factor required for initiating ribosome-directed polypeptide synthesis.
insertion sequence (**IS**) **element** The simplest transposable genetic element found in prokaryotes, it is a mobile segment of DNA that contains genes required for the process of insertion of the DNA segment into a chromosome and for the mobilization of the element to different locations.
interference Effect of one crossing-over event in (usually) decreasing the probability that a second crossover will take place nearby.

intervening sequence (**ivs**)(**intron**) A nucleotide sequence in eukaryotes that must be excised from a gene transcript to convert the transcript into a mature RNA molecule containing (in the case of mRNA molecules) only coding sequences that can be translated into the amino acid sequence of a polypeptide. Introns have also been found in tRNA and rRNA genes.

intron See **intervening sequence.**

inversion A chromosome mutation that results when a segment of a chromosome is excised then reintegrated in an orientation 180 degrees from the original orientation.

karyotype The chromosome complement of an individual as defined by the number and morphology of the chromosomes in the nucleus.

Klinefelter syndrome A human clinical syndrome that results from disomy for the X chromosome in a male, resulting in a 47,XXY male. Many of the affected males are mentally deficient, have underdeveloped testes, and are taller than average.

leader sequence One of the three main parts of the mRNA molecule. The leader sequence is located at the 5′ end of the mRNA and contains the coded information that the ribosome and special proteins read to tell it where to begin the synthesis of the polypeptide. The leader sequence is not translated into an amino acid sequence.

lethal allele An allele that results in the death of the organism when alone (if dominant) or when homozygous (if recessive).

linkage A term describing genes located on the same chromosome.

linked genes Genes located on the same chromosome.

locus (pl. **loci**) A gene's position on a genetic map.

Lyon hypothesis An hypothesis proposed by Mary Lyon in 1961 who said that the Barr body is an inactive or mostly inactive X chromosome and explained the survival of individuals with X chromosome aberrations.

lyonization A mechanism in mammals that allows them to compensate for X chromosome differences between males (one X) and females (two Xs) and for X chromosomes in excess of the normal complement. The X chromosomes in excess of one are cytologically condensed and inactivated, and they do not play a role in much of the development of the individual.

lysogenic A term describing a bacterium that contains a temperate phage in the prophage state. The bacterium is said to be lysogenic for that phage. Upon induction, phage reproduction is initiated, progeny phages are produced, and the bacterial cell lyses.

lysogenic pathway A path, besides the lytic cycle, that a phage can follow. The chromosome does not replicate; instead, it inserts itself physically into a specific region of the host cell's chromosome.

lysogeny The phenomenon of the insertion of a temperate phage chromosome into a bacterial chromosome, where it replicates when the bacterial chromosome replicates. In this state the phage genome is repressed and is said to be in the prophage state.

lytic cycle A type of phage life cycle in which the phage takes over the bacterium and directs its growth and reproductive activities to express the phage's genes and to produce progeny phages.

map distance The average number of exchanges occurring between two gene loci.

map unit (**mu**) A unit of measurement used for the distance between two gene pairs on a genetic map. A crossover frequency of 1% between two genes equals 1 map unit.

maternal effect Phenotype determined by the maternal nuclear genome and transmitted through the maternal cytoplasm.

maternal inheritance A phenomenon in

which the mother's phenotype is expressed exclusively.

meiosis Two successive nuclear divisions of a diploid nucleus that result in the formation of haploid gametes or meiospores having one-half the genetic material of the original cell.

meiosis I The first meiotic division that results in the reduction of the number of chromosomes. This division consists of four stages: prophase I, metaphase I, anaphase I, and telophase I.

meiosis II The second meiotic division, resulting in the separation of the chromatids.

Mendel, Gregor Johann (1822–1884) An Austrian priest whose breeding experiments with garden peas established the foundation of modern genetics.

Mendelian population An interbreeding group of individuals sharing a common gene pool; the basic unit of study in population genetics.

Mendel's first law, principle of segregation The law that two members of a gene pair (alleles) segregate from each other during the formation of gametes. As a result, one-half of the gametes carry one allele and the other half carry the other allele.

Mendel's second law, principle of independent assortment The law that genes for different traits assort independently of one another. In other words, genes on different chromosomes behave independently in the production of gametes.

messenger RNA (mRNA) One of the three classes of RNA molecules involved in protein synthesis; the RNA molecule that contains the coded information for the amino acid sequence of a polypeptide.

metaphase A stage in mitosis or meiosis in which chromosomes become aligned along the equatorial plane of the spindle.

missense mutation A gene mutation in which a base-pair change in the DNA causes a change in an mRNA codon, with the result that a different amino acid is inserted into the polypeptide in place of the one specified by the wild-type codon.

mitochondrion Organelle found in the cytoplasm of all aerobic animal and plant cells that is the principal source of ATP generation.

mitosis The process of nuclear division in haploid or diploid cells producing daughter nuclei that contain identical chromosome complements and that are genetically identical to one another and to the parent nucleus from which they arose.

mitotic crossing-over (mitotic recombination) A process during mitosis of a diploid cell that produces progeny cells with combinations of genes that differ from that of the diploid parental cell that entered the mitotic cycle.

moderately repetitive sequence A DNA sequence that is reiterated from a few to as many as 10^3 to 10^5 times in the genome.

monoploidy An aberrant state in a normally diploid cell or organism in which only one complete set of chromosomes is present instead of two.

monosomy An aberrant, aneuploid state in a normally diploid cell or organism in which one chromosome is missing, leaving one chromosome with no homolog.

multiple alleles Many alternative forms of a single gene.

mutagen Any physical or chemical agent that significantly increases the frequency of mutational events above a spontaneous mutation rate.

mutation Any detectable and heritable change in the genetic material not caused by genetic recombination.

mutational equilibrium A balance between mutation occurring in one direction and that occurring in the other direction in a population in Hardy-Weinberg equilibrium.

negative assortative mating Mating that occurs between dissimilar individuals more often than it does between randomly chosen individuals.

nonhistones Acidic proteins associated with DNA in eukaryotic chromosomes.

nonhomologous chromosomes The chromosomes containing dissimilar genes that do not pair during meiosis.

nonparental ditype (NPD) From a cross involving allelic differences at two loci, a meiotic tetrad in which the four members genotypically consist of two of each recombinant (nonparental) type.

nonsense mutation A gene mutation in which a base-pair change in the DNA causes a change in an mRNA codon from an amino acid-coding codon to a chain-terminating (nonsense) codon. As a result, polypeptide chain synthesis is terminated prematurely and the polypeptide is therefore either nonfunctional or, at best, partially functional.

nuclease An enzyme that catalyzes the degradation of a nucleic acid by breaking phosphodiester bonds. Nucleases may be specific for DNA (deoxyribonucleases) or for RNA (ribonucleases).

nucleosome Basic structure of eukaryotic chromosomes consising of DNA wound around a core of histones.

nucleotide A monomeric molecule of RNA and DNA that consists of three parts: a pentose sugar (ribose in RNA, deoxyribose in DNA), a nitrogenous base, and a phosphate group.

nucleus A discrete structure within a eukaryotic cell that is bounded by a nuclear membrane. It contains most of the genetic material of the cell.

nutritional mutation See **auxotrophic mutation**.

Okazaki fragments The relatively short, single-stranded DNA fragments in discontinuous DNA replication that are synthesized during DNA replication and subsequently covalently joined to make a continuous strand.

one gene–one enzyme hypothesis An hypothesis, based on Beadle and Tatum's studies in biochemical genetics, that each gene controls the synthesis of one enzyme.

operon A cluster of genes whose expressions are regulated together by operator-regulator protein interactions, plus the operator region itself and the promoter.

ordered tetrads Meiotic tetrads in which the four meiotic products are linearly arranged in a way that reflects exactly the orientation of the four chromatids of each tetrad at the metaphase plate of meiosis I.

paracentric inversion An inversion in which the inverted segment occurs on one chromosome arm and does not include the centromere.

parasexual cycle A cycle of events (e.g., in the fungus *Aspergillus*) involving the formation of a heterokaryon in a multinucleate mycelium, the rare fusion of haploid nuclei differing in genotype within a heterokaryon, mitotic crossing-over within diploid fusion nuclei, and the subsequent haploidization of the diploid fusion nuclei.

parental ditype (PD) From a cross involving allelic differences at two loci, a meiotic tetrad in which the four members genotypically consist of two of each parental type.

pedigree analysis A family tree investigation that involves the careful compilation of phenotypic records of the family over several generations.

penetrance The frequency with which a dominant or homozygous recessive gene manifests itself in the phenotype of an individual.

peptide bond A covalent bond in a polypeptide chain that joins the α-carboxyl group of one amino acid to the α-amino group of the adjacent amino acid.

pericentric inversion An inversion in which the inverted segment includes the parts of both chromosome arms and therefore includes the centromere.

phage A virus whose host is a bacterium.

phenotype The physical manifestation of a genetic trait that results from a specific genotype and its interaction with the environment.

photoreactivation (light repair) One way by which pyrimidine dimers can be repaired. The dimers are reverted directly to the original form by exposure to visible light in the wavelength range 320–370 nm.

plaque A round, clear area in a lawn of bacteria on solid medium that results from the lysis of cells by repeated cycles of phage lytic growth.

plasmid A genetic element within a cell that replicates autonomously, that is, independently of the host cell's chromosome(s).

point mutation A mutation caused by a substitution of one base pair for another.

polarity Referring to a bacterial operon that codes for a polygenic mRNA, the phenomenon whereby certain nonsense mutations not only result in the loss of activity of the enzyme encoded by the gene in which they are located but also reduce significantly or abolish the synthesis of enzymes coded by structural genes on the operator-distal side of the mutation. The mutations are called polar mutations.

poly(A) tail A sequence of 50 to 200 adenine nucleotides that is added as a posttranscriptional modification at the 3′ end of most eukaryotic mRNAs.

polycistronic mRNA (polygenic mRNA) Referring to prokaryotic operons, a single mRNA transcript of two or more adjacent structural genes that specifies the amino acid sequences of the corresponding polypeptides.

polypeptide A polymeric, covalently bonded linear arrangement of amino acids joined by peptide bonds.

polyploidy The condition of a cell or organism that has more than its normal number of sets of chromosomes.

polytene chromosome A special type of chromosome representing a bundle of numerous chromatids that have arisen by repeated cycles of replication of single chromatids without nuclear division. This type of chromosome is characteristic of various tissues of Diptera (two-winged flies, e.g., *Drosophila melanogaster*).

positive assortative mating Mating that occurs more frequently between individuals who are phenotypically similar than it does among randomly chosen individuals.

precursor mRNA (pre-mRNA) The initial transcript of a gene that is modified and/or processed to produce the mature functional mRNA molecule.

precursor rRNA (pre-RNA) A primary transcript of adjacent rRNA genes (16, 23, and 5S in prokaryotes; 18, 5.8, and 28S in eukaryotes) plus flanking and spacer DNA that must be processed to release the mature rRNA molecules.

precursor tRNA (pre-tRNA) A primary transcript of a tRNA gene whose bases must be extensively modified and that must be processed to remove extra RNA sequences to produce the mature tRNA molecule.

Pribnow box A part of the promoter sequence in prokaryotic genomes that is located about 10 base pairs upstream from the transcription starting point. The consensus sequence for the Pribnow box is TATAAT. The Pribnow box is often referred to as the TATA box.

primase The enzyme in DNA replication that catalyzes the synthesis of a short nucleic acid primer.

primer In DNA synthesis, a growing point for the addition of new deoxyribonucleotides.

prokaryote Bacteria and blue-green algae, which do not have a "true" nucleus.

promoter site (promoter sequence, promoter) A specific regulatory nucleotide sequence in the DNA to which RNA polymerase binds for the initiation of transcription.

prophage A temperate bacteriophage integrated into the chromosome of a lysogenic bacterium. It replicates with the replication of the host cell's chromosome.

propositus The individual through which the pedigree is discovered.

prototroph A strain that is a wild type for all nutritional requirement genes and thus requires no supplements in its growth medium.

Punnett square A matrix that describes all the possible gametic fusions that will give rise to the zygotes that will produce the next generation.

quantitative (continuous) traits Traits that show a continuous variation in phenotype over a range.

recombinant DNA molecule A new type of DNA sequence that has been constructed or engineered in the test tube from two or more distinct DNA sequences.

recombinant DNA technology A collection of experimental procedures that allow molecular biologists to splice a DNA fragment from one organism into DNA from another organism and to clone the new, recombinant DNA molecule. Development and application of particular molecular techniques, including **biotechnology** or genetic engineering. This technology is important in the production of antibiotics, hormones, and other medical agents used in the diagnosis and treatment of certain genetic diseases.

recombinants The individuals or cells that have nonparental combinations of genes as a result of the processes of genetic recombination.

recombination frequency Fraction of haploid cells that are recombinant for two gene loci.

replica plating The procedure for transferring the pattern of colonies from a master plate to a new plate, typically using a velveteen pad on a cylinder to pick up a few cells from each colony on the master plate and to inoculate them in the same pattern onto the new plate.

replication fork A Y-shaped structure formed when a double-stranded DNA molecule unwinds to expose the two single-stranded template strands for DNA replication.

replicon (replication unit) The stretch of DNA in eukaryotes from the origin of replication to the two termini of replication on each side of the origin.

repressible operons Amino acid biosynthesis operons, which are controlled when a chemical (such as an amino acid) is added.

repressor gene A regulatory gene whose product is a protein that controls the transcriptional activity of a particular operon.

repressor molecule The protein product of a repressor gene.

repulsion An arrangement in which each homologous chromosome carries the wild-type allele of one gene and the mutant allele of the other one.

restriction enzyme (restriction endonuclease) An enzyme that recognizes a specific sequence in DNA and, depending on the enzyme, either cleaves the DNA within or near the recognition sequence.

reverse mutation (reversion) Mutation that occurs in the mutant to wild-type direction.

ribonuclease (RNase) An enzyme that specifically catalyzes the degradation of RNA.

ribonucleic acid (RNA) A usually single-stranded polymeric molecule consisting of ribonucleotide building blocks. RNA is chemically very similar to DNA.

ribonucleotide A nucleotide that is a building block of RNA.

ribosomal RNA (rRNA) The RNA molecules of discrete sizes that, along with ribosomal proteins, comprise ribosomes.

ribosome A complex cellular particle consisting of ribosomal protein and rRNA molecules that is the site of amino acid polymerization during protein synthesis.

ribosome-binding site The nucleotide sequence on an mRNA molecule on which the ribosome becomes oriented in the correct reading frame for the initiation of protein synthesis.

RNA Shorthand for ribonucleic acid, the genetic material in some organisms, and intermediate molecules in the production of protein coded for in the DNA.

RNA ligase An enzyme that splices together the RNA pieces once the intervening sequence is removed from a pre-tRNA.

RNA polymerase An enzyme that catalyzes the synthesis of RNA from a DNA template in a process called transcription.

satellite DNA The DNA that forms a band in an equilibrium density gradient that is distinct from the band constituting the majority of the genomic DNA as a result of a different buoyant density.

semiconservative DNA replication Replication process in which the two parental strands unwind and each serves as a template for the synthesis of a new DNA strand. When replication is complete, each progeny DNA double helix consists of one old and one new DNA strand.

sense strand One of the two DNA strands whose complementary base sequence is produced when an RNA molecule is produced by transcription.

sex chromosome A chromosome in eukaryotic organisms that is represented differently in the two sexes. In many organisms one sex possesses a pair of visibly different chromosomes; one is an X chromosome and the other is a Y chromosome. Commonly the XX sex is female and the XY sex is male.

sex linkage The linkage of genes located on the sex chromosomes.

sexual reproduction The reproduction involving the fusion of haploid gametes produced by meiosis.

signal hypothesis The hypothesis that the secretion of proteins from a eukaryotic cell occurs through the binding of a hydrophobic amino terminal extension of the protein to the membrane and the subsequent removal and degradation of the extension in the cisternal space of the endoplasmic reticulum.

sister chromatid A chromatid derived from replication of one chromosome during interphase of the cell cycle.

somatic cell hybridization Fusion of two genetically different somatic cells of the same or different species to generate a somatic hybrid for genetic analysis.

spontaneous mutation The mutations that occur without the use of chemical or physical mutagenic agents.

stringent response A rapid shutdown of essential cellular activities.

structural gene A gene that codes for an mRNA molecule and hence for a polypeptide chain.

suppressor mutation A mutation at a second site that totally or partially restores a function lost because of a primary mutation at another site.

synapsis The specific pairing of homologous chromosomes during the zygonema stage of meiosis.

synkaryon A fusion nucleus produced following the fusion of cells with genetically different nuclei.

syntenic The genes that are localized to a particular chromosome by using an experimental approach.

telophase A stage during which the migration of the daughter chromosomes to the two poles is completed.

temperate bacteriophage The class of

bacteriophages that, when they infect a bacterial cell, are capable of evoking the lysogenic response. That is, the phage genome integrates into the bacterial chromosome and replicates along with the bacterial chromosome in the integrated state. The integrated phage genome is called the prophage.

template In DNA synthesis, a DNA strand that serves to specify the synthesis of a complementary strand.

testcross Cross of an individual with an individual that is homozygous recessive for all genes involved.

tetrad The four homologous chromatids (two from each homologous pair of chromosomes) synapsed during the first meiotic prophase and metaphase. The term is also used for the four haploid products of a single meiotic cycle.

tetrad analysis Genetic analysis of all the products of a single meiotic event. Tetrad analysis is possible in those organisms in which the four products of a single nucleus that has undergone meiosis are grouped together in a single structure.

tetratype (**T**) From a cross involving allelic differences at two loci, a meiotic tetrad in which the four members genotypically consist of two parentals (one of each type) and two recombinants (one of each type).

thymine (**T**) A pyrimidine base found in DNA but not in RNA. In double-stranded DNA, thymine pairs with adenine.

topoisomerases A class of enzymes that catalyze the supercoiling of DNA.

totipotency The capacity of a nucleus to direct events through all the stages of development and therefore to produce a normal adult.

trailer sequence The sequence of the mRNA molecule beginning at the end of the amino acid-coding sequence and ending at the 3′ end of the mRNA. The trailer sequence is not translated and varies in length from molecule to molecule.

transcription The transfer of information from a double-stranded DNA molecule to a single-stranded RNA molecule. It is also called RNA synthesis.

transcription-controlling sequence The base-pair sequence found around the beginning and end of each gene that is involved in the regulation of gene expression.

transcription terminator sequence A transcription regulatory sequence located at the distal end of a gene that signals the termination of transcription.

transducing phage The phage that is the vehicle by which genetic material is shuttled between bacteria.

transduction A process by which bacteriophages mediate the transfer of bacterial genetic information from one bacterium (the donor) to another (the recipient); a process whereby pieces of bacterial DNA are carried between bacterial strains by a phage.

transfer RNA (**tRNA**) One of three classes of RNA molecules produced by transcription and involved in protein synthesis; molecules that bring amino acids to the ribosome, where they are matched to the transcribed message on the mRNA.

transformant The genetic recombinant generated by the transformation process.

transformation A process in which genetic information is transferred from one cell to another by means of extracellular pieces of DNA.

transition mutation A specific type of base-pair substitution mutation that involves a change in the DNA from one purine-pyrimidine base pair to the other purine-pyrimidine base pair at a particular site (e.g., AT to GC).

translation (**protein synthesis**) The conversion in the cell of the mRNA base sequence information into the amino acid sequence of a polypeptide.

translocation A chromosome mutation in which there is a change in position of a chromosome segment or segments in the gene sequences they contain.

transposable genetic elements (**TGE**) Genetic elements of chromosomes of both prokaryotes and eukaryotes that have the capacity to mobilize themselves and move from one location to another in the genome.

transposon (**Tn**) A mobile DNA segment that contains genes for the insertion of the DNA segment into the chromosome and for the mobilization of the element to other locations on the chromosome.

transversion mutation A specific type of base-pair substitution mutation that involves a change in the DNA from a purine-pyrimidine base pair to a pyrimidine-purine base pair at a particular site (e.g., AT to TA or GC to TA).

trisomy In a normally diploid cell or organism, an aberrant, aneuploid state in which there are three copies of a particular chromosome instead of two copies.

trisomy-21 See **Down syndrome.**

true-breeding or pure-breeding strain A strain allowed to self-fertilize for many generations to ensure that the traits to be studied are inherited and unchanging.

Turner syndrome A human clinical syndrome that results from monosomy for the X chromosome in the female, which gives a 45,X female.

uniparental inheritance A phenomenon usually exhibited by extranuclear genes, in which all progeny have the phenotype of only one parent.

unique (**single-copy**) **sequence** A class of DNA sequences that has one to a few copies per genome.

unordered tetrads Meiotic tetrads in which the four meiotic products are located within a common structure in a random arrangement.

uracil (U) A pyrimidine base found in RNA but not in DNA.

virulent phage A phage like T4 that always follows the lytic cycle when it infects bacteria.

virus A noncellular, obligate, intracellular parasite that is incapable of autonomous replication and hence that can only replicate within a living host cell.

visible mutation A mutation that affects the morphology or physical appearance of an organism.

X chromosome A sex chromosome present in two copies in the homogametic sex and in one copy in the heterogametic sex.

X-linked Referring to genes located on the X chromosome.

X-linked dominant inheritance A trait due to a dominant mutant gene carried on the X chromosome. Since females have twice the number of X chromosomes that males have, X-linked dominant traits are more frequent in females than in males.

X-linked recessive inheritance A trait due to a recessive mutant gene carried on the X chromosome.

Y chromosome A sex chromosome that, when present, is found in one copy in the heterogametic sex, along with an X chromosome, and is not present in the homogametic sex.

Y-linked (**holandric**) **inheritance** A trait due to a mutant gene that is carried on the Y chromosome but has no counterpart on the X chromosome.

zygote The cell produced by the fusion of the male and female gametes.

Index